**Modern Crop Protection
Compounds**

Edited by
Wolfgang Krämer and
Ulrich Schirmer

1807–2007 Knowledge for Generations

Each generation has its unique needs and aspirations. When Charles Wiley first opened his small printing shop in lower Manhattan in 1807, it was a generation of boundless potential searching for an identity. And we were there, helping to define a new American literary tradition. Over half a century later, in the midst of the Second Industrial Revolution, it was a generation focused on building the future. Once again, we were there, supplying the critical scientific, technical, and engineering knowledge that helped frame the world. Throughout the 20th Century, and into the new millennium, nations began to reach out beyond their own borders and a new international community was born. Wiley was there, expanding its operations around the world to enable a global exchange of ideas, opinions, and know-how.

For 200 years, Wiley has been an integral part of each generation's journey, enabling the flow of information and understanding necessary to meet their needs and fulfill their aspirations. Today, bold new technologies are changing the way we live and learn. Wiley will be there, providing you the must-have knowledge you need to imagine new worlds, new possibilities, and new opportunities.

Generations come and go, but you can always count on Wiley to provide you the knowledge you need, when and where you need it!

William J. Pesce
President and Chief Executive Officer

Peter Booth Wiley
Chairman of the Board

Modern Crop Protection Compounds

Edited by
Wolfgang Krämer and Ulrich Schirmer

Volume 1

WILEY-VCH Verlag GmbH & Co. KGaA

The Editors

Dr. Wolfgang Krämer
Rosenkranz 25
51399 Burscheid
Germany

Dr. Ulrich Schirmer
Berghalde 79
69126 Heidelberg
Germany

1st Edition 2007
    1st Reprint 2007

Library of Congress Card No.: applied for

British Library Cataloguing-in-Publication Data
A catalogue record for this book is available from the British Library.

Bibliographic information published by the Deutsche Nationalbibliothek
The Deutsche Nationalbibliothek lists this publication in the Deutsche National-bibliografie; detailed bibliographic data is available in the Internet at ⟨http://dnb.d-nb.de⟩.

© 2007 WILEY-VCH Verlag GmbH & Co. KGaA, Weinheim

Printed in the Federal Republic of Germany
Printed on acid-free paper

Composition   Asco Typesetters, Hong Kong
Printing   betz-druck GmbH, Darmstadt
Bookbinding   Litges & Dopf GmbH, Heppenheim
Cover Design   Adam Design, Weinheim
Wiley Bicentennial Logo   Richard J. Pacifico

ISBN 978-3-527-31496-6

Contents

Modern Crop Protection Compounds. Edited by W. Krämer and U. Schirmer
Copyright © 2007 WILEY-VCH Verlag GmbH & Co. KGaA, Weinheim
ISBN: 978-3-527-31496-6

Preface

Modern market economies are not able to abandon modern crop protection as a basis for efficient and economical agriculture to nourish their growing population and the growing population in developing countries. Instead, they have to intensify the use of modern science to improve R&D processes for new cultivars or new crop protection compounds that fulfill worldwide registration demands and have no cross resistance to older ones, and to develop ecological and safety standards for real risk assessment.

To meet this challenge, scientists in universities, multinational organizations, like UNO and OECD, governmental authorities and in agricultural chemical companies and agricultural seed companies are working together to invent, develop and bring forward new solutions for effective, sustainable crop protection and the production of high-quality food.

This book aims to stimulate these processes by bringing together knowledge gained by modern biology, including genetics, biochemistry and chemistry, in crop protection during the last two decades, and by discussing the invention and development of modern crop protection compounds, whether unique in their chemistry or mode of action or as substance classes with a similar mode of action and similar or different chemistries.

Therefore, the contributions on new crop protection compounds are arranged not only under the headings new herbicides, fungicides and insecticides but also in respect of their biochemical mode of action.

Each of the main Sections, "Herbicides", "Fungicides", and "Insecticides", is introduced by a contribution of authors of the respective Resistance Committee, reflecting the common responsibilities of the crop protection industry for maintaining the efficacy of marketed crop protection compounds and supporting sustainable agriculture and improved public health.

These introductions allow us to mention, in a short overview, those compounds and compound classes that are not described in detail because they are dealt with *in extenso* in standard books such as *Chemistry of Plant Protection* (Springer, Berlin, Heidelberg, New York, Tokyo) and *Chemistry of Pesticides* (John Wiley and Sons, New York).

Our general target for "new" crop protection compounds was to include such compounds that have come to the market in the 15 years between 1990

Modern Crop Protection Compounds. Edited by W. Krämer and U. Schirmer
Copyright © 2007 WILEY-VCH Verlag GmbH & Co. KGaA, Weinheim
ISBN: 978-3-527-31496-6

and 2005 along with the development compounds of the "new millennium" up to 2006.

This book would not have been realized without the support of all major agricultural chemical companies and their research and development divisions, nor without the highly committed authors from them and from universities. We appreciate the tremendous work involved in collecting the literature and in writing and then submitting excellent manuscripts on time. We also especially appreciate that all of the chapter are written in a very individual manner, reflecting that all authors are, over many years, inventors or researchers or developers of plant protection compounds. As readers can see from the literature, many of the authors are the inventors of the compound or compound class they describe.

Consequently, we express our deepest gratitude to all our authors and their companies for their excellent contributions. We are sure the readers will enjoy this book and will use it as a compendium on plant protection research, in much the same way as we ourselves have experienced research on crop protection for about 30 years: stimulating, enjoyable, but also challenging. During the time period covered, tremendous market changes through new genetically modified crops have taken place, which have influenced research targets. The high demands of the public and registration authorities for safer, more ecologically compatible compounds, with tremendous increasing costs in research and development, have led to a strong concentration in the agricultural industry. From 20 companies with their own research and development in the 1980s only about five continue to carry out research in all main application fields (herbicides, fungicides, insecticides) and, additionally, in the seed business. In parallel, a deeper understanding of the efficacy and activity of crop protection compounds, their side effects on the basis of physicochemical properties and mode of action, has led to lower application rates, more selectivity of activity, and special uses, together with improved formulations.

We hope this book will also contribute to a better understanding between biologists, chemists, biochemists, agronomists, geneticists and conservationists dealing with plant protection science.

Note

The authors have named products/compounds preferably by their common names. Sometimes registered trademarks are cited. Their use is not free for everyone.

In view of the number of trademarks it was not possible to indicate each particular case in each table and contribution.

We accept no liability for this.

January 2007

Wolfgang Krämer
Ulrich Schirmer

List of Contributors

Peter Ackermann
Syngenta Crop Protection AG
Research Chemistry
WRO-1060.1.38
P.O. Box
4002 Basel
Switzerland

John T. Andaloro
DuPont Crop Protection
Stine-Haskell Research Center
Newark, DE 19711
USA

Nigel Armes
BASF Corporation
Global Insecticide Biology
P.O. Box 13528
Research Triangle Park, NC 27709
USA

Thomas Bretschneider
Bayer CropScience AG
Research Chemistry Insecticides
Bldg. 6220
Alfred Nobel Straße 50
40789 Monheim am Rhein
Germany

Heinrich Buchenauer
University of Hohenheim
Institute of Phytomedicine
Otto-Sander-Str. 5
70593 Stuttgart
Germany

Marco Busch
Bayer CropScience AG
Research Herbicides
Industriepark Hoechst, Bldg. H872N
65926 Frankfurt
Germany

Claire A. CaJacob
Monsanto Biotechnology Research
700 Chesterfield Pkwy. W.
Chesterfield, MO 63017
USA

Daniel Cordova
DuPont Crop Protection
Stine-Haskell Research Center
Newark, DE 19711
USA

Andrew. J. Corran
Syngenta Bioscience
Jealott's Hill
International Research Centre
Bracknell
Berkshire, RG42 6EY
UK

Modern Crop Protection Compounds. Edited by W. Krämer and U. Schirmer
Copyright © 2007 WILEY-VCH Verlag GmbH & Co. KGaA, Weinheim
ISBN: 978-3-527-31496-6

Gary D. Crouse
Dow AgroSciences LLC
9330 Zionsville Road
Indianapolis
IN 46268
USA

Denise P. Cudworth
Dow AgroSciences LLC
9330 Zionsville Road
Indianapolis, IN 46268
USA

Carl V. DeAmicis
Insect Management Biology
Dow AgroSciences LLC
9330 Zionsville Road
Indianapolis, IN 46268
USA

Gerrit J. deBoer
Dow Agrosciences
Discovery Research
9330 Zionsville Road
Indianapolis, IN 46268
USA

Mark A. Dekeyser
Chemtura Canada Co./Cie.
Guelph Technology Centre
120 Huron Street
Guelph
Ontario N1E 5L7
Canada

Tarlochan Singh Dhadialla
Dow AgroSciences LLC
9330 Zionsville Road
Indianapolis, IN 46268
USA

Jochen Dietz
BASF AG
Global Research Agricultural Products
GVA/FO – A30
Carl-Bosch-Straße
67056 Ludwigshafen
Germany

Günter Donn
Bayer CropScience AG
Biology Herbicides
Industriepark Höchst
Bldg. H 872 N
65926 Frankfurt am Main
Germany

Mark Drewes
Bayer CropScience AG
Research Department
Alfred Nobel Str. 50
40789 Monheim am Rhein
Germany

James E. Dripps
Dow AgroSciences LLC
9330 Zionsville Road
Indianapolis, IN 46268
USA

Fergus G. P. Earley
Syngenta
Jealott's Hill
International Research Centre
Bracknell
Berkshire, RG42 6EY
UK

Peter Eckes
Bayer CropScience AG
Scientific Computing
Industriepark Hoechst, Bldg. H872N
65926 Frankfurt
Germany

Andrew J. F. Edmunds
Syngenta Crop Protection AG
WRO-1060.3.38
Schwarzwaldallee 215
4002 Basel
Switzerland

Josef Ehrenfreund
Syngenta Crop Protection AG
WRO – 1060.144
P.O. Box
4002 Basel
Switzerland

Alfred Elbert
Bayer CropScience AG
Agronomic Development
Alfred Nobel Straße 50
40789 Monheim
Germany

Paul C. C. Feng
Monsanto Company
700 Chesterfield Pkwy. W.
Chesterfield, MO 63017
USA

Reiner Fischer
Bayer CropScience AG
Research Insecticides
Bldg. 6510
Alfred-Nobel-Straße 50
40789 Monheim am Rhein
Germany

Barja Francois
Laboratoire de Bioénergétique et
de
Microbiologie
University of Geneva
10 Ch. Embrouchis
1254 Jussy-Lullier
Switzerland

Roger E. Gast
Dow Agrosciences
Discovery Research
9330 Zionsville Road
Indianapolis, IN 46268
USA

Ulrich Gisi
Syngenta Crop Protection
Research Biology, WST 540
Schaffhauserstraße 101
4332 Stein
Switzerland

Toshio Goto
Bayer CropScience K.K.
Yuki Research Center
R&D Division
9511-4 Yuki, Yuki City
Ibaraki 307-0001
Japan

Hans Ulrich Haas
Syngenta Crop Protection
Münchwilen AG
Disease Control Research
Schaffhauserstr.
4332 Stein
Switzerland

Hiroshi Hamaguchi
Nihon Nohyaku Co. Ltd.
R&D Division
5th Floor Eitaro Building
1-2-5 Nihonbashi, Chu-Ku
Tokyo 103-8236
Japan

Gerhard Hamprecht
Rote Turmstraße 28
69469 Weinheim
Germany

Ryo Hanai
Kumiai Chemical Industry
Co., Ltd.
Life Science Research Institute
Kikugawa
Shizuoka 439-0031
Japan

Makoto Hatakoshi
Sumitomo Chemical Co. Ltd.
2-1 Takatsukasa 4-chome
Takarazuka
Hyogo 665-8555
Japan

Timothy R. Hawkes
Syngenta, Bioscience Dept.
Jealott's Hill
International Research Centre
Bracknell
Berkshire RG42 6EY
U.K.

Stefan Hillebrand
Bayer CropScience AG
BCS-R-CF, Bldg. 6550
Alfred Nobel Straße 50
40789 Monhcim am Rhcin
Germany

Takashi Hirooka
Nihon Nohyaku Co. Ltd.
R&D Division
5th Floor Eitaro Building
1-2-5 Nihonbashi, Chu-Ku
Tokyo 103-8236
Japan

Keith A. Holmes
BASF Agricultural Products
26 Davis Drive
Research Triangle Park, NC 27709
USA

Stefan Jansens
Bayer BioScience N.V.
Technologiepark 38
9052 Gent
Belgium

Peter Jeschke
Bayer CropScience AG
BCS-R-Cl Building 6240
Alfred-Nobel-Straße 50
40789 Monheim am Rhein
Germany

Timothy C. Johnson
Dow Agrosciences
Discovery Research
9330 Zionsville Road
Indianapolis, IN 46268
USA

Gertrude Knauf-Beiter
Syngenta Crop Protection AG
Research Biology
WST-540.2.65
4332 Stein
Switzerland

Helmut Köcher
Bayer CropScience AG
Industriepark Höchst
R&D, Bldg. H 872
65926 Frankfurt am Main
Germany

Karl-Heinz Kuck
Bayer CropScience AG
BCS-R-BF, Bldg. 6240
Alfred Nobel Straße 50
40789 Monheim am Rhein
Germany

David Kuhn
BASF Corporation
Global Insecticide Biology
P.O. Box 13528
Research Triangle Park, NC 27709
USA

Yoshio Kurahashi
Meiji University
Faculty of Agriculture
1-1-1 Higashi-mita, Tama-ku
Kawasaki-shi
214-8571, Kanagawa
Japan

George P. Lahm
DuPont Crop Protection
Stine-Haskell Research Center
Newark, DE 19711
USA

Clemens Lamberth
Syngenta Crop Protection
Research Chemistry WRO-1060
Schwarzwaldallee 215
4058 Basel
Switzerland

Darin W. Lickfeldt
Dow AgroSciences LLC
9330 Zionsville Road
Indianapolis, IN 46268
USA

John Lohrenz
Bayer CropScience AG
Research Department
Alfred Nobel Str. 50
40789 Monheim am Rhein
Germany

Daniel D. Loughner
Dow AgroSciences LLC
9330 Zionsville Road
Indianapolis, IN 46268
USA

Peter Maienfisch
Research and Technology
Syngenta Crop Protection AG
4002 Basel
Switzerland

Richard K. Mann
Dow Agrosciences
Discovery Research
9330 Zionsville Road
Indianapolis, IN 46268
USA

Lowell D. Markley
Dow AgroSciences LLC
9330 Zionsville Road
Indianapolis, IN 46268
USA

Alan McCaffery
Syngenta
Jealott's Hill
International Research Centre
Bracknell
Berkshire RG42 6EY
U.K.

Stephen F. McCann
DuPont Crop Protection
Stine-Haskell Research Center
Newark, DE 19711
USA

Andreas Mehl
Bayer CropScience AG
BCS-R-F-BF
Alfred-Nobel-Straße 50
40789 Monheim am Rhein
Germany

Hubert Menne
Bayer CropScience AG
Global Biology Herbicides
Industriepark Höchst
Bldg. H 872 N
65926 Frankfurt am Main
Germany

Koichi Moriya
Bayer CropScience K.K.
Yuki Research Centre
Chemical Research
9511-4 Yuki, Yuki City
Ibaraki 307-0001
Japan

Klaus-Helmut Müller
Bayer CropScience AG
R&D Insecticides
Bldg. 6510, 2.37
Alfred Nobel Straße 50
40789 Monheim am Rhein
Germany

Urs Müller
Drosselstrasse 6
4142 Münchenstein
Switzerland

Karl-Wilhelm Münks
Bayer CropScience AG
Portfolio Management
Bldg. 6100, C 3.81
Alfred Nobel Straße 50
40789 Monheim am Rhein
Germany

Ralf Nauen
Bayer CropScience AG
Research Insecticides Biology
Alfred Nobel Straße 50
40789 Monheim
Germany

Yukio Nezu
KI Chemical Research Institute
Co., Ltd.
Iwata
Shizuoka 437-1213
Japan

Thierry Niderman
Syngenta Crop Protection AG
WRO-1060.2.34
Schwarzwaldallee 215
4002 Basel
Switzerland

Nailah Orr
Dow AgroSciences LLC
9330 Zionsville Road
Indianapolis, IN 46268
USA

Oswald Ort
Bayer CropScience AG
Alfred-Nobel-Straße 50
40789 Monheim am Rhein
Germany

Stephen R. Padgette
Monsanto Company
700 Chesterfield Pkwy. W.
Chesterfield, MO 63017
USA

Thomas Pitterna
Syngenta Crop Protection AG
WRO-1060.1.34
Schwarzwaldallee 215
4002 Basel
Switzerland

Steven E. Reiser
Monsanto Company
700 Chesterfield Pkwy. W.
Chesterfield, MO 63017
USA

Joachim Rheinheimer
BASF AG
A30 – GVA
67056 Ludwigshafen
Germany

Beffa Roland
Department of Biochemistry
Bayer CropScience SA
14–20 rue Pierre Baizet
69009 Lyon
France

Chris Rosinger
Bayer CropScience AG
Industriepark Höchst
R&D, Bldg. H 872
65926 Frankfurt am Main
Germany

Ronald Ross
Dow AgroSciences LLC
9330 Zionsville Road
Indianapolis, IN 46268
USA

Shigeru Saito
Sumimoto Chemical Co. Ltd.
2-1, Takatsukasa 4-chome
Takarazuka
Hyogo 665-8555
Japan

Noriyasu Sakamoto
Sumimoto Chemical Co. Ltd.
2-1, Takatsukasa 4-chome
Takarazuka
Hyogo 665-8555
Japan

Vincent L. Salgado
BASF Agricultural Products
BASF Corporation
26 Davis Drive
Research Triangle Park, NC 27709
USA

Hubert Sauter
Alte Reichenbacher Straße 113
72270 Baiersbronn
Germany

Haruko Sawada
Bayer CropScience K.K.
Yuki Research Centre
Research/Fungicide
9511-4 Yuki, Yuki City
Ibaraki 307-0001
Japan

Michael Schindler
Bayer CropScience AG
BCS-R-Discovery, Bldg. 6500
Alfred Nobel Straße 50
40789 Monheim am Rhein
Germany

Klaus-Jürgen Schleifer
BASF AG
Computational Chemistry and
Biology, A030
67056 Ludwigshafen
Germany

Paul R. Schmitzer
Dow Agrosciences
Discovery Research
9330 Zionsville Road
Indianapolis, IN 46268
USA

Stefan Schnatterer
Bayer CropScience AG
Frankfurt Industriepark Hoechst
Research Chemistry Frankfurt
Bldg. G 836, L118
65926 Frankfurt am Main
Germany

Thomas Seitz
Bayer CropScience AG
BCS-R-F-CF
Alfred-Nobel-Straße 50
40789 Monheim am Rhein
Germany

Dale L. Shaner
Water Management Research
Unit
Natural Resources Research
Center
2150 Centre Avenue, Building D,
Suite 320
Fort Collins, CO 80526-8119
USA

Joel J. Sheets
Dow AgroSciences LLC
Biochemistry/Molecular Biology
9330 Zionsville Road
Indianapolis, IN 46268
USA

Tsutomu Shimizu
Kumiai Chemical Industry
Co., Ltd.
Life Science Research Institute
Kakegawa
Shizuoka 436-0011
Japan

Bijay Singh
BASF Corporation
26 Davis Drive
Research Triangle Park, NC 27709-3528
USA

Catherine Sirven
Department of Biochemistry
Bayer CropScience SA
14–20 rue Pierre Baizet
69009 Lyon
France

Thomas C. Sparks
Insect Management Biology
Dow AgroSciences LLC
9330 Zionsville Road
Indianapolis, IN 4628
USA

Klaus Stenzel
Bayer CropScience AG
Research
Alfred Nobel Str. 50
40789 Monheim am Rhein
Germany

Mark Stidham
Rx3 Pharmaceuticals
6310 Nancy Ridge Drive, Suite 105
San Diego, CA 92121
USA

George Theodoridis
Agricultural Product Group
FMC Corporation
P.O. Box 8
Princeton, NJ 08543
USA

Mark E. Thompson
DuPont Crop Protection
Discovery Research
Stine-Haskell Research Center
Newark, DE 19711
USA

Klaus Tietjen
Bayer CropScience AG
Research Department
Alfred Nobel Str. 50
40789 Monheim am Rhein
Germany

Isao Ueyama
Bayer CropScience K. K.
Yuki Research Center
9511-4 Yuki, Yuki City
Ibaraki 307-0001
Japan

Toquin Valerie
Department of Biochemistry
Bayer CropScience SA
14–20 rue Pierre Baizet
69009 Lyon
France

Jeroen Van Rie
Bayer BioScience N. V.
Technologiepark 38
9052 Gent
Belgium

Andreas van Almsick
Bayer CropScience AG
Industriepark Höchst
Chemistry Frankfurt, G 836
65926 Frankfurt am Main
Germany

Jean-Pierre Vors
Bayer CropScience SA
BCS-R-CF, La Dargoire
14–20 rue Pierre Baizet – BP 9163
69263 Lyon cedex 09
France

Clive Waldron
Dow AgroSciences LLC
9330 Zionsville Road
Indianapolis, IN 46268
USA

Frank Walker
University of Hohenheim
Institute of Phytomedicine (360)
Otto-Sander-Str. 5
70593 Stuttgart
Germany

Harald Walter
Syngenta Crop Protection AG
Research Chemistry
WRO-1060.202
4002 Basel
Switzerland

Yukiyoshi Watanabe
Bayer CropScience K.K.
Yuki Research Center
R&D Division
9511-4 Yuki, Yuki City
Ibaraki 307-0001
Japan

Jean Wenger
Syngenta Crop Protection AG
WRO-1060.2.34
Schwarzwaldallee 215
4002 Basel
Switzerland

William G. Whittingham
Syngenta Crop Protection
Jealott's Hill
International Research Centre
Bracknell
Berkshire RG42 6EY
UK

Matthias Witschel
BASF AG
Agricultural Products Global
Research
GVA/HC – B009
67056 Ludwigshafen
Germany

Akihiko Yanagi
Bayer CropScience K.K.
Yuki Research Center
R&D Division
9511-4 Yuki, Yuki City
Ibaraki 307-0001
Japan

Fumitaka Yoshida
KI Chemical Research Institute
Co., Ltd.
Iwata
Shizuoka 437-1213
Japan

David Young
Dow AgroSciencees LLC
Discovery Research
9330 Zionsville Road
Indianapolis, IN 46268
USA

Ronald Zeun
Syngenta Crop Protection AG
Research Biology
WST-540.E.67
4332 Stein
Switzerland

Jean-Luc Zundel
Bayer CropScience SA
La Dargoire Research Centre
14–20 rue Pierre Baizet
BP 9163
69263 Lyon Cedex 09
France

Part I
Herbicides

Modern Crop Protection Compounds. Edited by W. Krämer and U. Schirmer
Copyright © 2007 WILEY-VCH Verlag GmbH & Co. KGaA. Weinheim
ISBN: 978-3-527-31496-6

Overview

Wolfgang Krämer and Ulrich Schirmer

The Section on *Herbicides* reflects not only the changes in herbicide markets worldwide but also the changes in importance of the different herbicide classes and modes of action for the market as well as for research and development.

With the invention of the aceto-hydoxy-acid synthesis inhibitors (AHAS) the dominance of herbicides that act as photosynthesis inhibitors was dramatically broken – as it was also by the development of genetically modified herbicide tolerant crops. These especially important areas of research and development, from the 1990s up to now, are exemplified in Chapters 3 and 7. The development of 12 new sulfonyl urea herbicides launched since 1995 and the invention of four development compounds of the same chemical class, after the introduction to the market of twenty compounds already between 1980 and 1995, reflects the importance of this biochemical mode of action for the herbicide market as well as the different chemistries found to be active at this target, such as imidazolinones, triazolo-pyrimidines, pyrimidinyl-carboxylates, and sulfonylaminocarbonyl-triazolinones.

One of the most important reasons for the success of the AHAS inhibitors is their extremely low application rates, in the range of 10 g-a.i. ha^{-1}, corresponding to 1 mg-a.i. m^{-2}, allowing farmers a flexible use of such herbicides with reduced market prices.

The success story of genetically modified herbicide crops, with market shares, e.g., $\geq$80% in the soybean herbicide markets of USA and Argentina, reflects two facts: The low manufacturing and application costs of the herbicides used in those crops and, due to the activity of these compounds as total herbicides, the extremely broad spectrum against nearly all weeds, either broad leaf weeds or grassy weeds and also perennial ones.

Even if public opinion, especially in Europe, hindered the introduction and use of genetically modified crops in EU countries, the ability of these methods to protect crops from weed competition will not be prohibited worldwide for long. The more active (i.e., broader efficacy and spectrum, higher selectivity) and cheapest solution in solving weed problems will be used by farmers trying to survive under the pressure of low selling prices for their goods such as cereals, corn, soybeans and other crops.

Modern Crop Protection Compounds. Edited by W. Krämer and U. Schirmer
Copyright © 2007 WILEY-VCH Verlag GmbH & Co. KGaA, Weinheim
ISBN: 978-3-527-31496-6

However, the introduction of new herbicides, either from AHAS biochemistry or others such as HPPD inhibitors, ACC-ase inhibitors and others, shows that selective herbicides, sometimes together with safeners, will find their markets when they are competitive with older solutions and when they offer advantages to farmers, such as one application a season.

Thus, the contributions in the *Herbicides* section also aim to discuss the importance of the different biochemical pathways in the search for new herbicides.

Chapter 5, entitled "Safeners for Herbicides", demonstrates the progress in this research field, bringing out new compounds that create highly competitive products for the farmers out of only partly selective herbicides having a very broad weed spectrum and very low application rates ("the chemical answer to genetically modified herbicide resistant crops").

Chapter 10, entitled "Photosynthesis Inhibitors", discusses also regulatory aspects and the reregistration process in Europe, along with compounds as an example of the impact of political and public requests for safer and ecobiologically more selective compounds, and the impact on markets for producers and sellers of generic products. This example demonstrates that markets for generic compounds not only grow through compounds losing patent protection but that they can also shrink by losing registration in countries who are opinion leaders in registration requests. Conversely, registration demands and the fulfilling of the regulatory requests by the producer/seller will protect the compounds/products longer than the patent protection lasts.

Chapter 1 ("HRAC Classification of Herbicides and Resistance Development") describes the importance of resistance development of weeds for the herbicide developers and users. This impact on herbicidal activity under field situations also shows the necessity of continuous further research for new herbicides with new modes of action, at least by the global-acting agrochemical companies.

Plant growth regulators – having their own market as growth retardants, fruit thinning agents for better quality and fruit size, sprout suppressants, defoliants, stress protectants and harvesting help, for example by suppressing dwarfing, – are only of research and development interest in a small group of agrochemical companies. However, because they influence the biochemistry of plants we have included them in this section.

1
HRAC Classification of Herbicides and Resistance Development

Hubert Menne and Helmut Köcher

1.1
Introduction

The first cases of herbicide resistance were reported around 1970. Since then resistance of mono- and dicotyledonous weeds to herbicides has become an increasing problem world-wide.

In March 2006 the International Survey of Herbicide-Resistant Weeds recorded 305 herbicide-resistant biotypes with 182 weed species – 109 dicotyledonous and 73 monocotyledonous weeds [1]. The relatively steady increase in the number of new cases of resistance since 1980 accounts for the increasing importance of herbicide resistance in weeds in the major agricultural regions (Fig. 1.1).

In the period 1970–1990 most documented cases were concerning triazine resistance. The introduction of new herbicides with different modes of action (MoA) resulted in a shift, so that more recently ALS- and ACCase resistant weeds have been reported (Fig. 1.2).

The rapid adoption of glyphosate resistant crops in North and South America and the use of glyphosate as a pre-sowing treatment in different cropping systems has resulted in increasing cases of glyphosate resistance [1]. The probability of resistance development to glyphosate had been expressed as being likely, though less frequently in comparison with most mode of action classes [2].

1.2
HRAC Classification System of Herbicides

The global HRAC group proposed a classification system for herbicides according to their target sites, modes of action, similarity of induced symptoms or chemical classes (Table 1.1).

It is the most comprehensive existing classification system of herbicides globally. With the WSSA Code System and Australian Code System two similar classification systems were developed earlier for regional needs. The usage of differ-

Modern Crop Protection Compounds. Edited by W. Krämer and U. Schirmer
Copyright © 2007 WILEY-VCH Verlag GmbH & Co. KGaA, Weinheim
ISBN: 978-3-527-31496-6

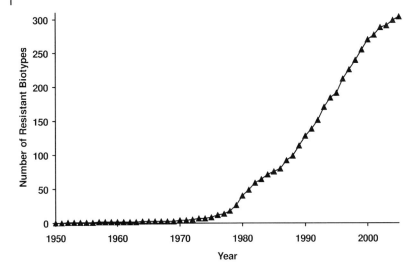

Fig. 1.1. Chronological increase in the number of herbicide-resistant weeds worldwide [3].

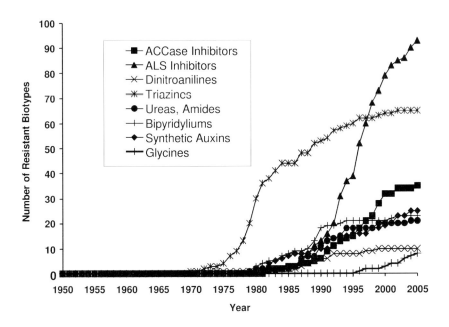

Fig. 1.2. Chronological increase in the number of herbicide-resistant weeds for the different herbicide classes [3].

Table 1.1 HRAC classification system in comparison to WSSA and Australian code system. (Adapted from Refs. [3–5].)

Mode of action	Chemical family	HRAC group	WSSA group[a]	Australian group[a]
Inhibition of acetyl CoA carboxylase (ACCase)	Aryloxyphenoxy-propionate, cyclohexanedione, phenylpyrazoline	A	1	A
Inhibition of acetolactate synthase ALS (acetohydroxyacid synthase AHAS)	Sulfonylurea, imidazolinone, triazolopyrimidine, pyrimidinyl(thio)benzoate, sulfonylaminocarbonyl-triazolinone	B	2	B
Inhibition of photosynthesis at photosystem II	Triazine, triazinone, triazolinone, uracil, pyridazinone, phenyl-carbamate	C1	5	C/K
	Urea, amide	C2	7	C
	Nitrile, benzothiadiazinone, phenyl-pyridazine	C3	6	C
Photosystem-I-electron diversion	Bipyridylium	D	22	L
Inhibition of protopor-phyrinogen oxidase (PPO)	Diphenyl ether, phenylpyrazole, N-phenylphthalimide, thiadiazole, oxadiazole, triazolinone, oxazolidinedione, pyrimidindione, other	E	14	G
Inhibition of the phytoene desaturase (PDS)	Pyridazinone, pyridinecarboxamide, other	F1	12	F
Inhibition of 4-hydroxyphenyl-pyruvate-dioxygenase (4-HPPD)	Triketone, isoxazole, pyrazole, other	F2	27	F
Inhibition of carotenoid biosynthesis (unknown target)	Triazole, diphenylether, urea (also C2)	F3	11	F
Inhibition of 1-deoxy-D-xylulose 5-phosphate synthase (DOXP synthase)	Isoxazolidinone	F4	13	F
Inhibition of EPSP synthase	Glycine	G	9	M

Table 1.1 (continued)

Mode of action	Chemical family	HRAC group	WSSA group[a]	Australian group[a]
Inhibition of glutamine synthetase	Phosphinic acid	H	10	N
Inhibition of DHP (dihydropteroate) synthase	Carbamate	I	18	K
Inhibition of microtubule assembly	Dinitroaniline, phosphoroamidate, pyridine, benzamide, benzoic acid	K1	3	D/K
Inhibition of mitosis/ microtubule organisation	Carbamate	K2	23	E
Inhibition of VLCFAs (inhibition of cell division)	Chloroacetamide, acetamide, oxyacetamide, tetrazolinone, other	K3	15	K
Inhibition of cell wall (cellulose) synthesis	Nitrile	L	20	K
	Benzamide	L	21	
	Triazolocarboxamide	L		
	Quinoline carboxylic acid	L	26	
Uncoupling (membrane disruption)	Dinitrophenol	M	24	
Inhibition of lipid synthesis – not ACCase inhibition	Thiocarbamate, phosphorodithioate	N	8	E
	Benzofuran	N	16	K
	Chloro-carbonic-acid	N	26	J
Action like indole acetic acid (synthetic auxins)	Phenoxy-carboxylic-acid, benzoic acid, pyridine carboxylic acid, quinoline carboxylic acid, other	O	4	I
Inhibition of auxin transport	Phthalamate, semicarbazone	P	19	
Unknown Note: While the mode of action of herbicides in Group Z is unknown it is likely that they differ in mode of action between themselves and from other groups	Arylaminopropionic acid	Z	25	K
	Pyrazolium	Z	26	
	Organoarsenical	Z	17	
	Other	Z	27	

[a] Not all chemical classes are classified.

ent numbers and letters in the different classification systems lead very often to confusion and misunderstanding on the global level. One common global system would be highly desirable for all users and for better understanding of differences between molecular classes. All single systems should give support and advice to all users of herbicides. This advice should state how to apply the individual active compounds to achieve the best results in terms of weed control and resistance management.

The classification system is describing not only the chemical family belonging to a specific mode of action but all compounds via their common names counted to each family, as shown in Table 1.2 for the modes of action such as "Inhibition of DHP (dihydropteroate) synthase", "Microtubule assembly inhibition", "Inhibition of mitosis/microtubule organization", "Inhibition of VLCFAs (Inhibition of cell division)" and "Inhibition of cell wall (cellulose) synthesis" as examples (not mentioned in other chapters of this book) (for a detailed table see www .plantprotection.org/HRAC/). The scheme "The World of Herbicides" available under this internet address also shows all chemical structures of the different herbicides belonging to the different chemical families.

1.3
Herbicide Resistance

In the weed population, herbicide resistance in weeds is a natural phenomenon that occurs at a low frequency and which has evolved over millions of years. Herbicide applications only select for these weeds in a population but they do not cause resistance. Increasing problems with herbicide resistant weed populations have predominantly occurred in countries with intensive agriculture cropping systems. The reliance on few of the available weed management tools with disregard of the principles of Integrated Weed Management (IWM) are closely related to changes in the weed population community. Changes in the farming environment and specifically the economic pressure on farmers are key factors that force farmers to change their practices to those that encourage resistance development.

The limitation in cropping systems, lack of rotation of herbicide chemistry or mode of action, limitation in weed control techniques, reduction of dose rates, etc. are major drivers for the selection of herbicide resistances. Regular country based surveys often make clear that farmers are aware of the problems and their causes. A survey in Germany in 2004 showed that 94% of the farmers are aware that the repeated use of the same herbicide, and 89% that the reduction of dose rates, causes the development of herbicide resistance. However, 86% of the farmers are forced to reduce their costs and they do not have a lot of scope with their weed management techniques [6].

As mentioned previously, the planting of herbicide resistant crops worldwide, which increased from 1.7 mio ha^{-1} in 1996 to around 90 mio ha^{-1} in 2005, has changed the farmers weed control tactic completely [7]. These systems have provided the farmers favorable economic advantages as well as more cropping flexi-

Table 1.2 Selected groups of the HRAC classification system with examples of the active ingredients, which are not mentioned in following chapters. (Adapted from Refs. [3–5].)

Mode of action	Chemical family	Active ingredient	HRAC group	WSSA group	Australian group[a]
Inhibition of DHP (dihydropteroate) synthase	Carbamate	Asulam	I	18	K
Microtubule assembly inhibition	Dinitroaniline	Benefin = benfluralin *Butralin* *Dinitramine* Ethalfluralin Oryzalin Pendimethalin Trifluralin	K1	3	D
	Phosphoro-amidate	*Amiprophos-methyl* *Butamiphos*			
	Pyridine	Dithiopyr Thiazopyr			
	Benzamide	Propyzamide = pronamide *Tebutam*			K
	Benzoic acid	DCPA = chlorthal-dimethyl		3	D
Inhibition of mitosis/microtubule organization	Carbamate	*Chlorpropham* *Propham* Carbetamide	K2	23	E
Inhibition of VLCFAs (inhibition of cell division)	Chloroacet-amide	Acetochlor Alachlor Butachlor *Dimethachlor* Dimethanamid Metazachlor Metolachlor *Pethoxamid* Pretilachlor Propachlor *Propisochlor* Thenylchlor	K3	15	K
	Acetamide	*Diphenamid* Napropamide *Naproanilide*			
	Oxyacetamide	Flufenacet Mefenacet			

Table 1.2 *(continued)*

Mode of action	Chemical family	Active ingredient	HRAC group	WSSA group	Australian group[a]
	Tetrazolinone	Fentrazamide			
	Other	Anilofos			
		Cafenstrole			
		Piperophos			
Inhibition of cell wall (cellulose) synthesis	Nitrile	Dichlobenil	L	20	K
		Chlorthiamid			
	Benzamide	Isoxaben		21	
	Triazolocarbox-amide	*Flupoxam*			
	Quinoline carboxylic acid	Quinclorac (for monocots) (also group O)		26	

ᵃ Not all chemical classes are classified.

bility. In Canada the adoption of herbicide resistant cropping systems has already reached 5.2 million ha (95%) out of 5.5 million ha for canola production [8].

The reliance on one herbicide has reduced the number of applications and the number of MoA used. In 2004, glyphosate was applied on 87% (62% in 2000, 25% in 1996) of the whole acreage of soybeans in the US [9]. No other herbicide was applied on more than 7% of the acreage (four herbicides with more than 10% in 2000) [9].

The continued use of herbicide resistant cropping systems with over-reliance on single weed management techniques selects for weeds that have already evolved resistance to the herbicide. Additionally, in the population, specific weed species can become dominant that were less frequent before but naturally resistant and, therefore, more difficult to control. It was suspected that a weed population shift will have a bigger impact on the cropping system than the selection of resistant weeds [10, 11]. Recent research studies and findings suggest that resistance in weeds and weed populations shifts are occurring more quickly than expected [12]. Statistical observations have shown that the use, dose rates and application frequency have already changed. In the US in 1996 glyphosate was applied in soybeans 1.1 times with 773 g-a.i. ha^{-1}. The usage increased to 1.3 applications and 1065 g-a.i. ha^{-1} in 2000, to 1.5 applications and 1211 g-a.i. ha^{-1} in 2004 [9]. Similar trends can be observed for corn and cotton and also for soybeans in other countries like Argentina and Brazil.

Intensive soil cultivation techniques and stubble burning were always common weed control techniques in agricultural areas in the past. The increasing limitation or ban of stubble burning caused increasing weed coverage, an increasing soil seed bank and the development of herbicide resistance in many agricultural regions. Different investigations have shown that the burning of straw drastically decreased weed densities, e.g., the number of waterplants (*Echinochloa* spp.) in comparison to the incorporation of rice straw into the soil [13]. Australian farmers in particular look for alternative weed control techniques during the harvest operation because of the limited choice of chemical solutions during the growing season. Balling of straw methods, such as trailing baler attached to the harvester or destroying of weed seeds physically during the harvesting operation ("Rotomill"), gives additional possibilities [14].

The economic pressure to farmers to produce at lowest costs and the changing environmental influences, like soil erosion or water availability, have led to the adoption of no-till practices in recent years. The use of no-till is expected to further increase globally. In most cases the shift to no-till systems causes an over reliance on herbicides. The price erosion of herbicides during the last years played a significant role in adoption of no-till practices. Survey studies showed that farmers are aware that no-till practices increase herbicides costs, herbicide resistance and in particular glyphosate resistance. Nevertheless, the acreage for no-till is expected to increase, especially in areas where no-till is still of low proportion [15]. However, growers with increasing herbicide resistance problems were planning to reduce the use of no-till.

Simulation studies showed that the risk of adopting no-till and the development of herbicide resistances can be reduced by alternating between minimum and no-tillage systems or by alternating between non-selective herbicides for pre-sowing weed control [16]. The most efficient weed control strategy for conserving susceptibility in no-tillage systems was the "double knockdown" pre-sowing application scheme of glyphosate and paraquat in sequence.

One of the most effective tools in the management of herbicide resistance and weed density is a diverse crop rotation practice. Weed species are typically associated with crops, and crop rotations compose their specific weed populations over time [17]. A high diversity provides the farmer more opportunities with more flexibility with respect to growing conditions, tillage practices and planting time, selecting of crop cultivars, rotating herbicides with different modes of action, varying the application timings of herbicides across years to a specific weed emergence period and/or including nonchemical management techniques etc. [18]. These practices give farmers opportunities to prevent or to slow down the selection and development of herbicide resistance. Selected resistance can remain in field populations for many years. They are stable until resistant weed seeds disappear from soil seed banks, which is very seldom. Investigations with triazine resistant weed strains showed that resistant weed seeds remained in soil despite changes in crop rotation and absence of triazine herbicides [19]. Similar results were obtained in studies that evaluated the effect of management practices on ACCase resistant *Alopecurus myosuroides* in the field [20]. The percentage of resistant plants did not change during a three-year period even without herbicide applications of ACCase inhibitors. The density of blackgrass plants was decreased, however, especially when spring crops were part of the crop rotation.

Neither cropping systems nor single weed management tactic can solve specific weed problems on a long-term basis. The use of all possible practices to prevent and to manage herbicide resistances in an integrated fashion should be the long-term goal for agricultural production.

As already mentioned, continuous application of a herbicide selects rare genotypes of weeds that are resistant to the herbicide and eventually at the same time already cross-resistant to other herbicides. These genotypes may already exist in a weed population in very low frequency before the introduction of the selecting herbicide.

1.3.1
Biochemistry of Herbicide Resistance

Resistance can be based on one of the following biochemical mechanisms [21]:

Target-site resistance is due to reduced or lost ability of the herbicide to bind to its target protein. This is usually an enzyme with a crucial function in metabolic pathways or the component of an electron transport system. As a further possibility, target-site resistance could also be caused by an overproduction of the herbicide-binding protein.

Nontarget-site resistance is caused by mechanisms that reduce the amount of herbicidally active compound reaching the target site. An important mechanism is enhanced metabolic detoxification of the herbicide in the weed, with the effect that only insufficient amounts of herbicidally active substance will reach the target site. Furthermore, reduced uptake and translocation or sequestration of the herbicide may lead to insufficient herbicide transport to the target site.

Cross-resistance means that a single resistance mechanism causes resistance to several herbicides. The term *target-site cross-resistance* is used when these herbicides bind to the same target site, whereas *nontarget-site cross-resistance* is due to a single nontarget-site mechanism (e.g., enhanced metabolic detoxification) that entails resistance across herbicides with different modes of action.

Multiple resistance is a situation where two or more resistance mechanisms are present within individual plants or within a population.

1.3.1.1 Target-site Resistance

Cases analyzed to date show that herbicide resistance is very frequently based on a target-site mutation. Within the past 35 years weed species have developed target-site resistance to most known herbicide chemistries. Those of major importance are discussed below.

Inhibitors of Photosystem II (PS II)   Early reports on resistance of weeds to PSII inhibitors of the triazine group appeared around 1970. Since then triazine resistance was reported for numerous, mainly dicotyledonous, weed species.

Research on the mechanism of resistance to triazines revealed that in most cases it is due to a mutation which results in a modification of the target site which is known to be the Qb site of the D1 protein in the PSII reaction center.

The triazine herbicides bind to this site and thus inhibit the photosynthetic electron flow. In the resistant mutants triazine binding is markedly reduced. As an example, the concentration of atrazine needed to obtain a 50% inhibition of photosynthetic electron flow in isolated chloroplasts of *Chenopodium album* was found to be at least 430× higher for chloroplasts from an atrazine-resistant mutant than for chloroplasts from wild-type plants [22].

In many cases, mutants of weed species with target-site resistance to triazines showed lower growth rate and ecological fitness than the susceptible wild type, when analyzed in absence of a triazine herbicide as selection agent. The quantum yield of CO_2 reduction in resistant biotypes was decreased. Furthermore, the electron transfer between the primary and secondary quinones in the PS II reaction center was slowed, which may have been the cause of increased susceptibility to photoinhibition in the resistant biotypes [23, 24].

The D1 protein is encoded by the chloroplast *psb*A gene, which is a highly conserved gene in higher plants, algae and cyanobacteria [25]. In almost all cases of investigated resistance of weed species in the field to triazines, resistance was attributed to a mutation in the *psb*A gene with a resultant serine 264 to glycine change in the herbicide binding niche of the D1 protein. Hence this resistance is usually maternally inherited. Though herbicides of the phenylurea group are also inhibitors of the PS II system, cross-resistance of atrazine-resistant mutants with a serine 264 to glycine change has not been observed to phenylureas. It was proposed that the binding sites of triazines and phenylureas are not identical but overlapping [26, 27]. Serine 264 provides a hydrogen bond to atrazine or other herbicides of the triazine group. Substitution of serine 264 by glycine removes this bond, which is important for binding the triazines. According to the concept of overlapping binding sites, hydrogen bonding to serine 264 is not important for phenylureas, due to a different binding geometry, hence phenylurea binding will not be affected by the serine 264 to glycine mutation.

In 1999 Masabni and Zandstra reported on a mutant of *Portulaca oleracea* with a resistance pattern to PS II inhibitors that was different to most triazine resistant weeds [28]. This mutant was resistant to the phenylureas linuron and diuron, but also cross-resistant to atrazine and other triazines. Sequencing of the D1 protein revealed that in the resistant biotype the serine 264 was replaced by threonine and not by glycine. This was the first report on a serine 264 to threonine mutation on a whole plant level. It was proposed that the serine-to-threonine mutation modified the conformation of the herbicide binding niche at the D1 protein in a way, which resulted in reduced binding of phenylureas and triazines as well.

Another novel mutant was identified, when field accessions of *Poa annua* with resistance to PS II inhibitors, collected in Western Oregon, were analyzed after amplification of the herbicide-binding region (933 base pair fragment) of the chloroplast *psb*A gene using PCR.

Sequence analysis of the fragment from a mutant with resistance to diuron and metribuzin (resistance factors between 10 and 20) revealed a substitution from valine 219 to isoleucine in the D1 protein encoded by the *psb*A gene. This amino acid substitution was previously identified after mutagenesis of laboratory cultures of algae and cell cultures of *Chenopodium rubrum*. The finding of a valine-219 to isoleucine substitution in *Poa annua*, however, was the first reported case of a weed species with resistance to PS II inhibitors in the field, due to a *psb*A mutation other than at position 264. As previously mentioned, electron transfer processes in the PS II reaction center of weeds with a mutation at position 264 were slowed and the ecological fitness of the mutants was reduced. In contrast, no effect on electron transfer in the PS II reaction center was found for the *Poa annua* mutant with the valine 219 to isoleucine change, and it was supposed that this mutant may be ecologically as fit as the wild type [29].

Inhibitors of Acetyl-CoA Carboxylase (ACCase) Acetyl-CoA carboxylase catalyzes the carboxylation of acetyl-CoA, which results in the formation of malonyl-CoA. In the plastids this reaction is the initial step of *de novo* fatty acid biosynthesis and hence of crucial importance in plant metabolism. Species of the Poaceae family (grasses) have in their plastids a homomeric, multifunctional form of ACCase with the following domains: biotin carboxy carrier protein (BCCP), biotin carboxylase (BC) and carboxyltransferase (CT). Other monocotyledonous species, so far examined, and most dicotyledonous species have in their plastids a heteromeric, multisubunit type of ACCase with the BCCP, BC and CT domains encoded on separate subunits. In addition all di- and monocotyledons, including the Poaceae, have a cytosolic ACCase, which belongs to the homomeric type. The ACCase-inhibiting herbicides inhibit only the plastidic homomeric ACCase in grasses (Poaceae), but not the plastidic heteromeric form of other mono- and dicotyledonous species nor the homomeric ACCase in the cytosol. Therefore, these herbicides selectively have a lethal effect only on grass species, while they are tolerated by other monocotyledonous and by dicotyledonous species. There are two different chemical groups of ACCase inhibitors, the aryloxyphenoxypropionates (APPs) and the cyclohexanediones (CHDs), which have developed in the past 15 to 20 years to a very important herbicide family with selective action on a broad spectrum of grass weed species.

Target-site resistance of biotypes to ACCase inhibitors has up to now been confirmed for quite a few grass weed species of economic importance. The earliest cases of target-site based resistance were reported for biotypes of *Lolium multiflorum* from Oregon, USA [30] and of *Lolium rigidum* from Australia [31].

ACCase prepared from the resistant *L. multiflorum* biotype, which had been selected by field use of diclofop, was inhibited by the APPs diclofop, haloxyfop and quizalofop with IC_{50}s (herbicide conc. needed for 50% enzyme inhibition) that were 28-, 9- and 10-times higher than for ACCase from a susceptible biotype. There was no cross-resistance to the CHD herbicides sethoxydim or clethodim [32]. ACCase resistance was subsequently also confirmed for *L. multiflorum* biotypes from other countries. In a resistant biotype selected by diclofop in Normandy, the resistance factor (ratio of the IC_{50} for ACCase from the resistant to the IC_{50} for ACCase from the susceptible biotype) was 19 for diclofop and 5 for haloxyfop, but only 2 for the CHDs clethodim and sethoxydim [33]. Interestingly, a different ACCase resistance pattern was found for the resistant *L. multiflorum* biotype Yorks A2, though field selection was apparently also mainly by diclofop. Resistance factors were 3 and 9, respectively, for the APPs diclofop and fluazifop, but 20 for the CHD herbicide cycloxydim [34].

First biotypes of *Lolium rigidum* with target-site resistance to ACCase inhibitors were identified in the early 1990s in Australia. Selection either with an APP or a CHD herbicide resulted in target-site cross-resistance to both herbicide groups. But, regardless of whether selection was by an APP or a CHD compound, the level of resistance in these biotypes was higher to APP than to CHD herbicides. ACCase resistance factors were 30–85 for diclofop, >10–216 for haloxyfop and 1–8 for sethoxydim [31, 35, 36].

Biotypes with target-site-based resistance to ACCase inhibitors were also selected in wild oat species (*Avena fatua*, *A. sterilis*). The resistance patterns were found to be variable. For example, the resistance factors for ACCase from the Canadian *A. fatua* biotype UM1 were 105 for sethoxydim, 10 for tralkoxydim, and 10 for diclofop and fenoxaprop, whereas for the *Avena fatua* biotype UM33 from Canada the ratios were 10.5 for fenoxaprop, 1.2 for diclofop, 5 for sethoxydim and 1.7 for tralkoxydim. It was proposed that this was due to different point mutations, each being associated with a characteristic resistance pattern [37]. Another reason could be the frequency of homozygote and heterozygote resistant and susceptible plants within a tested population.

During resistance studies with *Alopecurus myosuroides* populations from the UK two biotypes, Oxford A1 and Notts. A1, were identified, which were highly resistant to fenoxaprop, diclofop, fluazifop and sethoxydim due to an insensitive ACCase. Genetic studies revealed that the target-site resistance in the two *A. myosuroides* biotypes was monogenic and nuclear inherited, with the resistant allele showing complete dominance [38].

Target-site based resistance to ACCase has also been reported for several other grass weeds, e.g., two biotypes of *Setaria viridis* from Manitoba, Canada, one of them (UM8) conferring high levels of ACCase insensitivity to fenoxaprop and sethoxydim, while the ACCase of biotype UM 131 was highly insensitive to sethoxydim, but only moderately to fenoxaprop (reviewed in Ref. [36]). Biotypes of *Setaria faberi* and *Digitaria sanguinalis*, obtained in a vegetable cropping system in Wisconsin, USA, had an ACCase highly insensitive to sethoxydim and moderately insensitive to clethodim and fluazifop [39].

Based on the patterns of target-site-based cross-resistance of weeds to APP and CHD herbicides it was postulated that the two classes of ACCase inhibitors do not bind in identical manner to the target site ("overlapping binding sites"), and that different point mutations at the target enzyme account for variable resistance patterns. Molecular research with chloroplastic ACCase from wheat indicated first that a 400-amino acid region in the carboxyl transferase (CT) domain is involved in insensitivity to both APP and CHD herbicides [40]. Follow-up research with chloroplastic ACCase of *Lolium rigidum* showed that resistance to ACCase inhibitors was due to a point mutation which resulted in an isoleucine to leucine change in the CT domain of the enzyme [41]. Tal and Rubin have investigated the molecular basis of ACCase resistance in a *Lolium rigidum* biotype from Israel with resistance to CHD and APP herbicides [42]. After amplification by PCR of a 276-bp DNA encoding the CT domain of ACCase they found also in this resistant biotype a substitution of a single isoleucine by leucine. Inheritance studies of the same authors suggested that the alteration of ACCase in *L. rigidum* was governed by a single nuclear co-dominant gene.

It was shown that a point mutation resulting in an isoleucine to leucine substitution within the chloroplastic ACCase CT domain is also responsible for target-site resistance of *Avena fatua* [43] and of *Alopecurus myosuroides* [44]. Furthermore, in *Setaria viridis* biotype UM 131 a point mutation resulting in an isoleucine to leucine change of ACCase was detected [45]. The mutant leucine ACCase allele in this species was characterized to be dominant. No negative effect was detected on ACCase function of the mutant. It was suggested that the change in ACCase conformation caused by the isoleucine to leucine mutation is only minor, but sufficient to prevent or strongly reduce herbicide binding to the enzyme. Finally, also in *Alopecurus myosuroides*, an isoleucine to leucine substitution in the ACCase is associated with resistance to ACCase inhibitors [44]. These authors also pointed to the very interesting fact that the leucine found in the plastidic homomeric ACCase of mutated resistant grass weeds is also found in the heteromeric plastidic enzyme of non-grass species and in the cytosolic homomeric enzymes that are "naturally" resistant to these herbicides. Hence the selective action of ACCase-inhibiting herbicides appears to reside at this enzyme site.

Further studies by Délye and coworkers with *Alopecurus myosuroides* accessions from different sites in France shed more light on the molecular basis of the different resistance patterns to ACCase inhibitors. The isoleucine to leucine mutation (position 1781) resulted in resistance to fenoxaprop, diclofop and cycloxydim, but not to clodinafop and haloxyfop, while a newly discovered mutation of isoleucine to asparagine in position 2041 conferred resistance to fenoxaprop, diclofop, clodinafop and haloxyfop, but not to cycloxydim. Both resistance alleles can occur in the same plant and both are dominant, thus giving rise to plants that are resistant to the total spectrum of the above-mentioned herbicides [46]. Meanwhile, additional point mutations were identified in *Alopecurus myosuroides* that gave rise to insensitive ACCase due to exchange of one amino acid: Trp to Cys (pos. 2027), Asp to Gly (pos. 2078) and Gly to Ala (pos. 2096). The resistance patterns originating from these mutations gave further support to overlapping binding sites for APP and CHD herbicides at the ACCase enzyme [47].

Recently PCR amplification and sequencing of plastidic ACCase domains involved in herbicide resistance has been employed to screen a spectrum of 29 grass species for target-site-based resistance to ACCase inhibitors by direct comparison of the sequences of plastidic ACCase around the critical codons [48]. The authors found that, in *Poa annua* and *Festuca rubra*, a leucine residue occurred at position 1781, while the wild types of all other grass species had an isoleucine in this position. *Poa annua* and *F. rubra* are already known from enzyme inhibition tests to possess a plastidic ACCase that is markedly less susceptible to ACCase inhibitors than the ACCase of other grass species. Thus, the leucine in position

1781 can clearly be regarded as the basis or a substantial part of the natural inherent tolerance of both species to ACCase-inhibiting herbicides.

A different mechanism of target-site resistance to ACCase inhibitors to be mentioned here was identified in a *Sorghum halepense* biotype from Virginia, USA, which was selected in the field by quizalofop applications. The resistance level of this biotype *in vivo* was relatively low, with resistance factors (based on ED_{50} values) ranging between 2.5 and 10 for quizalofop, sethoxydim and fluazifop. No difference was found between herbicide susceptibility of ACCase from the resistant biotype and a susceptible standard. However, the specific activity of ACCase in the resistant biotype was found to be $2-3\times$ greater than in susceptible plants. The results suggested that an overproduction of ACCase was the mechanism that conferred a moderate level of resistance to these herbicides. Owing to the enzyme overproduction the resistant biotype was, presumably, able to sustain a level of malonyl-CoA production necessary for survival of herbicide treatment. This was so far the only reported case for this mechanism in a naturally occurring biotype [49].

Inhibitors of Acetolactate Synthase (ALS/AHAS)   The enzyme acetolactate synthase (ALS) plays in plants an essential role in branched-chain amino acid biosynthesis. In the pathway leading to valine and leucine, ALS catalyzes the formation of 2-acetolactate from two pyruvate molecules, and in the pathway to isoleucine the formation of 2-acetohydroxybutyrate from 2-ketobutyrate and pyruvate. Due to this double function the enzyme is also called with a more general term acetohydroxyacid synthase. ALS is inhibited by several groups of herbicides, mainly the sulfonylureas (SUs), imidazolinones (IMIs), triazolopyrimidines (TPs), pyrimidinylthiobenzoates(PTBs) and sulfonylaminocarbonyltriazolinone (SCTs) (see Chapter 2.1, M. E. Thompson).

Resistant biotypes being reported in the early 1990s were selected by chlorsulfuron or metsulfuron-methyl in wheat-growing areas or by sulfometuron-methyl in non-crop areas. While resistance of *Lolium rigidum* to ALS-inhibitors was attributed to enhanced herbicide metabolism [50] it was shown, for *Lolium perenne* and dicotyledonous species like *Stellaria media*, *Kochia scoparia*, *Salsola iberica* and *Lactuca serriola*, that resistant biotypes had a mutated ALS with reduced susceptibility to ALS-inhibiting herbicides [51–53]. The IC_{50}s for sulfonylureas, which were determined *in vitro* with ALS isolated from *Stellaria media*, *Salsola iberica* and *Lolium perenne*, increased 4- to 50-fold in the resistant biotypes. Smaller increases, about 2- to 7-fold, were determined in the same biotypes for the imidazolinone herbicide imazapyr [53].

Later ALS-inhibitors were developed for selective use in rice and led to the selection of resistant rice weed biotypes. A biotype of *Monochoria vaginalis*, discovered in Korea, showed high levels of cross-resistance to bensulfuron-methyl, pyrazosulfuron-ethyl and flumetsulam. Resistance factors determined for ALS *in vitro* were 158 to bensulfuron-methyl and 58 to flumetsulam, but only 1.6 to imazaquin [54]. In rice fields in Japan a biotype of *Scirpus juncoides* was selected, which exhibited a high degree of resistance to imazasulfuron (resistance factor of 271, calculated from ED_{50}s for growth inhibition). Inhibition tests with isolated ALS revealed an IC_{50} of 15 nM for the enzyme from susceptible plants, but of more than 3000 nM for ALS isolated from the resistant biotype, suggesting that resistance was due to an altered ALS enzyme [55].

It appears that reduced sensitivity of the target enzyme is the predominant cause of resistance to ALS inhibitors, and that resistance is conferred by a single, dominant or at least partial dominant, nuclear-encoded gene. Molecular studies revealed that resistance is caused by single substitution of one of five highly conserved amino acids in the ALS enzyme. These are the following (amino acid number standardized to the *Arabidopsis thaliana* sequence): Pro197, Ala122 and Ala205, located at the amino-terminal end, Trp574 and Ser653, located near the carboxy-terminal end [56]. For more details see also Chapter 2.1 (M. E. Thompson).

In the ALS of a *Lactuca serriola* biotype, highly resistant to SUs and moderately resistant to IMIs, Pro197 was substituted by His. The pyruvate binding domain on the ALS enzyme was not found to be altered by the mutation [57]. From *Kochia scoparia* it was reported that several substitutions of Pro197 by another amino acid (Thr, Arg, Leu, Gln, Ser, Ala) will confer resistance to sulfonylureas [58]. In the same species, it was found later that a substitution of Trp574 by Leu will also cause resistance to sulfonylureas and in addition cross-resistance to imidazolinones [59]. The latter substitution was also detected in resistant biotypes of several other dicotyledonous weed species.

In a biotype of *Amaranthus retroflexus* from Israel, resistance was caused by a change of Pro197 to Leu. This biotype exhibited cross-resistance to sulfonylureas, imidazolinones, triazolopyrimidines and to pyrithiobac-sodium *in vivo* and on the ALS enzyme level [60]. In mutations of *Amaranthus rudis*, Ser653 was found to be exchanged by Thr or Asn. These were only resistant to imidazolinones [61].

From the multiplicity of amino acid substitutions it was concluded that the herbicide-binding site of the ALS can tolerate substitutions of each of the five conserved amino acids without major consequences to normal catalytic functions. It was, therefore, speculated that the herbicide-binding site and the active site of ALS are different, though they are probably in close proximity. In absence of herbicide selection, the weed biotypes with mutated ALS showed, in most cases, no reduction or only negligible reduction of fitness (reviewed in Ref. [56]).

Glyphosate   Resistance to glyphosate has now appeared in several weed species. In resistant accessions of *Eleusine indica* from Malaysia it was found to be due to point mutations of the target enzyme EPSP synthase. By PCR amplification and sequence analysis of an EPSP synthase fragment an exchange of Pro106 by Ser was found in two resistant accessions, and an exchange of Pro106 by Thr in a third resistant accession [62].

1.3.1.2   Nontarget-site Resistance by Enhanced Metabolic Detoxification

Crop and weed species dispose of enzyme systems that catalyze the metabolic conversion of herbicides. The metabolites, which are usually more polar than the parent compound, are either non-phytotoxic at all or have a reduced phytotoxic potential. Among the various enzyme systems involved in metabolic herbicide detoxification, two are of particular importance in weeds and crops. One is the cytochrome-P450 monooxygenase system, which catalyzes oxidative transformations of the herbicide molecule (e.g., hydroxylations and oxidative dealkylations). Actually, it is a large enzyme family consisting of multiple cytochrome-P450 monooxygenases with diverse substrate specificities. The other enzyme is the glutathione-S-transferase (GST) family, catalyzing conjugation reactions that result in a nucleophilic displacement of aryloxy moieties, chlorine or other substituents by the tripeptide glutathione. Also the GSTs occur in various isoforms which differ in their catalytic properties.

The herbicide tolerance of crop species has been found to be based frequently on differential rates of metabolic herbicide detoxification in crop and weed species: while rates of herbicide detoxification in the weed species are too low to prevent binding of a lethal herbicide dosage at the target site, the tolerant crop is able to metabolically detoxify the herbicide with such a high rate that binding of the herbicide at the target site in sufficient amounts to cause irreversible herbicidal effects will be prevented. If weed biotypes with an improved ability for herbicide detoxification, comparable to the tolerant crop species, occur in a population they will survive herbicide application and will thus be selected.

To date quite a few weed biotypes have been described for which herbicide resistance was related to enhanced metabolic herbicide detoxification. Several cases have been published for *Lolium rigidum*. An early paper of Christopher et al. reported that excised shoots of biotype SLR 31 from Australia, which was resistant to diclofop, exhibited cross-resistance to the sulfonylureas chlorsulfuron, metsulfuron-methyl and triasulfuron [63]. The metabolite pattern of chlorsulfuron was identical in the resistant biotype and a susceptible standard, but the resistant biotype metabolized the herbicide more rapidly. The pathway of chlorsulfuron detoxification in *Lolium rigidum* was similar to the one described for wheat, ring hydroxylation being followed by glucose conjugation. The time course of chlorsulfuron metabolism in the *Lolium rigidum* biotype SR 4/84 (resistant to diclofop and cross-resistant to chlorsulfuron) was analyzed separately in shoots and roots. The half-life of chlorsulfuron in susceptible plants was longer in the roots (13 h) than in the shoots (4 h) and was reduced in the resistant biotype to 3 and 1 h, respectively. Detoxification of the herbicide by ring hydroxylation, likely catalyzed by a cytochrome-P450 monooxygenase with subsequent glucose conjugation was enhanced in the resistant biotype [50].

Two other *Lolium rigidum* biotypes from Australia (WLR2 and VLR69) developed metabolism-based resistance to PSII inhibitors. WLR2 came from a field with selection pressure by atrazine and amitrole, but never by phenylureas, and VLR69 from a field with selection pressure by diuron and atrazine. Both biotypes were resistant to triazines, and, despite the field selection by atrazine, resistance was more pronounced to the structurally related simazine. Furthermore, both biotypes were resistant to chlorotoluron, though only VLR69 was previously exposed to phenylureas. Analytical work revealed that in both resistant biotypes metabolism of chlorotoluron and simazine was enhanced, and that the main route of metabolism was via N-dealkylation reactions. This type of reaction and the fact that herbicide metabolism was inhibited by 1-aminobenzotriazole (ABT), an inhibitor of cytochrome-P450 monooxygenases, suggested increased activity of cytochrome-P450 monooxygenases in the resistant biotypes [64, 65]. The mechanism of phenylurea resistance of *Lolium rigidum* biotypes from Spain has been studied [66]: A biotype (R3) selected in the field by applications of diclofop plus isoproturon or plus chlorotoluron had *in vivo* resistance factors (ED_{50} R/ED_{50} S) of about 9.3 and 5.5 to chlorotoluron and isoproturon, respectively, and was also resistant to a broad spectrum of other phenylureas. Metabolism studies with chlorotoluron, in absence and presence of the cyochrome-P450 monooxygenase inhibitor 1-aminobenzotriazole, suggested that resistance was due to enhanced ability to degrade the molecule to non-toxic ring-alkylhydroxylated intermediates suitable for follow-up conjugation reactions. Several biotypes of *Lolium multiflorum* from the UK with resistance to diclofop have been analyzed [34]. While one biotype had an insensitive ACCase, resistance of three other biotypes could be attributed to enhanced metabolism of this herbicide.

The resistance of the grass weed *Phalaris minor* to isoproturon and of the dicotyledonous weed species *Abutilon theophrasti* to atrazine has also been attributed to enhanced metabolism. GST was the enzyme responsible for atrazine detoxification in *A. theophrasti* [67], whereas in *P. minor* the cytochrome P450 monooxygenase was probably involved in the enhanced detoxification of isoproturon [68].

The increasing occurrence of *Alopecurus myosuroides* resistance to herbicides in several European countries prompted research on resistance mechanisms also in this species. Aside from target-site-based resistance, cases of resistance due to enhanced herbicide metabolism had also been reported. Two biotypes, Peldon A1 and Lincs. E1, with *in vivo* resistance factors to isoproturon of 28 and 2.6, respectively, metabolized this herbicide faster than a susceptible standard. The rate of metabolism was higher in Peldon than in Lincs. Addition of the cytochrome-P 450 monooxygenase inhibitor 1-aminobenzotriazole decreased the rate of chlorotoluron metabolism and correspondingly increased phytotoxicity, suggesting the involvement of the cytochrome-P450 monooxygenase system in the detoxification of the herbicide. The major detoxification reaction in these biotypes appeared to be the formation of a hydroxymethylphenyl metabolite [69].

The same biotypes, Peldon A1 and Lincs. E1, are also resistant to the graminicide fenoxaprop, which is used for selective control of *A. myosuroides* and other grass weeds in cereals, mainly wheat. On a whole plant level, Lincs. E1 was more resistant than Peldon A1. The selectivity of this herbicide has been attributed to rapid detoxification by GST-catalyzed conjugation in the cereal species. In both resistant *A. myosuroides* biotypes GST activities towards fenoxaprop were found to be increased to a similar degree, when compared with a susceptible biotype. This was due to increased expression of a constitutive GST and to expression of two novel GST isoenzymes. Furthermore, glutathione levels were increased in the resistant biotypes, in Peldon more than in Lincs. The data pointed to an involvement of GST activity and glutathione levels in the resistance to fenoxaprop, though the lack of correlation to whole plant resistance of these biotypes did not permit definite conclusions [70]. A range of European *A. myosuroides* biotypes with resistance to fenoxaprop has been investigated [71]. Several of these biotypes, in particular one from Belgium, detoxified this herbicide with increased rates. The biotype from Belgium had also the highest GST activity towards the unspecific substrate CDNB, but GST activity towards the herbicide was not tested.

Studies on the mode of inheritance of metabolic herbicide resistance in *Alopecurus myosuroides* did not result in a uniform picture. It was reported that a single gene was responsible for metabolism-based resistance in a biotype resistant to fenoxaprop and flupyrsulfuron [72], while in another biotype resistance to chlorotoluron was attributed to more than one gene [73].

Different to the cases described above, the herbicide propanil is detoxified in rice and weed species by the action of an aryl acylamidase (aryl-acylamine amidohydrolase). High activity of this enzyme in rice confers crop tolerance. In Colombia, a biotype of *Echinochloa colona* was found that is resistant to propanil. Enzyme tests with extracts from this biotype revealed an about three-fold higher activity of aryl acylamidase in the resistant than in a susceptible biotype. It was concluded that resistance of the *E. colona* biotype is based on enhanced propanil detoxification [74].

1.3.1.3 Nontarget-site Resistance by Altered Herbicide Distribution

Cases of nontarget-site resistance by altered herbicide distribution have been reported for two important herbicides, paraquat and glyphosate.

Intensive use of the herbicide paraquat has resulted in the evolution of resistance in various weed species. Intensive research on the resistance mechanisms was mainly carried out with resistant biotypes from *Hordeum* spp. and *Conyza* spp., and altered distribution of the herbicide in the resistant weeds was suggested as the cause – or at least the partial cause – of resistance. In resistant *Conyza canadensis* it was supposed that a paraquat inducible protein may function by carrying paraquat to a metabolically inactive compartment, either the cell wall or the vacuole. This sequestration process would prevent the herbicide from getting in sufficient amounts into the chloroplasts as the cellular site of paraquat action. Inhibitors of membrane transport systems, e.g., *N,N*-dicyclohexylcarbodiimide (DCCD), caused a delay in the recovery of photosynthetic functions of the paraquat-resistant biotype, when given after the herbicide. These transport inhibitor experiments supported the involvement of a membrane transporter in paraquat resistance [75].

Translocation studies with two paraquat-resistant biotypes of *Hordeum leporinum* revealed that the basipetal transport of paraquat in resistant *H. leporinum* was much reduced compared with susceptible plants. It was concluded that the resistance to paraquat was the result of the reduced herbicide translocation out of the treated leaves [76]. One can suppose that also in this species herbicide sequestration may have been the primary cause for the altered long-distance transport.

Independent populations of *Lolium rigidum* with resistance to glyphosate have been reported from different locations in Australia. One of them, with a ca. 10-fold *in vivo* resistance to glyphosate, was used for intensive investigation of the mechanism of resistance. Neither a modification of the target enzyme EPSP synthase nor of herbicide metabolism contributed to the resistance in this case. Translocation studies after foliar application revealed, however, that in the resistant biotype glyphosate accumulated preferentially in the leaf tips, while in susceptible plants accumulation was stronger in the leaf bases and the roots. This result suggested a shift of glyphosate transport in the resistant plants from the phloem to the xylem system. It was speculated that the resistant biotype might have lost in efficiency to load glyphosate into the symplast. Thus more of the herbicide would remain in the apoplast and be translocated acropetally with the transpiration stream, while the concentration of glyphosate in the plastids of the sensitive meristematic tissues at the shoot base and in the roots would be reduced [77].

1.3.1.4 Multiple Resistance

As defined above, multiple resistance means that more than one resistance mechanism occurs in a weed population or an individual plant. This can either mean that a target site-based and a nontarget-site based mechanism occur in the same biotype, or that a biotype is resistant to herbicides with different mechanisms of action. Multiple resistance can result in resistance of a weed biotype to a very broad range of herbicide chemistries. Multiple resistance has been reported for several weed species, particularly for *Lolium rigidum*, *Alopecurus myosuroides*, *Kochia scoparia*, *Conyza canadensis* and *Amaranthus rudis*. It developed to a serious extent particularly in Australian biotypes of *Lolium rigidum*, probably as a result of agricultural conditions paired with biological characteristics of this weed (cross pollinating species with high genetic variability and seed production and high plant numbers per area).

Multiple resistance can develop by selection with a single herbicide or by selection with several herbicides, which are either used sequentially or simultaneously. Furthermore, cross-pollinating species can become multiple resistant, when two individuals, each with a different resistance mechanism, cross. An example for the selection of multiple resistance by a single herbicide (the ALS inhibitor chlorsulfuron) is the *Lolium rigidum* biotype WLR1. This biotype had as main mechanism of resistance an ALS with reduced sensitivity to chlorsulfuron, sulfometuron and imazamethabenz, and as additional mechanism enhanced metabolism of chlorsulfuron [78]. Extreme cases of multiple resistance, due to an application history of many herbicides, were reported from Australia for several *Lolium rigidum* biotypes. As an example, biotype VLR69 possessed the following mechanisms: enhanced metabolism of ACCase-inhibiting herbicides, resistant form of the ACCase enzyme, enhanced metabolism of the ALS-inhibitor chlorsulfuron, and in addition a resistant form of the ALS enzyme in 5% of the population [36].

Selection of multiple resistance after sequential use of different herbicides has been described for a biotype of *Kochia scoparia* from North America. Many years of triazine usage resulted in the selection of a biotype with target-site resistance of the D1 protein in photosystem II. By subsequent usage of ALS inhibitors, a point mutation in the gene encoding for ALS was selected in addition, which made this biotype also target-site-resistant to sulfonylureas and imidazolinones [59].

Obviously, multiple resistance leads to complex patterns of broad herbicide resistance, particularly in cross-pollinating weed species. This seriously restricts the remaining options of chemical weed control in agricultural practice.

References

1 I. Heap, personal communication. http://www.weedscience.com, **2006**.

2 I. Heap, H. LeBaron, In: *Herbicide Resistance and World Grains*. eds. S. B. Powles, D. L. Shaner CRC Press, Boca Raton, FL, **2001**, 1–22.

3 HRAC, *Classification of Herbicides According to Mode of Action*, http://www.plantprotection.org/HRAC/, **2005**.

4 C. A. Mallory-Smith, E. J. Retzinger, *Weed Technol.*, **2003**, 17, 605–619.

5 CropLife Australia, in: *Managing resistance, Herbicide Mode of Action Groups*, http://www.croplifeaustralia.org.au, update **2005**, 1–3.

6 BCS, Internal communication, **2004**.

7 C. James, *The International Service for the Acquisition of Agri-biotech Applications*, http://www.ISAAA.org, **2005**.

8 H. J. Beckie, *Can. J. Plant Sci.*, **2006**, in press.

9 National Agricultural Statistics Service, Agricultural Chemical Usage – **1996/2000/2004** Field Crops Summary (U.S. Dept. Agric. Washington DC).

10 S. O. Duke, *Proc. Workshop of Ecological Effects of Pest Resistance Genes in Managed Ecosystems*, Bethesda, MD, **1999**.

11 M. D. K. Owen, *Proc. Bright. Crop Prot. Conf.*, **1997**, 3, 955–963.

12 M. D. K. Owen, *Integr. Crop Manag. Conf.*, Iowa State Univ., **2005**, 55–59.

13 J. A. Bird, A. J. Eagle, W. R. Horwath, M. W. Hair, E. E. Zilbert, C. v. Kessel, *Calif. Agric.*, **2002**, 02, 69–75.

14 M. Walsh, *Austr. Farm J.*, **2003**, 03, 40–41.

15 F. H. D'Emden, R. S. Llewellyn, M. P. Burton, *Tech. Forec. Soc.*, **2003**. 03, 40–41.

16 P. Neve, A. J. Diggle, F. P. Smith, S. B. Powles, *Weed Res.*, **2003**, 43, 418–427.

17 T. Hyvönen, J. Salonen, *Plant Ecol.*, **2002**, 154, 73–91.

18 B. E. Valverde, in: *Weed Management for Developing Countries*, ed. R. Labrada, FAO, Rome, **2003**.

19 J. Gasquez, personal communication, **2003**.

20 B. Chauvel, J. P. Guillemin, N. Colbach, J. Gasquez, *Crop Prot.*, **2001**, 20, 127–137.

21 S. B. Powles, D. L. Shaner (eds.), *Herbicide Resistance and World Grains*. CRC Press, Boca Raton, FL, **2001**.

22 P. Böger, *Biol. Z.*, **1983**, 13.Jahrg., Nr.6, 170–177.

23 D. R. Ort, W. H. Ahrens, B. Martin, E. W. Stoller, *Plant Physiol.*, **1983**, 72, 925–930.

24 C. Sundby, W. S. Chow, J. M. Anderson, *Plant Physiol.*, **1993**, 103, 105–113.

25 G. Zurawski, H. Bohnet, P. Whitfeld, W. Bottomley, *Proc. Natl. Acad. Sci. U.S.A.*, **1982**, 79, 7699–7703.

26 A. Trebst, in: *Herbicide Resistance in Weeds and Crops*, eds. J. C. Caseley, G. W. Cussans, R. K. Atkin, Butterworth-Heinemann, Oxford, **1991**, 145–164.

27 A. Trebst, in: *Molecular genetics and evolution of pesticide resistance*, ed. T. M. Brown, *ACS Symposium series*, ACS, Washington DC, **1996**, 645, 44–51.

28 J. G. Masabni, B. H. Zandstra, *Weed Sci.*, **1999**, 47, 393–400.

29 L. W. Mengistu, G. W. Mueller-Warrant, A. Liston, R. E. Barker, *Pest Manag. Sci.*, **2000**, 56, 209–217.

30 C. E. Stanger, A. P. Appleby, *Weed Sci.*, **1989**, 37, 350–352.

31 J. A. M. Holtum, S. B. Powles, *Bright. Crop Prot. Conf. – Weeds*, **1991**, 1071–1078.

32 J. W. Gronwald, C. V. Eberlein, K. J. Betts, R. J. Baerg, N. J. Ehlke, D. L. Wyse, *Pestic. Biochem. Physiol.*, **1992**, 44, 126–139.

33 R. De Prado, J. González-Gutiérrez, J. Menéndez, J. Gasquez, J. W. Gronwald, R. Giménez-Espinosa, *Weed Sci.*, **2000**, 48, 311–318.

34 K. M. Cocker, D. S. Northcroft, J. O. D. Coleman, S. R. Moss, *Pest Manag. Sci.*, **2001**, 57, 587–597.

35 F. J. Tardif, J. A. M. Holtum, S. B. Powles, *Planta*, **1993**, 190, 176–181.

36 S. B. Powles, C. Preston, *The Herbicide Resistance Action Committee Monograph Number 2*, **1995**.

37 M. D. Devine, *Pestic. Sci.*, **1997**, 51, 259–264.

38 S. R. Moss, K. M. Cocker, A. C. Brown, L. Hall, L. M. Field, *Pest Manag. Sci.*, **2003**, 59, 190–201.

39 D. Volenberg, D. Stoltenberg, *Weed Res.*, **2002**, 42, 342–350.

40 T. Nikolskaya, O. Zagnitko, G. Tevzadze, R. Haselkorn, P. Gornicki, *Proc. Natl. Acad. Sci. U.S.A.*, **1999**, 96, 14647–14651.

41 O. Zagnitko, J. Jelenska, G. Tevzadze, R. Haselkorn, P. Gornicki, *Proc. Natl. Acad. Sci. USA*, **2001**, 98, 6617–6622.

42 A. Tal, B. Rubin, *Pest Manag. Sci.*, **2004**, 60, 1013–1018.

43 M. J. Christoffers, M. L. Berg, C. G. Messersmith, *Genome*, **2002**, 45, 1049–1056.

44 A. C. Brown, S. R. Moss, Z. A Wilson, L. M. Field, *Pestic. Biochem. Physiol.*, **2002**, 72, 160–168.

45 C. Délye, T. Wang, H. Darmency, *Planta*, **2002**, 214, 421–427.

46 C. Délye, C. Straub, A. Matéjicek, S. Michel, *Pest Manag. Sci.*, **2003**, 60, 35–41.

47 C. Délye, X.-Q. Zhang, S. Michel, A. Matéjicek, S. B. Powles, *Plant Physiol.*, **2005**, 137, 794–806.

48 C. Délye, S. Michel, *Weed Res.*, **2005**, 45, 323–330.

49 K. W. Bradley, J. Wu, K. K. Hatzios, E. S. Hagood Jr., *Weed Sci.*, **2001**, 49, 477–484.

50 J. C. Cotterman, L. L. Saari, *Pestic. Biochem. Physiol.*, **1992**, 43, 182–192.

51 C. A. Mallory-Smith, D. C. Thill, M. J. Dial, *Weed Technol.*, **1990**, 4, 163.

52 L. L. Saari, J. C. Cotterman, M. M. Primiani, *Plant Physiol.*, **1990**, 93, 55.

53 L. L. Saari, J. C. Cotterman, W. F. Smith, M. M. Primiani, *Pestic. Biochem. Physiol.*, **1992**, 42, 110–118.

54 I. T. Hwang, K. H. Lee, S. H. Park, B. H. Lee, K. S. Hong, S. S. Han, *Pestic. Biochem. Physiol.*, **2001**, 71, 69–76.

55 Y. Tanaka, *Pestic. Biochem. Physiol.*, **2003**, 77, 147–153.

56 P. J. Tranel, T. R. Wright, *Weed Sci.*, **2002**, 50, 700–712.

57 M. J. Guttieri, C. V. Eberlein, C. A. Mallory-Smith, D. C. Thill, D. L. Hoffmann, *Weed Sci.*, **1992**, 40, 670–676.

58 M. J. Guttieri, C. V. Eberlein, D. C. Thill, *Weed Sci.*, **1995**, 43, 175–178.

59 M. J. Foes, L. Liu, G. Vigue, E. W. Stoller, L. M. Wax, P. J. Tranel, *Weed Sci.*, **1999**, 47, 20.

60 M. Sibony, A. Michel, H. U. Haas, B. Rubin, K. Hurle, *Weed Res.*, **2001**, 41, 509–522.

61 W. L. Patzold, P. J. Tranel, *Proc. N. Cent., Weed Sci. Soc.*, **2001**, 56, 67.

62 C. H. Ng, R. Wickneswari, S. Salmijah, Y. T. Teng, B. S. Ismail, *Weed Res.*, **2003**, 43, 108–115.

63 J. T. Christopher, S. B. Powles, D. R. Liljegreen, J. A. M. Holtum, *Plant Physiol.*, **1991**, 95, 1036–1043.

64 M. W. M. Burnet, B. R. Loveys, J. A. M. Holtum, S. B. Powles, *Pestic. Biochem. Physiol.*, **1993**, 46, 207–218.

65 M. W. M. Burnet, B. R. Loveys, J. A. M. Holtum, S. B. Powles, *Planta*, **1993**, 190, 182–189.

66 R. De Prado, J. L. De Prado, J. Menendez, *Pestic. Biochem. Physiol.*, **1997**, 57, 126–136.

67 M. P. Anderson, J. W. Gronwald, *Plant Physiol.* **1991**, 96, 104–109.

68 S. Singh, R. C. Kirkwood, G. Marshall, *Pestic. Biochem. Physiol.*, **1998**, 59, 143–153.

69 L. M. Hall, S. R. Moss, S. B. Powles, *Pestic. Biochem. Physiol.*, **1995**, 53, 180–192.

70 I. Cummins, S. Moss, D. J. Cole, R. Edwards, *Pestic. Sci.*, **1997**, 51, 244–250.

71 K. M. Cocker, S. R. Moss, J. O. D. Coleman, *Pestic. Biochem. Physiol.*, **1999**, 65, 169–180.

72 A. Letouzé, J. Gasquez, *Theor. Appl. Genet.*, **2001**, 103, 288–296.

73 B. Chauvel, Ph.D. Thesis, **1991**, University of Paris-Orsay.

74 J. M. Leah, J. C. Caseley, C. R. Riches, B. Valverde, *Pestic. Sci.*, **1994**, 42, 281–289.

75 K. Halász, V. Sóos, B. Jóri, I. Rácz, D. Lásztity, Z. Szigeti, *Acta Biol. Szegediensis*, **2002**, 46, 23–24.

76 C. Preston, C. J. Soar, I. Hidayat, K. M. Greenfield, S. B. Powles, *Weed Res.*, **2005**, 45, 289.

77 D. F. Lorraine-Colwill, S. B. Powles, T. R. Hawkes, P. H. Hollinshead, S. A. J. Warner, C. Preston, *Pestic. Biochem. Physiol.*, **2003**, 74, 62–72.

78 J. T. Christopher, S. B. Powles, J. A. M. Holtum, *Plant Physiol.*, **1992**, 100, 1909–1913.

2
Acetohydroxyacid Synthase Inhibitors (AHAS/ALS)

2.1
Biochemistry of the Target and Resistance

Mark E. Thompson

2.1.1
Acetohydroxyacid Synthase (AHAS)

The first committed step in the biosynthetic pathway of the branched chain amino acids is catalyzed by the enzyme acetohydroxyacid synthase (AHAS, EC 2.2.1.6), which is also referred to as acetolactate synthase (ALS). As depicted in Fig. 2.1.1, the pathway leading to valine and leucine begins with the condensation of two molecules of pyruvate accompanied by loss of carbon dioxide to give (S)-2-acetolactate. A parallel reaction leading to isoleucine involves the condensation of pyruvate with 2-ketobutyrate to afford (S)-2-aceto-2-hydroxybutyrate after loss of carbon dioxide. Both reactions are catalyzed by AHAS, which requires the cofactors thiamin diphosphate (ThDP) and flavin adenine dinucleotide (FAD). A divalent metal ion, most commonly Mg^{2+}, is also required. Several excellent reviews of AHAS have appeared that describe the biochemistry, genetics, inhibition, and active site modeling of the enzyme [1–4].

Many authors have used the acronym ALS to refer to the enzyme that catalyzes the reaction of Fig. 2.1.1. However, since ALS refers specifically to the pathway leading to valine and leucine through the intermediate (S)-2-acetolactate, the designation AHAS better describes all products of the reaction. AHAS is present in bacteria, fungi, and plants. Many of the early kinetic, mechanistic, and structural studies were carried out with AHAS isolated and purified from enteric bacteria such as *Escherichia coli* and *Salmonella typhimurium*. Eukaryotic AHAS has proven more difficult to isolate and purify because of its reduced stability. Three AHAS isozymes have been characterized in bacteria – AHAS I, II, and III – whereas only one isozyme is known in fungi and plants.

Figure 2.1.2 shows the mechanism of the AHAS reaction [5, 6]. The first step involves removal of the proton attached to the 2-carbon atom of the thiazolium ring of ThDP to form an ylide. This ionization is followed by addition of the thiazolium 2-carbanion to the carbonyl group of pyruvate to give lactyl-ThDP, which

Modern Crop Protection Compounds. Edited by W. Krämer and U. Schirmer
Copyright © 2007 WILEY-VCH Verlag GmbH & Co. KGaA, Weinheim
ISBN: 978-3-527-31496-6

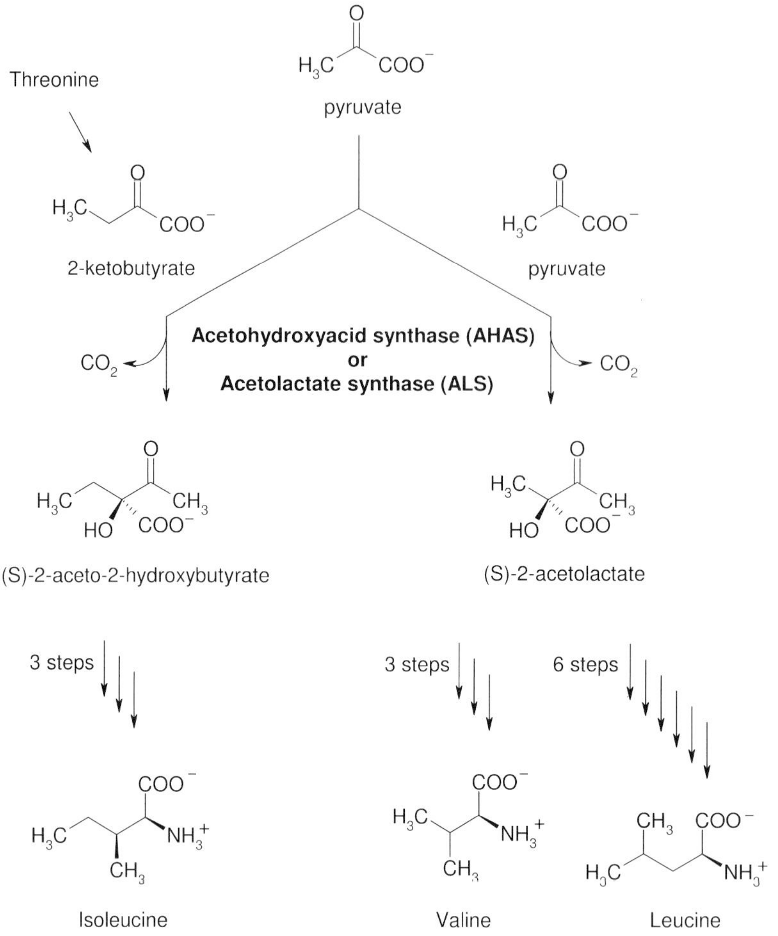

Fig. 2.1.1. Branched chain amino acid biosynthetic pathway.

loses carbon dioxide to generate the intermediate, hydroxyethyl-ThDP (HEThDP). The deprotonation is the only common step for all ThDP-dependent enzymes, and crystallographic studies combined with site-directed mutagenesis have identified a highly conserved glutamate residue as a key feature in ThDP catalysis [7]. NMR spectroscopic analysis of ThDP-ylide formation in yeast pyruvate decarboxylase (PDC) has provided convincing evidence that interaction of the glutamate with the 1′-nitrogen atom of the ThDP pyrimidine ring activates the 4′-amino group to facilitate removal of the thiazolium ring C-2 proton in an intramolecular process [8]. The exact details of the AHAS mechanism have not been completely resolved, especially with respect to the formation of the reactive ylide in the first step and whether it is a discrete or concerted process.

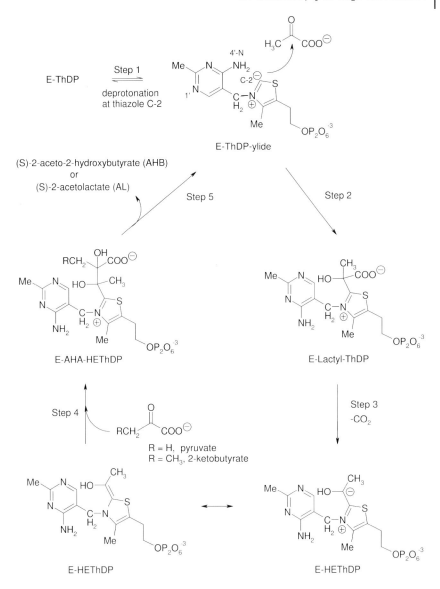

Fig. 2.1.2. Mechanism of the AHAS-catalyzed reaction.

The HEThDP intermediate reacts with either a second molecule of pyruvate or with 2-ketobutyrate to give acetolactate (AL) or acetohydroxybutyrate (AHB), respectively. In-depth studies of the kinetics of this reaction have been conducted on *E. coli* AHAS III [9]. Comparison of these results with earlier data for AHAS I and II indicate that all three bacterial isozymes catalyze similar mechanisms. However, AHAS II and III show a much more pronounced preference for reaction with 2-ketobutyrate in step 4. The substrate specificity in step 4 is an intrinsic property of the enzyme and is unaffected by pH or feedback inhibitors such as valine [10]. Recent work using a new NMR method for detecting covalent reaction intermediates has enabled the calculation of microscopic rate constants for both wild-type and mutant AHAS II from *E. coli*. These studies showed that addition of the ThDP C-2 anion to pyruvate is the rate-limiting step and that all other steps in the reaction are comparatively fast [6, 11].

The dependence of AL formation on pyruvate concentration in all three bacterial AHAS isozymes obeys Michaelis–Menten kinetics in the absence of 2-ketobutyrate. Such behavior implies that there is an irreversible step between the addition of the first and second pyruvate molecules [9, 12, 13]. Steady-state experiments confirmed that the rate-determining and product-determining steps in the mechanism are different. The observation that a wide range of substrate concentrations, changes in pH, or the presence of feedback inhibitors do not affect the specificity of the enzyme supports the idea that the first steps in the mechanism – preceding the binding of the second substrate – are rate determining [9]. Modulation of these first steps would be expected to affect the turnover rate of the enzyme without affecting the choice of products.

The role of FAD in the AHAS reaction is not fully understood since no oxidation or reduction occurs. Several hypotheses have been put forth, but so far no experimental evidence has conclusively supported any single explanation. One possibility is that FAD plays a structural role only and is likely an evolutionary remnant from a pyruvate oxidase (POX)-like ancestor [14]. The divalent metal ion does not play a direct role in the reaction, but serves to anchor the ThDP molecule to the protein by coordinating the diphosphate group and certain amino acid side chains [15].

The genes *ilv*BN, *ilv*GM, and *ilv*IH, which code for *E. coli* AHAS I, II, and III, respectively, have been cloned and sequenced [16–19]. Use of bacterial clones has enabled the production of significant quantities of each isozyme and purification essentially to homogeneity. The bacterial holoenzymes are heterotetramers composed of two types of subunits: Two large, identical catalytic subunits of approximately 60 kDa, which contain all of the catalytic machinery, and two small, identical regulatory subunits of molecular weight 9–17 kDa [12, 20, 21]. The catalytic subunit alone possesses low or no activity, but reconstitution *in vitro* with the regulatory subunit restores full enzymatic activity [22]. While AHAS I and III of *E. coli* are regulated by feedback inhibition from branched chain amino acids, especially valine, isozyme II is insensitive. The binding sites for the branched chain amino acid feedback modulators are located in the AHAS regulatory subunits [23].

Cloning and sequencing of the yeast AHAS gene, *ilv*2, and comparison of the amino acid sequence with that of *E. coli* AHAS showed a great deal of homology [24]. However, the subunit composition of eukaryotic AHAS is not as well characterized. In *Saccharomyces cerevisiae*, the *ilv*2 gene encodes a peptide that is homologous to the bacterial large (catalytic) subunit. Expression of *ilv*2 in *E. coli* gave a putative catalytic subunit, which showed considerably diminished enzymatic activity after isolation and purification [25]. Expression of the *ilv*6 gene from *S. cerevisiae* in *E. coli* and reconstitution with the yeast large subunit substantially enhanced the catalytic activity and conferred sensitivity to valine inhibition, thereby providing the first evidence for a eukaryotic small (regulatory) subunit [26].

Early studies in plants demonstrated that AHAS is a nuclear-encoded, chloroplast-localized enzyme [27, 28]. The first genes from higher plants encoding AHAS were isolated from *Arabidopsis thaliana* and *Nicotiana tabacum* using a yeast AHAS gene as a heterologous hybridization probe [29]. Comparison of the DNA and amino acid sequences from the two plants showed approximately 70% nucleotide homology and 85% homology in the mature proteins. Alignment of the plant DNA sequences with those from *E. coli* and yeast showed many regions of high homology interspersed with regions of divergence. One region of low homology between the plant proteins occurred in the first 85 amino acids, which were presumed to comprise N-terminal chloroplast transit peptides.

The first isolation of a regulatory subunit from plant AHAS was reported by Hershey et al. [30]. A cDNA clone was used to express the peptide from *Nicotiana plumbaginifolia* in *E. coli*. Based on homology with various bacterial AHAS small subunits and the observation of enhanced enzymatic activity when reconstituted with the large subunit of AHAS from either *N. plumbaginifolia* or *A. thaliana*, the authors concluded that they had isolated the regulatory subunit. Identification of the regulatory subunit of *A. thaliana* was subsequently reported by Lee and Duggleby [22], who showed that *in vitro* reconstitution not only enhanced the activity of the catalytic subunit, but also conferred sensitivity to regulation by all three branched-chain amino acids.

Several early structural models of AHAS were proffered based on homology to other known ThDP-dependent enzymes, such as POX from *Lactobacillus plantarum*, and carefully planned site-directed mutagenesis studies. Many of the features of the early models were borne out by the first X-ray crystal structure at 2.6-Å resolution of the dimeric catalytic subunit of AHAS from *S. cerevisiae* [15, 31]. Pang et al. thus confirmed that AHAS shares many structural features in common with other ThDP-dependent enzymes. Figure 2.1.3 depicts one of the monomers of the AHAS protein and the organization of the α-, β-, and γ-domains. The positions of the cofactors, ThDP, FAD, and Mg^{2+} are clearly defined. In this representation, the second ThDP molecule has been included to show the location of the enzyme active site, which is formed at the interface of the α-domain of one monomer and the γ-domain of the second monomer. The crystal structure shows that the active site without bound substrate or inhibitor is quite accessible to solvent. The two rings of ThDP are held in a bent conformation by interactions with Met525, Met555, Tyr113, Gly523, and Ala551 residues and by two hydrogen bonds [32]. ThDP is anchored to the γ-domain through its phosphate groups, which interact with the Mg^{2+} and several amino acids. FAD is most closely associated with, and binds with high affinity to, the β-domain. The cofactor is secured through numerous hydrogen bonds and van der Waals

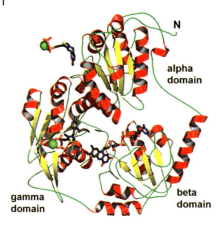

Fig. 2.1.3. Structure of a single monomer of the yeast AHAS catalytic subunit. ThDP and FAD molecules are represented as stick models. Mg^{2+} anchored to the ThDP is shown as a green sphere. ThDP of the γ-domain of the partner subunit has been added to more clearly illustrate the position of the enzyme active site at the interface of the dimer. (Reproduced with modification from Ref. [15], Figure 3, with permission from the publisher, Elsevier.)

interactions in an extended planar conformation [25, 32]. The crystal structure sheds no light on the functional role of FAD in the AHAS reaction, but does indicate that the cofactor appears to be too far removed from ThDP to be a direct participant.

2.1.2
Herbicides that Target AHAS

The discovery that certain synthetic small organic compounds inhibit AHAS and cause plant death has contributed significantly to the attention garnered by this enzyme. The first class of herbicides known to inhibit AHAS was the sulfonylureas (SUs) [33, 34]. The first commercial example of a sulfonylurea was chlorsulfuron, which was introduced by DuPont in 1982 under the trade name Glean®. This product provided highly effective control of dicotyledonous weeds in postemergence applications with excellent selectivity toward wheat [35]. Almost simultaneously, researchers at American Cyanamid discovered a structurally distinct family of herbicides, the imidazolinones (IMIs), which were also shown to inhibit the AHAS enzyme [36, 37]. Since then, three additional classes of AHAS-inhibiting herbicides have been discovered and commercial products introduced: triazolopyrimidines (TPs) from Dow AgroSciences [38], pyrimidinyl(thio)benzoates from Kumiai [39, 40], and sulfonylaminocarbonyltriazolinones from Bayer CropScience [41]. The AHAS inhibitors have proven to be among the most successful and widely used herbicides, with more than 50 active ingredients commercialized to date. This tremendous success can be attributed to the compounds' generally high bioefficacy, low field application rates, selectivity to many agronomically important crops, favorable environmental profiles, and ultra-low mammalian toxicity [42].

Early mode-of-action studies on the SUs revealed that treatment of plants with chlorsulfuron resulted in rapid cessation of cell division [43]. A further key observation in the laboratory was that growth inhibition of *S. typhimurium* caused by the SU herbicide, sulfometuron methyl, could be reversed by the addition of isoleucine, which pointed to the branched chain amino acid pathway as the biological process that was being disrupted. The mode of action of SUs in bacteria was confirmed by LaRossa and Schloss who showed that AHAS activity in extracts from wild-type *S. typhimurium* was completely inhibited by sulfometuron methyl [44]. These findings were subsequently confirmed in plants by Ray, who reported I_{50}s for chlorsulfuron inhibition of the AHAS enzyme from several plant species ranging from 18.5 (wheat) to 35.9 nM (Johnsongrass) [45]. Chaleff and Mauvais isolated tobacco mutants resistant to chlorsulfuron and sulfometuron methyl from tissue cell cultures and conclusively demonstrated that inhibition of the AHAS enzyme was the mode of herbicidal action in plants [46]. At about the same time, Shaner et al. reported that the phytotoxic effects of three IMI herbicides on corn tissue culture could be reversed by addition of valine, leucine, and isoleucine. The authors also showed that IMIs were potent inhibitors of the AHAS enzyme from *Zea mays*, with K_i values ranging from 1.7 to 12 μM [36].

2.1.3
Binding Site for AHAS-inhibiting Herbicides

The inhibition of AHAS by herbicidal compounds is a time-dependent process that is complex and not well understood [4, 47]. Since AHAS II is most similar to the enzyme in higher plants with respect to its sensitivity to various herbicides, most enzymological studies on the effects of synthetic small molecule inhibitors have been carried out on that isozyme [14, 48]. Early experiments showed that sulfometuron methyl exhibited slow, tight-binding inhibition of AHAS II from *S. typhimurium* with an initial, apparent K_i of 1.7 μM (50 mM pyruvate), followed by a time-dependent increase in potency to a final, steady-state K_i of 82 nM. The final, steady-state rate in the presence of excess herbicide indicated that the inhibition process is reversible. Although sulfometuron methyl binds to AHAS II in the absence of pyruvate, it only forms the reversible, tight-binding complex observed under turnover conditions and is competitive with pyruvate at both initial and final inhibition levels. Thus, pyruvate is required for the slowly reversible form of inhibition, but competes with sulfometuron methyl for binding to the enzyme. This observation has been explained in the context of the AHAS reaction mechanism by assuming that sulfometuron methyl binds most tightly to the enzyme following addition of the ThDP-ylide to the first molecule of pyruvate and decarboxylation (Fig. 2.1.2, steps 1–3) [49]. Evidence in support of this hypothesis was obtained from chemical quench experiments with AHAS II using ^{14}C-labeled pyruvate and ThDP, which showed that the level of HEThDP obtained by quenching steady-state reaction mixtures increased in the presence of sulfometuron methyl. Thus, while the SU virtually eliminated the enzymatic reaction, it increased the level of the HEThDP intermediate by inhibiting the binding and condensation of the second molecule of pyruvate [50].

The IMI herbicides also exhibit complex interactions with AHAS. When enzyme activity was measured over an extended period in the presence of various concentrations of imazapyr, inhibition increased with time, thereby suggesting that the equilibrium between the herbicide and AHAS was reached slowly, a characteristic of tight-binding inhibitors [51]. In contrast to SUs, substrate–inhibitor studies suggested that inhibition by imazapyr is uncompetitive with respect to pyruvate, which implies that the synthetic molecule binds to AHAS only after formation of the ternary enzyme–pyruvate–ThDP complex [52]. However, noncompetitive binding has also been reported for the IMIs, which underscores the complexity of the kinetics of AHAS inhibition [49].

The lack of obvious structural similarities among the AHAS inhibitors and the substrates or intermediates in the reaction catalyzed by AHAS suggested early on that the herbicides might not bind at the active site of the enzyme. Furthermore, the lack of obvious structural similarities among the different classes of AHAS-inhibiting herbicides, coupled with the differences in binding kinetics, has led to the speculation that the various classes of herbicides bind to different, albeit overlapping sites in the enzyme [4]. Because of the similarities between AHAS and POX, and the fact that the former enzyme requires FAD even though the reaction it catalyzes does not involve oxidation or reduction, Schloss et al. proposed that the SUs bind at a site that is distinct from the active site and is an evolutionary vestige of the ubiquinone binding site [14].

Early attempts to elucidate the herbicide binding site of AHAS were based on the similarity of AHAS to other ThDP-dependent enzymes for which X-ray crystallographic data existed. For example, a herbicide binding site structural model was postulated on the basis of homology between AHAS and POX, and an IMI molecule was positioned in the binding pocket using structure–activity information [53]. A significant milestone that greatly advanced the state of knowledge of the AHAS herbicide binding site was the publication by Pang et al. of the crystal structure at 2.8-Å resolution of yeast AHAS bound with chlorimuron ethyl, a commercial SU herbicide that is a potent inhibitor of the enzyme [54]. This crystal structure showed that the overall features of the AHAS•SU complex are quite similar to those of the free enzyme. The location and bent conformation of ThDP, which defines the enzyme active site at the interface of the α-domain of one monomer and the γ-domain of the second monomer, remain essentially unchanged vis-à-vis the free enzyme. Chlorimuron ethyl is positioned near FAD, which is in the same general location as in the free enzyme. However, the flavin ring of FAD has been displaced by several angstroms to avoid unfavorable steric interactions with the herbicide molecule.

One noteworthy difference between the crystal structures of the AHAS•SU complex and the free enzyme is that the volume of the protein in the region in which the active and herbicide binding sites are located has been reduced. A second, more significant difference is that a new substrate access channel has been formed in the AHAS•SU complex as a result of the ordering of two relatively short sequences of amino acids near the active and herbicide binding sites. As a result, ThDP exposure is substantially reduced with only the C-2 position of the thiazolium ring readily accessible to solvent. Numerous hydrophobic interactions between chlorimuron ethyl and amino acid residues along with four hydrogen bonds to the molecule's "bridge" ($-SO_2NHCONH-$ moiety) anchor the SU in the substrate access channel in such a way that the herbicide completely blocks access to the enzyme active site. The authors showed through molecular modeling that a cavity in the herbicide binding site that is normally occupied by a single water molecule will accommodate the reaction intermediate, HEThDP, with no unfavorable interactions and a stabilizing hydrogen bond to the 4′-amino group of the ThDP. This structural feature is consistent with the hypothesis that SUs bind most tightly to the AHAS enzyme following addition of the ThDP-ylide to the first pyruvate molecule and decarboxylation.

Crystal structures of four additional SUs bound to yeast AHAS were subsequently published by McCourt et al. [55]. Figure 2.1.4 shows the chemical structure of chlorsulfuron and key contact points with various amino acids in the binding site. While the conformations of all four bound SUs were similar, the authors were able to relate certain differences to structural features of the molecules and their respective binding affinities. For example, structure–activity studies had previously demonstrated the importance of the substituent in the *ortho*-position of the phenyl ring adjacent to the SU bridge for optimal herbicidal activity [34]. In the case of chlorsulfuron, this substituent is chloro. The crystal structures show that the chlorine atom of chlorsulfuron does not fit as tightly into the binding site as the carboxylic ester *ortho* groups of the other three SUs. This observation could account for the 39-fold lower binding affinity of chlorsulfuron for yeast AHAS and would be consistent with the earlier finding that the size of the *ortho* group is the most important attribute in determining the potency of SU inhibition of the enzyme [56]. The four SUs in this study differ somewhat in the nature of the extensive hydrophobic interactions that each makes with the highly conserved amino acid side-chains lining the substrate access channel. Mutation of several of the residues disrupts those interactions and has previously been shown to confer resistance to the four SUs although there was considerable variability [57]. The authors point out that, of the 13 amino acid residues making contact with the SUs, 11 are highly conserved across bacterial, fungal, and plant AHAS sequences, thereby suggesting that they play important roles in the AHAS reaction.

Twenty-two years after the introduction of the first commercial AHAS inhibiting herbicide, a preliminary crystal structure at 3.0-Å resolution of the AHAS catalytic subunit from a plant, *A. thaliana*, was published by Pang et al. [58]. This was followed two years later by crystal structures of the same enzyme complexed with five SUs and one IMI at 2.5- and 2.8-Å resolution, respectively [59]. The latter achievement by McCourt et al. represented the first reported X-ray crystal

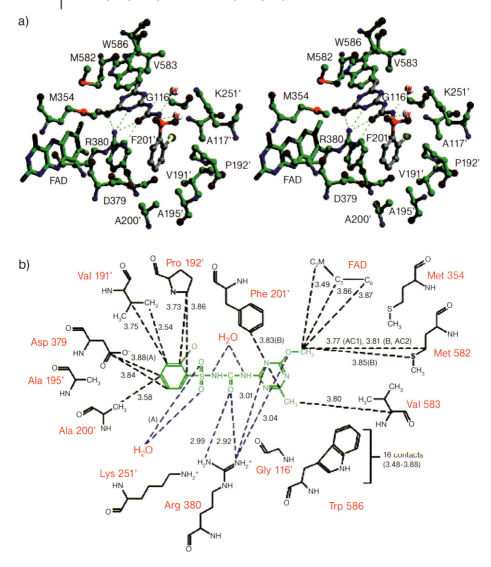

Fig. 2.1.4. Chlorsulfuron in the herbicide binding site of yeast AHAS. (a) The herbicide and nearby amino acids are shown as ball-and-stick models. The isoalloxazine ring of FAD is shown as a stick model. (b) Key contact distances (a) from chlorsulfuron to nearby amino acids. Hydrophobic contacts are broken black lines and hydrogen bonds are broken blue lines. Prime numbers denote residues from the other monomer. (Reproduced from Ref. [55], Figure 5, with permission from the publisher, American Chemical Society.)

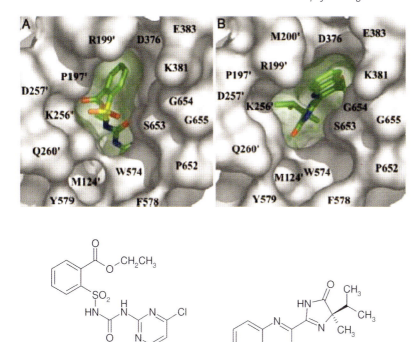

Chlorimuron ethyl   (R)-Imazaquin

Fig. 2.1.5. Herbicides bound to the AHAS protein of *Arabidopsis thaliana*, showing blockage of the channel leading to the enzyme active site. The herbicides are shown as stick models and the residues lining the channel are gray surfaces. Prime numbers denote residues from the other monomer. (Reproduced from Ref. [59], Figure 3, with permission from the publisher, National Academy of Sciences, USA.)

structures of plant protein–herbicide complexes. The overall structural features of the AHAS•SU and AHAS•IMI enzymes are similar, including a region consisting of short amino acid sequences in the vicinity of the active site that the authors hypothesize becomes more ordered in the presence of the inhibitors to form the substrate access channel in analogy to yeast AHAS. In addition, the conformations of the SU molecules bound to the plant AHAS are similar to those in yeast.

Figure 2.1.5(A) shows chlorimuron ethyl bound in the AHAS substrate access channel. The phenyl ring is located at the entrance to the channel with both the ortho-carboxylic ester group and the SU bridge pointing toward the enzyme active site. The pyrimidine ring is barely visible and is inserted deeper into the active site. Key interactions with several amino acid residues are apparent, including Trp574, Pro197, and one that is not present in yeast, Ser653. Figure 2.1.5(B) shows the IMI imazaquin positioned in the AHAS substrate access channel, with the imidazolinone ring directed toward the enzyme active site, as is the carboxylate substituent in the 3-position of the quinoline ring. Although racemic imazaquin was used in the crystallization, only the (R)-enantiomer was observed bound to the enzyme, which is consistent with the known higher herbicidal efficacy of this isomer versus the (S)-enantiomer [52].

The crystal structures provide excellent insight into the known higher AHAS binding affinity of SUs versus IMIs. For example, the apparent K_i values for inhibition of AHAS from *A. thaliana* for chlorimuron ethyl and chlorsulfuron are 10.8 and 54.6 nM, respectively, while that for imazaquin is 3 µM [60]. These differences can be attributed to the significantly greater number of van der Waal contacts and hydrogen bonds between the SUs and the enzyme versus imazaquin, and by the fact that the SUs are positioned closer to ThDP in the active site. While many of the residues interacting with the SUs and imazaquin are the same, there are six that only make contact with SUs and two that interact only with imazaquin. Thus, the crystal structures confirmed earlier suppositions that the two classes of herbicides occupy partially overlapping, but different, binding sites.

2.1.4
Molecular Basis for Resistance to AHAS Inhibitors

Several plants and cultured plant cells resistant to AHAS-inhibiting herbicides have been generated using both conventional mutation breeding and tissue culture cell selection. Mutants resistant to chlorsulfuron and sulfometuron methyl were first isolated from cultured cells of *Nicotiana tabacum* that were grown in the presence of one of the herbicides [61]. Crosses of fertile plants from several isolates established that resistance resulted from single dominant or semidominant nuclear mutation and the isolates were cross-resistant to both compounds. Co-segregation of resistance to the herbicides demonstrated that both resistances resulted from the same, or closely linked, mutations. Tobacco plants regenerated from the sulfometuron methyl-derived mutant cell lines showed resistance to high concentrations of chlorsulfuron.

Cloned yeast and bacterial genes were used to investigate the molecular basis for resistance to the SU herbicides [62]. Spontaneous mutations that conferred resistance to sulfometuron methyl were obtained in cloned genes for AHAS from *S. cerevisiae* and *E. coli*. The DNA sequences of the mutant AHAS genes showed single nucleotide differences from their wild-type counterparts, resulting in single amino acid substitutions in the corresponding proteins. The yeast mutant, Pro192Ser, resulted in reduced levels of enzyme activity, reduced sensitivity to sulfometuron methyl, and unaltered resistance to feedback inhibition from valine. The bacterial mutant, Ala26Val, resulted in unaltered levels of enzyme activity, greatly reduced sensitivity to sulfometuron methyl, and slightly reduced sensitivity to valine.

In yeast AHAS, spontaneous mutations at ten separate sites have each been shown to confer resistance to SUs [63, 64]. The X-ray crystal structure of yeast AHAS bound with chlorimuron ethyl revealed that nine of those residues make direct contact with the herbicidal molecule [54]. The authors studied the effects of several mutations on SU sensitivity in the context of molecular interactions in the herbicide active site. For example, the crystal structure showed that the indole ring of Trp586 is involved in aromatic π-orbital stacking interactions with the pyrimidine ring of chlorimuron ethyl. The mutation Trp586Leu in yeast AHAS is known to result in a >6000-fold reduction in sensitivity to chlorimuron ethyl, which can be understood in terms of the total disruption of the aromatic ring interactions.

A systematic study was carried out in which ten active mutants of yeast AHAS were constructed by mutagenesis and the resultant enzymes evaluated for their resistance to six SU and three IMI herbicides. The results were interpreted in terms of the herbicide binding site that was revealed by the X-ray crystal structure of AHAS from *S. cerevisiae* [57]. All ten mutants were resistant to some degree to the six SUs, although the levels of resistance spanned a range of nearly 10^4 and there was considerable variability in several mutants. The most consistent and highest levels of resistance were observed with Trp586Leu. The Pro192Ser mutant also displayed relatively high levels of resistance to all six Sus, and the crystal structure of yeast AHAS supports this observation in that Pro192 interacts with the phenyl rings of all of the bound herbicides. Eight of the mutants were resistant to the IMI, imazethapyr, although several of these were only barely affected and Asp379Asn was more sensitive than the wild-type enzyme. As represented schematically in Fig. 2.1.6, the positions in AHAS from various sources where

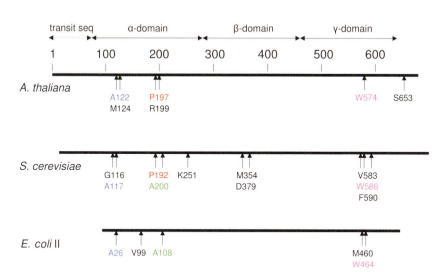

Fig. 2.1.6. AHAS mutations conferring herbicide resistance. Arrows point to positions in the sequences of AHAS from plant (*Arabidopsis thaliana*), yeast (*Saccharomyces cerevisiae*), and bacterial (*Escherichia coli*, isozyme II) sources where spontaneous or induced mutations result in an herbicide-insensitive enzyme. Colors designate substitutions occurring in more than one species.

mutations are known to confer resistance to one or more herbicides are distributed across the three domains of the protein [4, 32, 65]. At some sites, virtually any amino acid substitution confers resistance, while at others only a few substitutions are permitted.

Genes that specify herbicide-resistant forms of the AHAS enzyme were isolated from mutant *N. tabacum*, sequenced, and characterized [66]. The authors showed that a single amino acid change, Pro196Gln, in one of two distinct tobacco AHAS genes conferred resistance to SU herbicides. In the other tobacco AHAS gene, a double mutation of Pro196Ala and Trp573Leu resulted in significantly enhanced resistance to SUs. Transgenic plants carrying these mutant genes were highly resistant to chlorsulfuron treatments. Hattori et al. later showed that a single mutation in plant AHAS can confer resistance to multiple classes of AHAS-inhibiting herbicides [67, 68]. Thus, tobacco lines transformed with a plant AHAS gene specifying a single mutation, Trp557Leu, were strongly resistant to SUs, TPs, and IMIs. The tryptophan residue is conserved in virtually all wild-type AHAS proteins.

2.1.5
Resistance to AHAS-inhibiting Herbicides in Weeds

Certain characteristics of the AHAS target have played roles in the development of weeds resistant to herbicides that inhibit the enzyme: Target-based resistance is inherited as a single, semidominant trait that is carried on a nuclear gene; AHAS is the single site of action; there are multiple sites in AHAS that can be mutated to confer resistance; and mutant AHAS enzymes can possess full catalytic activity, which leads to resistant weeds that are fit [69]. Certain characteristics of the molecules themselves also contributed to resistance development, such as their potency and the relatively long soil residual of some of the early products. The first examples of resistance to chlorsulfuron occurring in the field were discovered in 1987 in the U.S. The fields containing those resistant weeds had been treated continuously with chlorsulfuron for five years. By 1992, there were numerous examples of weeds that had developed resistance to AHAS-inhibiting herbicides. Many additional observations of weeds resistant to AHAS-inhibiting herbicides have been reported since then, and there are now 93 known weed species with confirmed resistance to one or more of the five chemical classes of these compounds [70]. In 1998, the AHAS-inhibiting herbicides surpassed all other classes of herbicides in terms of the number of weed species for which at least one resistant population had been reported. At least two excellent reviews of resistance in weeds to AHAS-inhibiting herbicides have appeared [71, 72].

Studies have shown that AHAS resistance-conferring mutations can have subtle effects on plant growth and development, but they do not consistently reduce plant fitness. For example, the catalytic efficiencies of AHAS enzymes isolated from both resistant and susceptible biotypes of *L. serriola*, *K. scoparia*, *S. iberica*, *S. media*, and *L. perenne* have been shown to be virtually identical [71, 73].

In weed biotypes where the mechanism of evolved resistance has been confirmed, the majority has been due to reduced sensitivity of AHAS to the herbicide. One exception is *Lolium rigidum*, which first developed metabolism-based resistance to the ACCase inhibitor, diclofop-methyl, and then showed metabolism-based cross resistance to SUs and other classes of herbicides [74]. The identities of specific mutations in weed species that had evolved resistance to AHAS inhibitors were first determined by Guttieri et al. [75]. A Pro173His mutation was identified in resistant *L. serriola* and a Pro173Thr mutation was identified in *K. scoparia*. Mutations of five amino acid residues are known to be involved in causing resistant weed species: Ala122, Pro197, Ala205, Trp574, and Ser653 (*A. thaliana* numbering) [76]. These five amino acid residues are highly conserved across all known plant AHAS sequences [72]. Multiple substitutions have been identified for Pro197 with resistance primarily to SUs and, to a lesser extent, TPs and IMIs. The mutation, Pro197Thr, results in resistance to at least one herbicide from all five classes of AHAS inhibitors in *Chrysanthemum coronarium*, although the levels are only moderate for IMIs and TPs. The substitution Pro197Leu confers high levels of resistance to four classes of AHAS inhibitors in *Amaranthus retroflexus*. Six different amino acid substitutions in Pro197 have been linked to resistance in *K. scoparia* alone. Substitutions of Ala122 or Ser653 result in resistance to IMI, but not SU, herbicides. The mutation Trp574Leu confers resistance to several plant species and the levels of resistance are all high against IMIs, SUs, and TPs. No other evolved Trp574 mutations have been reported.

The patterns of mutation that confer evolved resistance to AHAS-inhibiting herbicides are quite understandable in light of the laboratory site-directed mutagenesis studies and the X-ray crystal structures of AHAS from yeast and *A. thaliana* that were described earlier. For example, Fig. 2.1.5 shows that Trp574 is strategically located at the opening of the herbicide binding site and interacts extensively with both chlorimuron ethyl and imazaquin. A leucine substitution alters many of those interactions and modifies the shape of the substrate access channel [59]. Figure 2.1.5 also shows that the Pro197 residue contacts the phenyl ring of chlorimuron ethyl, but is further removed from imazaquin. This structural feature explains why nearly any Pro197 replacement will hinder SU access to the channel and confer resistance while only the most bulky amino acid substitutions will displace IMIs. Conversely, Ser653 lies in close proximity to the quinoline ring of imazaquin so that replacement with a bulky amino acid would be expected to interfere with the compound's binding in the channel and confer resistance whereas this residue does not interact as strongly with chlorimuron ethyl. Not only is there wide variability in cross resistance to the different classes of AHAS-inhibiting herbicides, but resistance to one compound within a particular class does not necessarily guarantee cross resistance to all members of that family. This is particularly true of the SUs for which differential resistance has been reported in several species [77–80].

Despite the evolution of resistance to AHAS-inhibiting herbicides, these products are still among the most efficacious and widely used weed control agents in the world [72]. Commercial AHAS-inhibiting herbicides accounted for approximately 17.5% of the total worldwide herbicide market in 1997 [81]. The total sales of AHAS-inhibiting herbicides grew from 1.86 to 2.56 $billion during the decade 1994–2004, and these products still account for about 17.5% of the total worldwide herbicide market [82]. Within the past five years, eight new active ingredients have been introduced with this mode of action. Innovations in delivery methods have also recently appeared with the advent of homogeneous blends, which allow for customized mixtures of two or more different granular herbicides [83].

In cases where resistance to AHAS inhibitors has been selected, it has typically been after five to eight years of repeated, if not continuous, use of herbicides with that mode of action. Resistance has generally not been selected where AHAS-inhibiting herbicides have been used as part of an integrated program [84]. Resistance can be effectively managed by following several well-documented best practices such as rotating herbicides or using mixtures of herbicides with different modes of action with the same spectrum of weeds controlled [71, 81, 85, 86]. For example, the judicious use of glyphosate in combination with AHAS inhibitors may provide a powerful tool for managing resistance to both classes of herbicides. With this approach, the rate of increase of AHAS-resistant weed species in the USA may slow [81]. By following a well-planned weed management program, growers should be able to use the environmentally friendly AHAS-inhibiting herbicides for many years to come to achieve effective, broad-spectrum weed control.

Acknowledgments

The author is grateful to Drs. Hugh Brown, Josephine Cotterman, Ronald Duggleby, Jerry Green, Steven Gutteridge, Jennifer McCourt, Robert Pasteris, and Leonard Saari for their critical review of the manuscript, to Drs. Ya-Jun Zheng and John Andreassi for assistance with Figs. 2.1.3–2.1.5, to Mr. Thomas Dougherty for access to the global sales figures for AHAS-inhibiting herbicides, and to Ms. Debbie Carman and Ms. Susan Titter for assistance with the literature references.

References

1 Umbarger, H. E. *Annu. Rev. Biochem.*, **1978**, 47, 533–606.

2 Chipman, D. M., Barak, Z., Schloss, J. V. *Biochim. Biophys. Acta*, **1998**, 1385, 401–419.

3 Duggleby, R. G., Pang, S. S. *J. Biochem. Mol. Biol.*, **2000**, 33, 1–36.

4 Duggleby, R. G., Guddat, L. W., Pang, S. S., Structure and Properties of Acetohydroxyacid Synthase in *Thiamine: Catalytic Mechanisms in Normal and Disease States*, Vol. 11, Marcel Dekker, New York, **2004**, 251–274.

5 Pang, S. S., Duggleby, R. G., Schowen, R. L., Guddat, L. W. *J. Biol. Chem.*, **2004**, 279, 2242–2253.

6 McCourt, J. A., Duggleby, R. G. *Trends Biochem. Sci.*, **2005**, 30, 222–225.

7 Wikner, C., Meshalkina, L., Nilsson, U., Nikkola, M., Lindqvist, Y., Sundstrom, M., Schneider, G. *J. Biol. Chem.*, **1994**, 269, 32144–32150.

8 Kern, D., Kern, G., Neef, H., Tittmann, K., Killenberg-Jabs, M., Wikner, C., Schneider, G., Hübner, G. *Science*, **1997**, 275, 67–70.

9 Gollop, N., Damri, B., Barak, Z., Chipman, D. M. *Biochemistry*, **1989**, 28, 6310–6317.

10 Barak, Z., Chipman, D. M., Gollop, N. *J. Bacteriol.*, **1987**, 169, 3750–3756.

11 Tittmann, K., Golbik, R., Uhlemann, K., Khailova, L., Schneider, G., Patel, M., Jordan, F., Chipman, D. M., Duggleby, R. G., Hübner, G. *Biochemistry*, **2003**, 42, 7885–7891.

12 Grimminger, H., Umbarger, H. E. *J. Bacteriol.*, **1979**, 137, 846–853.

13 Schloss, J. V., Van Dyk, D. E., Vasta, J. F., Kutny, R. M. *Biochemistry*, **1985**, 24, 4952–4959.

14 Schloss, J. V., Ciskanik, L. M., Van Dyk, D. E. *Nature*, **1988**, 331, 360–362.

15 Pang, S. S., Duggleby, R. G., Guddat, L. W. *J. Mol. Biol.*, **2002**, 317, 249–262.

16 Squires, C. H., DeFelice, M., Devereux, J., Calvo, J. M. *Nucleic Acids Res.*, **1983**, 11, 5299–5313.

17 Lawther, R. P., Calhoun, D. H., Adams, C. W., Hauser, C. A., Gray, J., Hatfield, G. W. *Proc. Natl. Acad. Sci. U.S.A.*, **1981**, 78, 922–925.

18 Wek, R. C., Hauser, C. A., Hatfield, G. W. *Nucleic Acids Res.*, **1985**, 13, 3995–4010.

19 Friden, P., Donegan, J., Mullen, J., Tsui, P., Freundlich, M., Eoyang, L., Weber, R., Silverman, P. M. *Nucleic Acids Res.*, **1985**, 13, 3979–3993.

20 Eoyang, L., Silverman, P. M. *J. Bacteriol.*, **1984**, 157, 184–189.

21 Ibdah, M., Bar-Ilan, A., Livnah, O., Schloss, J. V., Barak, Z., Chipman, D. M. *Biochemistry*, **1996**, 35, 16282–16291.

22 Lee, T.-Y., Duggleby, R. G. *Biochemistry*, **2001**, 40, 6836–6844.

23 Mendel, S., Vinogradov, M., Vyazmensky, M., Chipman, D. M., Barak, Z. *J. Mol. Biol.*, **2003**, 325, 275–284.

24 Falco, S. C., Dumas, K. S. *Genetics*, **1985**, 109, 21–35.

25 Poulsen, C., Stougaard, P. *Eur. J. Biochem.*, **1989**, 185, 433–439.

26 Pang, S. S., Duggleby, R. G. *Biochemistry*, **1999**, 38, 5222–5231.

27 Miflin, B. J. *Plant Physiol.*, **1974**, 54, 550–555.

28 Jones, A. V., Young, R. M., Leto, K. *Plant Physiol.*, **1985**, 77, S293.

29 Mazur, B. J., Chui, C.-F., Smith, J. K. *Plant Physiol.*, **1987**, 85, 1110–1117.

30 Hershey, H. P., Schwartz, L. J., Gale, J. P., Abell, L. M. *Plant Mol. Biol.*, **1999**, 40, 795–806.

31 Pang, S. S., Guddat, L. W., Duggleby, R. G. *Acta Crystallog. Sect. D*, **2001**, 57, 1321–1323.

32 Pang, S. S., Duggleby, R. G., Schowen, R. L., Guddat, L. W. *J. Biol. Chem.*, **2004**, 279, 2242–2253.

33 Beyer, E. M., Duffy, M. J., Hay, J. V., Schlueter, D. D. Sulfonylureas in *Herbicides: Chemistry, Degradation, and Mode of Action*, Marcel Dekker, New York, **1988**, 117–189.

34 Levitt, G. Discovery of the Sulfonylurea Herbicides in *Synthesis and Chemistry of Agrochemicals II*, Baker, D. R., Fenyes, J. G., Moberg, W. K. (Eds.), American Chemical Society, Washington, D.C., **1991**, 16–31.

35 Sweetser, P. B., Schow, G. S., Hutchison, J. M. *Pestic. Biochem. Physiol.*, **1982**, 17, 18–23.

36 Shaner, D. L., Anderson, P. C., Stidham, M. A. *Plant Physiol.*, **1984**, 76, 545–546.

37 Shaner, D. L., O'Connor, S. L. (Eds.) *The Imidazolinone Herbicides*, CRC Press, Boca Raton, FL, **1991**.

38 Kleschick, W. A., Gerwick, B. C., Carson, C. M., Monte, W. T., Snider, S. W. *J. Agric. Food Chem.*, **1992**, 40, 1083–1085.

39 Shimizu, T. *J. Pestic. Sci.*, **1997**, 22. 245–256.

40 Shimizu, T., Nakayama, I., Nagayama, K., Miyazawa, T., Nezu, Y. Acetolactate Synthase Inhibitors in *Herbicide Classes in Development*, Böger, P., Wakabayashi, K., Hirai, K. (Eds.), Springer-Verlag, Berlin, **2002**, 1–41.

41 Pontzen, R. *Pflanz.-Nachrichten Bayer,* **2002**, 55, 37–52.

42 Russell, M. H., Saladini, J. L., Lichtner, F. *Pesticide Outlook,* **2002**, 166–173.

43 Ray, T. B. *Pestic. Biochem. Physiol.,* **1982**, 17, 10–17.

44 LaRossa, R. A., Schloss, J. V. *J. Biol. Chem.,* **1984**, 259, 8753–8757.

45 Ray, T. B. *Plant Physiol.,* **1984**, 75, 827–831.

46 Chaleff, R. S., Mauvais, C. J. *Science,* **1984**, 224, 1443–1445.

47 Kishore, G. M., Shah, D. M. *Annu. Rev. Biochem.,* **1988**, 57, 627–663.

48 Schloss, J. V., Van Dyk, D. E. *Methods Enzymol.,* **1988**, 166, 445–454.

49 Schloss, J. V., Aulabaugh, A. Acetolactate Synthase and Ketol-Acid Reductoisomerase: A Search for Reason and a Reason for Search, in *Biosynthesis of Branched Chain Amino Acids,* Barak, Z., Chipman, D. M., Schloss, J. V. (Eds.), VCH Verlagsgesellschaft, Weinheim, **1990**, 329–356.

50 Schloss, J. V., Ciskanik, L. M. *Biochemistry,* **1985**, 24, 3357 (Abstract).

51 Muhitch, M. J., Shaner, D. L., Stidham, M. A. *Plant Physiol.,* **1987**, 83, 451–456.

52 Stidham, M. A., Singh, B. K. Imidazolinone-Acetohydroxyacid Synthase Interactions in *The Imidazolinone Herbicides,* Shaner, D. L., O'Connor, S. L. (Eds.), CRC Press, Boca Raton, FL, **1991**, 71–90.

53 Ott, K.-H., Kwagh, J.-G., Stockton, G. W., Sidorov, V., Kakefuda, G. *J. Mol. Biol.,* **1996**, 263, 359–368.

54 Pang, S. S., Guddat, L. W., Duggleby, R. G. *J. Biol. Chem.,* **2003**, 278, 7639–7644.

55 McCourt, J. A., Pang, S. S., Guddat, L. W., Duggleby, R. G. *Biochemistry,* **2005**, 44, 2330–2338.

56 Andrea, T. A., Artz, S. P., Ray, T. B., Pasteris, R. J. Structure-Activity Relationships of Sulfonylurea Herbicides in *Rational Approaches to Structure, Activity, and Ecotoxicology of Agrochemicals,* Draber, W., Fujita, T. (Eds.), CRC Press, Boca Raton, FL, **1992**, 373–395.

57 Duggleby, R. G., Pang, S. S., Yu, H., Guddat, L. W. *Eur. J. Biochem.,* **2003**, 270, 2895–2904.

58 Pang, S. S., Guddat, L. W., Duggleby, R. G. *Acta Crystallogr., Sect. D,* **2004**, 60, 153–155.

59 McCourt, J. A., Pang, S. S., King-Scott, J., Guddat, L. W., Duggleby, R. G. *Proc. Natl. Acad. Sci. U.S.A.,* **2006**, 103, 569–573.

60 Chang, A. K., Duggleby, R. G. *Biochem. J.,* **1998**, 333, 765–777.

61 Chaleff, R. S., Ray, T. B. *Science,* **1984**, 223, 1148–1151.

62 Yadav, N., McDevitt, R. E., Bernard, S., Falco, S. C. *Proc. Natl. Acad. Sci. U.S.A.,* **1986**, 83, 4418–4422.

63 Falco, S. C., McDevitt, R. E., Chui, C.-F., Hartnett, M. E., Knowlton, S., Mauvais, C. J., Smith, J. K., Mazur, B. J. *Dev. Ind. Microbiol.,* **1989**, 30, 187–194.

64 Bedbrook, J. R., Chaleff, R. S., Falco, S. C., Mazur, B. J., Somerville, C. R., Yadav, N. S. U.S. Patent 5,013,659, **1991**.

65 Mazur, B. J., Falco, S. C. *Annu. Rev. Plant Physiol. Plant Mol. Biol.,* **1989**, 40, 441–470.

66 Lee, K. Y., Townsend, J., Tepperman, J., Black, M., Chui, C. F., Mazur, B., Dunsmuir, P., Bedbrook, J. *EMBO J.,* **1988**, 7, 1241–1248.

67 Cole, D. J., Rodgers, M. W. Plant Molecular Biology for Herbicide-Tolerant Crops and Discovery of New Herbicide Targets in *Herbicides and Their Mechanisms of Action,* Cobb, A., Kirkwood, R. C. (Eds.), Sheffield Academic Press, Sheffield, UK, **2000**, 239–278.

68 Hattori, J., Brown, D., Mourad, G., Labbé, H., Ouellet, T., Sunohara, G., Rutledge, R., King, J., Miki, B. *Mol. Gen. Genetics,* **1995**, 246, 419–425.

69 Hartnett, M. E., Chui, C.-F., Mauvais, C. J., McDevitt, R. E., Knowlton, S., Smith, J. K., Falco, S. C., Mazur, B. J. Herbicide-Resistant Plants Carrying Mutated Acetolactate Synthase Genes in *Managing Resistance to Agrochemicals,* Green, M. B., LeBaron, H. M., Moberg, W. K. (Eds.), American Chemical Society, Washington, D.C., **1990**, 459–473.

70 Heap, I. M. The International Survey of Herbicide Resistant Weeds. Online. Internet. February 10, 2006. Available http://www.weedscience.com.

71 Saari, L. L., Cotterman, J. C., Thill, D. C. Resistance to Acetolactate Synthase Inhibiting Herbicides, in *Herbicide Resistance in Plants*, Powles, S. B., Holtum, J. A. M. (Eds.), CRC Press, Boca Raton, FL, **1994**, 83–139.

72 Tranel, P. J., Wright, T. R. *Weed Sci.*, **2002**, 50, 700–712.

73 Holt, J. S., Thill, D. C. Growth and Productivity of Resistant Plants in *Herbicide Resistance in Plants: Biology and Biochemistry*, Powles, S. B., Holtum, J. A. M. (Eds.), CRC Press, Boca Raton, FL, **1994**, 299–316.

74 Christopher, J. T., Powles, S. B., Holtum, J. A. M. *Plant Physiol.*, **1992**, 100, 1909–1913.

75 Guttieri, M. J., Eberlein, C. A., Mallory-Smith, D. C., Thill, D. C., Hoffman, D. L. *Weed Sci.*, **1992**, 40, 670–677.

76 Tranel, P. J., Wright, T. R., Heap, I. M. ALS mutations from herbicide-resistant weeds. Online. Internet. February 10, 2006. Available http://www.weedscience.com.

77 Devine, M. D., Marles, M. A. S., Hall, L. M. *Pestic. Sci.*, **1991**, 31, 273–280.

78 Hart, S. E., Saunders, J. W., Penner, D. *Weed Sci.*, **1993**, 41, 317–324.

79 Saari, L. L., Cotterman, J. C., Smith, W. F., Primiani, M. M. *Pestic. Biochem. Physiol.*, **1992**, 42, 110–118.

80 Sibony, M., Michel, A., Haas, H. U., Rubin, B., Hurle, K. *Weed Res.*, **2001**, 41, 509–522.

81 Shaner, D. L., Heap, I. Herbicide Resistance in North America: The Case for Resistance to ALS Inhibitors in the United States in *Agrochemical Resistance*, ACS Symposium Series, 808, American Chemical Society, Washington, D.C., **2002**, 161–167.

82 Phillips, M. Phillips McDougall. Personal communication. January 2006.

83 Geigle, W. L., Gleich, S. I. U.S. Patent 6,270,025, **2001**.

84 Shaner, D. L., Feist, D. A., Retzinger, E. J. *Pest. Sci.*, **1997**, 51, 367–370.

85 Shaner, D. L. *J. Weed Sci. Technol.*, **1999**, 44, 405–411.

86 Preston, C., Mallory-Smith, C. A. Biochemical Mechanisms, Inheritance, and Molecular Genetics of Herbicide Resistance in Weeds in *Herbicide Resistance and World Grains*, Powles, S. B., Shaner, D. L. (Eds.), CRC Press, Boca Raton, FL, **2001**, 23–60.

2.2
Newer Sulfonylureas

Oswald Ort

2.2.1
Introduction

Most commercial sulfonylurea herbicides are characterized by the typical sulfonylated urea bridge connecting a nitrogen-containing heterocycle with an ortho-substituted aryl or heteroaryl moiety (Fig. 2.2.1). To date, sulfonylurea herbicides have been developed and commercialized worldwide in over 80 countries, in all major agronomic crops and for many specialty uses (e.g., rangeland/pasture, forestry, vegetation management applications).

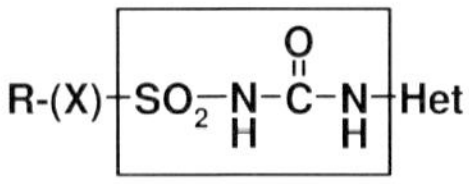

typical sulfonylurea bridge

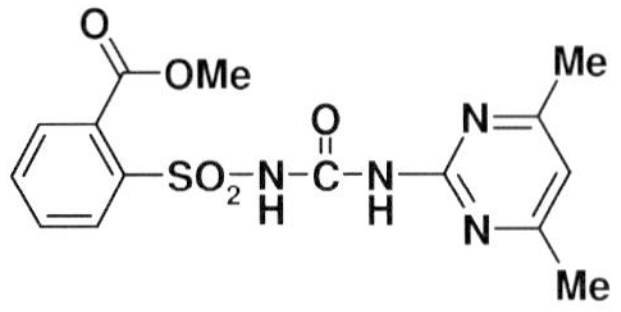

Sulfometuron-methyl (DuPont) 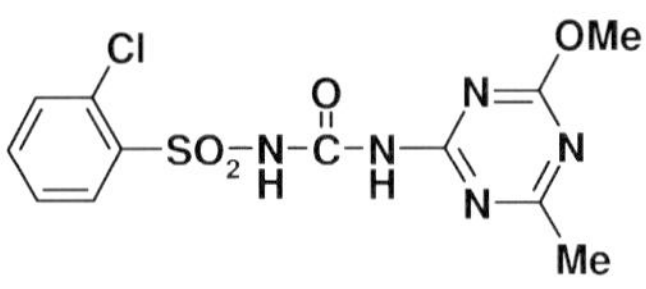   Chlorsulfuron (DuPont)

Fig. 2.2.1. First generation sulfonylurea herbicides.

Herbicidal sulfonylureas have a unique mode of action: they interfere with a key enzyme required for plant cell growth – acetohydroxyacid synthase (AHAS, EC 2.2.1.6) [1, 2, 3] (see also Mark E. Thompson in this volume, Chapter 2.1 "Biochemistry of the Target and Resistance"). AHAS is the enzyme responsible for the synthesis of the branched-chain amino acids valine, leucine and isoleucine. Inhibition of this enzyme disrupts the plant's ability to manufacture proteins, and this disruption subsequently leads to the cessation of all cell division and eventual death of the plant.

The visible signs of herbicide action after postemergent application of sulfonylurea herbicides are an almost immediate arrest of growth, followed by leaf yellowing (chlorosis), stimulation of anthocyanin production (leading to the typical reddish coloration of weed leaves), and finally, progressive shoot death. Depending on the weed species and environmental conditions, plant death will usually occur between seven and twenty days after herbicide application.

Since the initial discovery of the first sulfonylurea herbicides by George Levitt at Du Pont in 1975, this compound class has attracted, and indeed continues to attract, very high interest and much activity within the agrochemical research domain.

This continued research interest can largely be attributed to the fact that the herbicidal activity levels demonstrated by this class of compounds remain unsurpassed today – with the most active compounds able to control undesired vegetation at rates lower than 10 g-a.i. ha^{-1} ($= 1$ mg-a.i. m^{-2}).

This, in turn, then contributes to a reduction in environmental burden by replacement of older higher-rate herbicides and provides an attractive return on investment to both the farmer and the producing company. In addition, the favorable environmental properties and low acute mammalian toxicology shown by the sulfonylureas usually provide a large margin of safety with regard to ecological and toxicological effects (cf. Tables 2.2.1 and 2.2.2).

Table 2.2.1 Acute toxicity for birds and water organisms (LD$_{50}$, LC$_{50}$ or EC$_{50}$).

Compound	Birds[a] (mg kg^{-1})	Fish (96 h, mg L^{-1})	Daphnia magna (48 h, mg L^{-1})
Azimsulfuron	>2250	>154[b]	941
Cyclosulfamuron	Not available	>100	Not available
Ethoxysulfuron	>2000	>78.4	307
Flucetosulfuron	Not available	>10[c]	>10
Flupyrsulfuron	>2250	470	721
Foramsulfuron	>2000	>100	>100
Iodosulfuron	>2000	>100	>100
Mesosulfuron	>2000	>100	>100
Oxasulfuron	>2250	>100	>89.4
Sulfosulfuron	>2250	>91	>96
Trifloxysulfuron	>2000	>120[c]	>108
Tritosulfuron	Not available	>100	>100

[a] Mallard duck/Bobwhite quail.
[b] Oncorhynchus mykiss.
[c] Cyprinus carprio.

Table 2.2.2 Acute toxicity for mammals (LD$_{50}$ or LC$_{50}$).

Compound	Oral rat (mg kg^{-1})	Dermal rat (mg kg^{-1})	Inhalation rat (mg L^{-1})
Azimsulfuron	>5000	>2000	Not available
Cyclosulfamuron	>5000	>4000[a]	>5.2
Ethoxysulfuron	>3270	>4000	3.6
Flucetosulfuron	>5000	>2000	>5.03
Flupyrsulfuron	>5000	>2000	>5.8
Foramsulfuron	>5000	>2000	>5.0
Iodosulfuron	2678	>2000	>2.8
Mesosulfuron	>5000	>5000	1.3
Orthosulfamuron	>5000	>5000	>2.2
Oxasulfuron	>5000	>2000	>5.1
Sulfosulfuron	>5000	>5000	>3.0
Trifloxysulfuron	>5000	>2000	>5.0
Tritosulfuron	3310	>2000	5.9

[a] Rabbit.

The objective of this Chapter is to give an overview of the sulfonylurea herbicides that either have been introduced to the market since 1995 or are currently in their later stages of development. These include flupyrsulfuron-methyl-sodium, sulfosulfuron, iodosulfuron-methyl-sodium, mesosulfuron-methyl, tritosulfuron and monosulfuron for use in cereals; ethoxysulfuron, azimsulfuron, cyclosulfamuron, flucetosulfuron, TH 547 and orthosulfamuron in rice; foramsulfuron in maize; oxasulfuron in soybeans; and trifloxysulfuron-sodium in sugarcane and cotton.

2.2.1.1 History and Development

Since George Levitt's landmark discovery of herbicidal sulfonylurea herbicides at Du Pont in 1975 many hundreds of patents have been granted to Du Pont and, in addition, to over twenty other agrochemical companies.

Numerous review articles about sulfonylurea herbicides are available. Amongst those particularly recommended for further reading is George Levitt's original description of his work [4]. This monograph contains seven more papers on sulfonylurea herbicides covering the literature up to 1991. Another, earlier standard text for sulfonylurea enthusiasts is that of Beyer et al. [5]. More recent sulfonylurea reviews can be found in Ref. [6].

The first generation of crop selective sulfonylurea herbicides, e.g., chlorsulfuron and metsulfuron-methyl, was found to be mainly active against broadleaf weeds. At that time it was thought that this herbicide class would be specifically applicable for controlling broadleaf weed species in a wide range of crops. This view changed with the appearance of the second generation of sulfonylureas bearing pyridylsulfonamide moieties, such as nicosulfuron and rimsulfuron, for use in maize. These compounds were active not only against broadleaf weeds but also against a broad spectrum of grasses. A further important break-through was achieved with the advent of the third generation of sulfonylurea herbicides: grass-killer experts such as mesosulfuron, blackgrass specialists such as flupyrsulfuron, and cross spectrum compounds such as iodosulfuron (cf. Table 2.2.6) and foramsulfuron (cf. Table 2.2.20).

Table 2.2.3 gives an overview of those sulfonylurea herbicides that were introduced prior to 1995.

Before going on to describe the compounds and their uses in more detail, the following section will provide a short overview of the most commonly applied synthesis methods that can be used in the production of these compounds.

Table 2.2.3 Sulfonylurea herbicides introduced before 1995 (in alphabetical order).

Chemical structure	Main crop	Common name (company)	Application rates (g-a.i. ha^{-1})	Ref.
CH$_3$-SO$_2$-N-SO$_2$-N-C-N— (pyrimidine with OCH$_3$, OCH$_3$); CH$_3$	Cereals	Amidosulfuron (Bayer CropScience)	30–60	6, 60
(benzene with C=O, OCH$_3$)-CH$_2$-SO$_2$-N-C-N— (pyrimidine with OCH$_3$, OCH$_3$)	Rice	Bensulfuron-methyl (Du Pont)	20–75	61
(benzene with C=O, OCH$_2$CH$_3$)-SO$_2$-N-C-N— (pyrimidine with OCH$_3$, Cl)	Soybeans	Chlorimuron-ethyl (Du Pont)	8–13	62
(benzene with Cl)-SO$_2$-N-C-N— (triazine with OCH$_3$, CH$_3$)	Cereals	Chlorsulfuron (Du Pont)	9–25	63
(benzene with O-CH$_2$-CH$_2$-OCH$_3$)-SO$_2$-N-C-N— (triazine with OCH$_3$, OCH$_3$)	Rice	Cinosulfuron (Syngenta)	20–40	6
(benzene with C=O, OCH$_3$)-SO$_2$-N-C-N— (triazine with NHCH$_3$, OCH$_2$CH$_3$)	Oilseed rape	Ethametsulfuron-methyl (Du Pont)	15–20	6, 64
(pyridine with CF$_3$)-SO$_2$-N-C-N— (pyrimidine with OCH$_3$, OCH$_3$)	Turf, vegetation management	Flazasulfuron (Ishihara)	25–100	6, 65
(pyrazole with Cl, C=O OCH$_3$, CH$_3$)-SO$_2$-N-C-N— (pyrimidine with OCH$_3$, OCH$_3$)	Maize, turf	Halosulfuron (Nissan)	18–35	6, 66

Table 2.2.3 (continued)

Chemical structure	Main crop	Common name (company)	Application rates (g-a.i. ha^{-1})	Ref.
	Rice, turf	Imazosulfuron (Takeda)	50–100	6
	Cereals, rice vegetation management	Metsulfuron-methyl (Du Pont)	3–8 14–168	6, 67
	Maize	Nicosulfuron (Du Pont/ Ishihara)	35–70	68, 69
	Maize	Primisulfuron-methyl (Syngenta)	20–40	70
	Cereals, maize	Prosulfuron (Syngenta)	20–40	6, 71
	Rice	Pyrazosulfuron-ethyl (Nissan)	15–30	6
	Maize	Rimsulfuron (Du Pont)	5–35	72
	Vegetation management	Sulfometuron-methyl (Du Pont)	26–420	73

Table 2.2.3 *(continued)*

Chemical structure	Main crop	Common name (company)	Application rates (g-a.i. ha^{-1})	Ref.
	Cereals, maize, soybeans	Thifensulfuron-methyl (Du Pont)	2–30	74
	Cereals	Triasulfuron (Syngenta)	10–30	75
	Cereals	Tribenuron-methyl (Du Pont)	9–18	6, 76
	Sugar beet	Triflusulfuron-methyl (Du Pont)	15–30	6, 77

2.2.1.2 Synthesis

The five synthetic approaches shown in Scheme 2.2.1 have been employed most widely for the construction of the typical sulfonylurea bridge [7, 8].

Of these methods, number 1 is the most commonly used for the commercial production of sulfonylureas. The reaction is high yielding and highly atom efficient, giving advantages for downstream waste processing. In addition, the required sulfonylisocyanates are readily accessed from the corresponding sulfonamides by reaction with phosgene under several conditions. The second method has the advantages of saving two reaction steps to prepare the sulfonylisocyanate and allowing the reaction to proceed in one pot. This method is particularly useful in cases where the sulfonylisocyanate is difficult to isolate or where formation of a saccharin as by-product is problematic. Whilst methods 3 and 4 result in good conversion into the targeted products, they suffer from the undesirable production of phenol as a by-product. This can be overcome by employing alkoxy N-heterocyclylcarbamates in the presence of AlMe$_3$, leading to the generation of more innocuous alcoholic by-products.

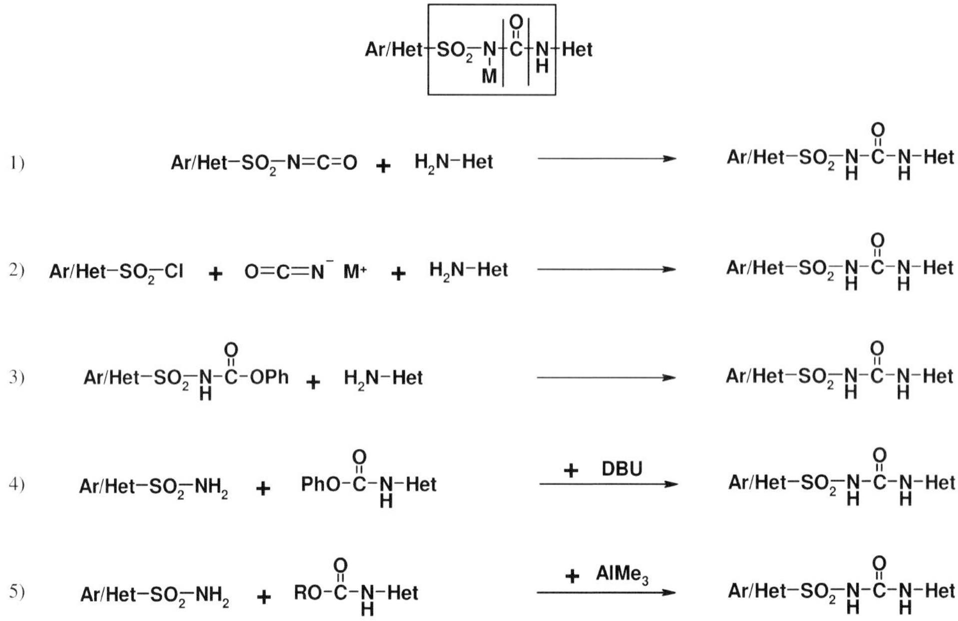

$$\boxed{\text{Ar/Het}-\text{SO}_2-\overset{\underset{\textstyle M}{|}}{\text{N}}-\overset{\overset{\textstyle O}{\|}}{\text{C}}-\underset{\underset{\textstyle H}{|}}{\text{N}}-\text{Het}}$$

1) $\text{Ar/Het}-\text{SO}_2-\text{N}{=}\text{C}{=}\text{O} \;+\; \text{H}_2\text{N}-\text{Het} \longrightarrow \text{Ar/Het}-\text{SO}_2-\underset{\text{H}}{\text{N}}-\overset{\overset{\text{O}}{\|}}{\text{C}}-\underset{\text{H}}{\text{N}}-\text{Het}$

2) $\text{Ar/Het}-\text{SO}_2-\text{Cl} \;+\; \text{O}{=}\text{C}{=}\text{N}^-\,\text{M}^+ \;+\; \text{H}_2\text{N}-\text{Het} \longrightarrow \text{Ar/Het}-\text{SO}_2-\underset{\text{H}}{\text{N}}-\overset{\overset{\text{O}}{\|}}{\text{C}}-\underset{\text{H}}{\text{N}}-\text{Het}$

3) $\text{Ar/Het}-\text{SO}_2-\underset{\text{H}}{\text{N}}-\overset{\overset{\text{O}}{\|}}{\text{C}}-\text{OPh} \;+\; \text{H}_2\text{N}-\text{Het} \longrightarrow \text{Ar/Het}-\text{SO}_2-\underset{\text{H}}{\text{N}}-\overset{\overset{\text{O}}{\|}}{\text{C}}-\underset{\text{H}}{\text{N}}-\text{Het}$

4) $\text{Ar/Het}-\text{SO}_2-\text{NH}_2 \;+\; \text{PhO}-\overset{\overset{\text{O}}{\|}}{\text{C}}-\underset{\text{H}}{\text{N}}-\text{Het} \xrightarrow{\;+\;\text{DBU}\;} \text{Ar/Het}-\text{SO}_2-\underset{\text{H}}{\text{N}}-\overset{\overset{\text{O}}{\|}}{\text{C}}-\underset{\text{H}}{\text{N}}-\text{Het}$

5) $\text{Ar/Het}-\text{SO}_2-\text{NH}_2 \;+\; \text{RO}-\overset{\overset{\text{O}}{\|}}{\text{C}}-\underset{\text{H}}{\text{N}}-\text{Het} \xrightarrow{\;+\;\text{AlMe}_3\;} \text{Ar/Het}-\text{SO}_2-\underset{\text{H}}{\text{N}}-\overset{\overset{\text{O}}{\|}}{\text{C}}-\underset{\text{H}}{\text{N}}-\text{Het}$

Scheme 2.2.1. Basic construction routes to the sulfonylurea bridge.

2.2.2
Agricultural Utility

In 2004 the global crop protection market (excluding seeds & biotechnology) amounted to 25.9 bio €, with herbicides accounting for roughly half of this value at 12.3 bio € sales. The herbicide subtotal is relatively evenly spread over the three main crops: cereals, maize and soybeans (19%, 18% and 16%, respectively); rice 8% and others 39% (including non-selective/non-crop use), with the main markets for herbicides in North America and Europe [9]. Table 2.2.4 gives an overview of the global acreage and production of the world's major arable crops.

The following sections give an overview, split by crop segment, of the new sulfonylurea herbicides that have either been introduced since 1995 or are currently in their later stages of development.

2.2.2.1 Cereals

Cereals (wheat, barley, sorghum, oats, rye and triticale) are the most important of the arable crops (Table 2.2.4). In 2005, global cereal production was approximately 870 mio tonnes on 340 mio hectares of land, with wheat (*Triticum aestivum*) being the most important cereal grain, accounting for more than two-thirds of the total production (Table 2.2.5).

Geographically, the largest cereal production areas are in regions with temperate conditions, such as Europe, North America and cooler parts of Australia and China.

Table 2.2.4 Major crops of the world, average 2004–05/source FAO [78].

Crop	Mio (ha)	Mio MT[a] production
Cereals[b]	333	882
Rice, paddy	152	613
Maize (grain)	147	709
Soybeans	91	208
Sugarcane	20	1309
Rapeseed	26	46
Sunflower	22	29
Potatoes	19	325
Sugar beets	6	246

[a] MT = metric tonnes.
[b] Wheat, rye, oats, triticale, barley, sorghum.

A significant area of cereal fields world-wide are infested with grass weeds. More than 40 major grass weed species can be found in cereal fields and these weeds are often highly competitive with the crop, causing substantial losses in both yield and quality. As mentioned above, sulfonylurea herbicides have been used in cereals, mainly as herbicides against broadleaf weeds, since they were first introduced in the early 1980s. Tank-mixtures with these first generation sulfonylureas or spraying programs involving different grass weed herbicides remained the most widely employed chemical control strategy up to the late 1990s. The introduction of the new generation of sulfonylurea herbicides, such as flupyrsulfuron-methyl-sodium, iodosulfuron-methyl-sodium or mesosulfuron-methyl with their broad-spectrum grass and broadleaf performance, provided the farmer with a single, innovative and easy one-pass solution saving both time and cost (Table 2.2.6).

In the following sub-sections, each of the compounds listed in Table 2.2.6 are described in more detail.

Table 2.2.5 Major cereal crops of the world, average 2004–05/Source FAO [78].

Crop	Mio (ha)	Mio MT production
Wheat	212	624
Barley	55	144
Oats	12	25
Rye	7	17
Triticale	4	14
Sorghum	43	58

Table 2.2.6 Cereal sulfonylurea herbicides in order of market introduction.

Chemical structure	Common name (company, launch year)	Agricultural utility	Application rate (g-a.i. ha^{-1})
	Flupyrsulfuron-methyl-sodium (Du Pont, 1997)	Grass weeds and select broadleaf weeds	8–10
	Sulfosulfuron (Takeda/ Monsanto, 1997)	Grass weeds and broadleaf weeds	10–35
	Iodosulfuron-methyl-sodium (Bayer CropScience, 1999)	Broadleaf and grass weeds	5–10 + safener Mefenpyr
	Mesosulfuron-methyl (Bayer CropScience, 2001)	Grass weeds and select broadleaf weeds	6–15 + safener Mefenpyr
	Tritosulfuron (BASF, 2004)	Broadleaf weeds	30–50

2.2.2.1.1 Flupyrsulfuron-methyl-sodium

Flupyrsulfuron-methyl-sodium (DPX-KE459) (Table 2.2.7) [10] is a postemergent cereal herbicide designed for the control of problem grass weeds, such as *Alopecurus myosuroides* and *Apera spica-venti*, and a wide range of broadleaf weeds with application rates of 8–10 g-a.i. ha^{-1}. It is commercialized by Du Pont [11] under the trade name "Lexus® 50DF" as a stand-alone product or as "Lexus® Class" in a 1:2 ratio in combination with carfentrazone-ethyl and was launched in 1997. At the 10 g-a.i. ha^{-1} rate, the following broadleaf weeds are well controlled: *Chenopodium album*, *Lamium purpureum*, *Matricaria* sp., *Polygonum aviculare*, *P. convolvulus*, *Senecio vulgaris* and *Sinapis arvensis*.

Ciral® is a combination of flupyrsulfuron-methyl-sodium (33.3%) and metsulfuron-methyl (16.7%) for the control of *Alopecurus myosuroides*, *Apera spica-venti*, *Poa annua* and annual broadleaf weeds (except *Galium aparine*) such as *Thlaspi arvense*, *Capsella bursa-pastoris*, *Galeopsis* spp., *Matricaria* spp., *Papaver rhoeas*, *Centaurea cyanus*, *Brassica napus*, *Viola arvensis*, *Lamium* spp., *Myosotis ar-*

Table 2.2.7 Physicochemical properties of flupyrsulfuron-methyl-sodium.

Common name (ISO)	Flupyrsulfuron-methyl-sodium
CAS-No.	144740-54-5
Code numbers	DPX-KE459
Melting point	Not determined (decomposition at 165–170 °C)
Vapor pressure	$<1 \times 10^{-9}$ Pa (20 °C)
Dissociation constant (at 20 °C)	pK_a 4.94 (93.4%)
Solubility in water (g L^{-1} at 20 °C)	0.06 (93.4%) (pH 5) 0.61 (pH 6) Instability of the solution (pH 7)
Solubility in organic solvents (g L^{-1} at 20 °C)	Acetone 3.1 (93.4%) Acetonitrile 4.3 Benzene 0.028 Dichloromethane 0.60 Hexane <0.001 Methanol 5.0 *n*-Octanol 0.19
Partition coefficient (log P_{OW}) in octanol–water (at 25 °C)	0.96 (pH 5.0) 0.11 (93.4%) (pH 6.0)

vensis and *Stellaria media*. Ciral® is a flexible product that can be applied in autumn and spring in wheat and in flax at a dose rate of 25 g ha^{-1} [12] of formulated product containing 8 g-a.i. flupyrsulfuron-methyl-sodium.

2.2.2.1.2 Sulfosulfuron

Sulfosulfuron (MON 37500) (Table 2.2.8) [13] is a postemergent herbicide for the control of grass (especially *Bromus* species) and broadleaf weeds in cereal crops at rates of 10–35 g-a.i. ha^{-1}. Sulfosulfuron was jointly developed [14, 15] by Monsanto Company and Takeda Chemical Industries and launched in 1997. Barley and oats are sensitive to applications of sulfosulfuron and so use in these crops is not recommended. At rates of 20–30 g-a.i. ha^{-1} the following weeds are controlled with at least 85% efficiency: *Elymus repens, Apera spica-venti, Agrostis*

Table 2.2.8 Physicochemical properties of sulfosulfuron.

Common name (ISO)	Sulfosulfuron	
CAS-No.	141776-32-1	
Code numbers	MON 37500 TKM 19	
Melting point	201.1–201.7 °C	
Vapor pressure	3.05×10^{-8} Pa (20 °C)	
Dissociation constant (at 20 °C)	pK_a 3.51	
Solubility in water (g L^{-1} at 20 °C)	0.018 (pH 5) 1.627 (pH 7) 0.482 (pH 9)	
Solubility in organic solvents (g L^{-1} at 20 °C)	Acetone	0.71
	Ethyl acetate	1.01
	Dichloromethane	4.35
	n-Heptane	<0.001
	Methanol	0.33
	Xylene	0.16
Partition coefficient (log P_{ow}) in octanol–water	0.73 (pH 5.0) −0.77 (pH 7.0) −1.44 (pH 9.0)	

stolonifera, Avena fatua (North America), *Bromus commutatus, B. japonicus, B. mollis, B. rigidus, B. secalinus, B. sterilis, B. tectorum, Poa bulbosa* and *Poa trivialis, Ambrosia artemisiifolia, Amsinckia lycopsoides, Atriplex patula, Brassica nigra, Capsella bursa-pastoris, Claytonia per, Descurainia pinnata, D. sophia, Fumaria officinalis, Galium aparine, Helianthus* sp., *Matricaria chamomilla, M. inodora, Polygonum aviculare, P. persicaria, Sinapis arvensis, Sisymbrium altissimum, Stellaria media, Thlaspi arvense* and *Viola arvensis.* In Europe sulfosulfuron is commercially available under the trade name "Monitor".

2.2.2.1.3 Iodosulfuron-methyl-sodium

At a rate of 2.5–10 g-a.i. ha^{-1}, iodosulfuron-methyl-sodium (AE F115008) (Table 2.2.9) [16] controls more than 50 different broadleaf weed species, including some very competitive weeds that cause a substantial reduction of cereal productivity, e.g., *Galium aparine, Matricaria chamomilla, Stellaria media, Raphanus* ssp.,

Table 2.2.9 Physicochemical properties of iodosulfuron-methyl-sodium.

Common name (ISO)	Iodosulfuron-methyl-sodium
CAS-No.	144550-36-7
Code numbers	AE F115008
Melting point	148–152 °C
Vapor pressure	2.6×10^{-9} Pa (20 °C) 6.7×10^{-9} Pa (25 °C)
Dissociation constant (at 20 °C)	pK_a 3.22 (under strong acidic conditions – pH 2 – formation of iodosulfuron-methyl)
Solubility in water (g L^{-1} at 20 °C)	0.16 (pH 5) 25.0 (pH 7) 65.0 (pH 9)
Solubility in organic solvents (g L^{-1} at 20 °C)	Acetonitrile 52 Ethyl acetate 23 *n*-Heptane 0.001 Methanol 12 2-Propanol 4.4 Toluene 2.1
Partition coefficient (log P_{OW}) in octanol–water (at 20 °C)	1.96 (pH 4.0) 1.22 (pH 9.0)

Cirsium arvense, Lamium ssp. Whilst application of iodosulfuron-methyl-sodium at the lower end of the suggested use-rate is usually sufficient for control of broadleaf weeds, a higher rate is needed for consistent grass weed control. Major grass weeds controlled with a 7.5–10 g-a.i. ha^{-1} dose rate applied at the three-leaf stage up to end of tillering are *Agrostis gigantea, Apera spica-venti, Lolium multiflorum, L. perenne, L. persicum, L. rigidum, Phalaris brachystachys, P. canariensis, P. paradoxa, Poa annua,* and *P. trivialis.*

Iodosulfuron-methyl-sodium was the first safened sulfonylurea herbicide on the market [17, 18] when introduced in 1999 and has been commercialized by Bayer CropScience for use both in cereals and maize. A safener such as mefen-pyr-diethyl (cf. Fig. 2.2.2) is a chemical that, when applied to crop plants, reduces the injury caused by herbicides to an acceptable level. A safener ideally does not

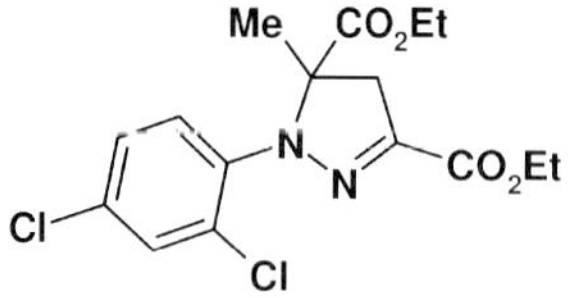

Fig. 2.2.2. Cereal safener mefenpyr-diethyl (AE F107892).

reduce activity against the target weeds. A series of experiments were conducted to compare the behavior of iodosulfuron-methyl-sodium with and without the safener mefenpyr-diethyl. The findings suggest that the safener acts by specific catalytic enhancement of herbicide degradation in cereals but not in target weeds such as wild oat. The topic concerning safeners is dealt with later in more detail (see Chris Rosinger and Helmut Koecher in this volume, Chapter 5 "Safener for Herbicides").

In cereals, iodosulfuron-methyl-sodium is commercialized under the trade name "Hussar" as a straight product in a 1:3 ratio with the safener mefen-

Table 2.2.10 Iodosulfuron-based products, formulations and composition.

Iodosulfuron-based products	Formulation type	Iodo-sulfuron	Meso-sulfuron	Fenoxa-prop-ethyl	Amido-sulfuron	Mefenpyr-diethyl
Hussar[R], Husar[R], Huzar[R], Huszar[R], Al Fares[R], Wipe[R]	WG[a]	50[b]				150[b]
Sekator[R], Grodyl[R] Ultra	WG[a]	12.5[b]			50[b]	125[b]
Sekator[R] OD	OD[c]	25[d]			100[d]	250[d]
Chekker[R], Hoestar[R] Super	WG[a]	12.5[b]			125[b]	125[b]
Hussar[R] OF Evolution	SC[e]	8[d]		64[d]		24[d]
Hussar[R] OD	OD[c]	100[d]				300[d]
Atlantis[R] WG	WG[a]	6[b]	30[b]			90[b]
Pacifica[R]	WG[a]	10[b]	30[b]			90[b]
Archipel[R], Cossack[R], Chevalier[R], Hussar maxx[R]	WG[a]	30[b]	30[b]			90[b]

[a] WG: water dispersible granules.
[b] Units: g-a.i. kg^{-1}.
[c] OD: oil dispersion.
[d] Units: g-a.i. L^{-1}.
[e] SC: suspension concentrate.

pyr-diethyl. The compound is also commercialized in various combinations with other mixing partners such as "Hussar[R] OF" (+ fenoxaprop-P-ethyl + mefenpyr-diethyl), "Sekator[R]"/"Chekker[R]" (+ amidosulfuron + mefenpyr-diethyl), "Cossack[R]" (+ mesosulfuron + mefenpyr-diethyl) and "Atlantis[R]" (+ mesosulfuron + mefenpyr-diethyl) (cf. Table 2.2.10).

2.2.2.1.4 Mesosulfuron-methyl

Mesosulfuron-methyl (AE F130060) (Table 2.2.11) [19] was the second safened sulfonylurea herbicide for cereal crops to be commercialized. This compound was introduced in 2001 and has been commercialized by Bayer CropScience [20, 21]. Its strength is broad-spectrum post-emergence grass weed control. Mesosulfuron-methyl, at a dose rate of 4.5–15 g-a.i. ha^{-1}, reliably controls 24 different grass weed species from 12 different families. Among the commercially

Table 2.2.11 Physicochemical properties of mesosulfuron-methyl.

Common name (ISO)	Mesosulfuron-methyl	
CAS-No.	208465-21-8	
Code numbers	AE F130060	
Melting point	195.4 °C (98.7% purity)	
Vapor pressure	1.1×10^{-11} Pa (25 °C)	
Dissociation constant (at 20 °C)	pK_a 4.35	
Solubility in water (g L^{-1} at 20 °C)	0.007 (pH 5) 0.483 (pH 7) 15.39 (pH 9)	
Solubility in organic solvents (g L^{-1} at 20 °C)	Acetone Ethyl acetate Dichloromethane *n*-Hexane Toluene	13.66 2.0 3.8 <0.0002 0.013
Partition coefficient (log P_{OW}) in octanol–water (at 25 °C)	1.39 (pH 5.0) −0.48 (pH 7.0) −2.06 (pH 9.0)	

important grass weed species, it provides good control of *Agrostis* spp., *Alopecurus myosuroides*, *Apera spica-venti*, *Avena* spp. *Lolium* spp., *Phalaris brachystachis*, *P. minor*, *P. paradoxa*, *Poa annua*, *Poa trivialis*, *Pucciniella* spp. and *Sclerochloa kengiana*. Additionally mesosulfuron-methyl controls, or has a strong suppressive effect on, some very persistent grass weed species, such as *Bromus catharticus*, *B. diandrus*, *B. erectus*, *B. japonicus*, *B. mollis*, *B. tectorum*, *B. secalinus*, *B. sterilis* and *Vulpia* spp.

The compound is applied on soft and durum wheat, triticale and rye, together with the safener mefenpyr-diethyl (Fig. 2.2.2) as the straight products "Atlantis® OF", "Silverado®" and "Osprey®" or in combination with iodosulfuron-methyl-sodium ("Atlantis® WG", "Cossack®", "Pacifica®"), diflufenican and propoxycarbazone-sodium (Table 2.2.12).

"Atlantis® WG" is positioned in market segments where grass weeds are the main target, whereas "Cossack®" is a cross spectrum product, active against grasses and against a large number of important broadleaf weeds. Mesosulfuron-methyl belongs to the group of modern OnePass® products. It predominantly acts via the leaves of treated weeds; however, highly susceptible grasses, such as

Table 2.2.12 Mesosulfuron based products, formulations and composition.

Mesosulfuron-based products	Formulation type	Mesosulfuron-methyl	Iodosulfuron-methyl sodium	Diflu-fenican	Propoxy-carbazone	Mefenpyr-diethyl
Atlantis® OD	OD[a]	30[b]				90[b]
Atlantis®	WG[r]	30[d]	6[d]			90[d]
Pacifica®	WG[c]	30[d]	10[d]			90[d]
Archipel®, Cossack®, Chevalier®, Hussar maxx®	WG[c]	30[d]	30[d]			90[d]
Silverado®	WG[c]	20[d]				120[d]
Osprey®	WG[c]	45[d]				90
Alister®	OD[a]	9[b]	3[b]	150[b]		27[b]
Othello®	OD[a]	7.5[b]	2.5[b]	50[b]		22.5[b]
Olympus flex®	WG[c]	45[d]			67.5[d]	90[d]

[a] OD: oil dispersion.
[b] Units: g-a.i. L^{-1}.
[c] WG: water dispersible granules.
[d] Units: g-a.i. kg^{-1}.

Apera and *Alopecurus*, are also successfully controlled by uptake of mesosulfuron-methyl via the soil and the roots.

The safener mefenpyr-diethyl, as with iodosulfuron-methyl-sodium, selectively accelerates the degradation of the active ingredient to non-phytotoxic compounds in cereals but not in weeds.

2.2.2.1.5 Tritosulfuron

Tritosulfuron (BAS-635) (Table 2.2.13) [22] is a broad-spectrum postemergent dicot herbicide mainly for use in cereals, rice, maize and turf with application rates of 40–75 g-a.i. ha^{-1}. In cereals it was commercialized by BASF in 2004 under the trade name "Biathlon®" [23] as a WG formulation containing 714 g kg^{-1} tritosulfuron and is applied at a rate of 50 g-a.i. ha^{-1}. The following weeds are well controlled: *Thlaspi arvense*, *Mercurialis annua*, *Urtica urens*, *Cirsium arvense*, *Veronica hederifolia*, *Chenopodium* spp., *Sinapis arvensis*, *Capsella bursa-pastoris*, *Galeopsis*

Table 2.2.13 Physicochemical properties of tritosulfuron.

Common name (ISO)	Tritosulfuron
CAS-No.	142469-14-5
Code numbers	BAS-635
Melting point	167–169 °C
Vapor pressure	1.0×10^{-5} Pa (20 °C)
Dissociation constant (at 20 °C)	pK_a 4.69
Solubility in water (g L^{-1} at 20 °C)	<0.001 (pH 1.7) 0.04 (pH 7.0) 78.32 (pH 10.2)
Solubility in organic solvents (g L^{-1} at 20 °C)	Acetone — Acetonitrile — Ethyl acetate 83.0 Dichloromethane 25.0 *n*-Heptane <0.001 Methanol 23.0 Toluene 4.2
Partition coefficient (log P_{OW}) in octanol–water (at 20 °C)	2.85 (pH 4.0) 0.62 (pH 7.0) −2.38 (pH 10.0)

tetrahit, *Matricaria* spp., *Galium aparine*, *Polygonum* spp., *Centaurea cyanus*, *Lamium* spp., *Myosotis arvensis*, *Stellaria media*, *Vicia* spp., *Convolvulus arvensis*, *Sonchus arvensis*, *Brassica napus*. Tritosulfuron acts mainly through the treated leaves and not via the soil. The compound has the advantage of having a short soil half-life, which allows re-cropping after 60 days without plowing [24, 25].

Tritosulfuron is selective in the following cereal crops: wheat, rye, barley, triticale, oat, durum wheat and spelt. The application window of tritosulfuron in all winter and summer cereals ranges from vegetation start up to ES 39. Sold in maize as "Tooler®", it can be applied from ES 12 to ES 18.

2.2.2.1.6 Cereals Development Candidates

Two compounds, monosulfuron and NPC-C9908, from research in China are currently in the early market introduction phase or late development stage in China. Thus there is only limited public knowledge available about these compounds.

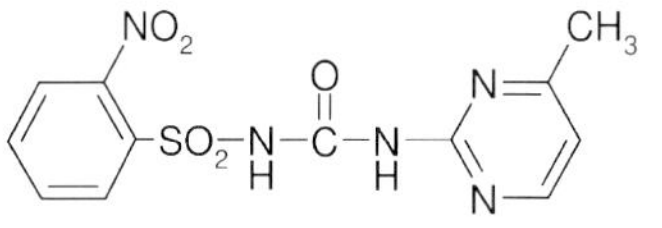

Fig. 2.2.3. Monosulfuron.

Monosulfuron (CAS-No.: 155860-63-2) (Fig. 2.2.3) is a new herbicide for the control of weeds in wheat (*Triticum aestivum*) and millet (*Panicum miliaceum*), with application rates ranging from 15–60 g-a.i. ha^{-1}. The molecule was discovered by Nankai University in 1993 [26] and recently registered in China. Monosulfuron provides effective control of various broadleaf and grass weeds, such as *Leptochloa chinensis*, *Amaranthus retroflexus*, *Chenopodium album*, *Abutilon theophrasti*, *Xanthium sibiricum* Patrin., *Portulaca oleracea*, *Acalypha australis*, *Solanum nigrum*, *Digitaria sanguinalis*, *Descurainia sophia*, *Echinochloa phyllopogon*, *Eriochloa villosa* and *Puccinellia distans*. Further properties and environmental data of monosulfuron are detailed in several papers by Fan [27].

HNPC-C9908 [2-(4-methoxy-6-methylthiopyrimidin-2-yl) carbamoyl sulfonyl benzoate] (CAS-No.: 441050-97-1) (Fig. 2.2.4) is a novel sulfonylurea herbicide [28, 29] discovered by the Hunan Branch of the National Pesticide R&D South Center, Changsha, China, and is reported to be effective in controlling various broadleaf weeds and some grasses in wheat.

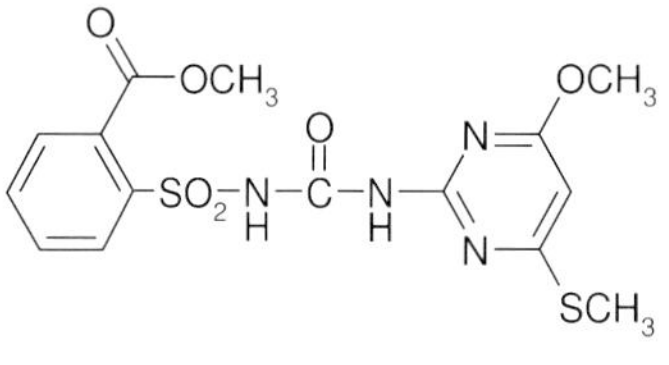

Fig. 2.2.4. HNPC-C9908 herbicide.

Table 2.2.14 Rice sulfonylurea herbicides in order of market introduction.

Chemical structure	Common name (company, launch year)	Agricultural utility	Application rate (g-a.i. ha^{-1})
(structure: O–C$_2$H$_5$, O–SO$_2$–N–C–N–, OCH$_3$, OCH$_3$)	Ethoxysulfuron (Bayer CropScience, 1996)	Annual and perennial broadleaf and sedge weeds	6–60
(structure: N–N–CH$_3$, SO$_2$–N–C–N–, OCH$_3$, OCH$_3$, N–N–CH$_3$)	Azimsulfuron (Du Pont, 1996)	Annual and perennial broadleaf and sedge weeds	6–25
(structure: cyclopropyl ketone, N–SO$_2$–N–C–N–, OCH$_3$, OCH$_3$)	Cyclosulfamuron (BASF, 1997)	Annual and perennial broadleaf and sedge weeds	10–60
(structure: CH$_2$OCH$_3$, CHFCH$_3$, SO$_2$–N–C–N–, OCH$_3$, OCH$_3$)	Flucetosulfuron (KRICT/LG Chem., 2004)	Annual and perennial broadleaf and sedge weeds	15–60

2.2.2.2 Rice

Around 60% of the global population, particularly in Asia, rely on rice (*Oryza sativa*) as a major food source. Rice is grown mainly in the humid and sub-humid tropics of the Far East. Rice production on ca. 154 mio hectares totaled 618 mio tonnes in 2005, with the biggest two producers, China and India, being responsible for more than half of the global total [78]. However, value-wise Japan is the largest rice market with >40% of the total market value.

It is estimated that, on average, weed infestation in tropical rice areas accounts for 10–20% of yield loss, but there are studies that show that some problem weed species, such as red rice (*Oryza sativa* ssp.) and barnyard grass (*Echinochloa crusgalli*), can cause even higher losses. Red rice (the term "red rice" is used synonymously for weedy rice because its grains frequently have a red pigmented pericarp) is in the same genus and species as cultivated conventional rice, which makes it very difficult to eliminate in rice fields. Fisher and Ramirez [30, 31] found that a 5% density of red rice decreased conventional rice yields by up to 40%.

Typically herbicides in rice are used in combinations of active ingredients and, as labor is becoming more expensive, the trend is towards single application products. These products usually contain sulfonylureas as the main active ingredient (Table 2.2.14).

The following sub-sections describe in more detail each of the compounds listed in Table 2.2.14.

2.2.2.2.1 Ethoxysulfuron

Ethoxysulfuron (HOE 095404) [32] is a very flexible herbicide for the control of broadleaf and sedge weed species (Table 2.2.15).

Although rice is the main use crop, the compound can also be applied in cereals and sugar cane [33]. Selectivity is achieved due to a differential metabolism in the target crops to that in the weeds [34]. With an application rate of 15–60 g-a.i. ha^{-1} a wide range of important annual and perennial rice weeds are controlled,

Table 2.2.15 Physicochemical properties of ethoxysulfuron.

Common name (ISO)	Ethoxysulfuron	
CAS-No.	126801-58-9	
Code numbers	HOE 095404	
Melting point	144–147 °C	
Vapor pressure	6.6×10^{-5} Pa (20 °C)	
Dissociation constant (at 20 °C)	pK_a 5.28	
Solubility in water (g L^{-1} at 20 °C)	0.026 (pH 5) 1.353 (pH 7) 9.628 (pH 9)	
Solubility in organic solvents (g L^{-1} at 20 °C)	Acetone	36.0
	Ethyl acetate	14.1
	Dichloromethane	107.0
	n-Hexane	0.006
	Methanol	7.7
	Poly(ethylene glycol)	22.5
	Toluene	2.5
Partition coefficient (log P_{OW}) in octanol–water (at 21 °C)	2.89 (pH 3.0) 0.004 (pH 7.0) −1.22 (pH 9.0)	

Table 2.2.16 Ethoxysulfuron-based products, formulations and composition.

Ethoxysulfuron-based products	Formulation type	Ethoxy-sulfuron	Anilofos	Fenoxaprop-ethyl	Safener
Gladium, Grazie, Hero, Skol, Sunrice	WG[a]	600[b]			
Sunrice, Sunstar	WG[a]	150[b]			
Ricestar®, Ricestar® Xtra, Tiller® Gold, Turob®	OD[c]	20[d]		69[d]	75[d]
Sunrice Plus	SC[e]	15[d]	300[d]		

[a] WG: water dispersible granules.
[b] Units: g-a.i. kg^{-1}.
[c] OD: oil dispersion.
[d] Units: g-a.i. L^{-1}.
[e] SC: suspension concentrate.

such as *Cyperus* spp., *Aeschynomene* spp., *Eleocharis* spp., *Sagittaria* spp., *Scirpus* spp., *Amannia* spp., *Lindernia* spp., *Ludwigia* spp. and *Monochoria vaginalis*. Ethoxysulfuron is fully selective in all types of seeded rice (dry drilled, pre-germinated wet seeded, pre-germinated water seeded) and all types of transplanted rice. The selectivity is not influenced by the rice growth stage at application time, the water management or other environmental factors. Ethoxysulfuron was introduced in rice in 1996 (Vietnam) and has been commercialized by Bayer CropScience as a straight product under the trade name "Sunrice® WG" and as SC formulated products in combination with anilofos as "Riceguard®", "Benefiter®", "Sunrice® Super" and "Sunrice® Plus" (Table 2.2.16).

2.2.2.2.2 Azimsulfuron

Azimsulfuron (DPX-A8947) [35] is a new rice herbicide introduced in 1996 by Du Pont [36] for the control of broadleaf weeds (including hard-to-control perennials) (Table 2.2.17). At rates of 8–20 g-a.i. ha^{-1} it gives superior weed control, including *Echinochloa crus-galli*, when compared with the first-generation sulfonylurea bensulfuron at 50–75 g-a.i. ha^{-1}. Azimsulfuron is targeted to replace or supplement bensulfuron in some applications. In Japan, in planted rice, azimsulfuron is used as a pre-mixture with bensulfuron (6 + 30 g-a.i. ha^{-1}) to boost the activity against perennial weeds. Good control has also been reported of other members of the Echinochloa family, such as *E. hispidula*, *E. oryzicola* and *E. oryzoides*. Other weeds controlled include *Alisma lanceolatum*, *A. plantago-aquatica*, *Butomus umbellatus*, *Cyperus difformis*, *Scirpus maritimus*, *S. mucronatus*, *S. supinus*, *Heteranthera limosa*, *Potamogeton nodosus*, *Ammannia coccinea*, *A. robusta*, *Bergia capensis* and *Lindernia dubia*.

Azimsulfuron is sold under the trade names "Gulliver" and "Azin".

Table 2.2.17 Physicochemical properties of azimsulfuron.

Common name (ISO)	Azimsulfuron	
CAS-No.	120162-55-2	
Code numbers	DPX-A8947	
Melting point	170 °C	
Vapor pressure	4.0×10^{-9} Pa (25 °C)	
Dissociation constant (at 20 °C)	pK_a 3.6	
Solubility in water (20 °C)	0.072 (pH 5)	
	1.050 (pH 7)	
	6.536 (pH 9)	
Solubility in organic solvents	Acetone	26.4
(g L^{-1} at 25 °C)	Acetonitrile	13.9
	Ethyl acetate	13.0
	Dichloromethane	65.9
	n-Hexane	<0.2
	Methanol	2.1
	Toluene	1.8
Partition coefficient (log P_{OW}) in octanol–water (at 25 °C)	0.646 (pH 5.0)	
	−1.367 (pH 7.0)	
	−2.076 (pH 9.0)	

2.2.2.2.3 Cyclosulfamuron

Cyclosulfamuron (AC 322,140) herbicide was launched in 1997 and is commercialized by BASF for control of a wide range of broadleaf weeds and sedge species in rice, wheat and barley [37] (Table 2.2.18). Rice weeds controlled with greater 90% efficiency at an application rate of 45–60 g-a.i. ha^{-1} include *Cyperus serotinus, C. difformis, Elatine triandra, Eleocharis congesta, E. kuroguwai, Lindernia annua, L. procumbens, Monochoria vaginalis, Rotala indica, Sagittaria pygmaea, S. trifolia* and *Scirpus juncoides*. Selectivity in the rice paddy is achieved due to various factors, including rapid metabolic degradation of the herbicide in rice shoots, placement of rice seedlings during transplanting and the compound's soil binding properties, which retain cyclosulfamuron in the upper soil layer of the paddy [38].

Table 2.2.18 Physicochemical properties of cyclosulfamuron.

Common name (ISO)	Cyclosulfamuron	
CAS-No.	136849-15-5	
Code numbers	AC 322,140	
Melting point	170–171 °C	
Vapor pressure	$<2.2 \times 10^{-5}$ Pa (20 °C)	
Dissociation constant (at 20 °C)	pK_a 5.04	
Solubility in water (25 °C)	0.001 (pH 5)	
	0.003 (pH 6)	
	0.006 (pH 7)	
	0.032 (pH 8)	
Solubility in organic solvents ($g L^{-1}$ at 20 °C)	Acetone	21.0
	Ethyl acetate	5.0
	Dichloromethane	50.0
	n-Hexane	<0.001
	Methanol	1.5
	Toluene	1.0
Partition coefficient (log P_{OW}) in octanol–water (at 25 °C)	1.58 (pH 3.0)	
	2.05 (pH 5.0)	
	1.69 (pH 6.0)	
	1.41 (pH 7.0)	
	0.70 (pH 8.0)	

In rice, cyclosulfamuron is commercialized under the trade name "Ichiyon-maru" and "Saviour". In combinations with daimuron and cafenstrole it is commercialized as "Nebiros" and in combination with pentoxazone as "Utopia". "Shakariki" is the trade name for the mixture with esprocarb.

At rates of 25–50 g-a.i. ha^{-1} cyclosulfamuron can also be used in cereal crops for pre- and postemergent control of several important broadleaf weeds, such as *Veronica persica*, *V. hederifolia*, *Galium aparine*, *Matricaria* spp. and *Polygonum convolvulus* [38].

Cyclosulfamuron cannot be synthesized by any of the general methods depicted in Scheme 2.2.1. From the methods published in the patent literature, Brady et al. [40] describe a straightforward reaction of 2-amino-4,6-

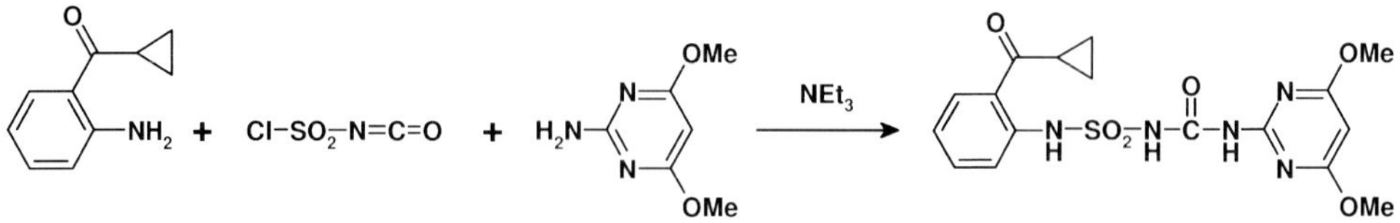

Scheme 2.2.2. CSI route to cyclosulfamuron.

dimethoxypyrimidine with chlorosulfonylisocyanate (CSI) at 0 °C with a mixture of 2-aminophenyl cyclopropyl ketone and triethylamine to yield 70% of the desired herbicide (Scheme 2.2.2). This synthesis method of cyclosulfamuron and its intermediate products can also be found in the paper by Tan from 2005 [41].

Table 2.2.19 Physicochemical properties of flucetosulfuron.

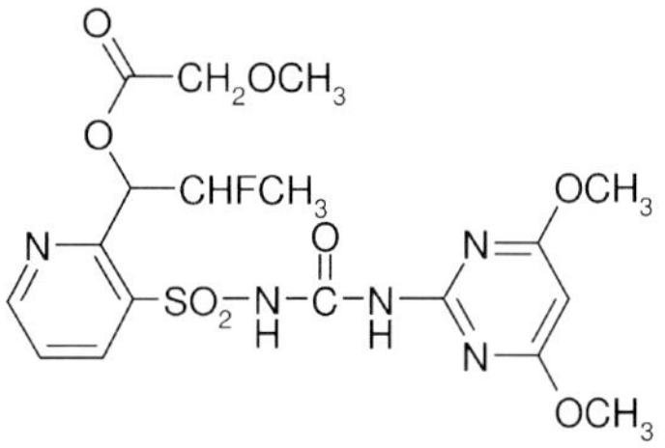

Common name (ISO)	Flucetosulfuron
CAS-No.	412928-75-7
Code numbers	LGC-42153
Melting point	178–182 °C
Vapor pressure	$<1.86 \times 10^{-5}$ Pa (25 °C)
Dissociation constant (at 20 °C)	pK_a 3.5
Solubility in water (25 °C)	0.114 g L^{-1} (pH 7)
Solubility in organic solvents (g L^{-1} at 20 °C)	Acetone 22.9
	Ethyl acetate 11.7
	Dichloromethane 113.0
	Dimethylformamide 265.0
	Dimethyl sulfoxide 211.7
	n-Hexane 0.006
	Methanol 3.8
Partition coefficient (log P_{OW}) in octanol–water (temperature not published)	n.a. (pH 3.0)
	1.05 (pH 7.0)
	n.a. (pH 9.0)

2.2.2.2.4 Flucetosulfuron

Flucetosulfuron (LGC-42153) [42] was presented at the BCPC Conference in 2003 by researchers from LG Life Sciences Ltd. and KRICT [43, 44] and was commercialized in 2004 (Table 2.2.19). It can be used for the control of broadleaf weeds, some grass weeds and also sedges in rice and cereal crops. In rice, flucetosulfuron provides excellent control of *Echinochloa crus-galli*, which is usually not controlled by other commercial rice sulfonylurea products. In addition, the following weeds are controlled at a rate of 10–20 g-a.i. ha^{-1}: *Alisma* spp., *Ammannia coccinea*, *Cyperus difformis*, *Fimbristylis* spp., *Lindernia* spp., *Monochoria vaginalis*, *Rorippa silvestri*, *Rotala indica*, *Scirpus juncoides*, *S. mucronatus* and *S. maritimus*. At a higher rate of 20–30 g-a.i. ha^{-1}, greater than 90% control of *Aeschymene indica*, *Butomus umbellatus*, *Eleocharis kuroguwai*, *Sagittaria pygmaea*, *S. trifolia* and *Sparganium erectum* is achieved by flucetosulfuron with a high crop safety margin when applied to soil or foliage in direct-seeded or transplanted rice.

In cereal crops, flucetosulfuron at a 20–30 g-a.i. ha^{-1} rate shows excellent activity against *Galium aparine* and other broadleaf weeds, such as *Capsella bursa-pastoris*, *Galeopsis tetrahit*, *Lamium purpureum*, *Matricaria* spp., *Myosotis arvensis*, *Papaver rhoeas*, *Raphanus raphanistrum*, *Senecio vulgaris*, *Sinapis arvensis*, *Stellaria media* and *Thlaspi arvense*, while being safe to use in wheat and barley at up to three times the recommended application rate.

2.2.2.2.5 Rice Development Candidates

In rice there are currently two compounds, TH 547 from research at Sumika-Takeda, and orthosulfamuron from Isagro, shortly before market introduction.

TH 547 (Fig. 2.2.5) is a new sulfonylurea under development by Sumika-Takeda and is currently in official trials in Japan. Although the structure has not been officially confirmed, it is believed to be related to the imazosulfuron class [45, 46]. The compound is expected to be introduced in 2008–2009 and is considered to be a new generation sulfonylurea for the control of annual and perennial broadleaf weeds and sedges, especially against ALS-resistant weed biotypes. At rates of 70 g-a.i. ha^{-1}, TH 547 controls *Cyperus serotinus*, *C. difformis*, *Elatine triandra*, *Eleocharis congesta*, *E. kuroguwai*, *Lindernia annua*, *L. procumbens*, *Monochoria vaginalis*, *Rotala indica*, *Sagittaria pygmaea*, *S. trifolia* and *Scirpus juncoides*. At a higher rate of 90 g-a.i. ha^{-1}, it gives total weed control, including *Echinochloa* spp.

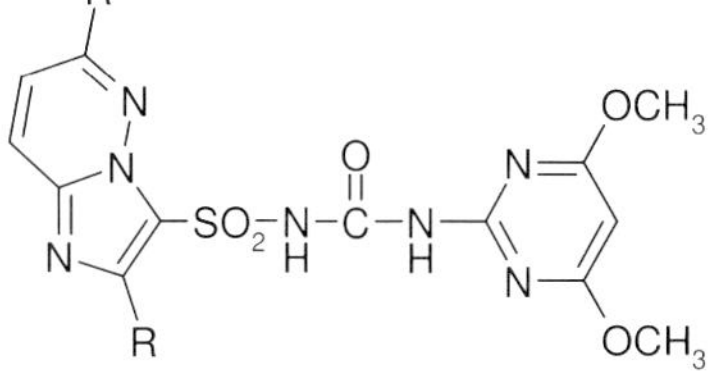

Fig. 2.2.5. Tentative structure of TH 547 (R = Cl; R′ = n-Pr).

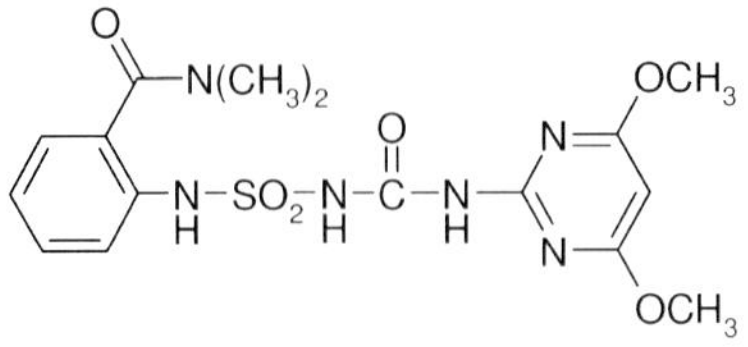

Fig. 2.2.6. Orthosulfamuron herbicide.

Orthosulfamuron (IR-5878, CAS-No.: 213464-77-8) (Fig. 2.2.6) [47], is a broad-spectrum pre- and postemergent rice herbicide developed by Isagro for the control of annual broadleaf and sedge weeds. There is only limited public knowledge available about this compound but it is assumed to be applied at a use rate of 25–150 g-a.i. ha^{-1} and could be on the market by 2007.

2.2.2.3 Maize

Approximately 700 mio tonnes of maize were produced worldwide in 2005 on more than 140 mio hectares land [78]. Maize (*Zea mays*), occupies third place in world production as a source of food, forage and processed products for industry. The main producing countries are the USA, China and Brazil, which together account for ca. two-thirds of global production. Maize is most commonly grown for animal feed use, although it is a dietary staple in some areas such as Mexico and other Latin American countries.

Maize, with its shallow root system, is particularly prone to competition by other plants in its early growth stages. While older generations of maize herbicides are predominantly used as preemergent herbicides, e.g., atrazine from the triazines class, there are now modern, postemergent sulfonylurea products available to the farmer for cost-effective and time-flexible weed control. The latest compound to be introduced after 1995 is discussed below (Table 2.2.20).

Table 2.2.20 Maize sulfonylurea herbicides.

Chemical structure	Common name (company, launch year)	Agricultural utility	Application rate (g-a.i. ha^{-1})
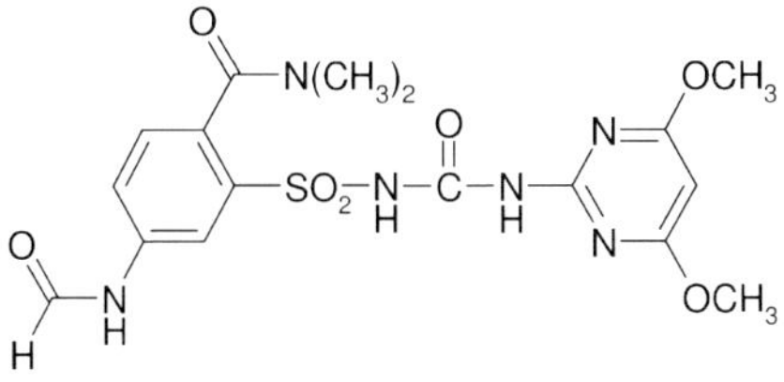	Foramsulfuron (Bayer CropScience, 2001)	Grass and broadleaf weeds	30–45 + Safener Isoxadifen

Table 2.2.21 Physicochemical properties of foramsulfuron.

Common name (ISO)	Foramsulfuron	
CAS-No.	173159-57-4	
Code numbers	AE F130360	
Melting point	194.5 °C (98.4% w/w)	
Vapor pressure (Pa)	4.2×10^{-11} (20 °C) 1.3×10^{-10} (25 °C)	
Dissociation constant (at 21.5 °C)	pK_a 4.6	
Solubility in water (g L^{-1} at 20 °C)	0.037 (pH 5) 3.293 (pH 7) 94.577 (pH 8)	
Solubility in organic solvents (g L^{-1} at 20 °C)	Acetone	1.925
	Acetonitrile	1.111
	1,2-Dichloroethane	0.185
	Ethyl acetate	0.362
	Heptane	<0.01
	Methanol	1.660
	p-Xylene	<0.01
Partition coefficient (log P_{OW}) in octanol–water (20 °C)	1.44 (pH 2.0) 0.60 (pH 5.5–5.7) −0.78 (pH 7.0) −1.97 (pH 9.0)	

2.2.2.3.1 Foramsulfuron

Foramsulfuron (AE F130360) [48] is a postemergence sulfonylurea herbicide for the control of major grass species and certain broadleaf weeds in maize (Table 2.2.21). It is applied with the safener isoxadifen-ethyl (AE F122006) (Fig. 2.2.7) and in some products in combination with small quantities iodosulfuron-methyl-sodium [49].

Introduced in 2001 and subsequently commercialized by Bayer CropScience, the three-way mixture of foramsulfuron with iodosulfuron-methyl-sodium and isoxadifen-ethyl is used for postemergent weed control in maize. Foramsulfuron,

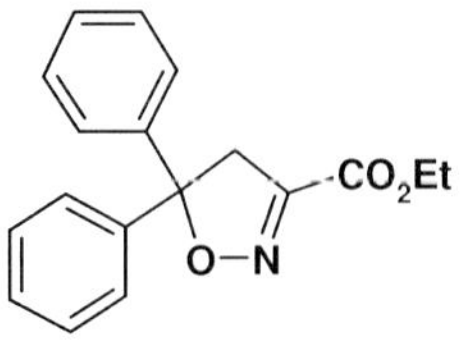

Fig. 2.2.7. Maize safener isoxadifen-ethyl (AE F122006).

at a dose rate of 30–45 g-a.i. ha^{-1}, offers a minimum of 90% weed control on most grassy weeds, such as *Echinochloa crus-galli*, *Setaria* spp., *Agropyron repens*, *Apera spica-venti*, *Alopecurus myosuroides*, *Lolium multiflorum*, *Panicum dichotomiflorum*, *Poa annua* and *Sorghum halepense*, and a wide selection of broadleaf weed species, such as *Abutilon theophrasti*, *Amaranthus* spp., *Galinsoga parviflora*, *Lamium purpureum*, *Solanum nigrum* and *Stellaria media* [50]. The addition of 1–2 g-a.i. ha^{-1} of iodosulfuron-methyl-sodium improves the level of weed control on broadleaf weed species, including *Chenopodium album*, *Galium aparine*, *Fallopia convolvulus*, *Ipomoea* spp., *Polygonum aviculare*, *P. lapathifolium*, *Sonchus arvensis* and *Xanthium strumarium*.

The basis of selectivity of foramsulfuron in the presence of the safener isoxadifen-ethyl is a more rapid rate of metabolic detoxification in maize compared with target weeds, in which little or no degradation of the parent sulfonylurea occurs [51]. Three main routes of metabolism have been established in maize – a hydrolytic cleavage of the sulfonylurea bridge, a deformylation of the amino group and oxidative metabolism of the dimethoxypyrimidine ring.

Foramsulfuron is commercialized with the safener isoxadifen-ethyl under the trade names "Option®" and "Equip®", whereas in combination with iodosulfuron-methyl-sodium the ternary mixture is sold as "MaisTer®", "Mester®", "Fortuna®" or "Equip® Plus" and "Option® 360" (Table 2.2.22). The

Table 2.2.22 Foramsulfuron-based products, formulations and composition.

Foramsulfuron-based products	Formulation type	Foramsulfuron	Iodosulfuron	Isoxadifen-ethyl
Option®WG	WG 70[a]	350[b]		350[b]
MaisTer®WG, Mester®, Fortuna®	WG 61[a]	300[b]	10[b]	300[b]
Equip®	OD 05[c]	22.5[d]		22.5[d]
Equip® Plus, Option® 360	WG 62[a]	300[b]	20[b]	300[b]

[a] WG: water dispersible granules.
[b] Units: g-a.i. kg^{-1}.
[c] OD: oil dispersion.
[d] Units: g-a.i. L^{-1}.

combination of the two herbicides probably make "MaisTer®" and "Mester®" the widest-spectrum maize herbicides used in Europe today.

2.2.2.4 Other Crops

Soybeans are the number one oilseed crop world-wide. In 2005, a total of 210 mio metric tonnes of soybean were produced. Relatively few countries produce soybeans: the USA accounts for more than 40% of the world production, with Brazil, Argentina and China together accounting for an additional 55%. In Europe, Italy, Russia and the Ukraine are the main producer countries. In the USA, Brazil and Argentina, the most widely planted soybeans are genetically modified varieties (GMO), which are tolerant against the herbicide glyphosate.

Sugarcane and cotton also represent important crops that benefit from newer sulfonylurea herbicides. Table 2.2.23 shows the most recent compounds introduced after 1995.

Table 2.2.23 Other sulfonylurea herbicides for use in soybeans, cotton and sugarcane.

Chemical structure	Common name (company, launch year)	Crop	Agricultural utility	Application rate (g-a.i. ha^{-1})
	Oxasulfuron (Syngenta, 1996)	Soybeans	Broadleaf weeds	45–90
	Trifloxysulfuron-sodium (Syngenta, 2001)	Sugar cane, cotton, turf	Sedges and broadleaf weeds	5–23

2.2.2.4.1 Oxasulfuron

Oxasulfuron (CGA 277476) (Table 2.2.24) [52] was launched in 1996 by Syngenta as a preemergent and postemergent herbicide. At application rates of 66–92 g-a.i. ha^{-1}, it provides greater than 80% control of *Abutilon theophrasti, Xanthium strumarium, Amaranthus* spp., *Ambrosia artemisiifolia, A. trifida, Bidens pilosa, Cyperus esculentus, Polygonum pensylvanicum, Sorghum bicolor, Echinochloa crus-galli, Helianthus annuus, Sesbania exaltata* and *Ipomoea* spp. [53] in soybeans. The observed selectivity is due to rapid metabolization in the target crop.

2.2.2.4.2 Trifloxysulfuron-sodium

Trifloxysulfuron-sodium (CGA 362622) (Table 2.2.25) [54] is a post-emergence herbicide commercialized by Syngenta in 2001 for use in all major cotton and

Table 2.2.24 Physicochemical properties of oxasulfuron.

Common name (ISO)	Oxasulfuron
CAS-No.	144651-06-9
Code numbers	CGA 277476
Melting point	158 °C (decomposition)
Vapor pressure	$<2 \times 10^{-6}$ Pa (25 °C)
Dissociation constant (temperature not published)	pK_a 5.10
Solubility in water (g L^{-1} at 25 °C)	0.052 (pH 5.1) 0.063 (pH 5.0, buffer solution) 1.70 (pH 6.8, buffer solution) 19.0 (pH 7.8, buffer solution)
Solubility in organic solvents (g L^{-1} at 25 °C)	Acetone 9.3 Ethyl acetate 2.3 Dichloromethane 69.0 n-Hexane 0.0022 Toluene 0.32
Partition coefficient (log P_{OW}) in octanol–water (25 °C)	0.75 (pH 5.0) -0.81 (pH 7.0) -2.2 (pH 9.0)

sugarcane production areas [55]. In cotton, it is formulated as a WG 75 and can be applied postemergent at 5–7.5 g-a.i. ha^{-1} in conventional or GMO cotton and at higher rates of 10–15 g-a.i. ha^{-1} postemergent directed. At the lower rates, the following weeds are controlled: *Acanthospermum hispidum, Ambrosia artemisiifolia, Bidens pilosa, Senna obtusifolia, Cassia occidentalis, Chenopodium album, Euphorbia heterophylla, Ipomoea* spp., *Melochia corchorifolia, Mollugo vertillata, Sesbania exalta, Trianthema portulacastrum, Xanthium strumarium*. With post-directed sprays and higher dosages, additional control is achieved of *Ageratum conyzoides, Amaranthus hybridus, A. palmeri, Cyperus esculentus* and *Tridax procumbens*. Application of trifloxysulfuron-sodium may be made after cotton (picker-type varieties only) has reached a minimum of five true leaves, with applications continuing until 60 days before harvest. Due to reduced crop tolerance, the product is not recommended as a postemergent over-the-top spray on stripper-type cotton varieties.

Table 2.2.25 Physicochemical properties of trifloxysulfuron-sodium.

Common name (ISO)	Trifloxysulfuron-sodium	
CAS-No.	199119-58-9	
Code numbers	CGA 362622	
Melting point	170.2–177.7 °C	
Vapor pressure	$<1.3 \times 10^{-6}$ Pa (25 °C)	
Dissociation constant (at 20 °C)	pK_a 4.76	
Solubility in water (g L^{-1}) (25 °C)	0.063 (pH 5.0) 5.016 (pH 7.0) 25.7 (pH 7.4)	
Solubility in organic solvents (g L^{-1} at 25 °C)	Acetone	17.0
	Ethyl acetate	3.8
	n-Hexane	<0.001
	Methanol	50.0
	Octanol	4.4
	Toluene	<0.001
Partition coefficient (log P_{OW}) in octanol–water (25 °C)	1.4 (pH 5.0) −0.43 (pH 7.0)	

In cotton and sugarcane trifloxysulfuron-sodium is commercialized under the trade name "Envoke®" as a straight product. In cotton it is used in combination with prometryn as "Suprend®" and in sugarcane in combination with ametryn as "Krismat®". In sugarcane "Envoke®" can be used for a maximum of three applications pre-spiking, post-emergence over-the-top, and/or post-emergence directed at a total rate of 78 g-a.i. ha^{-1} per season. The product may be applied to sugarcane at a plant height of 45–60 cm up to 100 days before harvest. At a dose rate of 16 g-a.i. ha^{-1} the following weeds are controlled with greater than 85% efficacy: *Alternanthera philoxeroides, Acanthospermum hispidum, Panicum adspersum, Mollugo vertillata, Xanthium strumarium, Cassia occidentalis, Gnaphalium pensylvanicum, Eupatorium cappilliforium, Desmodium tortuosum, Trianthema portulacastrum, Sesbania exaltata, Rottboellia cochinchinensis, Chenopodium album, Ipomoea* spp., *Cyperus esculentus, C. rotundus, Amaranthus* spp., *Ambrosia artemisiifolia, Melochia corchorifolia, Senna obtusifolia, Bidens bipinnata, Linaia canadensis, Abutilon theophrasti* and *Euphorbia heterophylla*.

2.2.3
Metabolic Fate and Behavior in the Soil

There is abundant knowledge about the metabolic fate of sulfonylurea herbicides. However, especially with regard to animal data to support product registrations, most of this information is as yet unpublished. For readers who are interested in more information on plant metabolism and crop selectivity, reference is given to articles by Brown et al. [56]. Another excellent review article on the metabolic fate of sulfonylurea herbicides is found in Part 1 of the *Metabolic Pathways of Agro-chemicals* series also authored by Brown et al. at Du Pont [57].

In the soil, there are two major pathways of sulfonylurea degradation [58]: (a) chemical hydrolysis and (b) microbial degradation. The breakdown of sulfonylureas in sterile soils is solely attributable to chemical hydrolysis, whereas breakdown in non-sterile soils is a combination of both microbial degradation and

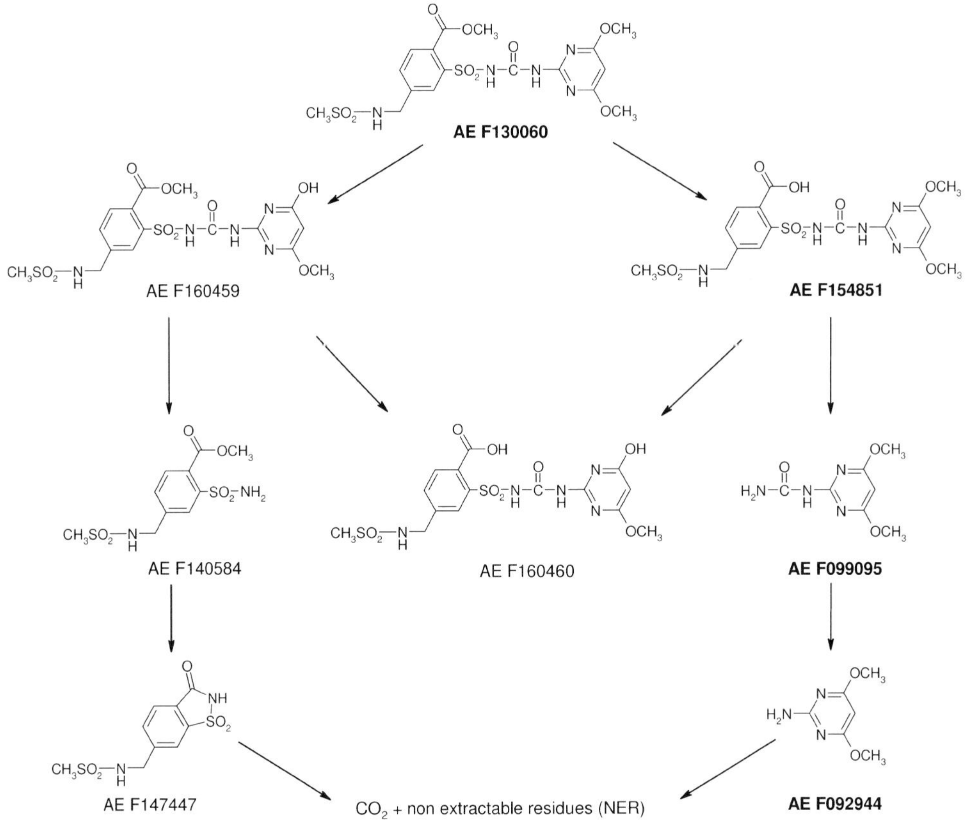

Scheme 2.2.3. Metabolic pathway of mesosulfuron-methyl in soil under aerobic and anaerobic conditions (compounds labeled in bold were detected at >10% of the applied radioactivity).

chemical hydrolysis. The relative importance of the microbial degradation can then be calculated from the differential rate.

The main soil degradation pathways of sulfonylurea herbicides are cleavage of the sulfonylurea bridge, *O*- and *N*-dealkylation reactions, aryl and aliphatic hydroxylation reactions, dehalogenation and ester hydrolysis. It is not within the scope of this chapter to discuss each of these in detail for all of the above-mentioned new sulfonylureas. Instead mesosulfuron-methyl is taken below as a general illustration of commonly found soil degradation pathways established within the sulfonylurea family.

Mesosulfuron-methyl is degraded in soil and water via hydrolysis and *O*-demethylation reactions. Its metabolites are also readily degraded to non-extractable-residues (NER) and CO_2 [59]. During soil metabolism studies with radiolabeled mesosulfuron-methyl, major metabolites found, representing more than 10% of the applied radioactivity, were mesosulfuron acid AE F154851, pyrimidinyl urea AE F099095 and aminopyrimidine AE F092944 (Scheme 2.2.3). *O*-Demethylation of the methyl ether in the pyrimidine moiety to yield hydroxypyrimidine derivative AE F160459 proved to be of minor relevance in soil. Other minor soil metabolites were the sulfonylurea AE F160460, the sulfonamide AE F140584 and the saccharin derivative AE F147447. Carbon dioxide and unidentified non-extractable residues (NER) bound to the soil matrix were the final products of the degradation in the soil.

2.2.4
Concluding Remarks

Twelve new sulfonylurea herbicides for all major crops have been commercialized since 1995 and four new compounds from this class are currently in their late development stage. These together with the 20 sulfonylurea products that already have been on the market prior to 1995 give a remarkable figure, outnumbering any other herbicidal class in modern crop protection. The reason for this is a combination of the environmental friendliness of the products, their versatility as regards applicable crops and timing flexibility and also their cost/benefit performance. It remains to be seen whether the market can accept yet further innovations from this class, and whether resistant weed development will one day become an issue despite hitherto successful resistance strategies employed by the agro-industry.

In conclusion, it is fascinating to see the development that began with George Levitt's pioneering work at Du Pont over 30 years ago. In "Gulliver's Travels" (*Voyage to Brobdingnag*, Ch. 6), the Irish author Jonathan Swift (1667–1745) wrote that

> Whoever could make two ears of corn or two blades of grass grow
> upon a spot of ground where only one grew before, would deserve
> better of mankind, and do more essential service to his country
> than the whole race of politicians put together.

It is against this background that the achievements of George Levitt and all other colleagues involved in the world's agrochemical industry should be viewed.

Acknowledgments

The author is pleased to acknowledge Drs. Darren Mansfield, Klaus-Helmut Mueller, Graham Holmwood, Arno Schulz and Shinichi Shirakura for their critical review of the manuscript and many helpful suggestions, and Mrs. Tong Lin and
Mr. Tetsuya Murata for translations of Chinese and Japanese publications.

References

1 R. S. Chaleff, C. J. Mauvais,
Acetolactate synthase is the site of
action of two sulfonylurea herbicides
in plants, *Science*, 224, 1443–1444
(**1984**).

2 R. A. LaRossa, J. V. Schloss, *J. Biol.
Chem.*, 259, 8753–8757 (**1984**).

3 D. Delfourne, J. Bastide, R. Badon,
A. Rachon, P. Genix, Specificity of
Plant Acetohydroxyacid Synthase:
Formation of Products and Inhibition
by Herbicides, *Plant Physiol. Biochem.*,
32, 473–477 (**1994**).

4 G. Levitt, Discovery of the Sulfonylurea Herbicides, in: D. R. Baker, J. G.
Fenyes, W. K. Moberg (eds) *Synthesis
and Chemistry of Agrochemicals II*, ACS
Symposium Series 443, 16–31 (**1991**).

5 E. M. Beyer, M. J. Duffy, J. V. Hay,
D. D. Schlueter, in: P. C. Kearney,
D. D. Kaufman (eds) *Herbicides –
Chemistry, Degradation and Mode of
Action*. Marcel Dekker, New York,
Volume 3, 117–189 (**1987**).

6 H. M. Brown, J. C. Cotterman:
Recent advances in sulfonylurea
herbicides, in: W. Ebing (editor-in-
chief) *Chemistry of Plant Protection*,
J. Stetter (volume editor) Herbicides
inhibiting branched-chain amino acid
biosynthesis. Springer, Berlin
Heidelberg, Volume 10, 47–81,
(**1994**).

7 S. K. Gee, J. V. Hay, Recent
developments in the chemistry of
sulfonylurea herbicides, in: W. Ebing
(editor-in-chief) *Chemistry of Plant
Protection*, J. Stetter (volume editor)
Herbicides inhibiting branched-chain
amino acid biosynthesis. Springer,
Berlin Heidelberg, Volume 10, 15–46,
(**1994**).

8 J. V. Hay, Chemistry of sulfonylurea
herbicides, *Pestic. Sci.*, 29, 247–261
(**1990**).

9 Cropnosis, Agrochemical Service,
Update of the Crops Section, Rice,
Cereals and Maize. November
2004.

10 T. A. Andrea, P. H. T. Liang, Preparation of Me 2-[[[[(4,6-dimethoxy-2-
primidinyl)amino]carbonyl]amino]-
sulfonyl]-6-trifluoromethyl-3-pyridine-
carboxylate and salts as herbicides,
Eur. Pat. Appl. EP 502740 A1
(**1992-09-09**).

11 S. R. Teaney, L. Armstrong, K.
Bentley, D. Cotterman, D. Leep, P. H.
Liang, C. Powley, J. Summers, S.
Cranwell, F. Lichtner, R. Stichbury,
Proc. BCPC Conference – Weeds 49–56
(**1995**).

12 M. Lechner, Ciral® – Ein neues
Nachauflaufherbizid zur Bekämpfung
von Ungräsern und Unkräutern im
Getreide, *Mitt. Biol. Bundesanst. Land-
und Forstwirtsch.*, 390, 243–244
(**2002**).

13 Y. Ishida, K. Ohta, H. Yoshikawa,
Preparation of sulfonylurea herbicides, Eur. Pat. Appl. EP 477808 A1
(**1992-04-01**).

14 S. K. Parrish, J. E. Kaufmann, K. A.
Croon, Y. Ishida, K. Ohta, S. Itoh,
Proc. BCPC Conference – Weeds 57–63
(**1995**).

15 G. Gibson, G. de Kerchove, *Proc.
BCPC Conference – Weeds* 87–92
(**1999**).

16 O. Ort, K. Bauer, H. Bieringer, Preparation of [(arylsulfonyl)ureido]azines
as herbicides and plant growth regulators, PCT Int. Appl. WO 9213845
A1 (**1992-08-20**).

17 E. Hacker, H. Bieringer, L. Willms, O. Ort, H. Koecher, H. Kehne, *Proc. BCPC Conference – Weeds* 15–22 (**1999**).

18 E. Hacker, H. Bieringer, L. Willms, W. Roesch, H. Koecher, R. Wolf, Mefenpyr-diethyl – A safener for fenoxaprop-P-ethyl and iodosulfuron in cereals, *Z. PflKrankh. PflSchutz, Sonderh.* XVII, 493–500 (**2000**).

19 K. Lorenz, L. Willms, K. Bauer, H. Bieringer, Phenylsulfonyl ureas, process for producing the same and their use as herbicides and plant growth regulators, PCT Int. Appl. WO 9510507 A1 (**1995-04-20**).

20 E. Hacker, H. Bieringer, L. Willms, K. Lorenz, H. Koecher, H. P. Huff, G. Borrod, R. Brusche, Mesosulfuron-methyl – a new active ingredient for grass weed control in cereals, *Proc. BCPC Conference – Weeds* 43–48 (**2001**).

21 Atlantis", *Pflanzenschutz-Nachrichten Bayer*, 58(2), 165–299, (**2005**), ISSN 0340-1723.

22 H. Mayer, G. Hamprecht, K.-O. Westphalen, H. Walter, M. Gerber, K. Grossmann, W. Rademacher, Herbicidal N-[(1,3,5-triazin-2-yl)-aminocarbonyl]benzenesulfonamides and their preparation Ger. Offen. (**1992**), DE 4038430 (Publ. 04.06.1992), PCT WO92/09608 (Publ. 11.06.1992).

23 P. Dombo, H. D. Brix, M. Landes, W. Nuyken, Der neue Herbizidwirkstoff Tritosulfuron – Charakterisierung und Einsatz im Ackerbau, *Mitt. Biol. Bundesanst. Land- und Forstwirtsch.* 390, 473–474 (**2002**).

24 A. Schönhammer, N. Pörksen, J. Freitag, W. Nuyken, Biathlon – das erste Tritosulfuron-haltige Herbizid zur Unkrautbekämpfung in monokotylen Kulturen, *Mitt. Biol. Bundesanst. Land- und Forstwirtsch.*, 390, 241–242 (**2002**).

25 J. Dressel, C. Beigel, Estimation of standardized transformation rates of a pesticide and its four soil metabolites from field dissipation studies for use in environmental fate modeling. *Proc. BCPC Conference – Weeds* 119–126 (**2001**).

26 Z. Li, G. Jia, L. Wang, *et al.*, Faming Zhuanli Shenqing Gongkai Shuo-mingshu (1994), Chinese Patent Application CN 93-101976 (**1993-02-27**).

27 Z. Fan, Y. Ai, C. Qian, Z. Li, *J. Environ. Sci.*, 17(3), 399–403 (**2005**) and references cited therein.

28 X. Ou, M. Lei, M. Huang, Y. Wang, X. Wang, D. Fan, *Nongyaoxue Xuebao* 5(3), 16–23 (**2003**).

29 M. Huang, L. Huang, C. Chen, L. Zhao, M. Lei, T. Wu, S. Yu, Faming Zhuanli Shenqing Gongkai Shuo-mingshu, Chinese Patent Application CN 2000-113423 (**2000-05-11**).

30 A. Diarra, R. J. Smith, Jr., R. E. Talbert, Interference of red rice (*Oryza sativa*) with rice (*O. sativa*), *Weed Sci.*, 33, 644–649 (**1985**). A. Fisher, A. Ramirez, Red rice (Oryza sativa): Competition studies for management decisions, *Int. J. Pest Manage.*, 39 (2), 133–138 (**1993**).

31 A. Ferrero (Weedy rice, biological features and control) in: R. Labrada (ed.), *Weed Management for Developing Countries (Addendum 1)*, Serie title: FAO Plant Production and Protection Papers – 120 Add.1 (**2003**), ISBN: 9251050198.

32 H. Kehne, L. Willms, K. Bauer, H. Bieringer, H. Buerstell, Heterocyclic 2-alkoxyphenoxysulfonylureas and their use as herbicides or as plant growth regulators, Eur. Pat. Appl. EP 342569 A1 (**1989-11-23**).

33 E. Hacker, K. Bauer, H. Bieringer, H. Kehne, L. Willms, HOE 095404: A new sulfonylurea herbicide for use in cereals, rice and sugarcane, *Proc. BCPC Conference – Weeds* 73–78 (**1995**).

34 M. Hess, E. Rose, HOE 095404: A new herbicide for broadleaf weed and sedge control in rice, *Proc. BCPC Conference – Weeds* 763–768 (**1995**).

35 G. Levitt, Tetrazole-containing sulfonylureas, their herbicidal compositions, and their use in weed control, U.S. Pat. US 4,746,353 A (**1988-05-24**).

36 T. Marquez, M. M. Joshi, T. Pappas Fader, W. Massasso, *Proc. BCPC Conference – Weeds* 65–72 (**1995**).

37 M. E. Condon, T. E. Brady, D. Feist, T. Malefyt, P. Marc, L. S. Quakenbush, S. J. Rodaway, D. L. Shaner, B. Tecle, AC 322,140 – a new broad-spectrum herbicide for selective weed control in rice and cereals *Proc. BCPC Conference – Weeds* 41–46 (**1993**).

38 S. J. Rodaway, B. Tecle, D. L. Shaner, *Proc. BCPC Conference – Weeds* 239–246 (**1993**).

39 T. E. Brady, M. E. Condon, P. A. Marc, U.S. Pat. US 5,009,699 (**1991-04-23**).

40 X. Tan, D. Wang, Huaxue Yu, *Shengwu Gongcheng* (**2005**), 22(2), 47–48.

41 S. J. Koo, J. H. Cho, J. S. Kim, S. H. Kang, K. G. Kang, D. W. Kim, H. S. Chang, Y. K. Ko, J. W. Ryu, Preparation of herbicidally active pyridylsulfonyl ureas, PCT Int. Appl. WO 2002030921 A1 (**2002-04-18**).

42 D. S. Kim, S. J. Koo, J. N. Lee, K. H. Hwang, K. G. Kim, K. G. Kang, K. S. Hwang, G. H. Joe, J. H. Cho, D. W. Kim, Flucetosulfuron: a new sulfonylurea herbicide, *Proc. BCPC Int. Congress – Crop Sci. Technol.*, 87–92 (**2003**).

43 D. S. Kim, J. N. Lee, K. H. Hwang, K. G. Kang, T. Y. Kim, S. J. Koo, Flucetosulfuron: a new tool to control Galium aparine and broadleaf weeds in cereal crops, *Proc. BCPC Int. Congress – Crop Sci. Technol.*, 941–946 (**2003**).

44 Y. Tanaka, Y. Kajiwara, M. Noguchi, T. Kajiwara, T. Tabuchi, Fused heterocyclic sulfonylurea compound, herbicide containing the same, and method of controlling weed with the same, PCT Int. Appl. WO 03061388 (**2003-07-31**).

45 Y. Tanaka, Y. Kajiwara, T. Nishiyama, Composition of herbicide, JP 2005-126415 (**2005-05-19**).

46 F. Bettarini, S. Massimini, G. Meazza, G. Zanardi, D. Portoso, E. Signorini, PCT Int. Appl. WO 9840361 A1 (**1998-09-17**).

47 G. Schnabel, L. Willms, K. Bauer, H. Bieringer, Acylated Aminophenylsulphonylureas, process for their preparation and their use as herbicides and plant-growth regulators, PCT Int. Appl. WO 9529899 (**1995-11-09**).

48 B. Collins, D. Drexler, M. Maerkl, E. Hacker, H. Hagemeister, K. E. Pallett, C. Effertz, Foramsulfuron – a new foliar herbicide for weed control in corn (maize), *Proc. BCPC Conference – Weeds* 35–42 (**2001**).

49 J. A. Bunting, C. L. Sprague, D. E. Riechers, Incorporating foramsulfuron into annual weed control systems for corn, *Weed Technol.* (**2005**), 19 (1), 160–167.

50 J. A. Bunting, C. L. Sprague, D. E. Riechers, Physiological basis for tolerance of corn hybrids to foramsulfuron, *Weed Technol.* (**2004**), 52 (5), 711–717.

51 W. Meyer, Sulfonylureas as herbicides, Eur. Pat. Appl. EP 496701 A1 (**1992-07-29**).

52 R. L. Brooks, A. Zoschke, P. J. Porpiglia, CGA-277476: a short residual herbicide for soybean weed control programs, *Proc. BCPC Conference – Weeds* 79–85 (**1995**).

53 W. Foery, Sulfonylurea salts as herbicides, PCT Int. Appl. WO 9741112 A1 (**1997-11-06**).

54 S. Howard, M. Hudetz, J.-L. Allard, Trifloxysulfuron-sodium: a new post emergence herbicide for use in cotton and sugarcane, *Proc. BCPC Conference – Weeds* 29–34 (**2001**).

55 H. M. Brown, T. P. Fuesler, T. B. Ray, S. D. Strachan, in: H. Frehse (editor), *Pestic. Chem.: Adv. Int. Res., Dev., Legis., Proc. Int. Congr. Pestic. Chem.*, 7th (**1991**), Meeting Date 1990, 257–266. Publisher: VCH, Weinheim.

56 H. M. Brown, V. Gaddamidi, P. W. Lee, in: T. R. Roberts (editor-in-chief), D. H. Hutson, P. W. Lee, P. H. Nicholls, J. R. Plimmer (contributing editors) *Metabolic Pathway of Agrochemicals, Part 1: Herbicides and Plant Growth Regulators*, 451–473, The Royal Society of Chemistry Information Services **1998**, ISBN 0-85404-494-9.

57 H. M. Brown, Mode of action, crop selectivity, and soil relations of the sulfonylurea herbicides, *Pestic. Sci.*, 29, 263–281 (**1990**).

58 H. Gildemeister, D. Schäfer, Behaviour of the herbicide mesosulfuron-methyl in the

environment, *Pflanzenschutz-Nachrichten Bayer*, 58 (2), 195–214 (**2005**), ISSN 0340-1723.

59 D. S. M. D'Souza, I. A. Black, R. T. Hewson, Amidosulfuron – a new sulfonylurea for the control of Galium aparine and other broadleaf weeds in cereals, *BCPC Conference – Weeds* 567–72 (**1993**).

60 T. Yuyama, S. Takeda, H. Watanabe, T. Asami, S. Peudpaichit, J. L. Malassa, P. Heiss, Proc. 9th Asian Pacific Weed Sci. Soc. Congr., 554–559 (**1983**), S. Takeda, D. L. Erbes, P. B. Sweetser, J. V. Hay, T. Yuyama, *Weed Res.* (Japan), 31, 157–163 (**1986**).

61 M. H. Russell, J. L. Saladini, F. Lichtner, *Pesticide Outlook* 166–173 (**2002**).

62 H. L. Palm, J. D. Riggleman, D. A. Allison, *Proc. BCPC Conference – Weeds* 1–6 (**1980**).

63 J. M. Hutchison, C. J. Peter, K. S. Amuti, L. H. Hageman, G. A. Roy, R. Stichbury, *Proc. BCPC Conference – Weeds* 63–67 (**1987**).

64 T. Haga, Y. Tsujii, K. Hayashi, F. Kimura, N. Sakashita, K. Fujikawa, in: D. R. Baker, J. G. Fenyes, W. K. Moberg (eds) *Synthesis and Chemistry of Agrochemicals II*, ACS Symposium Series 443, 107–119 (**1991**).

65 K. Suzuki, T. Nawamaki, S. Watanabe, S. Yamamoto, T. Sato, K. Morimoto, *Proc. BCPC Conference – Weeds* 31–37 (**1991**).

66 R. I. Doig, G. A. Carraro, N. D. McKinley, *Proc. 10th Int. Congress of Plant Protection* 324–331 (**1983**).

67 S. Murai, T. Haga, K. Fujikawa, N. Sakashita, F. Kimura, in: D. R. Baker, J. G. Fenyes, W. K. Moberg (eds)

Synthesis and Chemistry of Agrochemicals II, ACS Symposium Series 443, 98–106 (**1991**).

68 F. Kimura, T. Haga, N. Sakashita, S. Murai, K. Fujikawa, *Proc. BCPC Conference – Weeds* 29–34 (**1989**).

69 W. Maurer, H. R. Gerber, J. Rufener, *Proc. BCPC Conference – Weeds* 41–48 (**1987**).

70 M. Schulte, K. Kreuz, N. Nelgen, M. Hudetz, W. Meyer, *Proc. BCPC Conference – Weeds* 53–59 (**1993**).

71 H. L. Palm, P. H. Liang, T. P. Fuesler, G. L. Leek, S. D. Strachan, V. A. Wittenbach, M. L. Swinchatt, *Proc. BCPC Conference – Weeds* 23–28 (**1989**).

72 R. H. Harding, C. B. Chumley, G. E. Cook, F. A. Holmes, DPX-5648 – a new herbicide for control of johnsongrass and many other weeds, *Proc. Western Soc. Weed Sci.*, 34, 120–121 (**1981**).

73 S. D. Sionis, H. G. Drobny, P. Lefebvre, M. E. Upstone, *Proc. BCPC Conference – Weeds* 49–54 (**1985**).

74 J. Amrhein, H. R. Gerber, *Proc. BCPC Conference – Weeds* 55–62 (**1985**).

75 D. T. Ferguson, S. E. Schehl, L. H. Hageman, G. E. Lepone, G. A. Carraro, *Proc. BCPC Conference – Weeds* 43–48 (**1985**).

76 K. A. Peeples, M. P. Moon, F. T. Lichtner, V. A. Wittenbach, T. H. Carski, M. D. Woodward, K. Graham, H. Reinke, *Proc. BCPC Conference – Weeds* 25–30 (**1991**).

77 UN Food & Agriculture Organisation (FAO), FAOSTAT data, 2006 (last updated 24 April **2006**). (http://faostat.fao.org).

2.3
Imidazolinone Herbicides

Dale L. Shaner, Mark Stidham, and Bijay Singh

2.3.1
Overview

The imidazolinone herbicides (Table 2.3.1) are a family of six compounds that were discovered and developed by American Cyanamid Corporation. Readers may obtain comprehensive and detailed information in *The Imidazolinone Herbicides* [1], a book authored by the researchers who discovered and developed the herbicides. The herbicides as a class are broad spectrum and are active both pre- and postemergence. Imidazolinones are absorbed and moved through both xylem and phloem, eventually accumulating in the meristematic tissue. Activity is characterized by rapid cessation of growth followed by plant death days or weeks after treatment. Selectivity is based most often on metabolic inactivation except for selection-developed target site based resistance.

Synthesis methodology for numerous imidazolinones is described in the patent literature [2–6]. Figure 2.3.1(A) shows a simple one-step method [7].

Imidazolinones are generally formulated as the amine salts. Perhaps because of their high potency, broad spectrum, and high water solubility, the imidazolinones have been co-formulated with many other herbicides.

Table 2.3.1 Structure of commercialized imidazolinones.

R1	Common Name
H	imazapyr
CH$_3$	imazapic
CH$_2$CH$_3$	imazethapyr
CH$_2$OCH$_3$	imazamox

Imazamethabenz-methyl

Imazaquin

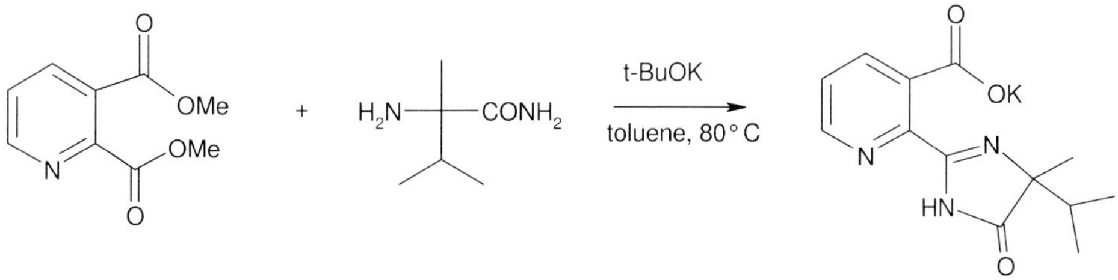

Fig. 2.3.1. Synthesis method for imidazolinones.

2.3.2
History of Discovery

The imidazolinone herbicides were discovered through a long process of observation, exploration, and optimization. The account here has been presented in greater detail elsewhere [2, 3]. The initial lead molecule **1** was synthesized in the 1950s by an American Cyanamid Medical Division chemist working on anticonvulsives (Fig. 2.3.2). The compound came years later to the Agricultural Division for random screening, where it showed herbicidal activity at 4 kg ha^{-1}, sufficient for additional synthesis effort. The mode of action was not known or even investigated at the time, but years after the discovery of the imidazolinones, this original phthalimide was shown to be an inhibitor of acetohydroxyacid synthase (AHAS).

Initial modifications did not improve the herbicidal activity, but derivative **2** showed interesting plant growth regulant activity similar to gibberellic acid (Fig. 2.3.2) [4, 5]. This new compound was further optimized for plant growth regulation, resulting in **3**.

Associated work to enable production of field trial samples produced a tricyclic compound, and in the spirit of comprehensive exploration (and thorough patent coverage), the same reaction was attempted on the original herbicide lead compound, resulting in **4** (Fig. 2.3.3). This compound showed broad-spectrum herbi-

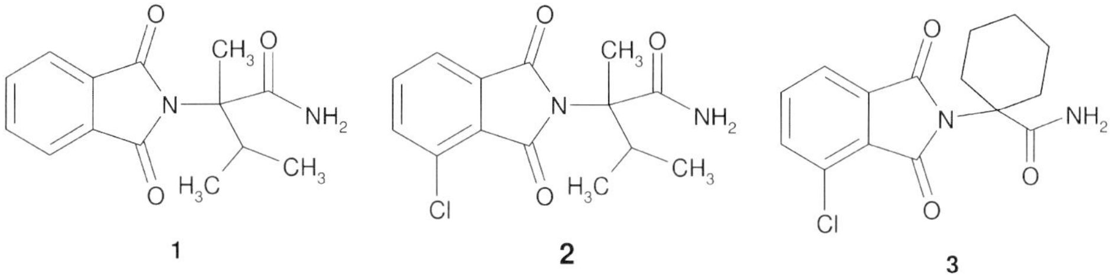

1	**2**	**3**

Fig. 2.3.2. Early lead compounds that led to the imidazolinones.

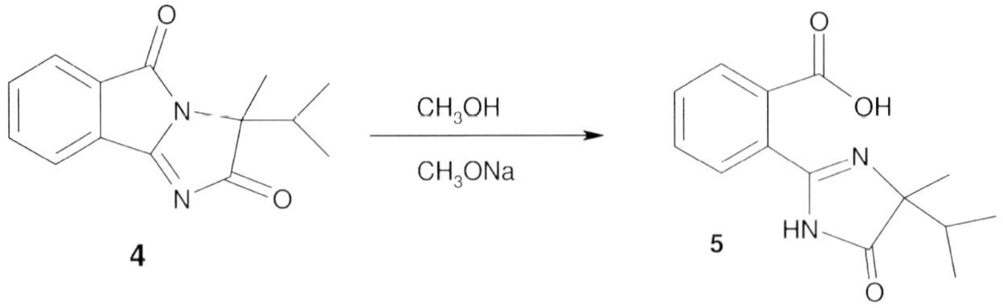

Fig. 2.3.3. Synthesis of first imidazolinone lead.

cidal activity, and continued exploration in the series resulted in the first imidazo-linone **5** (Fig. 2.3.3). This compound had markedly improved herbicidal spectrum and potency with some selectivity in rice.

Work continued in this program, eventually resulting in **6**, the isomer mixture imazamethabenz-methyl, a wheat-selective herbicide (Fig. 2.3.4).

A quantum leap in herbicidal potency and spectrum occurred when the benzene ring was replaced with a pyridine ring. The resulting compound had pre- and postemergence activity at doses in the range $10–100$ g ha^{-1} in greenhouse tests. Exploration of this new series demonstrated that the picolinic acid and iso-nicotinic acid had far less herbicide activity than the nicotinic acid. Also, high activity is maintained only in derivatives with substituents at the 5- and 6-position of the pyridine ring. Thus, unlike the other major classes of AHAS-inhibiting herbicides, the imidazolinones have a relatively narrow structure–activity pattern for weed control [8].

2.3.3
Physical Chemical Properties

The imidazolinone salts have high water solubility, ranging from $>57\%$ (imazapyr/isopropylamine salt) to 17% (imazaquin ammonium salt). Imazapyr has two sites for protonation, namely the imidazolinone secondary nitrogen and

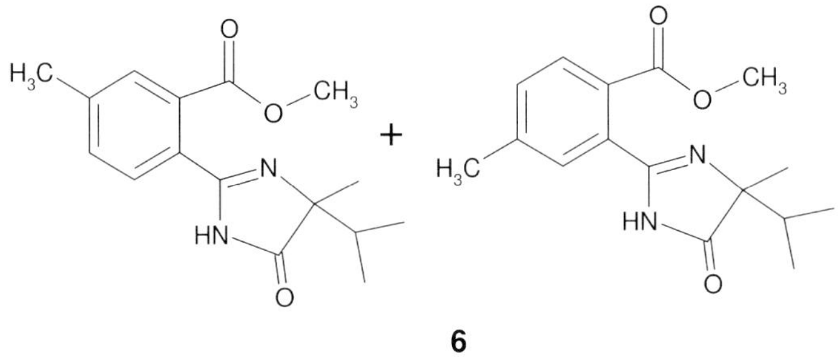

6

Fig. 2.3.4. Structure of imazamethabenz-methyl.

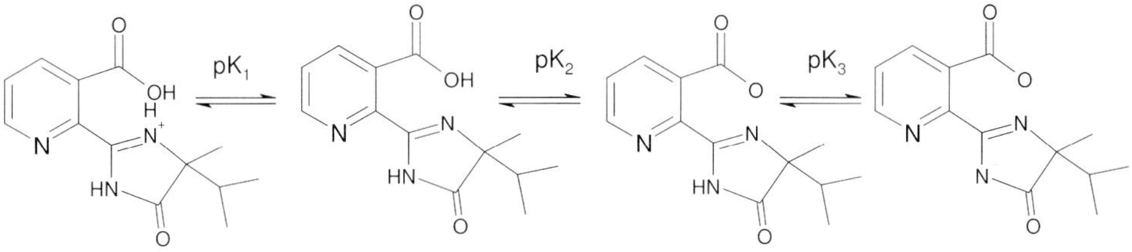

Fig. 2.3.5. Ionization states of imidazolinones.

the carboxylic acid substituent on the pyridine ring. The ionization constants are relatively similar for the pyridine imidazolinone herbicides; for imazapyr, pK_1 is 1.9 and pK_2 is 3.6 (Fig. 2.3.5). A third ionization on the primary imidazolinone nitrogen occurs at pH $\sim$ 11 (Fig. 2.3.5) [9]. The pK_2 is important for concentrating the herbicide inside the cell through a weak acid trapping mechanism. Outside the cell in the apoplast, a relatively low pH allows a substantial proportion of the imidazolinone to exist in an uncharged state, with enough lipophilicity to passively cross cell membranes. Once inside the cell, the pH is much higher and the charged form predominates, effectively trapping the herbicides inside the cell [10].

2.3.4
Structural Features of Herbicidal Imidazolinones

The structural features of imidazolinones important for target site and herbicide activity have been summarized [11–14]. The orientation of the imidazolinone ring ortho to the acid equivalent is critical. Derivatives of the acid equivalent are herbicidally active if they can be metabolized to the acid either in the soil or in the plant. Likewise, tricyclic derivatives such as **4** are pro-herbicides that must be metabolized to the acid-imidazolinone form.

The commercial herbicides are a mixture of R and S isomers at the chiral center where the methyl and isopropyl substituents are placed, but the (R)-isomer is approximately ten-fold more potent both as an enzyme inhibitor and as an herbicide. Substituents other than methyl and isopropyl are substantially weaker enzyme inhibitors an herbicides [14].

The aromatic ring component illustrates the relative contributions of enzyme inhibition and physicochemical properties to herbicidal activity. The benzene imidazolinones are approximately ten-fold more potent than the corresponding pyridine derivatives as enzyme inhibitors but are less potent as herbicides.

The primary factor that determines the biological activity of the imidazolinones besides the inhibition of AHAS is their ability to translocate to meristematic tissue. AHAS, the target site for these herbicides, functions primarily in rapidly dividing tissue and decreases rapidly as tissue matures [15]. Thus, the difference in herbicidal activity among the six commercial imidazolinones depends on differences in their ability to be absorbed and translocate within the plant. Imazaquin is primarily used as a soil applied herbicides. This imidazolinone is the most lip-ophilic of all the commercial herbicides and is the most readily absorbed by roots and translocated to the shoot [16, 17]. Imazapyr and imazamox, in contrast, are the least lipophilic and are also the most active when applied to the foliage [16, 17]. Imazethapyr and imazapic fall in between these two extremes.

The differences in herbicidal activity among these analogs appear to be related to their ability to be trapped in the phloem. The imidazolinones are absorbed into phloem via an ion-trapping mechanism as described previously. Thus, all imida-zolinones can penetrate the phloem and will be carried to meristematic tissue. However, the concentration of the herbicide that actually reaches the meristems is a function of how rapidly the chemicals diffuse out of the phloem as it moves through the plant. Imazaquin will diffuse out of the phloem more readily than imazamox or imazapyr because it is more lipophilic. Thus, it will not be carried as far and has limited postemergent activity compared with imazamox or imaza-pyr. As mentioned previously, the benzene imidazolinones are not as herbicidally active as the pyridine imidazolinones, although the former analogs are more po-tent inhibitors of AHAS. The benzene imidazolinones are more lipophilic than the pyridine imidazolinones and hence are not trapped in the phloem as well.

There may be other factors governing the herbicidal activity of the imidazoli-nones. The position of the nitrogen in the pyridine ring in relation to the carbox-ylic and imidazolinone ring substitution is critical, although inhibition of AHAS is unaffected by the relative position of the nitrogen in the imidazolinone ring in relation to the substitutions [16]. Cellular uptake of the imidazolinones is affected by the relative position of the nitrogen to the carboxylic acid moiety. Hawkes et al. have shown that if the nitrogen in the pyridine ring is not ortho to the carboxylic acid group, the compound is not absorbed by the cell [18]. The mechanism of this differential uptake is not known. If an imidazolinone is not absorbed or translo-cated well within the plant, it is not herbicidal.

2.3.5
Mode of Action of Imidazolinones

Although plant growth stops soon after application of the imidazolinone herbi-cides, death of the whole plant may take 2–3 weeks. Meristematic tissues exhibit chlorosis and necrosis first followed by slow necrosis of the mature tissues. Phys-iological changes resulting from herbicide treatment include changes in metabo-lite concentrations [19, 20], reduction of assimilate transport [21, 22], inhibition of DNA synthesis [23, 24] and cell division [23, 25]. These physiological effects in plants result from inhibition of AHAS, the first enzyme in the biosynthesis of branched chain amino acids, valine, leucine and isoleucine. Supplementation of plants with branched chain amino acids reverses the effects of herbicide [21] which suggests that starvation for branched chain amino acids is the primary cause of plant death [26].

The I_{50} for various commercial imidazolinone herbicides *in vitro* AHAS assays varies between 0.1 and 10 μM depending upon the assay conditions [27]. Under *in vivo* conditions, binding of imidazolinones seems to cause irreversible loss of AHAS activity [28]. The level of AHAS activity extracted from plants treated with lethal dose of an imidazolinone herbicide is reduced more than 80%. This effect of inhibitors can be discerned within an hour after treatment and the loss of AHAS activity is proportional to the concentration of the inhibitor in the plant tissue. There are several possible reasons for the loss of extractable AHAS activity in the imidazolinone treated plants. The imidazolinones may interact with the enzyme in such a way *in vivo* that the herbicide does not easily separate from the enzyme during the extraction procedure; the herbicide causes a change in the protein structure such that it is enzymatically inactive; or the inhibitor bound enzyme is easily degraded by the proteases. The last possibility was ruled out by immunoassay studies (Bijay Singh, unpublished). Binding of imazethapyr with AHAS appears to stabilize the AHAS protein in relation to other proteins that are degraded after the herbicide treatment [29].

2.3.6
Imidazolinone-tolerant Crops

Owing to many desirable properties of imidazolinone class of chemistry, development of imidazolinone-tolerant crops began in early 1980s, the same time when different imidazolinone herbicides were being discovered and developed for commercialization. This example is probably the first in which selection for a herbicide tolerant crop began so early in the development of a class of herbicides. During this research, Anderson and Georgeson [30] were successful in obtaining imidazolinone-tolerant maize plants through tissue culture selection and regeneration. Subsequent research showed that resistance at the whole plant was a semidominant trait that resulted from an alteration in the gene encoding AHAS. This early work not only proved that imidazolinone-tolerant crops could be selected, but it also led to the discovery of the site of action of this class of herbicides and to the development of other imidazolinone-tolerant crops.

Plants tolerant to imidazolinones have been produced by both transgenic and non-transgenic mechanisms. However, all of the imidazolinone-tolerant crops currently being sold have been developed by non-transgenic methods. The first imidazolinone-tolerant crop (maize) was introduced in 1992. Subsequently, four additional imidazolinone-tolerant crops (canola, rice, wheat and sunflower) have been commercialized [31]. All of the imidazolinone-tolerant crops are being sold under the CLEARFIELD^{TM} trade name.

These imidazolinone tolerance traits in different crops were developed by various methods. These methods included tissue culture selection (maize), pollen mutagenesis (maize), microspore selection (canola), seed mutagenesis (wheat and rice) and incorporation of resistance trait from a weedy relative (sunflower). Details of these methods have been previously reviewed [31, 32]. In all of these cases, the basis of tolerance is due to the presence of an altered form of AHAS that is resistant to inhibition by imidazolinones. The resistant enzyme is produced due to a single base pair change in the gene encoding the large subunit of AHAS that results in a single amino acid change in the mature protein. Several mutations in the gene encoding the large subunit of AHAS have been identified that confer tolerance to imidazolinones [31, 33]. Specifically, the amino acid changes identified in different imidazolinone-tolerant crops are Ala205Val (sunflower; amino acid number in reference to AHAS sequence from *Arabidopsis thaliana*), Trp574Leu (maize and canola), Ser653Asn (maize, canola, wheat and rice), and Gly654Glu (rice). The amino acid changes that confer tolerance to imidazolinones are distributed over the entire primary structure of the AHAS protein. However, these amino acids reside in a pocket of the folded protein in the quaternary structure of the enzyme [34, 35].

From the imidazolinone family, four different molecules, imazapyr, imazapic, imazethapyr and imazamox, have been registered for weed control in various imidazolinone-tolerant crops in different regions of the world. These herbicides are applied alone or in combination with other imidazolinones or with other classes of herbicides for a broad spectrum, season-long weed control. A combination of different imidazolinone tolerance traits and multiple herbicide options provides an effective weed management tool for farmers around the world.

2.3.7
Commercial Uses of the Imidazolinone Herbicides

Six imidazolinones are commercially available. These herbicides have extremely low toxicity or are non-toxic to mammals, birds, invertebrates and fish [36]. The crops on which these herbicides are registered and whether or not they are applied to foliage or to the soil is determined by the structure of the chemical (Table 2.3.2). When applied to the foliage of plants, a non-ionic surfactant or oil adjuvant is required for maximum activity. The addition of either urea or another form of nitrogen can also increase herbicidal activity.

Imazamethabenz methyl is strictly applied postemergent to most major varieties of wheat (spring and winter), barley (spring and winter) and rye as well as some varieties of winter triticale and sunflower and safflower. Imazamox is used postemergent in leguminous crops, including soybeans, alfalfa and edible beans, as well as in imidazolinone resistant wheat, sunflower, rice, and canola. Imazaquin, though, is primarily a soil applied herbicide that is used in soybeans, established bermudagrass, centipedegrass, St. Augustinegrass, zoysiagrass, and selected landscape ornamentals. Imazethapyr is used both postemergent and preemergent in soybeans, edible beans, alfalfa, peanut, and imidazolinone resistant maize, rice, and canola. Imazapic is also applied both to the foliage and the soil in peanuts, rangeland, sugarcane, and imidazolinone resistant canola, maize, wheat, and rice. Imazapyr controls the broadest spectrum of weeds of the imidazolinones, but has selectivity on many coniferous species as well as date and oil palms. It is used for weed control and site preparation in pines and date and oil palms. It is also used in non-crop sites for control of weedy vegetation and/or maintenance of bare ground as well as in imidazolinone resistant maize and sunflower.

Table 2.3.2 Registered uses of imidazolinone herbicides in the U.S.A.

Imidazolinone	Application	Crop	Imidazolinone-resistant crop
Imazamethabenz methyl	Foliar	Barley, wheat, sunflower	
Imazethapyr	Foliar and soil	Edible beans, peas, soybean, lentils, alfalfa, peanuts, clover, birdsfoot trefoil, crown vetch, lupine, switchgrass, wheatgrass, little bluestem, orchardgrass, western wheatgrass, big bluestem, canarygrass	Maize, rice, canola
Imazamox	Foliar	Soybeans, chicory, peas, edible beans, alfalfa, clover	Canola, wheat, sunflowers
Imazapyr	Foliar and soil	Forest lands, wetlands, noncrop areas, roadsides, bahiagrass, bermudagrass	Maize
Imazaquin	Foliar and soil	Soybeans, yucca, hosta, bermudagrass, centipedegrass, mondo grass, pachysandra, St. augustinegrass, zoysiagrass, liriope, crape myrtle, gardenia, Indian hawthorn, wax-myrtle, dwarf yaupon, holly, Fraser photinia, Pfitzer juniper	
Imazapic	Foliar and soil	Peanuts, sugar cane, pastures, rangeland, ornamental turf, ditch banks, conservation reserve program land, noncrop areas	

2.3.8
Mechanisms of Selectivity

Crop selectivity of the imidazolinones is primarily dependent on differential metabolism of the herbicide between the crops and targeted weeds. For the 5'-substituted imidazolinones (i.e., imazethapyr, imazamox, imazapic, and imaza-methabenz-methyl) detoxification of the herbicides is through a mixed function oxidase that hydroxylates the substitution followed by conjugation of the metabolite to glucose through the hydroxyl group [36] (Fig. 2.3.6). Imazapyr and imaza-quin are metabolized via a different route in which there is a condensation between the carboxylic acid on the aromatic ring to nitrogen in the imidazolinone ring followed by cleavage of the imidazolinone ring (Fig. 2.3.6). The half-life of

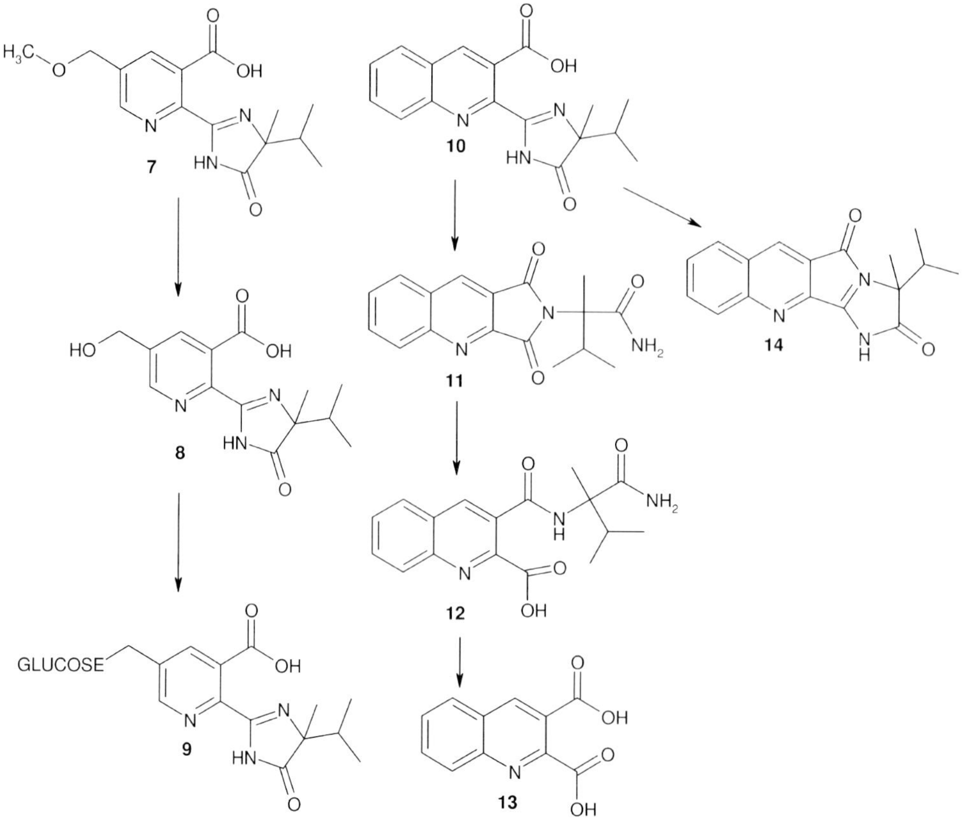

Fig. 2.3.6. General routes of metabolism of imidazolinones [36].

imidazolinones in naturally tolerant crops is less than 24 h [37]. Imidazolinone resistant crops contain a natural mutation in the AHAS gene which encodes an enzyme that no longer binds these herbicides, although metabolism may play a role in determining the level of tolerance of the resistant crop.

The weed spectrum of the imidazolinones is dependent on differential metabolism. Imazethapyr controls many broadleaf weeds and some grasses, but has limited activity on legumes and many composites. Imazamox, in contrast, has much better activity than imazethapyr on grasses. Imazapyr controls the broadest spectrum of weeds of all the imidazolinones, although it is not as active on legumes and composites. The reason for these differences is due to the ability of weeds to metabolize the herbicides. The half-life of imazethapyr in many grasses is less than 24 h because they can rapidly hydroxylate the 5′-ethyl substituent of imazethapyr [37]. However, most grasses are unable to rapidly hydroxylate the 5′-methoxyethyl substituent on imazamox [37]. Legumes and many composites can hydroxylate the 5′ substituent of both imazethapyr and imazamox. Since imazapyr does not have any substituents on the pyridine ring, most species are unable to metabolize the herbicide. However, legumes and some composites can metabolize imazapyr via the mechanism described above [37].

2.3.9
Conclusion

The imidazolinone herbicides have been and continue to be highly successful products. The ability to mix and match different imidazolinones to take advantage of their differing weed control spectrum and pre- and postemergent activity has proved invaluable in designing products for imidazolinone-resistant crops throughout the world. Although the number of imidazolinone analogs that were commercialized is extremely small compared with other ALS inhibiting herbicides, these compounds fill vital niches in many weed management programs.

References

1 Shaner, D. L., O'Connor, S. L. (Eds.) *The Imidazolinone Herbicides*, CRC Press, Inc. Boca Raton, FL, 1991.

2 Los, M. Preparation of Imidazolinyl Benzoic Acids, U.S. Patent 4,608,437, **1996**.

3 Los, M. Herbicidal 2-(2-Imidazolin-2-yl) fluoroalkoxy-, alkenyloxy- and alknyloxypyridines, U.S. Patent 4,647,301, **1987**.

4 Los, M., Ladner, D. W., Cross, B. (2-Imidazolin-2-yl)thieno- and -furo[2,3-*b*] and -[3,2-*b*]pyridines and Intermediates for the Preparation thereof, and Use of Said Compounds as Herbicidal Agents, U.S. Patent 4,650,514, **1987**.

5 Los, M. Herbicidal 2-(2 Imidazolin-2-yl)fluoroalkoxy-, alkenyloxy- and alkynyloxyquinolines, U.S. Patent 4,772,311, **1988**.

6 Los, M., Ladner, D. W., Cross, B. (2-Imidazolin-2-yl)thieno- and -furo[2,3-*b*]pyridines and Use of Said Compounds as Herbicidal Agents, U.S. Patent 4,752,323, **1988**.

7 Ciba-Geigy, 2-Imidazolinyl-pyridine- and -quinolinecarboxylic Acid Production by Reaction of Pyridine or Quinoline-2,3-dicarboxylic Acid Esters with a 2-Amino-alkanoic Acid Amide, EP 233-150A, **1986**.

8 Los, M. Synthesis and Biology of the Imidazolinone Herbicides, in *Pesticide Science and Biotechnology*, Greenhalgh, R., Roberts, T. R. (Eds.), Blackwell Scientific Publications, Oxford, **1987**.

9 Ladner, D. W. Structure–Activity Relationships among the Imidazolinone Herbicides, in *The Imidazolinone Herbicides*, Shaner, D. L., O'Connor, S. L. (Eds.), CRC Press, Inc. Boca Raton, FL, **1991**.

10 Van Ellis, M. R., Shaner, D. L. *Pestic. Sci* **1988**, 23, 25–34.

11 Los, M. Discovery of the Imidazolinone Herbicides, in *The Imidazolinone Herbicides*, Shaner, D. L., O'Connor, S. L. (Eds.), CRC Press, Inc. Boca Raton, FL, **1991**.

12 Los, M., Kust, C. A., Lamb, G., Diehl, R. E. *HortScience* **1986**, 15, 22–28.

13 Suttle, J. C., Schreiner, D. R. *J. Plant Growth Regul.* **1982**, 1, 139–145.

14 Ladner, D. W. *Pestic. Sci.* **1990**, 29, 317–325.

15 Stidham, M. A., Singh, B. K. Imidazolinone-Acetohydroxyacid Synthase Interactions, in *The Imidazolinone Herbicides*, Shaner, D. L., O'Connor, S. L. (Eds.), CRC Press, Inc. Boca Raton, FL, **1991**.

16 Wepplo, P. J. Chemical and Physical Properties of the Imidazolinone

Herbicides, in *The Imidazolinone Herbicides*, Shaner, D. L., O'Connor, S. L. (Eds.), CRC Press, Inc. Boca Raton, FL, **1991**.

17 Little, D. L., Shaner, D. L., Ladner, D. W., Tecle, B., Ilnicki, R. D. *Pestic. Sci.* **1994**, 41, 161–169.

18 Hawkes, T. R. *Monograph: British Crop Protection Council.* **1989**, 42, 131–138.

19 Rhodes, D., Hogan, A. L., Deal, L., Jamieson, G. C., Haworth, P. *Plant Physiol.* **1987**, 84, 775–780.

20 Singh, B. K., Shaner, D. L. *Plant Cell* **1995**, 7, 935–944.

21 Shaner, D. L., Singh, B. K. How does inhibition of amino acid biosynthesis kill plants? In Biosynthesis and Molecular Regulation of Amino Acids in Plants, Singh, B. K., Flores, H. E., Shannon, J. C. (Eds), American Society of Plant Physiologists, Rockville, MD, **1992**.

22 Kim, S., Vanden Born, W. H. *Pestic. Biochem. Physiol.* **1996**, 56, 141–148.

23 Rost, T. L., Gladish, D., Steffen, J., Robbins, J. *J. Plant Growth Regul.* **1990**, 9, 227–232.

24 Shaner, D. L. Sites of action of herbicides in amino acid metabolism: primary and secondary physiological effects. In *Plant Nitrogen Metabolism*, Poulton, J. E., Romeo, J. T., Conn, E. E. (Eds.), Plenum Press, New York, **1989**.

25 Pillmoor, J. B., Caseley, J. C. *Pestic. Biochem. Physiol.* **1987**, 27, 340–349.

26 Shaner, D. L., Singh, B. K. *Plant Physiol.* **1993**, 103, 1221–1226.

27 Shaner, D. L., Singh, B. K. Acetohydroxyacid synthase inhibitors, in *Herbicide Activity: Ttoxicology, Biochemistry and Molecular Biology*,

Roe, R. M., Burton, J. D., Kuhr, R. J. (Eds), IOS Press, Washington DC, **1997**.

28 Shaner, D. L., Singh, B. L., Stidham, M. A. *J. Agric. Food Chem.* **1990**, 38, 1279–1282.

29 Shaner, D. L., Singh, B. K. *Plant Physiol.*, **1991**, 97, 1339–1341.

30 Anderson, P. C., Georgeson, M. *Genome* **1989**, 31, 994–999.

31 Tan, S., Evans, R. R., Dahmer, M. L., Singh, B. K., Shaner, D. L. *Pest Manag. Sci.* **2005**, 61, 246–257.

32 Shaner, D. L., Bascomb, N. F., Smith, W. Imidazolinone-resistant crops: Selection, characterization, and management. In *Herbicide Resistant Crops*, Duke, S. O. (Ed), Lewis Publishers, Boca Raton, FL, **1996**.

33 Tranel, P. J., Wright, T. R. *Weed Sci.* **2002**, 50, 700–712.

34 Ott, K. H., Kwagh, J. G., Stockton, G. W., Sidorov, V., Kakefuda, G. *J. Mol. Biol.* **1996**, 263, 359–368.

35 McCourt, J. A., Pang, S. S., King-Scott, J., Guddat, L. W., Duggleby, R. G. *Proc. Natl. Acad. Sci. U.S.A.* **2006**, 103, 569–573.

36 Shaner, D. L. Imidazolinone Herbicides in *Encyclopedia of Agrochemicals*, J. Plimmer, J. (Ed.) John Wiley and Sons, New York, **2003**.

37 Shaner, D. L., Tecle, B. Designing Herbicide Tolerance Based on Metabolic Alteration: The Challenges and the Future, in *Pesticide Biotransformation in Plants and Microorganisms*, Hall, J. C., Hoagland, R. E., Zablotowicz, R. M. (Eds.) ACS Symposium Series 777, American Chemical Society, Washington, DC, **2001**.

2.4
Triazolopyrimidines

Timothy C. Johnson, Richard K. Mann, Paul R. Schmitzer,
Roger E. Gast, and Gerrit J. deBoer

2.4.1
Introduction

Triazolopyrimidine sulfonamides and related compounds have been studied extensively since their discovery in the early 1980s. The initial lead was discovered while examining bioisosteric relationships to the sulfonyl ureas [1]. Further investigations of structure–activity relationships around this lead eventually led to the triazolo[1,5-*a*]pyrimidine sulfonanilides and the discovery of flumetsulam (**1**) and metosulam (**2**) (Table 2.4.1). Flumetsulam was developed for use in maize and soybeans and metosulam was developed for use in maize and cereals. Studies have shown the triazolopyrimidine sulfonamides to be competitive with the amino acid leucine for binding to acetohydroxyacid synthase (AHAS) isolated from cotton (*Gossypium hirsutum*) [2]. The same study showed similar results for the sulfonylurea and imidazolinone herbicides. In addition, analysis of *Arabidopsis thaliana* mutants with resistance to AHAS-inhibiting herbicides identified a mutation that conferred resistance to triazolopyrimidine sulfonanilide and sulfonylurea herbicides but not to the imidazolinone herbicides [3]. Since those discoveries, additional work has led to the development of diclosulam (**3**) and cloransulam-methyl (**4**) for broadleaf weed control in soybeans and florasulam (**5**) for broadleaf weed control in cereals. Research efforts of new N-aryl-triazoloazinyl sulfonamides, which include the triazolo[1,5-*a*]pyridine, the triazolo[1,5-*a*]pyrazine, N-triazolo[1,5-*c*]pyrimidine and N-triazolo[1,5-*a*]pyrimidine sulfonamides, led to the discovery of penoxsulam (**6**) and DE-742 (**7**). Penoxsulam was developed for broadleaf, grass and sedge weed control in rice and DE-742 is being developed for broadleaf and grass weed control in wheat.

2.4.2
N-Triazolo[1,5-*c*]pyrimidine Sulfonanilides

2.4.2.1 Synthesis
Synthetic routes leading to triazolo[1,5-*c*]pyrimidine sulfonanilides have been reviewed [4]. Scheme 2.4.1 shows a general synthetic route to the triazolo[1,5-*c*]pyrimidine sulfonanilides [5]. An appropriately substituted 4-hydrazino-2-methylthiopyrimidine is reacted with carbon disulfide followed by benzyl chloride to afford 3-benzylthio-5-methylthio-1,2,4-triazolo[4,3-*c*]pyrimidine (**8**). Compound **8** is then treated with methoxide to afford 2-benzylthio-5-methoxy-1,2,4-triazolo[1,5-*c*]pyrimidine (**9**). The benzyl sulfide (**9**) is oxidized to the sulfonyl chloride (**10**) by treatment with chlorine and water. The sulfonyl chlorides

Table 2.4.1 Commercial and developmental triazolopyrimidine sulfonamides.

	Chemical structure	Common name	Launch date	Log P	Melting point (°C)
1		Flumetsulam	1994	−0.68	251–253
2		Metosulam	1994	0.98	210–211
3		Diclosulam	1997	0.98	216–218
4		Cloransulam-methyl	1997	0.85	218–221
5		Florasulam	1999	−1.22	193–230
6		Penoxsulam	2004	−0.35	212
7		DE-742		1.83	194–195

are then reacted with N-trimethylsilylanilines in the presence of a catalytic amount of dimethyl sulfoxide or with anilines in the presence of a catalytic amount of dimethyl sulfoxide and pyridine to afford the desired sulfonanilides (**11**).

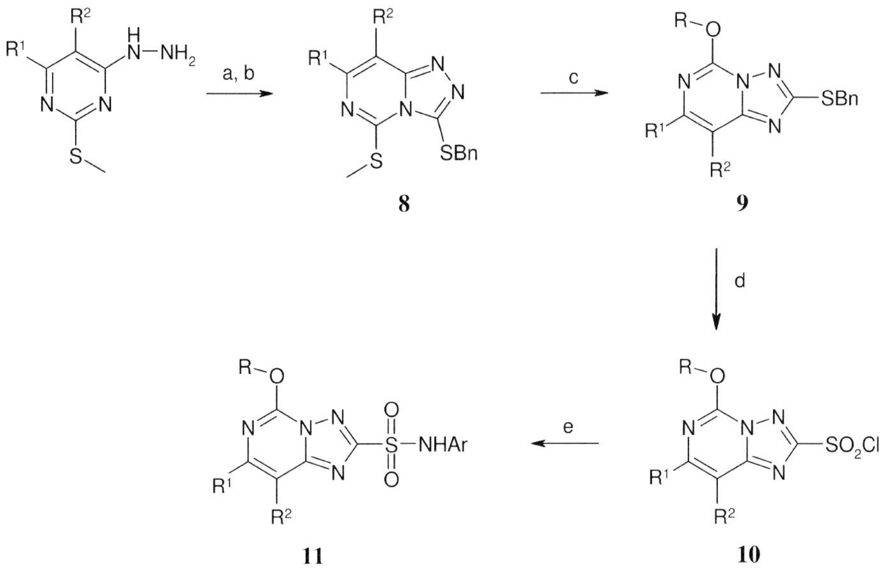

Scheme 2.4.1. (a) CS$_2$, dioxane, Et$_3$N; (b) BnCl; (c) NaOMe, MeOH, ethyl acrylate; (d) Cl$_2$, H$_2$O; (e) ArNHSi(Me)$_3$, DMSO (cat), CH$_3$CN or ArNH$_2$, pyridine, DMSO (catalytic), CH$_3$CN.

2.4.2.2 Biology

Unless otherwise noted, the *in vivo* greenhouse screening data presented in the following sections is a tabulation of postemergence foliar applied results and expressed as a "percent in growth reduction" (GR) for treated plants compared with untreated plants, where the rate identified provides the level of weed control or crop injury. The broadleaf weed activity (BW) is given as an average percent reduction in growth at a given concentration, as indicated, over five to eight broadleaf weeds chosen from the following: *Xanthium strumarium*, *Datura stramonium*, *Chenopodium album*, *Helianthus* spp., *Ipomoea* spp., *Amaranthus retroflexus*, *Abutilon theophrasti*, *Veronica heteraefolia*, *Ipomoea hederacea*, *Stellaria media* and *Polygonum convolvulus*. The grass weed activity (GW) is averaged over five weeds chosen from *Alopecurus* spp., *Echinochloa crus-galli*, *Setaria fabarii*, *Sorghum halapense*, *Digitaria sanguinalis* and *Avena fatua* and expressed in a manner similar to broadleaf weeds.

The general structure–activity relationships (SAR) for triazolo[1,5-*c*]pyrimidine sulfonanilides (**11**) have been described [4]. The SAR identified compounds with alkoxy in the 5-position (**11**, OR) and halogen or alkoxy in the 7- and 8-position (**11**, R^1 and R^2) as having the highest levels of activity. Further investigation identified compounds with halogen in the 7-position as having good levels of activity on broadleaf weeds and selectivity to soybeans. In addition, compounds with halogen in the 8-position were identified as having good activity on broadleaf weeds with selectivity to wheat.

2.4.2.2.1 Cloransulam-methyl and Diclosulam Crop Utility

Cloransulam-methyl and diclosulam are members of the triazolo[1,5-c]pyrimi-dine sulfonanilide family of AHAS-inhibiting herbicides. Both compounds show excellent crop selectivity, broad-spectrum broadleaf weed control and low toxicity. The herbicidal utility of cloransulam-methyl in soybeans was first presented in 1994 [6, 7] and further described in 1995 [8, 9] and 1996 [10–12]. Diclosulam was first described for use in soybeans and peanuts in 1997 [13] with additional description in 1998 [14] and 1999 [15–17].

Cloransulam-methyl was commercialized in the United States under the trade name FirstRate (Trademark of Dow Agrosciences, LLC) herbicide for the control of annual broadleaf weeds and certain perennial sedges in soybeans. Applications can be made preplant surface, preplant incorporated, preemergence and poste-mergence for the control of broadleaf weed species. Postemergence applications of cloransulam-methyl at 17.5 g-a.i. ha^{-1} or soil-applied treatments at rates of 35–44 g-a.i. ha^{-1} provide control of a large number of soybean relevant weeds. Cloransulam-methyl does not provide control of annual and perennial grass weeds or certain broadleaf weeds such as *Solanum* spp. [18].

Diclosulam is registered in the United States and in Latin America for use in peanuts and soybeans. Applications can be made preplant surface, preplant incor-porated and preemergence at rates of 17.5–26 g-a.i. ha^{-1} for the control of numer-ous broadleaf weed species. Diclosulam does not provide control of annual and perennial grass weeds or certain broadleaf weeds such as *Solanum* spp.

2.4.2.2.2 Florasulam Crop Utility

Florasulam (**5**) provides excellent postemergence selectivity in turf and small grain cereal crops such as wheat, barley, oats, rye and triticale [19, 20]. The Euro-pean and North American cereal markets are of primary commercial interest for florasulam due to its specialized spectrum of broadleaf weed control. Florasulam is highly active on economically important species in the Compositae, Caryophyl-laceae, Cruciferae, Rubiacea and Leguminosae plant families at a typical use rate of 5 g-a.i. ha^{-1} [21]. Owing to the relatively short half-life in soil, only postemer-gence applications are used in commercial practice [22].

2.4.2.3 Mechanism of Crop Selectivity

2.4.2.3.1 Cloransulam-methyl and Diclosulam Mechanism of Crop Selectivity

The metabolism of triazolopyrimidine sulfonanilides (**1–4**) in plants has been re-viewed [23, 24]. It has been shown that diclosulam (**3**) and cloransulam-methyl (**4**) are rapidly metabolized in soybeans by facile conjugation with homogluta-thione which displaces the 7-fluoro substituent (Fig. 2.4.1) [25]. This mechanism was found to only occur in soybeans for **3** and **4**. Oxidation at the 4-position of the aniline ring occurs rapidly in maize for **3** and **4**. In wheat, **4** undergoes O-dealkylation of 5-ethoxy followed by glucose conjugation and oxidation at the 4-position of the aniline ring occurs for **3** [25].

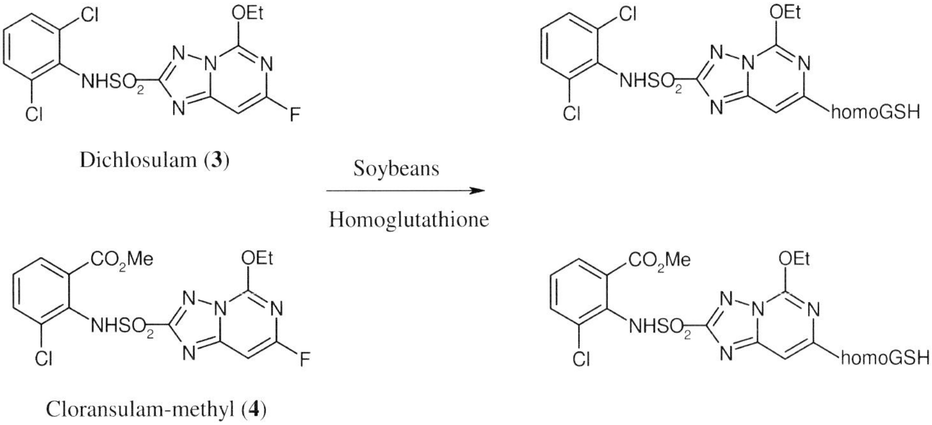

Fig. 2.4.1. Metabolism of diclosulam (**3**) and cloransulam-methyl (**4**) in soybeans (*Glycine max*).

2.4.2.3.2 Florasulam Mechanism of Crop Selectivity

The selectivity of florasulam (**5**) to wheat and the high level of herbicidal activity on important weeds are related primarily to the difference in rates of metabolism [26]. Florasulam has a half-life in wheat of 2.4 h as compared with a half-life in *Galeopsis tetrahit* L., *Polygonum papathifolium*, and *Galium aparine* L. of 19.8 h, 43.6 h and >48 h, respectively (Table 2.4.2). In wheat, florasulam has been shown to undergo rapid metabolism at the 4-position of the phenyl ring to give the 4-hydroxy metabolite which, in turn, is conjugated to glucose (Fig. 2.4.2). In contrast, slow metabolism is observed in *Galeopsis tetrahit* L. and *Polygonum papathifolium* with little degradation of florasulam observed in *Galium aparine* L., even at 48 h after treatment. Similar differences in the rate of metabolism in wheat compared with broadleaf weeds accounted for the sensitivity of broadleaf weeds to closely related analogs [24].

Table 2.4.2 Herbicidal activity and metabolism of **5**.

Species	GR$_{50}$ (g ha^{-1})	$T_{1/2}$[a] (h)
Triticum aestivum	≫32	2.4
Galium aparine L.	≪2.5	>48 (202.4)
Galeopsis tetrahit L.	<2.5	19.8
Polygonum papathifolium	<2.5	43.6

[a] Time required for plants to metabolize 50% of the applied compound.

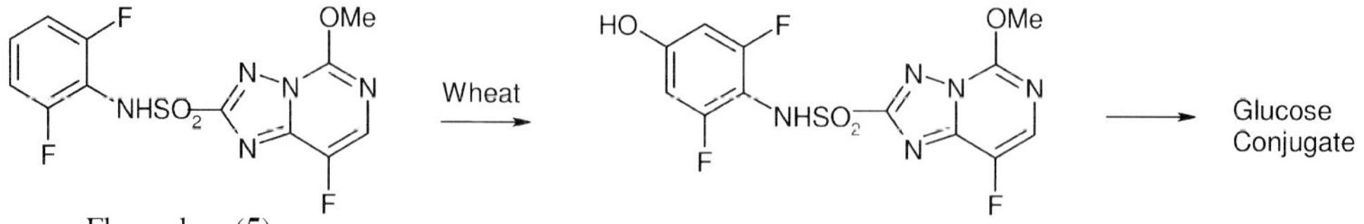

Fig. 2.4.2. Metabolism of florasulam (**5**) in wheat (*Triticum aestivum*).

2.4.2.4 Environmental Degradation, Ecotox and Tox

2.4.2.4.1 Cloransulam-methyl and Diclosulam Environmental Degradation

Metabolism in aerobic soils is a significant dissipation mechanism for both cloransulam-methyl (**4**) and diclosulam (**3**). Analysis of soil samples from bareground applications of cloransulam-methyl and diclosulam gave half-life ranges of 3–11 days and 13–43 days, respectively [27, 28]. Organic matter content and soil temperature were found to be the two factors that most influenced the soil degradation rates. Dealkylation of the 5-ethoxy on the triazolopyrimidine ring to form the associated 5-hydroxytriazolopyrimidine is a metabolic manipulation shared by both compounds. The shared aminosulfonyl triazolopyrimidine (**12**) was a metabolite identified in soil degradation studies for both compounds [29, 30]. Both compounds underwent additional, unique metabolic manipulations in the soil. Unextractable residues accounted for a significant amount of the final metabolite distribution for both cloransulam-methyl and diclosulam. Photolysis in water was also shown to be a significant avenue of degradation for cloransulam-methyl, with a half-life of less than 1 h [31]. Diclosulam does not show significant degradation by photolysis (half-life > 100 days).

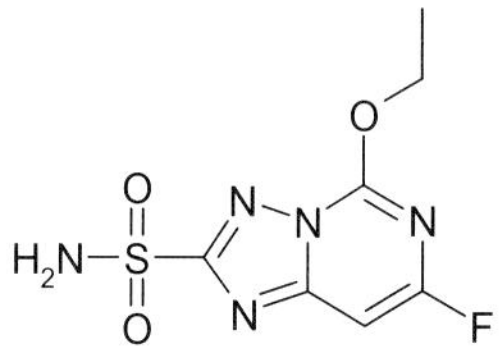

12

Both cloransulam-methyl and diclosulam have low acute toxicology profiles and no indication of any chronic toxicology issues. Both compounds do show slight toxicity to *Daphnia* but are considered practically non-toxic to birds, insects, aquatic organisms and earthworms.

2.4.2.4.2 Florasulam

Florasulam (**5**) dissipates primarily through microbial degradation [22]. Other patterns of degradation or dissipation contribute minimally to the loss of florasulam in the agricultural field environment. As florasulam degrades in the soil, several metabolites of the herbicide are formed. The primary soil metabolite, the 5-hydroxytriazolo[1,5-c]pyrimidine sulfonanilide analog of **5** (**5**-OH), has been shown to have limited plant activity (a factor of 100× or greater) relative to the parent while other metabolites are inactive. The field half-life of florasulam in soil ranges from 2 to 18 days. While soil pH, texture and level of organic matter influence rate of degradation, temperature has the greatest impact on soil half-life. In natural sediment and surface water in the dark at 20 °C, florasulam is degraded to the **5**-OH metabolite with a half-life of 9 to 29 days. In anaerobic conditions, the half-life was approximately 13 days. In water, the aqueous photolytic half-life was 4.9 days.

The overall toxicological profile of florasulam is very favorable. It is not acutely toxic, does not pose an inhalation hazard, nor is it a skin sensitizer. No evidence of mutagenic or carcinogenic potential was obtained from any study. It showed no teratogenic effects in either rats or rabbits. Tests also indicate that florasulam is not a reproductive hazard or concern.

2.4.3
N-Triazolo[1,5-c]pyrimidine Sulfonamides

The N-triazolo[1,5-c]pyrimidine sulfonamides and related compounds differ from their sulfonanilide counterparts by the orientation of the linkage between the triazoloazine and the aryl or heteroaryl ring. However, in most cases synthesis of sulfonamides is similar to the sulfonanilides, as the target molecules are formed by reaction of a sulfonyl chloride and an amine in the final step. Notably, the synthesis of arylsulfonyl chlorides account for much of the diversity in these molecules [32–36].

2.4.3.1 Synthesis
Scheme 2.4.2 outlines a straightforward and general route for the synthesis of N-triazolo[1,5-c]pyrimidine sulfonamides (**13**) [32, 35]. 4-Hydrazino-2-methyl-thiopyrimidines (**14**) are reacted with cyanogen bromide to give the 3-amino-5-methylthiotriazolo[4,3-c]pyrimidines (**15**), usually as the hydrogen bromide salt. Treatment of **15** with sodium methoxide affords the 2-amino-5-methoxy-triazolo[1,5-c]pyrimidine ring system (**16**). The sulfonamides (**13**) are prepared by reacting **16** with arylsulfonyl chlorides (**17** Scheme 2.4.3) in the presence of pyridine and a catalytic amount of dimethyl sulfoxide.

Several substituted benzene and pyridine sulfonyl chlorides from which to prepare sulfonamides have been investigated. However, with respect to crop selectivity, most interest has focused on 2,6-disubstituted benzenesulfonyl chlorides and 2,4-disubstituted pyridine-3-sulfonyl chlorides. A general method for the preparation of various benzene and pyridine sulfonyl chlorides is via ortho directed metalation [32, 37]. The sulfonyl chlorides (**17**) can be prepared directly from the aryl lithium species by reacting with sulfur dioxide followed by sulfuryl chloride (Scheme 2.4.3). Alternatively, reaction of the aryl lithium species with a disulfide,

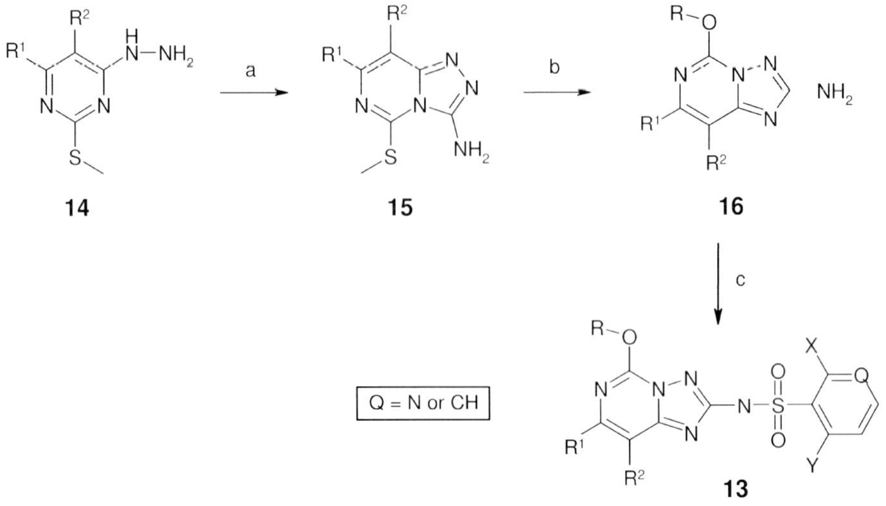

Q = N or CH

Scheme 2.4.2. (a) BrCN, *i*-PrOH; (b) NaOR, ethyl acrylate; (c) ArSO$_2$Cl, pyridine, DMSO (catalytic), CH$_3$CN.

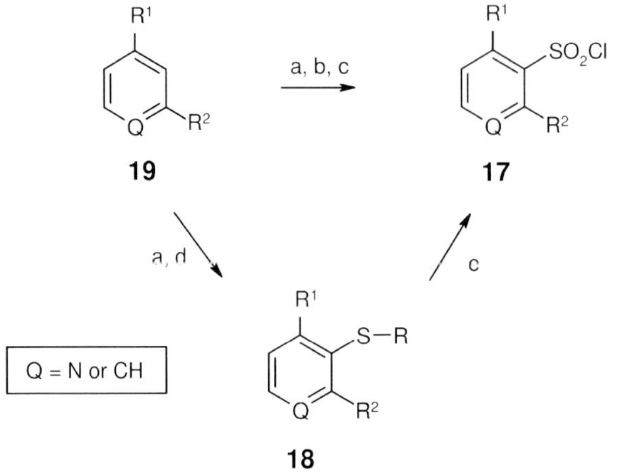

Q = N or CH

Scheme 2.4.3. (a) Excess 19, BuLi, TMEDA, i-Pr$_2$NH, THF or Et$_2$O; (b) SO$_2$, Et$_2$O; (c) SO$_2$Cl$_2$; (d) (n-PrS)$_2$; (e) Cl$_2$, H$_2$O, HOAc.

most commonly propyl disulfide, gives an alkyl aryl sulfide (**18**: R = *n*-Pr) which can be converted into the sulfonyl chloride using chlorine and water. The later method is commonly used when further manipulation on the aryl ring is required.

2.4.3.2 Biology

The structure–activity trends for triazolo[1,5-*c*]pyrimidine sulfonamides (**13**) have been studied extensively [32–36]. Table 2.4.3 summarizes the herbicidal activity

Table 2.4.3 Herbicidal activity for analogs of 13 (R = Me) (structure shown in Scheme 2.4.2).

X	Y	Q	R^2	R^1	Average GR$_{80}$ BW (ppm)	Average GR$_{80}$ GW (ppm)
Cl	Cl	CH	OMe	H	3	11
Cl	Cl	CH	OEt	H	216	>500
Cl	Cl	CH	Me	H	<15	>500
Cl	Cl	CH	Cl	H	1	15
Cl	Cl	CH	H	OMe	>1000	>1000
OMe	CF$_3$	CH	OMe	H	<0.2	1
OMe	OMe	CH	OMe	H	0.5	<0.1
OMe	F	CH	OMe	H	1	3
OMe	Me	CH	OMe	H	1	3
OMe	CO$_2$Me	CH	OMe	H	10	2
OMe	Cl	N	OMe	H	12	15
OMe	OMe	N	OMe	H	<1	2
OMe	CF$_3$	N	OMe	H	4	16
OEt	CF$_3$	N	OMe	H	2	64
CF$_3$	OMe	N	OMe	H	>250	>250

for **13** with various substitutions on the triazolo[1,5-c]pyrimidine ring and 2,6-disubstitutions on the aryl ring. Analogs with substitution in the 8-position of the triazolopyrimidine ring (R^2) are more active than those with substitution in the 7-position (R^1). The 8-methoxy analog has the best activity on both grass and broadleaf weeds. Halogen substitutions in the 8-position have good levels of activity on broadleaf species with somewhat reduced levels of activity on grass species. High levels of activity are achieved with 2,6-disubstitutions on the phenyl ring, especially when one of the substituents is methoxy. The highest levels of activity are achieved when both the 2- and 6-positions are methoxy, although good levels of activity are achieved with various substituents in the 6-position when there is a methoxy in the 2-position. For substitutions on the pyridine ring, good levels of activity are achieved when at least one of the substituents is methoxy. The best levels of activity on both grass and broadleaf species is gained with the dimethoxy analog (**13**, Q = N, X = Y = OMe, R = Me, R^2 = OMe). The 4-methoxy analog (**13**, Q = N, X = CF$_3$, Y = OMe) has very little activity on either grass or broadleaf weeds.

Several 2-trifluoromethylphenyl analogs of **13** have been prepared with various alkoxy and substituted alkoxy groups in the 6-position of the phenyl ring and some of these molecules demonstrated trends for selectivity toward rice (*Oryza sativa*) with activity on barnyard grass (*Echinochloa crus-galli*) [32, 35]. Tables 2.4.4 and 2.4.5 summarize the activity observed on rice and key rice weeds when applied as a water-injected treatment (Table 2.4.4) and as a postemergence foliar treatment (Table 2.4.5) in the greenhouse to 1–3 lf rice and weeds for 2-alkoxy-6-trifluoromethylphenyl substituted analogs that were identified as having activity of interest. Particularly noteworthy are the 2,2-difluoroethoxyphenyl (**13**, Q = CH,

Table 2.4.4 Herbicidal activity on transplanted paddy rice and weeds for analogs of **13** (Q = CH, R = Me, R^1 = H, R^2 = OMe).

X	Y	*Oryza sativa* GR_{20} (g-a.i. ha^{-1})	*Echinochloa crus-galli* GR_{80} (g-a.i. ha^{-1})	*Monochoria vaginalis* GR_{80} (g-a.i. ha^{-1})	*Scirpus juncoides* GR_{80} (g-a.i. ha^{-1})	*Cyperus difformis* GR_{80} (g-a.i. ha^{-1})
CF_3	OCH_2CH_2F	14	10	5.2	9	8
CF_3	OCH_2OMe	124	16	4	15	18
CF_3	OCH_2CF_3	51	19	5	21	36
CF_3	OCH_2CF_2H	75	12	<2	12	14
CF_3	$OCH(CH_2F)_2$	140	14	1	9	31

Table 2.4.5 Herbicidal activity on direct-seeded rice and weeds as a postemergence foliar application for **13** (Q = CH, R = Me, R^1 = H, R^2 = OMe).

X	Y	*Oryza sativa* GR_{20} (g-a.i. ha^{-1})	*Echinochloa crus-galli* GR_{80} (g-a.i. ha^{-1})	*Scirpus juncoides* GR_{80} (g-a.i. ha^{-1})
CF_3	$O(CH_2)_2F$	4	1	3
CF_3	OCH_2OMe	>70	12	5
CF_3	OCH_2CF_3	>70	9	–
CF_3	OCH_2CF_2H	>140	20	30
CF_3	$OCH(CH_2F)_2$	>70	10	–

X = CF_3, R = Me, R^2 = OMe) and 2-fluoroethoxyphenyl (**13**, Q = CH, X = CF_3, R = Me, R^2 = OMe) analogs which showed high levels of activity on all weeds species, particularly barnyard grass, with good selectivity to rice.

2.4.3.3 Penoxsulam Crop Utility

Based on the above greenhouse results, several 2-trifluoromethyl-6-alkoxyphenyl analogs of **13** were tested in key rice growing countries from 1997 to 1999 to characterize their activity. From these analogs, the 2,2-difluoroethoxyphenyl analog of **13** (**6**) was identified as having good rice tolerance, broad spectrum weed control (*Echinochloa* spp. and many key broadleaf and sedge weeds) and providing good residual weed control depending on the rates applied. Other analogs tested were not selected for several reasons, such as being too injurious to rice, providing poor weed control, or having short residual activity, when compared with **6**. Additionally, it was discovered that **6** could be co-applied with the grass herbicide cyhalofop-butyl which can not be tank-mixed with commercially available ALS or auxin mode of action herbicides without antagonizing the control of *Echinochloa* spp. Based on the ability to meet many of the commercial rice herbicide needs (crop tolerance, broad-spectrum weed control, residual weed control activity, and tank-mix ability) in transplanted rice, direct-seeded rice and water-seeded rice in over 25 rice countries, **6** was identified for development as a new rice herbicide with the code number DE-638 and the common name "penoxsulam" [38–46].

2.4.3.4 Penoxsulam Mechanism of Crop Selectivity
Metabolism studies conducted on **6** showed that *O*-dealkylation of one heterocycle methoxy group was occurring in rice (Fig. 2.4.3) [47]. A comparison of metabolic degradation rates and activity on indica rice, japonica rice and barnyardgrass for **6** indicates that degradation rates explain the major differences observed in activity (Table 2.4.6). Other factors, such as site of uptake and transport, which are modulated by plant structure and metabolism, may contribute to additional rice selectivity observed for penoxsulam.

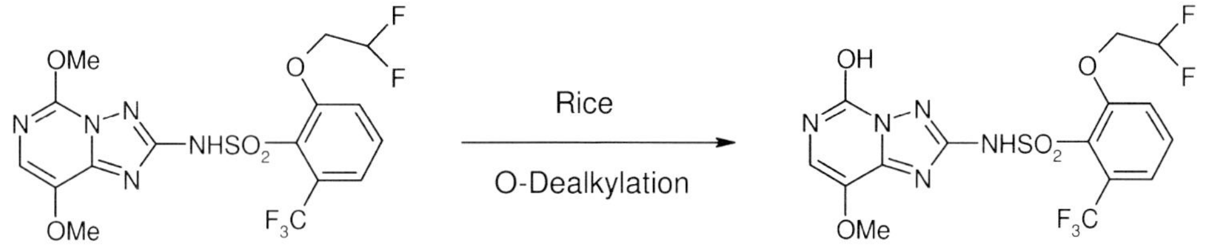

Fig. 2.4.3. Metabolism of penoxsulam (**6**) in rice (*Oryza sativa*).

Table 2.4.6 Herbicidal activity and metabolism of **6**.

Species	GR$_{80}$ (ppm)	$T_{1/2}$[a] (h)
Indica rice	>250	14.4
Japonica rice	>250	38.4
Echinochloa crus-galli	0.24	106

[a] Time required to metabolize 50% of the applied compound.

2.4.3.5 Penoxsulam Environmental Degradation, Ecotox and Tox
Dissipation of penoxsulam occurs primarily through microbial degradation. Other patterns of degradation or dissipation (e.g., photolysis, volatility, leaching and chemical hydrolysis) contribute to the loss of penoxsulam in the agricultural field environment. As penoxsulam degrades in the soil, several metabolites of the herbicide are formed. The primary soil metabolite, the 5-hydroxytriazolo[1,5-*c*]pyrimidine sulfonamide analog of **6**, has been shown to have very limited plant activity (a factor of >100×) relative to the parent, while other metabolites are inactive. The half-life of penoxsulam under field conditions averaged 6.5 days (4 to 10 days) under flooded water-seeded rice conditions, and averaged 14.6 days (13

to 16 days) under non-flooded, dry-seeded rice conditions. Soil pH, texture and level of organic matter will influence the rate of degradation. Half-lives for aerobic aquatic conditions averaged 25 days (11–34 days) while half-lives for anaerobic conditions averaged 7 days (5–11 days). The major route of degradation in water is photolysis (half-life in water from photolysis was 2 days; summer sunlight, 40° N latitude).

Penoxsulam has a toxicological profile similar to other triazolopyrimidine sulfonamides. There was no indication of acute or chronic toxicity issues to mammalian and non-target organisms such as fish, fresh water invertebrates, honey bees, earthworm and beneficial arthropods.

2.4.4
N-Triazolo[1,5-*a*]pyrimidine Sulfonamides

2.4.4.1 Synthesis
Scheme 2.4.4 outlines a general route for the synthesis of N-triazolo[1,5-*a*]pyrimidine sulfonamides (**20**) [48, 49]. 2-Amino-4,6-dimethoxypyrimidine (**21**) is reacted with thiocarbonyldiimidazole followed by hydroxylamine in the presence of a base to give **23**, which is then reacted with various substituted benzene and pyridine sulfonyl chlorides (**17** Scheme 2.4.3) in the presence of a catalytic amount of dimethyl sulfoxide and a base to give **20**.

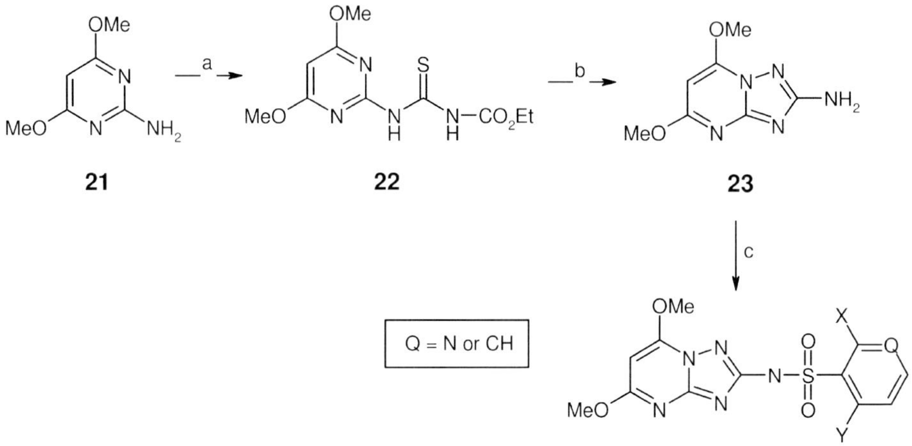

Scheme 2.4.4. (a) SCNCO$_2$Et; (b) HONH$_2$, Et(*i*-Pr)$_2$N; (c) ArSO$_2$Cl, pyridine, DMSO (catalytic).

2.4.4.2 Biology
Structure–activity trends for analogs of **20** have been reported previously [1]. For substitutions on the triazolo[1,5-*a*]pyrimidine ring, these studies showed that better herbicidal activity was achieved when the 5- or 7-position is substituted with

Table 2.4.7 Herbicidal activity for analogs of **20** (structure shown in Scheme 2.4.4).

X	Y	Q	Average GR$_{80}$ BW (ppm)	Average GR$_{80}$ GW (ppm)	*Alopecurus myosuroides* GR$_{80}$ (ppm)	*Triticum aestivum* GR$_{20}$ (ppm)
Cl	Cl	CH	4	4	4	<1
OMe	OMe	CH	8	8	4	<1
OMe	CF$_3$	CH	15	1	<1	<1
O(CH$_2$)F	CF$_3$	CH	2	16	27	13
OCH$_2$CF$_2$H	CF$_3$	CH	2	30	28	>250
OCH$_2$CF$_3$	CF$_3$	CH	4	62	240	46
OMe	CF$_3$	N	2	2	2	2
OMe	CF$_2$CF$_3$	N	>250	125	164	3
OMe	I	N	31	8	1	<1
OEt	CF$_3$	N	15	62	10	62

methoxy than when the 5- or 7-position is substituted with alkyl, halogen or halo alkyl. Recent efforts have shown that when both the 5- and 7-position are substituted with methoxy superior levels of herbicidal activity are achieved. With respect to the phenyl ring, previous efforts showed that 2,6-disubstitutions have superior herbicidal Activity over other substitution patterns. Much of the recent work has focused on 2,6-disubstituted phenyl and 2,4-disubstituted 3-pyridyl analogs of **20** with a 5,7-dimethoxy substitution on the triazolo[1,5-*a*]pyrimidine ring [48]. Table 2.4.7 summarizes general trends in activity on grass weeds, broadleaf weeds, blackgrass (*Alopecurus myosuroides*) and wheat (*Triticum aestivum*) for phenyl and pyridyl analogs of **20**. Good levels of herbicidal activity are achieved on both grass and broadleaf weeds with 2,6-substitutions on the phenyl ring (**20**, Q = CH) and in particular when one of the substituents is methoxy. Superior levels of activity on grass species is achieved with the 2-methoxy-6-trifluormethyl-phenyl analog of **20** (Q = CH, X = OMe, Y = CF$_3$). However, these analogs cause significant injury to wheat. With higher alkoxy substitutions (e.g., **20**, Q = CH, X = OMe, Y = OCH$_2$CH$_2$F) the activity on blackgrass decreases. With 2-methoxy-4-trifluoromethyl substitution on the pyridyl ring of **20** (Q = N, X = OMe, Y = CF$_3$) excellent activity is achieved on both grass and broadleaf weeds. Replacing methoxy with ethoxy (**20**, Q = N, X = OEt, Y = CF$_3$) results in a loss of herbicidal activity which is more significant on grass than broadleaf species. Good levels of activity on blackgrass are observed for 4-trifluoromethylpyridyl analogs of **20** (Q = N, X = OMe, Y = CF$_3$) and this analog shows a trend for wheat selectivity. Based on grass and broadleaf weed control in field studies, this analog was identified for development as a new herbicide for wheat with the code name DE-742.

2.4.4.3 DE-742 Crop Utility
DE-742 (**7**) requires addition of a safener to achieve commercial levels of postemergence selectivity in small grain cereal crops, the main target market. Com-

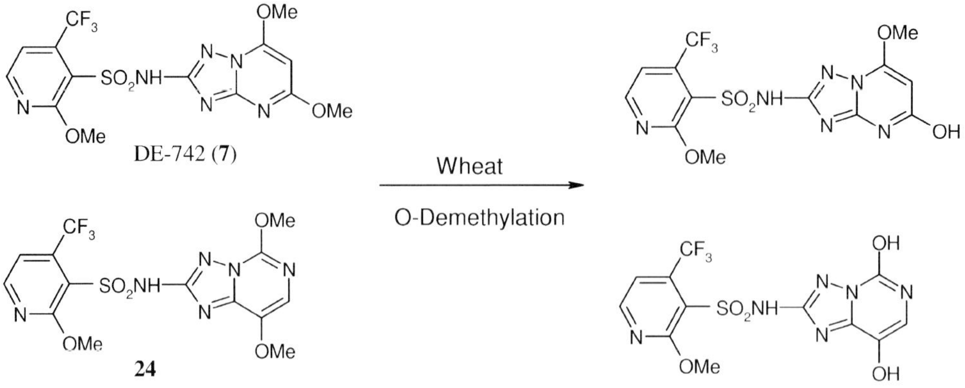

Fig. 2.4.4. Metabolites identified for DE-742 (**7**) and **24** in wheat (*Triticum aestivum*).

mercial selectivity is limited to wheat, rye and triticale varieties. DE-742 is broadly active on annual grass and broadleaf weeds, with some activity on certain perennial weed species. It derives most of its activity from foliar application but has some ability to provide soil residual control of emerging weeds. DE-742 controls economically important grass and broadleaf weed species in the global cereals markets.

2.4.4.4 DE-742 Mechanism of Crop Selectivity

Metabolism studies have been conducted on **7** and the closely related triazolo[1,5-c]pyrimidine analog **24** (Fig. 2.4.4). The metabolites identified in wheat (*Triticum aestivum*) for **7** and **24** are shown in Fig. 2.4.4. These studies showed that O-dealkylation of one heterocycle methoxy groups was occurring with **7** in wheat (Fig. 2.4.4). In comparison, both methoxy groups on the heterocycle of **24** underwent O-dealkylation. Table 2.4.8 compares the metabolism rates and activity on wheat and blackgrass (*Alopecurus myosuroides*) for **5**, **7** and **24**. The order of rank-

Table 2.4.8 Herbicidal activity and metabolism of **5**, **7** and **24**.

Herbicide	*Triticum aestivum*		*Alopecurus myosuroides*	
	GR_{20} (ppm)	$T_{1/2}$ (h)[a]	GR_{80} (ppm)	$T_{1/2}$ (h)[a]
Florasulam (**5**)	7.8	2.4	15.6	NA
24	12.2	5.7	4.4	51.6
DE-742 (**7**)	2.9	14	0.31	46

[a] Time required for plants to metabolize 50% of the applied compound.

ing is **5** > **24** > **7**, with respect to rate of metabolism in wheat, and **7** > **5** > **24**, with respect to activity on wheat. The slower rate of metabolism in wheat along with the higher levels of activity most likely account for the injury observed when **7** is not used in conjunction with a herbicide safener, such as cloquintocet. However, **7** is significantly more active on blackgrass than either **5** or **24**, with a rate of metabolism in blackgrass comparable to **24**.

2.4.4.5 DE-742 Environmental Degradation, Ecotox and Tox

Under aerobic conditions, laboratory studies have shown that DE-742 degrades rapidly in soil. The average laboratory half-life was 4 days at 20 °C across 20 different soils from Europe, the United States, and Canada. The principal metabolites are 7-hydroxytriazolo[1,5-*a*]pyrimidine sulfonamide (**25**), 5-hydroxytriazolo[1,5-*a*]pyrimidine sulfonamide (**26**), 7-hydroxy-6-chlorotriazolo-[1,5-*a*]pyrimidine sulfonamide (**27**), 5,7-dihydroxytriazolo[1,5-*a*]pyrimidine sulfonamide (**28**), and the corresponding sulfonic acid of DE-742 (**29**) (Fig. 2.4.5). All soil metabolites have little or no phytotoxicity compared with DE-742. DE-742 degrades at a moderate rate under anaerobic conditions, with a half-life of 47 days determined on a single soil. DE-742 does not photodegrade at a measurable rate on soil surfaces. Studies indicate overall toxicological profile of DE-742 is very favorable and similar to other triazolopyrimidine sulfonamides.

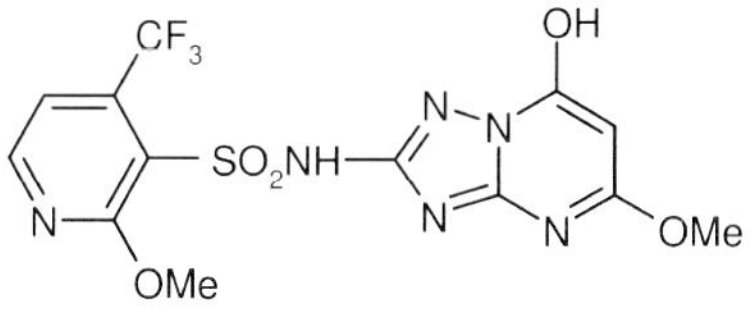

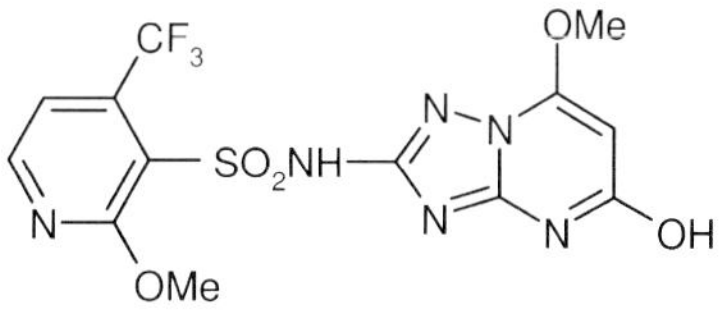

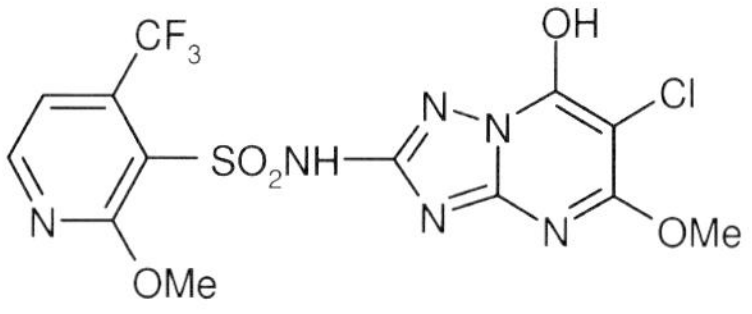

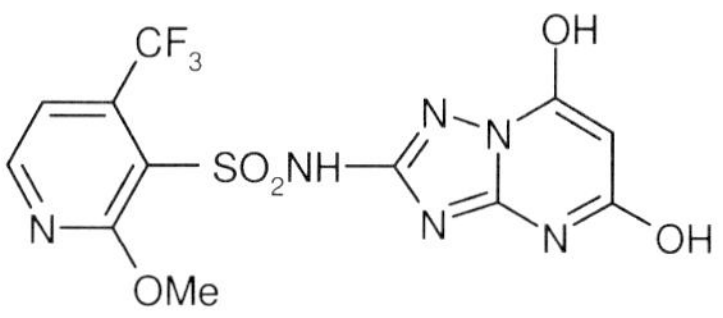

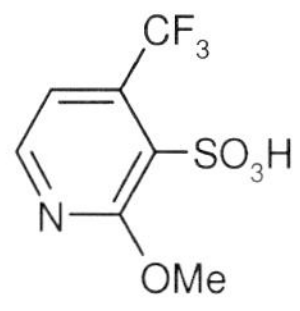

Fig. 2.4.5. Soil metabolites of DE-742 (**7**).

2.4.5
Other Systems

2.4.5.1 Synthesis

2.4.5.1.1 N-Aryl-triazolo[1,5-*a*]pyridine Sulfonanilides

The triazolo[1,5-*a*]pyridine-2-sulfonanilides (**30**) can be prepared by the general route as outlined in Scheme 2.4.5 [49, 50]. The intermediate 2-benzylthio-triazolo[1,5-*a*]pyridine (**31**) is prepared starting from an appropriately substituted 2-aminopyridine. The 2-aminopyridine is reacted with O-mesitylenesulfonylhydroxylamine to give the N-aminopyridinium mesitylate (**32**). Compound **32** is then reacted with thiocarbonyldiimidazole followed by benzyl chloride to give **31**. Conversion of the benzyl sulfide, **31**, into the corresponding sulfonyl chlorides (**33**) is accomplished with chlorine and water. The sulfonyl chlorides are converted into the desired sulfonanilide (**30**) by reaction with aniline in the presence of pyridine and a catalytic amount of dimethyl sulfoxide.

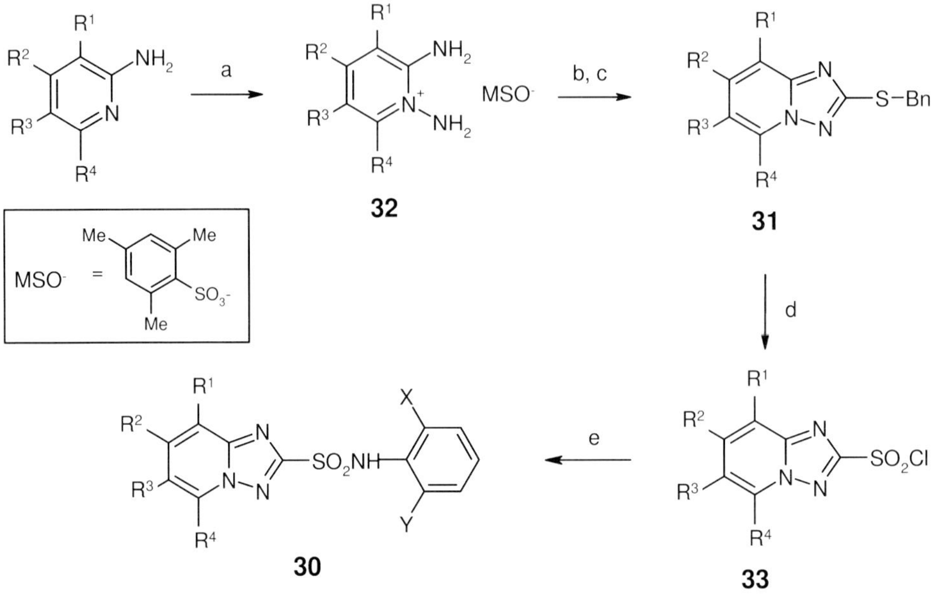

Scheme 2.4.5. (a) O-Mesitylenesulfonylhydroxylamine;
(b) 1,1′-thiocarbonyldiimidazole; (c) BnCl; (d) Cl_2, H_2O; (e) $ArSO_2Cl$,
pyridine, DMSO (catalytic).

2.4.5.1.2 N-Aryl-triazolo[1,5-*a*]pyrazine Sulfonanilides

The triazolo[1,5-*a*]pyrazines sulfonanilides (**34**) can be prepared in a manner analogous to the triazolo[1,5-*a*]pyridines, starting from 2-aminopyrazines (Scheme 2.4.6) [51]. 2-Aminopyrazines are reacted with O-mesitylenesulfonylhy-

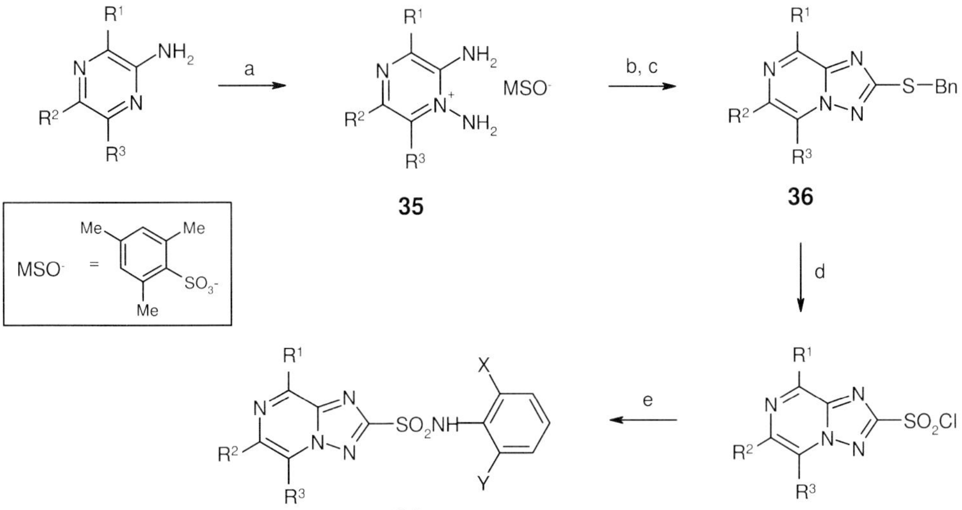

Scheme 2.4.6. (a) O-Mesitylenesulfonylhydroxylamine;
(b) 1,1'-thiocarbonyldiimidazole; (c) BnCl, BuOH; (d) Cl₂, H₂O;
(e) ArNH₂, pyridine, DMSO (catalytic).

droxylamine to yield the N-aminopyrazinium mesitylate (**35**). Compound **35** is then converted into the intermediate 2-benzylthiotriazolo[1,5-a]pyrazine (**36**) by first reacting with thiocarbonyldiimidazole followed by reaction with benzyl chloride in hot butanol. Further manipulation of the triazolo[1,5-a]pyrazine ring can take place at this stage to introduce additional functionality. Compound **36** is converted, by reaction with chlorine in water, into the sulfonyl chloride, which is then reacted with substituted anilines to give **34**.

2.4.5.1.3 N-Triazolo[1,5-a]pyridine Sulfonamides

The N-triazolo[1,5-a]pyridine sulfonamides (**37**) are prepared by the general methods outlined in Scheme 2.4.7 [32, 36]. A substituted 2-aminopyridine (**38**) is reacted with ethoxycarbonylisothiocyanate to give the thiourea (**39**). Reaction of **39** with hydroxylamine in the presence of a base yields the 2-aminotriazolo[1,5-a]-pyridines (**40**). Compound **40** is then reacted with substituted sulfonyl chloride in the presence of a catalytic amount of dimethyl sulfoxide and pyridine to give **37**.

2.4.5.2 Biology

2.4.5.2.1 Triazolo[1,5-a]pyridine Sulfonanilides

The structure–activity relationships, with respect to substitutions on the phenyl ring, for the triazolo[1,5-a]pyridine sulfonanilides (**30**) are similar *in vivo* against broadleaf and grass weeds to the triazolo[1,5-a]pyrimidine sulfonanilides [4, 49, 50]. Compounds with substitutions in the 2- and 6-position of the phenyl ring

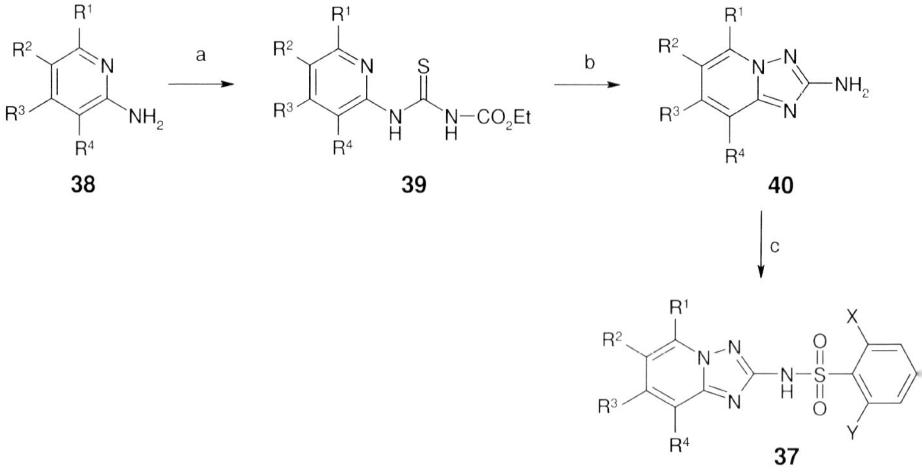

Scheme 2.4.7. (a) $SCNCO_2R$; (b) $HONH_2$, $Et(i-Pr)_2N$; (c) $ArSO_2Cl$, pyridine, DMSO (catalytic).

are more active than those with substitution in the 3- and 4-position. Substitutions on the triazolo[1,5-*a*]pyridine ring have been studied more extensively than those on the phenyl [4, 5]. Table 2.4.9 presents a compilation of herbicidal activity for substitutions on the fused heterocyclic ring. With a methoxy in the 5-position (**30**, R^1 = OMe) there are good levels of activity on both broadleaf and grass weeds. When the substituent in the 8-position is methoxy (**30**, R^4 = OMe) there

Table 2.4.9 Herbicidal activity for **30** (X = Y = Cl or X = Y = F) (structure shown in Scheme 2.4.5).

X	Y	R^1	R^2	R^3	R^4	Average GR_{80} BW (ppm)	Average GR_{80} GW (ppm)
Cl	Cl	Cl	H	Me	H	31	>500
Cl	Cl	OMe	H	Me	H	1	8
Cl	Cl	OMe	H	OMe	H	16	62
Cl	Cl	H	Me	H	OMe	62	125
Cl	Cl	OMe	H	Br	H	1	62
F	F	OMe	H	H	H	2	8
F	F	OMe	H	H	OMe	1	125
F	F	OMe	H	Me	H	1	8
F	F	OEt	H	Me	H	8	31
F	F	Cl	H	Cl	H	>2000	>2000
F	F	OMe	H	Cl	H	1	16
F	F	OMe	H	H	Br	15	31

Table 2.4.10 Herbicidal activity for **34** (X = Y = Cl) (structure shown in Scheme 2.4.6).

R^1	R^2	R^3	Average GR$_{80}$ BW (ppm)	Average GR$_{80}$ GW (ppm)
OMe	H	H	>250	>125
H	H	OMe	36	>125
Br	H	OMe	11	600
OMe	H	OMe	<4	27
H	Cl	OMe	4	375

are good levels of activity on broadleaf weeds, but the activity on grass species is somewhat less. The best levels of activity, on both grass and broadleaf species, are observed when the 5-position is methoxy (**30**, R^1 = OMe) and the 7-position (R^3) is methyl or chlorine.

2.4.5.2.2 Triazolo[1,5-*a*]pyrazine Sulfonanilides

The structure–activity trends for the triazolo[15-*a*]pyrazines have not been studied as extensively as other members of triazolopyrimidine sulfonanilides [51]. Table 2.4.10 shows the activity on broadleaf and grass species for a series of substitutions on the fused heterocyclic portion of **34**. The highest levels of activity on grass and broadleaf species are observed when both 5- and 8-positions are substituted with methoxy (**34**, R^1 = R^3 = OMe). However, the herbicidal activity observed for **34** is weaker than that for the triazolo[1,5-*a*]pyrimidine sulfonamides.

2.4.5.2.3 N-Triazolo[1,5-*a*]pyridine Sulfonamides

The structure–activity relationships for substitutions on the phenyl ring of the triazolo[1,5-*a*]pyridine sulfonamides (**37**) are similar to **13**. Disubstitutions are

Table 2.4.11 Herbicidal activity for phenyl analogs of **37** (X = Y = OMe) (structure shown in Scheme 2.4.7).

R^1	R^2	R^3	R^4	Average GR$_{80}$ BW (ppm)	Average GR$_{80}$ GW (ppm)
H	H	H	OMe	57	118
Cl	H	H	OMe	<4	17
OMe	H	H	Cl	15	<62
OMe	H	H	OMe	2	10
OMe	H	Me	H	>1000	>1000
OMe	H	Cl	H	>62	>62
H	OMe	H	OMe	216	>250

more active than mono-substitutions and 2,6-disubstituted analogs, especially when both are methoxy, give rise to the best levels of activity on both grass and broadleaf weeds. Structure–activity relationships for substitutions on the triazolo[1,5-*a*]pyridine ring of **37** have been studied more than those on the phenyl ring [32, 36]. Table 2.4.11 presents a compilation of activity for **37** with various substitutions on the triazolo[1,5-*a*]pyridine ring. The 5,8-dimethoxy analog (**37**, $R^1 = R^4 = OMe$) has the best level of activity on both grass and broadleaf weeds. However, these compounds are weaker herbicides than the triazolopyrimidine sulfonamides.

2.4.6
Conclusion

The triazolopyrimidine class of ALS inhibitors has grown to include several fused triazole ring systems containing a bridgehead nitrogen. Members have demonstrated control of grass, broadleaf, and sedge weeds in several agronomically important crops. The discovery of molecules with crop safety and favorable environmental profiles has led to the development of seven new herbicides for use in many crops, including corn, peanuts, soybeans, wheat, barley, oats, rye, triticale, rice, sugarcane, sorghum and turf.

References

1 Kleschick WA, Costales MJ, Dunbar JE, Meikle RW, Monte WT, Pearson NR, Snider SW, Vinogradoff AP (**1990**) *Pest. Sci.* 29, 341.

2 Subramanian MV, Loney-Gallant V, Dias JM, Mireles LC (**1991**) *Plant Physiol.* 96, 310–313.

3 Mourad G, King J (**1992**) *Planta* 188, 491–497.

4 Kleschick WA, Triazolopyrimidine Sulfonanilides and Related Compounds. In *Herbicides Inhibiting Branch Chain Amino Acid Biosynthesis*, Stetter, J. Ed., Spinger-Verlag, Germany, **1994**, Vol. 10, pp 119–143.

5 Van Heertum JC, Gerwick BC, Kleschick WA, Johnson TC (**1992**) US 5,163,995.

6 Hunter JJ, Schultz ME, Mann RK, Cordes RC, Lassiter RB (**1994**) *Proc. N Cent. Weed Sci. Soc.* 49, 124.

7 Jachetta JJ, Van Heertum JC, Gerwick BC (**1994**) *Proc. N Cent. Weed Sci. Soc.* 49, 123–124.

8 Jachetta JJ, Van Heertum JC, Gerwick BC, Barrentine JL (**1995**) *Proc. S Weed Sci. Soc.* 48, 199.

9 Hunter JJ, Langston VB, Grant DL, McCormick RW, Barrentine JL, Braxton LB (**1995**) *Proc. S Weed Sci. Soc.* 48, 201.

10 Braxton LB, Barrentine JL, Dorich RA, Geselius TC, Grant DL, Langston VB, Redding KD, Richburg JS (**1996**) *Proc. S Weed Sci. Soc.* 49, 170–171.

11 Choate JH, Wilcut JW, York AC (**1996**) *Proc. S Weed Sci. Soc.* 49, 193.

12 Stabler GF, Murdock EC, Keeton A, Isgett TD (**1996**) *Proc. S Weed Sci. Soc.* 49, 18.

13 Sheppard BR, Braxton RL, Barrentine JL, Geselius TC, Grant DL, Langston VB, Redding KD, Richburg JS, Roby DB (**1997**) *Proc. S Weed Sci. Soc.* 50, 161.

14 Arnold JC, Shaw DR, Bennett (**1998**) *Proc. S Weed Sci. Soc.* 51, 2.

15 Bailey WA, Wilcut JW, Jordan DL, Swann CW, Langston VB (**1999**) *Weed Technol.* 13, 450–456.

16 Bailey WA, Wilcut JW, Jordan DL, Swann CW, Langston VB (**1999**) *Weed Technol.* 13, 771–776.

17 Shaw DR, Bennett AC, Grant DL (**1999**) *Weed Technol.* 13, 791–798.

18 Nelson KA, Renner KA (**1998**) *Weed Technol.* 12, 293–299.

19 Thompson AR, McReath AM, Carson CM, Ehr RJ, DeBoer GJ (**1999**) *Proc. Br. Crop Protect. Conf.* 1, 73–80.

20 Lepiece D, Rijckaert G, Thompson A (**2000**) *Proc 52nd Int. Symp. Crop Protec.*, Gent 65(2a), 141–149.

21 Daniau P, Prove P (**2001**) *Phytoma* 534, 49–51.

22 Jackson R, Ghosh D, Paterson G (**2000**) *Pest Manag. Sci.* 56(12), 1065–1072.

23 Gerwick BC, Deboer GJ, Schmitzer PR, Mechanism of Tolerance to Triazolopyrimidine Sulfonamide. In *Herbicides Inhibiting Branch Chain Amino Acid Biosynthesis*, Stetter, J. Ed., Spinger-Verlag, Germany, **1994**, Vol. 10, pp 145–160.

24 Owen WJ, deBoer GJ. Plant Metabolism and Design of New Selective Herbicides. In, *Eighth International Congress of Pesticide Chemistry: Options 2000.* Ragsdale, N. N., Kearney, P. C., Plimmer, J. R. Eds., ACS Conference Proceedings Series, American Chemical Society, Washington, DC, **1994**, pp 257–268.

25 Owen WJ. Herbicide Metabolism as a Basis for Plant Selectivity. In *Metabolism of Agrochemicals in Plants.* Roberts, T, Ed., John Wiley & Sons Ltd., London, **2000**, pp 240–249.

26 DeBoer GJ, Ehr RJ, Thornburgh S (**2006**) *Pest Manag. Sci.* 62, 316–324.

27 van Wesenbeek IJ, Zabik JM, Wolt DW, Bormett GA, Roberts DW (**1997**) *J. Agric. Food Chem.* 45, 3299–3307.

28 Zabic JM, van Wesenbeeck IJ, Peacock AL, Kennard LM, Roberts DW (**2001**) *J. Agric. Food Chem.* 49, 3284–3290.

29 Wolt JD, Smith JK, Sims JK, Duebelbeis DO (**1996**) *J. Agric. Food Chem.* 44, 324–332.

30 Yoder RN, Huskin MA, Kennard LM, Zabik JM (**2000**) *J. Agric. Food Chem.* 48, 4335–4340.

31 Krieger M, Yoder R, Stafford L, Batzer F, Smith J, Cook W, Lewer P (**2000**) Book of Abstracts, 219th ACS National Meeting, San Francisco.

32 Johnson TC, Ehr RJ, Johnston RD, Kleschick WA, Martin TP, Pobanz MA, Van Heertum JV, Mann RK (**1999**) US 5,858,924.

33 Johnson TC, Ehr RJ, Martin TP, Pobanz MA, Van Heertum JV, Mann RK (**2000**) US 6,130,335.

34 Johnson TC, Ehr RJ, Martin TP, Pobanz MA, Van Heertum JV, Mann RK (**2001**) US 6,303,814.

35 Johnson TC, Ehr RJ, Johnston RD, Kleschick WA, Martin TP, Pobanz MA, Van Heertum JV, Mann RK (**1999**) US 6,005,108.

36 Johnson TC, Ehr RJ, Kleschick WA, Pobanz MA, Van Heertum JV, Mann RK (**1999**) US 5,965,490.

37 Smith MG, Pobanz MA, Roth GA, Gonzales MA (**2002**), US 6,462,240.

38 Larelle D, Mann R, Cavanna S, Bernes R, Duriatti A, Mavrotas C (**2003**) *Proc. Int. Congr. Br. Crop Protect. Conf – Crop Sci. Technol.*, Glasgow, UK, Nov. 10–12, 1, 75–80.

39 Mann RK, Lassiter RB, Haack AE, Langston VB, Simpson DM, Richburg JS, Wright TR, Gast RE, Nolting SP (**2003**) *Proc. Weed Sci. Soc. Am.* 43, 40.

40 Mann RK, Haack AE, Langston VB, Lassiter RB, Richburg JS (**2005**) *Proc. Weed Sci. Soc. Am.* 45, 308.

41 Mann RK, Mavrotas C, Huang YH, Larelle D, Patil V, Min YK, Shiraishi I, Nguyen L, Nonino HL, Morell M (**2005**) *Proc. 20th Asian-Pacific Weed Sci. Soc.*, Vietnam, pp 289–294.

42 Min YK, Mann RK (**2004**) *Korean J. Weed Sci.* 24(3), 192–198.

43 Min YK, Mann RK (**2004**) *Korean J. Weed Sci.* 24(3), 199–205.

44 Shiraishi I (**2005**) *J. Pestic. Sci.* 30(3), 265–268.

45 Wang CL, Lee MS, Li YW, Yao ZW, Shieh JN, Mann RK, Huang YH (**2004**) *15th Int. Plant Protect. Congr. Abstracts.* Beijing, China, pp 598.

46 Mann RK, Huang YH, Larelle D, Mavrotas C, Min YK, Morell M, Nonino H, Shiraishi I (**2003**). *Proc. 3rd Int. Temperate Rice Conf.*, Punta

del Este, Uruguay, March 10–13, abstract WD055, pp 68.

47 Deboer GJ, Thornburgh S (**2005**) Abstract Agro 062, 229th ACS National Meeting, San Diego, CA.

48 Johnson TC, Pobanz MA, Van Heertum JV, Ouse DG, Arndt KE, Walker DK (**2003**) US 6,559,101.

49 VanHeertum JV, Kleschick WA, Arndt

KA, Costales MJ, Ehr RJ, Bradley KB, Reifschneider W, Benko Z, Ash ML, Jachetta JJ (**1997**) US 5,700, 940.

50 VanHeertum JV, Kleschick WA, Arndt KA, Costales MJ, Ehr RJ, Bradley KB, Reifschneider W, Benko Z, Ash ML, Jachetta JJ (**1996**) US 5,571,775.

51 Benko Z, Jachetta JJ, Costales MJ, Arndt KE (**1997**) US 5,602,075.

2.5
Pyrimidinylcarboxylates

Fumitaka Yoshida, Yukio Nezu, Ryo Hanai, and Tsutomu Shimizu

2.5.1
Introduction

Pyrimidinylsalicylates is the class of ALS-inhibiting herbicides disclosed in the late 1990s by Kumiai Chemical Industry and Ihara Chemical Industry. Their biological activities are as potent as those of the sulfonylureas (SUs). Further research led to the pyrimidinylglycolates, which are experimental herbicides. Since both types have the carboxyl moiety in their chemical structure, they are called pyrimidinylcarboxylates (PCs) or pyrimidinyl carboxy (PC) herbicides.

2.5.2
Discovery of the Pyrimidinylcarboxylates

The discovery of the PC herbicides started with attempts to synthesize new herbicides that incorporate a dimethoxypyrimidine [1]. During the synthetic and bioassay project, phenoxyphenoxypyrimidine **1** was found to show potent herbicidal activity with symptoms similar to Hill reaction inhibitors. In elaborating the structure to develop more systemic herbicides, a carboxylate group was introduced to give the compound **2** [2]. While the ethyl ester **2** was inactive, its "regio-isomer" **3** exhibited moderate activity against broadleaf weeds with pre- as well as post-emergent treatments [3]. Symptoms observed on plants after treatment with **3** were similar to those of the SUs; ALS-inhibiting herbicides and unlike those of the Hill reaction inhibitor **1**. By removing the second phenoxy group from **3**, the resulting *O*-pyrimidinylsalicylate **4** was found to exhibit highly potent herbicidal activity, characteristic of ALS inhibition (Fig. 2.5.1).

Using the skeletal structure of **4** as a new lead compound, various derivatives were synthesized to optimize the herbicidal activity. Among conventional substituents on the benzene ring, the carboxylate group ortho to the pyrimidinyloxy

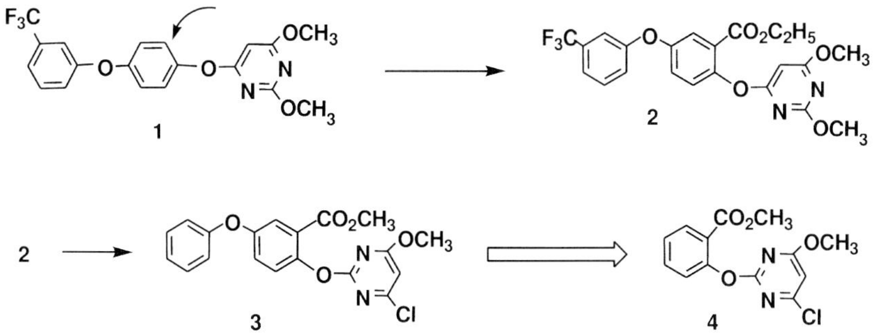

Fig. 2.5.1. Structural modification pathway towards the lead compound **4**.

one was essential for potentiating activity. A pyrimidine ring was better than any other nitrogen heterocycle. The most favorable substitution pattern was the 4,6-dimethoxy-2-pyrimidinyl **5**. At the rate of 1 kg-a.i. ha^{-1}, it controlled various grass- and broad-leave weed species pre- as well as post-emergently with phyto-toxic symptoms similar to those of the SUs. Unfortunately, the safety margin of **5** for crops was narrow and unacceptable as a selective-herbicide despite a marked increase in herbicidal activity (Fig. 2.5.2) [4]. Thus, the ALS inhibitory activities of **4** and **5** were assessed. As shown in Table 2.5.1, the free acid of **5** was potent in terms of the I_{50} (nM) of ALS inhibitory activity. This compound exhibited a much higher ALS inhibitory activity than imazapyr (a representative IMI) [5].

This study of ALS inhibition demonstrated that the PCs are a novel class of ALS-inhibiting herbicides, differing from both the SUs and the IMIs. The PCs and the SUs are structurally unrelated, but possess common structural parts of a weakly acidic proton and an N-containing heterocyclic ring. On a two-dimensional hexagonal grid template, the common parts in both molecules overlap (Fig. 2.5.3) [6]. This suggested that a weakly acidic proton and an N-containing heterocyclic ring, appropriately located in a molecule are requisites for inhibiting ALS. Further modifications based on this hypothesis led to the highly active pyrimidinylglycolates in which a carboxylic and a pyrimidinyloxy group were not directly connected with a benzene ring (Fig. 2.5.4) [7].

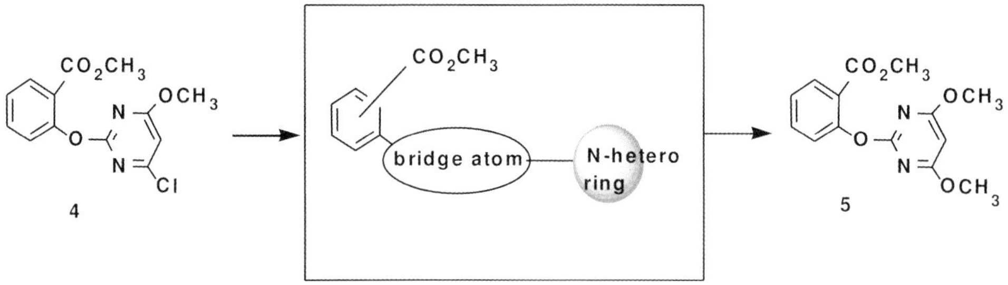

Fig. 2.5.2. Optimization from the lead compound **4**.

Table 2.5.1 ALS inhibitory activity of the acids of compounds **4** and **5**.

Compound	ALS inhibitory activity I_{50} (nM)[a]
4 (COOH)	4600
5 (COOH)	250
Imazapyr	9100
Chlorsulfuron	27

[a] I_{50}, molar concentration required for 50% inhibition of the ALS activity. ALS sample prepared from etiolated pea seedlings.

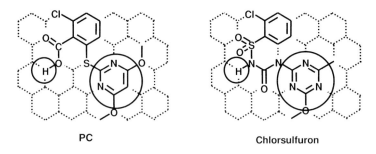

Fig. 2.5.3. Comparison of PC and chlorsulfuron on a hexagonal grid template.

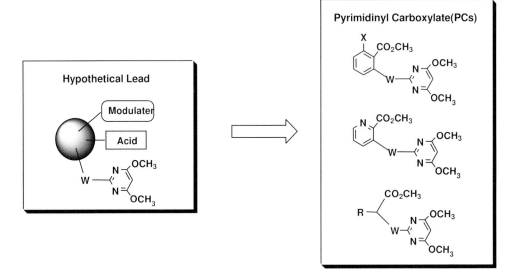

Fig. 2.5.4. Pyrimidinylcarboxylates – a new class of ALS inhibitors.

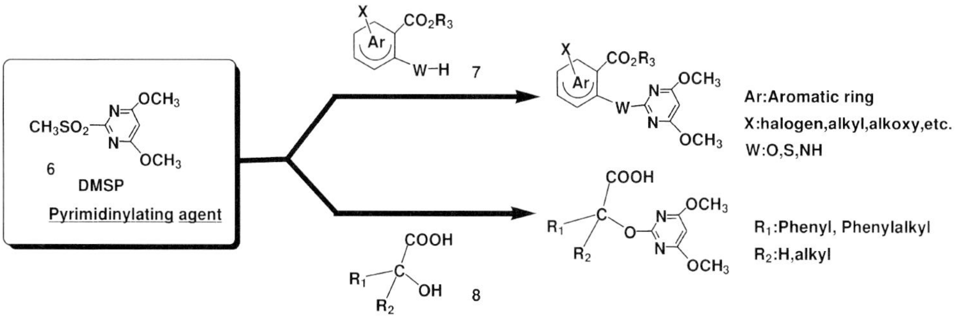

Fig. 2.5.5. Synthesis via 2-methanesulfonylpyrimidinyl intermediate **6**.

2.5.3
Structure–Activity Relationships

The first PC compound was prepared using the 2-methanesulfonylpyrimidine (OMSP) **6** and the salicylic acid ester in DMF. The sulfonyl compound **6** was a very efficient intermediate to synthesize 2-substituted pyrimidines [8]. The methanesulfonyl group in **6** is easily replaced by nucleophilic reagents like **7** and **8** (Fig. 2.5.5) [7, 9, 10]. This method was generally employed to synthesize numerous analogues aiming at new herbicides, not only with a high potency but also with an enhanced crop safety. Structural modifications were first made with the skeletal structure **9** (Fig. 2.5.6) [3].

2.5.3.1 Effects of Benzene Ring Substituents in the *O*-Pyrimidinylsalicylic Acids

The herbicidal activity of the pyrimidinylsalicylates (PSs) varied with the structure and position of ring substituents (X) (Table 2.5.2). First, the position-specific effect of ring substituents was examined. For Cl derivatives, the 6-Cl was obviously more potent than 3-Cl and 5-Cl (Table 2.5.2) [4, 11]. Also in other cases such as the methyl and fluoro derivatives, a similar positional pattern was shown. Therefore, the sequence of effects of the ring substituents in position was 6 > H (unsubstituted) > 3 > 5 ≫ 4. Thus, only substitution at the 6-position was favorable for enhancing the herbicidal activity of unsubstituted compound **5**. Among the 6-substituted derivatives, the halogeno, methyl, acetyl, phenyl, CF_3 and lower alkoxy derivatives exhibited extremely high activity at both the pre- and post-emergent

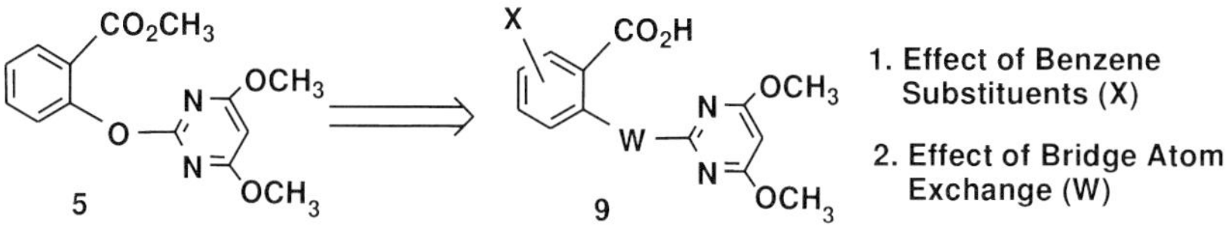

Fig. 2.5.6. Optimization for both herbicidal activities and safety to crops.

Table 2.5.2 Post-emergence herbicidal and ALS inhibitory activities of the dimethoxypyrimidinyl salicylic acids.[a]

X	pI$_{50}$[b]	Ech[c]	Dig[d]	Pol[e]	Ama[f]	Che[g]	Cyp[h]
3-Cl	6.3	2	1	5	5	1	5
4-Cl	4	0	0	0	0	0	0
5-Cl	5.4	0	0	2	5	3	4
6-Cl	7.6	5	5	5	5	5	5
H	6.6	5	4	5	5	5	4

[a] Applied at the dose of 250 g-a.i. ha^{-1}, and assessed with 6 grades from 0 (no effect) to 5 (complete kill).
[b] pI$_{50}$, ALS inhibitory activity ($-\log$ I$_{50}$).
[c] Ech, *Echinochloa crus-galli*.
[d] Dig, *Digitaria adscendens*.
[e] Pol, *Polygonum nodosum*.
[f] Ama, *Amaranthus retroflexus*.
[g] Che, *Chenopodium album*.
[h] Cyp, *Cyperus iria*.

applications. These compounds controlled weeds completely at a dose of 250 g-a.i. ha^{-1}, but their phytotoxicity to crops could not be improved up to practical use [4].

2.5.3.2 Effect of a Bridge Atom in the Pyrimidinylsalicylates

Fixing the 6-substituent as Cl, the effect of the bridge between the two rings was examined (Table 2.5.3). The *S*-bridge derivative showed excellent herbicidal activity, and the *CH$_2$*-bridge derivative was moderately active, whereas the *NH*-bridge and *SO*-bridge derivatives exhibited poor activity even at the application rate of 250 g-a.i. ha^{-1}. In an *in vitro* study of the ALS inhibition, the *S*-bridge and *SO*-bridge derivatives have inhibitory activities comparable to that of the *O*-bridge derivative, whereas those of the *NH*-bridge and *CH$_2$*-bridge derivatives decreased [4].

2.5.3.3 Pyrimidinylglycolates

As discussed in the Section 2.5.2, the pyrimidinylglycolates shown in Fig. 2.5.4, in which carboxylic and pyrimidinyloxy groups were not directly connected with a benzene ring also have high herbicidal activity [7, 9]. We attempted to examine favorable distances among important substructures, namely a carboxylic group, a pyrimidine and benzene rings (Fig. 2.5.7). First, compounds **10, 11** were synthesized, because the distances between the carboxylic group and pyrimidine ring in these compounds were supposed to be close to that in the pyrimidinylsalicylates

Table 2.5.3 Herbicidal activities of 6-Cl pyrimidinylsalicylic acid analogues in which the O-bridge is modified.[a]

W	Post-emergence[b]				Pre-emergence			
	Ech	Dig	Pol	Ama	Ech	Dig	Pol	Ama
O	5	4	5	5	4	5	2	4
S	5	3	5	5	5	5	5	5
NH₂	0	0	0	3	0	0	0	2
SO	3	0	3	5	1	3	1	5
CH₂	4	4	4	5	4	4	5	5

[a] Applied at the dose of 250 g-a.i. ha^{-1}, and assessed with 6 grades from 0 (no effect) to 5 (complete kill).
[b] *Ech, Echinochloa crus-galli; Dig, Digitaria adscendens; Pol, Polygonum nodosum; Ama, Amaranthus retroflexus.*

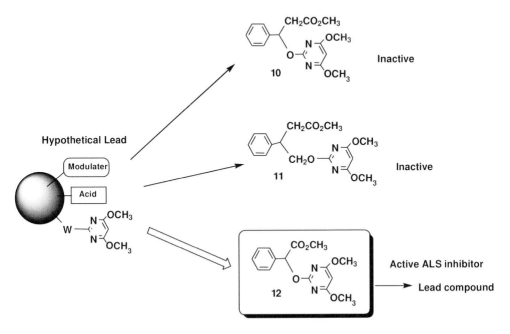

Fig. 2.5.7. Discovery of a new lead compound.

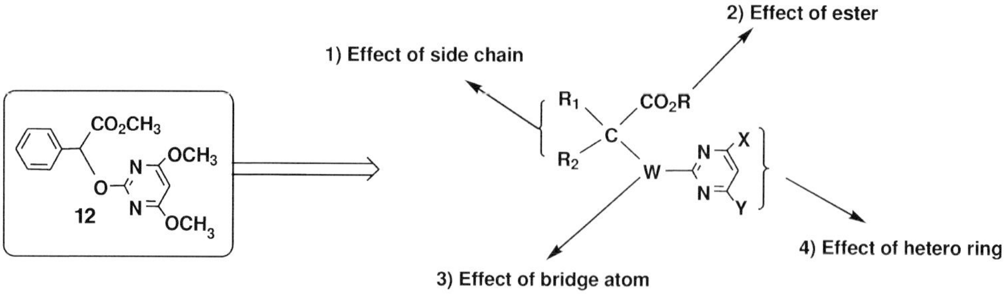

Fig. 2.5.8. Structural modifications from lead compound **12**.

having strong activity [12]. However, they were totally inactive. Next, compound **12**, with the carboxylic group placed at α position of the pyrimidinyloxy one, showed encouraging ALS inhibitory and herbicidal activities. Starting from **12**, structural modifications were made as shown in Fig. 2.5.8 to discern the effects of side chains (R_1, R_2), ester residues (R), bridge atoms (W) and pyrimidine substituents (X,Y) on herbicidal activity. As a result, the bridge atom (W) were fixed as oxygen and the pyrimidine substituents (X,Y) as methoxy in subsequent examinations. These options led to the most active compounds.

The herbicidal activities varied with substituents R_1, R_2 and R (Table 2.5.4). The α-hydrogen atom is, probably, essential for the herbicidal activity because the di-

Table 2.5.4 Herbicidal efficacy of the pyrimidinylglycolates.

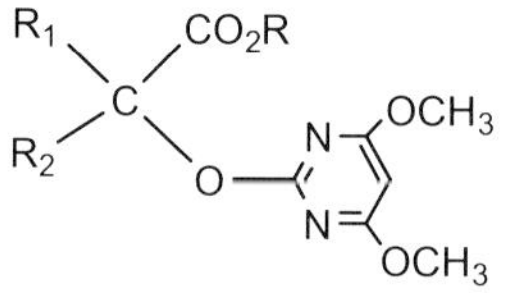

R_1	R_2	R	Pre-emergence (ED_{90}[a])		Post-emergence (ED_{90}[a])	
			Ech[b]	Pol[c]	Ech[b]	Pol[c]
Ph	H	H	B	A	A	B
PhCH$_2$	H	H	A	A	A	A
PhC$_2$H$_4$	H	H	B	A	A	B
Ph	CH$_3$	H	C	C	C	C
Ph	H	C$_2$H$_5$	C	C	A	C
PhCH(CH$_3$)	H	H	A	A	A	A
tert-C$_4$H$_9$	H	H	A	A	A	A

[a] ED_{90} (kg-a.i. h^{-1}): A, 1 or less; B, 1–4; C, 4 or more.
[b] *Ech, Echinochloa crus-galli.*
[c] *Pol, Polygonum nodosum.*

substituted compound (R_1 = Ph, R_2 = CH_3) completely lost activity. In contrast, two compounds (R_1 = $PhCH_2$, R_2 = H) and (R_1 = PhC_2H_4, R_2 = H) extended by methylene length(s) in R_1 almost retained the herbicidal activities of the starting compound **12** (R_1 = Ph, R_2 = H). Furthermore, the phenyl group in **12** was replaced by straight alkyl ones. The optimal length of the alkyl chain R_1 is around C_3 but with lower herbicidal activity. If the phenyl group in **12** is replaced by α-branching alkyl groups, *tert*-alkyl groups are more active than *sec*-alkyl groups, but again with lower herbicidal activity than **12**. The free acid is more active than the esters (e.g. R = ethyl).

2.5.3.4 Commercialized PC Herbicides

Further optimization of the pyrimidinylsalicylates led to three useful herbicides: (1) The sodium salt of 6-chloro-2-[(4,6-dimethoxypyrimidin-2-yl)thio]benzoic acid **13** (pyrithiobac-sodium) was selected as one of the best cotton herbicides [13]. (2) The sodium salt of 2,6-bis[(4,6-dimethoxypyrimidin-2-yl)oxy]benzoic acid **14** (bispyribac-sodium) is used as a post-emergent herbicide for the control of a wide range of weeds with excellent selectivity on direct-seeded rice and for the vegetative growth reduction [14]. (3) Methyl 6-[1-(methoxyimino)ethyl]-2-[(4,6-dimethoxypyrimidin-2-yl)oxy]benzoate **15** (pyriminoibac-methyl) is a selective herbicide with outstanding efficacy on *Echinochloa* spp. in paddy rice (Fig. 2.5.9) [15].

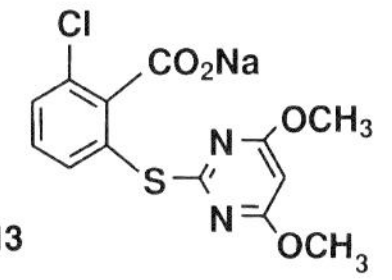

13
"Pyrithiobac-sodium" Staple(R)
1996 market introduction (USA)

M.p., 233.8-234.2 °C (decomp.); V.p., 4.80×10^{-9} Pa; logP=0.6 (pH 5); -0.84 (pH 7)
Solubility: In water, 264 g/l (pH 5); 705 g/l (pH 7); 690 g/l (pH 9);In methanol, 270 g/ml; In n-hexane, 10 mg/ml
Acute oral: LD_{50} for male rats, 3300 mg/kg; LD_{50} for female rats, 3200 mg/kg
Eye irritation for rabbits, Irritating
TLm (96 hr) for rainbow trout, >1000 ppm; TLm (48 hr) for daphnia, >1100 ppm

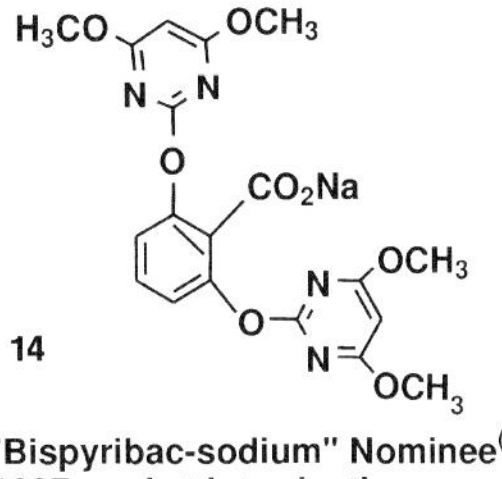

14
"Bispyribac-sodium" Nominee(R)
1997 market introduction

M.p., 223-224 °C ; V.p., 5.05×10^{-9} Pa; logP=-1.03
Solubility: In water, 73.3 g/l; In methanol, 26.3 g/100l; In n-hexane, 3.56 mg/l Acute oral: LD_{50} for male rat, 4111 mg/kg; LD_{50} for female rats, 2635 mg/kg
Eye irritation for rabbits: Slightly irritating
TLm (96 hr) for rainbow trout, >100ppm;
TLm (48 hr) for daphnia, >100ppm

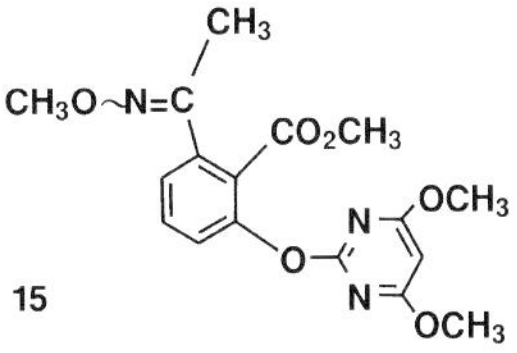

15
"Pyriminobac methyl" Prosper(R)
1997 market introduction

M.p.: Tech. 105 °C ; E-isomer, 107-109 °C ; Z-isomer, 70 °C
V.p.: E-isomer, 3.5×10^{-5} Pa ; Z-isomer, 2.681×10^{-5} Pa
logP of E-isomer= 2.98; logP of Z-isomer= 2.70
Solubility: E-isomer in water, 9.25 mg/ml; E-isomer in methanol, 14.6 g/ml; Z-isomer in water, 175 mg/l; Z-isomer in methanol, 14.0 g/l
Acute oral, LD_{50} for rats, > 5000 mg/kg
Eye irritation for rabbits: slightly irritating
TLm (96 hr) for rainbow trout, 21.2ppm;
TLm (24 hr) for daphnia, >20ppm

Fig. 2.5.9. Commercialized pyrimidinylcarboxy herbicides.

2.5.4

"Pyrithiobac-sodium" – Cotton Herbicide

2.5.4.1 Discovery

The effect of the 6-substituents in the thiosalicylate moiety was fine-tuned (Table 2.5.5). Halogeno and thioalkyl derivatives had good post-emergent herbicidal activity against broadleaf weeds, but were weak on *Echinochloa crus-galli*, whereas the acetyl derivative showed good herbicidal activity both against *Abutilon theophrasti* and *Echinochloa crus-galli*. This suggested that hydrophobic substituents such as halogens are favorable for killing broadleaf weeds, whereas the hydrophilic properties of the acetyl group affect positively the activity against grass weeds [4, 11]. Based on the good safety margin of the sodium salt of **13** (pyrithiobac-sodium) it was selected to be developed as a cotton herbicide against broadleaf weeds such as *Abutilon theophrasti* and *Ipomoea lacunosa* [16].

Table 2.5.5 Post-emergence herbicidal activities of 6-substituted pyrimidinylthiosalicylic acids.[a]

X	Crops[b]			Weeds[c]			
	Zea	*Gly*	*Gos*	*Ech*	*Abu*	*Ipo*	*Xan*
F	7	10	3	3	6	7	5
Cl	9	8	0	4	9	6	7
Br	9	7	0	1	7	4	6
I	10	7	0	2	9	6	8
CH$_3$	8	6	1	0	4	0	4
CF$_3$	8	5	0	0	6	2	2
COCH$_3$	10	4	1	10	9	8	4
OCH$_3$	9	9	0	2	5	1	6
OC$_3$H$_7$-*i*	2	4	1	0	2	0	0
SCH$_3$	9	8	4	0	10	10	9
NO$_2$	3	6	2	1	6	3	1

[a] Applied at the dose of 16 g-a.i. ha^{-1}, and assessed with 11 grades from 0 (no effect) to 10 (complete kill).
[b] Crops: *Zea*, Zea mays; *Gly*, Glycine max; *Gos*, Gossypium hirsutum.
[c] Weeds: *Ech*, Echinochloa crus-galli; *Abu*, Abutilon theophrasti; *Ipo*, Ipomoea lacunosa, *Xan*, Xanthium strumarium.

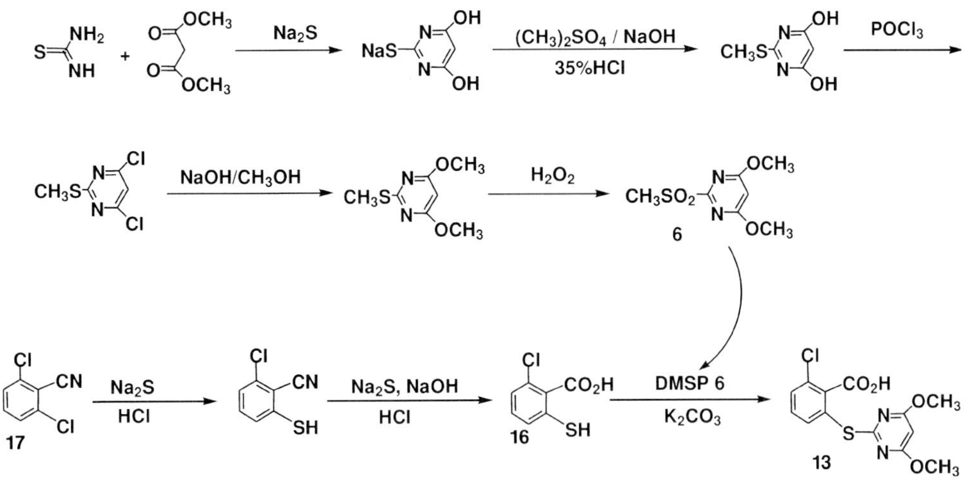

Fig. 2.5.10. Synthetic route for pyrithiobac.

2.5.4.2 Synthesis
Pyrithiobac-sodium is prepared by the condensation of 2-chloro-6-mercap-tobenzoic acid **16** and DMSP **6**. Compound **16** was synthesized through two steps starting from 2,6-dichlorobenzonitrile **17**. The industrial process of DMSP **6** synthetic scheme (Fig. 2.5.10) has been optimized [17, 18].

2.5.4.3 Biology
Pyrithiobac-sodium is a herbicide for controlling a wide range of weeds in cotton [11, 16, 19]. This compound provides excellent control of troublesome weeds such as *Ipomoea* spp., *Xanthium strumarium*, *Ahutilon theophrasti*, *Sida spinosa*, *Seshania exaltata*, and *Sorghum halepense*. It can be applied pre- or post-emergently. Soil or foliar treatment with pyrithiobac-sodium at 35–105 g-a.i. ha^{-1} provides excellent control of weeds. Adjuvants such as non-ionic surfactants or some petroleum-based adjuvant oils play an important role in achieving consistent performance on several weed species when applied post-emergent. A good safety margin for cotton at rates that are effective on weeds has been observed with pre-emergence treatment in both the greenhouse and the field.

2.5.5
"Bispyribac-sodium" – Herbicide in Direct-seeded Rice

2.5.5.1 Discovery
Our previous studies on the PCs showed that substitution at the 6-position of the salicylate moiety was preferable for herbicidal and ALS inhibitory activities. Some PCs showed a strong activity against various weeds even at rate of around

10 g-a.i. ha^{-1}, but rice injury was severe. In structural modification of the 6-substitutent on the benzene ring, compounds with halogeno, alkyl or alkoxy group did not improve rice safety. With a bulky substituent such as phenoxy group, the herbicidal activity was somewhat decreased, but rice injury was significantly alleviated [20, 21]. Starting from the 6-phenoxy compound as a basic structure, various substituents Y were introduced on the 6-phenoxy. Unfortunately, no compounds gave acceptable rice safety and strong herbicidal activity at the same time (Table 2.5.6). The severe rice injury was attributed to the hydrophobic property of the phenoxy group. Thus, more hydrophilic substituents of heterocycle-oxy groups were introduced in its place. Among five- or six-membered hetero-rings, pyrimidinyloxy groups exhibited the most suitable performances as a rice herbicide in both aspects of activity and rice safety. PCs with 2 or 4-(substituted)-

Table 2.5.6 Effect of substituent(s) Y on the benzene ring of 6-phenoxypyrimidinylsalicylic acids on herbicidal activity and rice phytotoxicity at post-emergence application.[a]

Y	Phytotoxicity Ory.[b]	Herbicidal activity[c]			
		Ech	Pol	Ama	Xan
H	4	9	10	10	7
2-Cl	3	5	7	8	2
3-Cl	0	0	7	8	2
4-Cl	6	4	4	8	0
2-F	5	4	10	9	2
3-F	0	3	9	9	4
4-F	2	0	8	8	2
2-CH$_3$	2	4	8	9	6
3-CH$_3$	0	4	9	9	6
4-CH$_3$	0	0	7	10	4
2-OCH$_3$	6	8	7	9	6
2-NO$_2$	4	4	9	9	0
3,5-(OCH$_3$)$_2$	4	7	6	7	8

[a] Applied at the dose of 16 g-a.i. ha^{-1}, and assessed with 11 grades from 0 (no effect) to 10 (complete kill).
[b] Ory, Oryza sativa.
[c] Weeds: Ech, Echinochloa crus-galli; Pol, Polygonum nodosum; Ama, Amaranthus retroflexus, Xan, Xanthium strumarium.

Table 2.5.7 Effect of substituents Rn on the pyrimidine ring of bis(pyrimidinyloxy)benzoic acids on herbicidal activity and rice phytotoxicity at post-emergence application.[a]

R_1 ... N ... W ... Z ... R_2 — O — (benzene ring with CO_2H) — O — (pyrimidine with OCH_3, OCH_3); N, N

R_1	R_2	Z	W	Phytotoxicity[b] Ory.	Herbicidal activity[c]			
					Ech	Pol	Ama	Xan
H	H	N	CH	8	4	8	6	0
Cl	CH_3	N	CH	3	4	7	9	2
Cl	OCH_3	N	CH	5	9	8	9	9
CH_3	CH_3	N	CH	5	8	8	8	6
CH_3	OCH_3	N	CH	4	9	9	9	8
OCH_3	OCH_3	N	CH	1	10	9	10	9
H	H	N	CCl	7	4	8	7	2
OCH_3	OCH_3	CH	N	5	8	9	8	8

[a] Applied at the dose of 16 g-a.i. ha^{-1}, and assessed with 11 grades from 0 (no effect) to 10 (complete kill).
[b] Ory, Oryza sativa.
[c] Weeds: Ech, Echinochloa crus-galli; Pol, Polygonum nodosum, Ama, Amaranthus retroflexus, Xan, Xanthium strumarium.

pyrimidinyloxy group as a 6-substituent on the benzene ring were, furthermore, synthesized and evaluated (Table 2.5.7).

In comparison with the unsubstituted compound ($R_1 = R_2 = H$), it was, consequently, revealed that the introduction of substituents into 4 and 6 positions of the pyrimidine was favorable for improving rice safety without decreasing herbicidal activity. In particular, the 4,6-dimethoxy compound, being a bis-pyrimidinyl compound, showed both a remarkable improvement in rice safety and excellent activity against Echinochloa spp. and broad-leave weeds. Finally, the sodium salt of 2,6-bis[(4,6-dimethoxypyrimidin-2-yl)oxy]benzoic acid (**14**, bispyribac-sodium) was selected to be commercialized as a herbicide on direct-seeded rice [22].

2.5.5.2 Synthesis
After double pyrimidinylation by the reaction of benzyl 2,6-dihydroxybenzoate with two equivalents of DMSP **6**, the benzyl group is removed by hydrogenation to yield bispyribac (Fig. 2.5.11) [20, 21].

2.5.5.3 Biology
Bispyribac-sodium is a post-emergent herbicide for the control of a wide range of weeds with excellent selectivity on direct-seeded Indica-type rice [22, 23]. The low

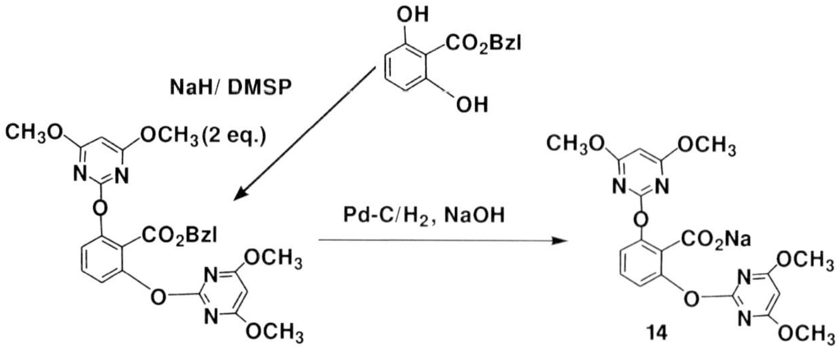

Fig. 2.5.11. Synthetic route for bispyribac-sodium.

application rate of 15–45 g-a.i. ha^{-1} with surfactant has provided outstanding efficacy on *Echinochloa* spp. and can be applied from the 1- to 7-leaf stage of the weed. It can control other troublesome weeds, including *Brachiaria* spp., *Cyperus* spp., *Scirpus* spp., *Polygonum* spp., *Sagittaria* spp., *Commelina* spp. and *Sesbania exaltata*. Adjuvants, such as non-ionic surfactants, silicon-type adjuvants or crop oil concentrate play an important role in enhancing the activity and achieving a consistent performance of this compound. Bispyribac-sodium has high selectivity between Indica-type rice and *Echinochloa oryzicola* by foliar application under dry-seeded conditions, suggesting that this compound can be used against a wide range of growth stages of *Echinochloa* spp. without rice crop injury. On the other hand, bispyribac-sodium at the rate of 150 g-a.i. ha^{-1} pre-mixed with a non-ionic surfactant reduced the vegetative growth of weeds such as *Imperata cylindrica*, *Digitaria adscendens*, *Miscanthus sinensis* and *Artemisia princes* [24]. The growth reduction persisted for 50 days after application of this compound when applied 5–10 days after mowing (at 10–20 cm plant height). Also, bispyribac-sodium controlled a wide range of weed species such as *Solidago altissima*, *Polygonum lapathifolium*, *Aeschynomene indica*, *Paspalum distichum* and *Echinochloa crus-galli* that grew in rice levees or on highway and railroad right-of-ways. The results indicated that bispyribac-sodium can reduce the frequency of mowing in paddy rice levees, and on highway and railroad right-of-ways.

2.5.6
"Pyriminobac-methyl" – Rice Herbicide

2.5.6.1 Discovery
Since ALS inhibitory and low-dose herbicides, including the SUs, were not commercially available for the effective control of *Echinochloa* spp. in transplanted rice when the pyriminobac methyl project was initiated, we focused our studies on

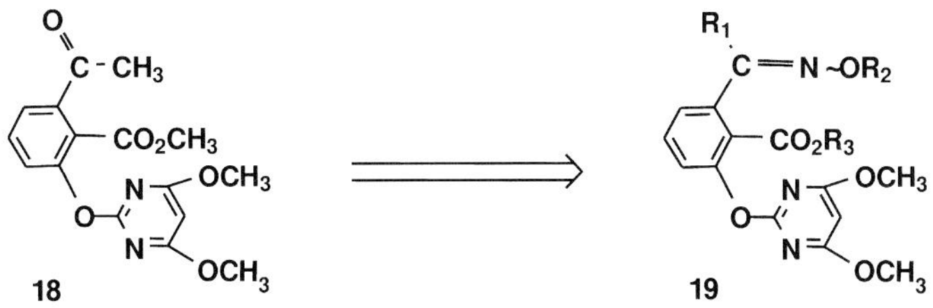

Fig. 2.5.12. Modification to an oxyimino group from an acyl one.

Table 2.5.8 Herbicidal activities of the 6-alkoxyiminosalicylate analogues against barnyard grass.

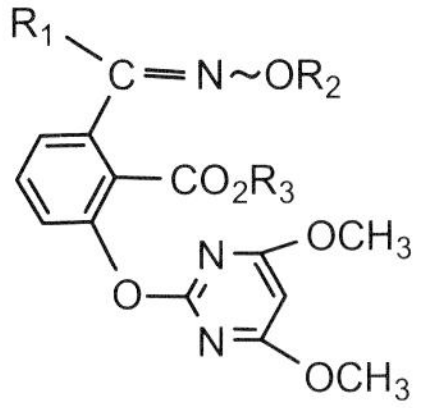

R1	R2	R3	Pre-emergence		Selectivity ED_{10}/ED_{90}
			Phytotoxicity $ED_{10}(Ory.^{[a]})$	Herbicidal activity $ED_{90}(Ech.)^{[b]}$	
H	CH_3	CH_3	63	16	4
CH_3	CH_3	CH_3	250	16	16
C_2H_5	CH_3	CH_3	63	63	1
C_3H_7	CH_3	CH_3	63	>1000	<1/16
CH_3	H	CH_3	4	16	1/4
CH_3	C_2H_5	CH_3	250	16	16
CH_3	C_3H_7	CH_3	250	16	16
CH_3	C_3H_7-i	CH_3	63	16	4
CH_3	C_4H_9	CH_3	63	16	4
CH_3	CH_3	C_2H_5	63	63	1
CH_3	CH_3	H	<4	16	<1/4

[a] Active ingredient amounts (g ha^{-1}) required for less than 10% phytotoxicity of *Oryza sativa* (*Ory.*)
[b] Active ingredient amounts (g ha^{-1}) required for more than 90% control of *Echinochloa oryzicola* (*Ech.*).

low-dose herbicides particularly effective in controlling *Echinochloa* spp. in the paddy rice. Our previous studies of pyrimidinylsalicylates had provided the following findings: substitution at the 6-position of the salicylate moiety was preferable for herbicidal and ALS inhibitory activities; electron-withdrawing groups contributed to ALS activity; and hydrophilic groups led to better activities against grass weeds than broad-leaf weeds. Therefore, 6-acyl compounds were specially interesting, since the acyl groups are both hydrophilic and electron-withdrawing. Compound **18** showed excellent control of *Echinochloa* spp., but caused unacceptable phytotoxicity to rice. However, the herbicidal profile of **18** satisfied our minimum requirements as a prototype for *Echinochloa* spp. herbicide [25].

To reduce rice injury, while keeping herbicidal activity of **18**, the introduction of an oxyimino group was attempted to give a hypothetical bio-isosteric analogue of **18**. The methoxyimino group has a similar $[\sigma_p]$ (acyl group: $\sigma_p = 0.4$, methoxyimino group: $\sigma_p = 0.3$) and a steric similarity to a carbonyl group, and the hydrophilicity of the oxyimino moiety can be varied by alkylation and acylation. Extensive synthetic modifications were then made to the 6-alkyl moiety (R_1), the alkoxyimino moiety (R_2) and the ester moiety (R_3) of **19** (Fig. 2.5.12).

Structure–activity relationships of the synthesized compounds were studied by examining their herbicidal activities against *Echinochloa oryzicola* in paddy rice at various growth stages, including pre-emergence (Table 2.5.8). Compounds with $R_1 = CH_3$, $R_3 = CH_3$ and $R_2 = $ alkyl showed the best selectivity/activity relationship, but compounds with $R_3 > CH_3$ had reduced herbicidal activity at a higher growth stage [25, 26].

According to a study of the mode of action (Table 2.5.9), the ALS inhibitory activities of the methyl compound **15** against both *Echinochloa oryzicola* and rice were almost identical and about 1000× lower than that of the carboxylic acid (**20**). Besides, the metabolic transformation of **15** into **20**, which is considered to be the metabolically activated form as an ALS inhibitor, was enhanced, particularly

Table 2.5.9 ALS inhibitory activity, herbicidal activity and phytotoxicity of 6-methoxyiminosalycilate and its acid.

Compound	ALS I_{50}[a] (µM)		Phytotoxicity $ED_{10}(Ory.)$[b]	Herbicidal activity $ED_{90}(Ech.)$[c]
	Rice	Barnyardgrass		
15	59	47	6.3	25
20	0.018	0.016	<0.4	0.4

[a] Concentration required for 50% inhibition.
[b] Maximum effective dosage (g-a.i. ha^{-1}) for less than 10% phytotoxicity against transplanted *Oryza sativa* (*Ory.*) at 3 cm in depth.
[c] Minimum effective dosage (g-a.i. ha^{-1}) required for more than 90% control against *Echinochloa oryzicola* (*Ech.*) in pre-emergence.

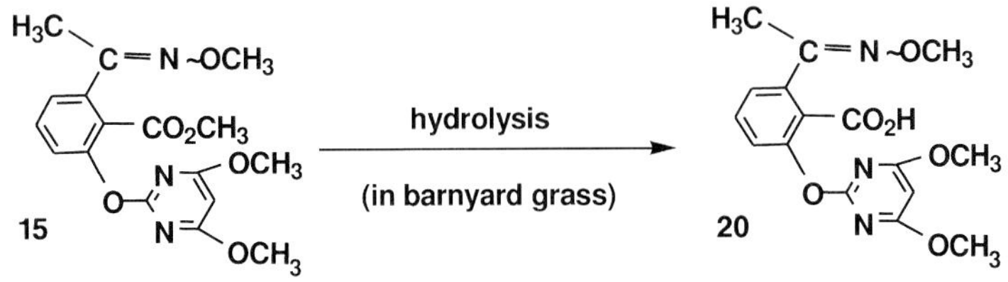

Fig. 2.5.13. Proposed metabolic pathway of 15 to the activated form 20 in barnyard grass.

in *Echinochloa oryzicola*, while not enhanced in rice (Tables 2.5.8 and 2.5.9 and Fig. 2.5.13) [26]. Methyl 6-[1-(methoxyimino)ethyl]-2-[(4,6-dimethoxypyrimidin-2-yl)oxy]-benzoate, (pyriminobac-methyl) (15) was, finally, selected as the candidate for commercialization as a novel barnyard-grass killer with an excellent selectivity for rice [27].

2.5.6.2 Synthesis

Figure 2.5.14 shows a synthetic route for pyriminobac-methyl [28]. The key step is ortho-lithiation reaction (step 3) of compound 21 protected by dimethylacetal and benzylation, followed by regioselective carbomethoxylation at the 2-position with methyl chloroformate via lithiated benzene prepared by *n*-butyllithium. Through several processes of deacetalization, methoxyimination of the acetyl group and debenzylation, compound 22 is condensed with DMSP 6 to give pyriminobac-methyl 15 [29].

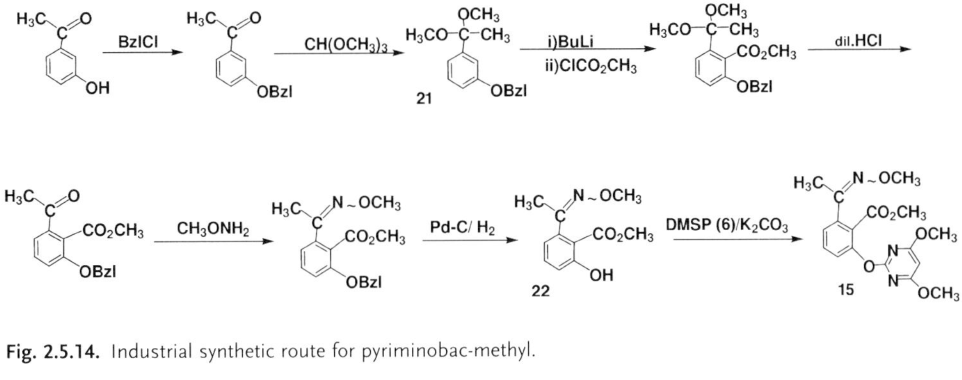

Fig. 2.5.14. Industrial synthetic route for pyriminobac-methyl.

2.5.6.3 Biology

Pyriminobac-methyl is a selective herbicide with outstanding efficacy on *Echinochloa* spp. in paddy rice [26, 27, 30]. This compound has a specific effectiveness against *Echinochloa* spp. during a wide range of growth stages, from pre- to late post-emergence, with an excellent crop safety in rice. The use rate of pyriminobac-methyl is extremely low in comparison with the recommended rate of molinate and thiobencarb. Pyriminobac-methyl has shown excellent safety on all eleven varieties tested of water-seeded rice and can be applied at any growth stage of rice. There was no observed significant difference in susceptibility to pyriminobac-methyl among rice varieties tested. Pyriminobac-methyl can be used alone or mixed with other rice herbicides such as bensulfuron-methyl. The residual activity of pyriminobac-methyl at 30 g-a.i. ha^{-1} was superior to thiobencarb at 3000 g-a.i. ha^{-1} under flooded conditions in the greenhouse.

2.5.7
"Pyribenzoxim and Pyriftalid" – Rice Herbicides

Pyribenzoxim (Fig. 2.5.15) is an oxime ester of bispyribac, which has been developed by LG Chemical Ltd. for use in rice [31]. This chemical compound has post-emergent activity on various grass and broadleaf weeds, including *E. crus-galli*, *Alopecurus myosuroides* and *Polygonum hydropiper*. *E. crus-galli* is controlled by pyribenzoxim at 30 to 40 g ha^{-1}, when applied alone in the field. Pyriftalid (Fig. 2.5.15) is an ALS-inhibiting herbicide categorized as a PC herbicide, which has been developed by Syngenta Crop Protection AG for use in rice [32]. This chemical compound controls grasses, especially *Echnochlora* spp., with an application rate of 100–300 g ha^{-1}. Herbicidal activities of pyriftarid are assumed to derive from conversion of the structure into the ring-open salicylic acid form.

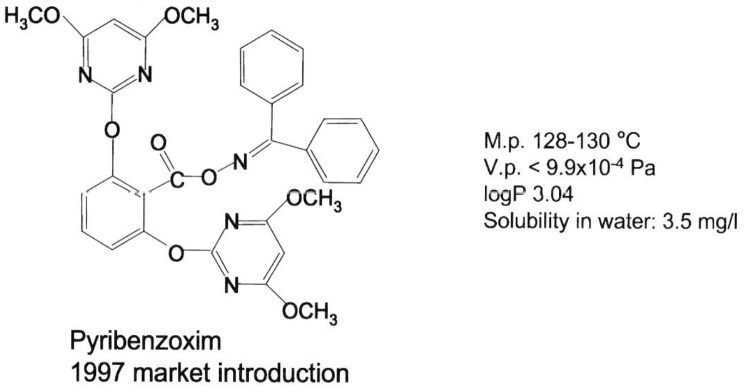
H3CO
OCH3
N
N
O
O
C
O
N
OCH3
N
O
N
OCH3
Pyribenzoxim
1997 market introduction
M.p. 128-130 °C
V.p. < 9.9x10-4 Pa
logP 3.04
Solubility in water: 3.5 mg/l

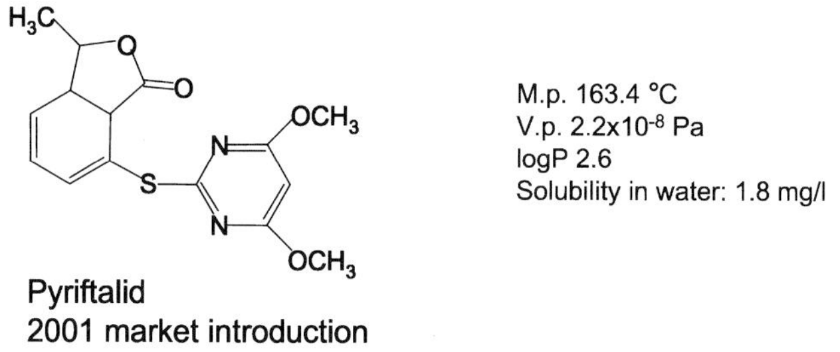
H3C
O
O
OCH3
N
S
N
OCH3
Pyriftalid
2001 market introduction
M.p. 163.4 °C
V.p. 2.2x10-8 Pa
logP 2.6
Solubility in water: 1.8 mg/l

Fig. 2.5.15. Other pyrimidinylcarboxy herbicides.

2.5.8
Mode of Action of the PC Herbicides

The growth inhibition of rice seedlings and chlorella by the PCs were alleviated by simultaneous application of three branched-chain amino acids [5]. The PSs, including pyrithiobac, bispyribac and pyriminobac, strongly inhibited ALS in various plant species at concentrations in the nanomolar range [33]. The SU [34] and the triazolopyrimidine (TP) [35] inhibit plant ALSs activity in the mixed-type with respect to pyruvate in the steady state analysis, while the IMI inhibits in the uncompetitive manner [36]. We have shown the following kinetic results in our studies [37, 38]. Pyrithiobac and bispyribac inhibited the ALS of etiolated pea seedlings in the mixed-type with respect to pyruvate by means of a 40-min steady-state analysis. This inhibition pattern was the same as that of a SU, chlorsulfuron, but different from that of a IMI, imazapyr. Imazapyr inhibited this enzyme in an uncompetitive manner. The inhibition pattern of pyrithiobac for ALS of *Pseudomonas aeruginosa* was non-competitive with respect to pyruvate as same as that of chlorsulfuron, whereas that of imazapyr was uncompetitive [39]. The inhibition patterns of these inhibitors are different from those by feedback inhibitors, whose inhibition patterns are partially competitive. The small ALS from etiolated pea seedlings, which lost its sensitivity to the feedback inhibition, was potently inhibited by the PCs. These results indicated that the binding sites of these inhibitors on the enzyme are different from those of feedback inhibitors. Imazapyr has been demonstrated to compete with a SU, sulfometuron-methyl for the binding to ALS of *Salmonella typhimurium* [40]. In our study, chlorsulfuron competed with bispyribac for the binding to ALS of etiolated pea seedlings. This competition was more potent than that of pyrithiobac [38]. The binding site of the PCs on ALS is located on the allosteric site in a wide sense near the catalytic center. Both the SUs and the TPs might share the binding site with the PCs. Whereas, the IMIs bind to the site that is somewhat distinct from but overlaps that of the SUs, the TPs and the PSs. These sites are not on the regulatory subunit, but are considered to be in the vestige of the ubiquinone binding site on the catalytic subunit [40] that lost its role in the enzymatic reaction during the evolutionary process.

Despite their reversible nature to the inhibition of ALS, the SU and the IMI are slow-binding inhibitors of plant ALS [41, 42], which inactivate ALS irreversibly after reaching the final steady inhibitions. Irreversible inactivation of the enzyme has been found in both the presence [43] and absence [44] of pyruvate. Pyrithiobac and bispyribac inhibited the ALS of etiolated pea seedlings with slow-binding properties. Pyrithiobac showed the mixed-type pattern with respect to pyruvate in the initial inhibition. The inhibition constants in the initial inhibition by pyrithiobac and bispyribac were about 20-fold larger than those in the final steady state. The maximal first-order rate constant (k_1, 0.069 min^{-1}) for transition from the initial to the final steady state inhibition of pyrithiobac [37] was nearly identical to those of the SU and IMI. However, the dissociation constant of bispyribac to the ALS of etiolated pea seedlings after reaching the final steady inhibition was nearly identical with the inhibition constant in the initial inhibition [33].

2.5.9
Mode of Selectivity of the PC Herbicides in Crops

Despite the high selectivity of pyrithiobac for cotton and bispyribac for rice, there were no differences in the sensitivities of ALSs to pyrithiobac between cotton and other plants, and to bispyribac between rice and other plants. The selectivities of pyrithiobac and bispyribac must be determined by other factors. As for pyrithiobac, there is no published paper on its selectivity for cotton. However, oxidative demethylation of the 3,5-dimethoxy moiety has been shown to account for the tolerance of tall morning-glory to pyrithiobac [45]. Thus, the same mechanism is assumed to be involved in its selectivity between cotton and other sensitive plants. Regarding bispyribac, translocation of the compound mainly accounts for its selectivity between rice and barnyard grass [unpublished data]. Since des-methyl bispyribac was detected in a rice plant treated with bispyribac [46], and application of P-450 inhibitors such as 1-aminobenzotriazol and piperonyl butoxide reduced selectivity of bispyribac for Indica-type rice [unpublished data], the oxidative detoxification metabolism, like that of pyrithiobac, is presumed to be another factor in the selectivity of bispyribac for the rice plant.

One of the methyl ester compounds of the PC (**5** in Fig. 2.5.2), which has the same herbicidal potency as its free acid, hardly inhibited the activity of ALS separated from esterase. However, this compound inhibited the ALS activity as potently as its free acid, when the esterase was added in the reaction mixture [47]. Thus, the active forms of ester compounds are their free acids. However, pyriminobac-methyl inhibited ALS less potently than its free acid even in the presence of esterase. Pyriminobac-methyl was hardly hydrolyzed by the esterase existing in the soluble fractions of both rice and barnyard grass, whereas it was hydrolyzed by the microsomal fraction of barnyard grass [unpublished data]. Also, the free acid of pyriminobac-methyl was detected in barnyard grass treated with this compound, but not in rice [48]. These results indicate that the selectivity of pyriminobac-methyl between rice and barnyard grass depends on the difference in substrate specificity of the enzyme having esterase activity in the membrane fraction of plants.

2.5.10
PC-resistant Plants and their Mutated ALS Genes

It was expected that novel mutated ALS genes that had different mutations from those reported [2] were obtained through the selection of plant cells under the pressure of the PC herbicides. First, the callus from rice seeds was induced. The calli were then cultured with 1 µM bispyribac-sodium for about 2 months so that the bispyribac-sodium resistant cells were generated. The cells were next cultured with higher concentrations of bispyribac-sodium. Finally, several kinds of spontaneous BS-resistant cells that could grow under the pressure of 100 µM bispyribac-sodium were obtained. A wild-type ALS gene and a mutated ALS gene have been cloned from the bispyribac-sodium resistant cells using the partial cDNA that is

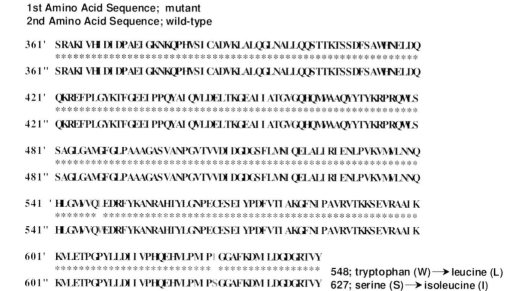

1st Amino Acid Sequence; mutant
2nd Amino Acid Sequence; wild-type

361' SRAKI VHI DI DPAEI GKNKQPHVSI CADVKLALQGLNALLQQSTTKISSDFSAWFNELDQ
      ****
361" SRAKI VHI DI DPAEI GKNKQPHVSI CADVKLALQGLNALLQQSTTKISSDFSAWFNELDQ

421' QKREFPLGYKIFGEEI PPQYAI QVLDELTKGEAI I ATGVGQHQMWAAQYYTYKRPRQWLS
      ****
421" QKREFPLGYKIFGEEI PPQYAI QVLDELTKGEAI I ATGVGQHQMWAAQYYTYKRPRQWLS

481' SAGLGAMGFGLPAAAGASVANPGVTVVDI DGDGSFLMNI QELALI RI ENLPVKVMVLNNQ
      ****
481" SAGLGAMGFGLPAAAGASVANPGVTVVDI DGDGSFLMNI QELALI RI ENLPVKVMVLNNQ

541 ' HLGMVVQI EDRFYKANRAHTYLGNPECESEI YPDFVTI AKGFNI PAVRVTKKSEVRAAI K
      ******* **
541" HLGMVVQVEDRFYKANRAHTYLGNPECESEI YPDFVTI AKGFNI PAVRVTKKSEVRAAI K

601' KMLETPGPYLLDI I VPHQEHVLPM PI GGAFKDM LDGDGRTVY
      **************************** ****************
601" KMLETPGPYLLDI I VPHQEHVLPM PSGGAFKDM LDGDGRTVY

548; tryptophan (W) → leucine (L)
627; serine (S) → isoleucine (I)

Fig. 2.5.16. Comparison of amino acid sequences between ALSs from the mutant and the wild type.

an expressed sequence tag obtained from the Ministry of Agriculture, Forestry and Fishery (MAFF) DNA bank of Japan as a homologous hybridization probe. Figure 2.5.16 shows a comparison of the deduced amino acid sequences between the wild-type ALS and one of the mutated ALSs. The first amino acid shows the sequences between position 361 and the C-terminal position 644 in the mutated ALS, and second amino acid sequence shows that in the wild-type ALS. The mutations involved the residues of tryptophan 548 to leucine and serine 627 to isoleucine. This double mutation on rice is a new combination of spontaneous mutations with the novel substitution at the serine position (DDBJ accession number, AB049823) [49].

One-point mutated ALS genes were then prepared to compare the sensitivities of their recombinant ALSs to the ALS-inhibiting herbicides with that of the two-point mutant. Each one-point mutant was prepared from the two-point mutant by PCR and the self-polymerase reaction. Recombinant ALSs from these ALS genes were expressed in *Escherichia coli* as GST-fused proteins and the proteins were examined for their sensitivities to herbicides. The ALS expressed from the wild-type gene showed a similar sensitivity to bispyribac-sodium and chlorsulfuron compared with that prepared from the natural source. Conversely, the ALS expressed from the two-point mutated ALS gene showed quite different sensitivities to the herbicides. This ALS showed a stronger resistance to bispyribac-sodium than to chlorsulfuron. Bispyribac-sodium had no effect on the enzyme even at 100 µM, which is an approximately 10 000-fold higher concentration than the I_{50} for the wild-type enzyme (Fig. 2.5.17). Notably, the two-point mutated gene imparted

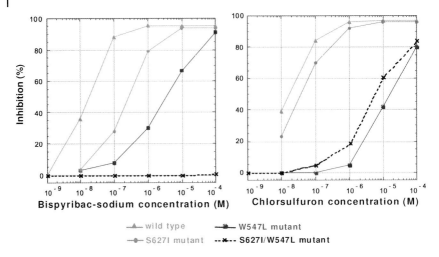

Fig. 2.5.17. Sensitivities of GST-fused ALSs to bispyribac-sodium and chlorsulfuron.

synergistic resistance to ALS against bispyribac-sodium that is stronger than the additive effect predicted from the degree of each resistance of the one-point mutated ALS [49].

2.5.11
Use of the Mutated ALS Genes for Genetic Transformation of Plants

As shown above, the novel mutated ALS gene from rice exhibited a high resistance to bispyribac-sodium. Thus we studied the use of this gene as a selectable marker for the genetic transformation of plants. Promoters and terminators derived from rice were used and a new binary vector was constructed. The two-point mutated ALS gene was driven with a rice callus specific promoter, and the GFP gene was driven with a constitutive promoter. Rice seeds were transformed with this vector by the Agrobacterium method and the transformed cells were selected by the pressure of bispyribac-sodium. As a result, fluorescence from GFP was detected only in selected cells, indicating that the two-point mutated ALS gene was an effective selection marker for rice transformation [50].

Transgenic rice plants were then generated to examine whether this gene works normally in the plant or not. The two-point mutated ALS gene was driven with a constitutive 35S promoter cassette with enhanced expression activity. Rice seeds were transformed with this vector and a transgenic rice plant was generated. This transgenic rice plant exhibited resistance to bispyribac-sodium and grew normally so that it was fertile. T_1 seeds were collected and the bispyribac-sodium resistant phenotype of T_1 plants were examined. The result showed that the phenotype was segregated by approximately 3:1 according to Mendel's law. The plants that exhibited resistance to bispyribac-sodium were cultivated on a large scale and several kinds of T_2 seeds were collected. Consequently, homozygotes for the

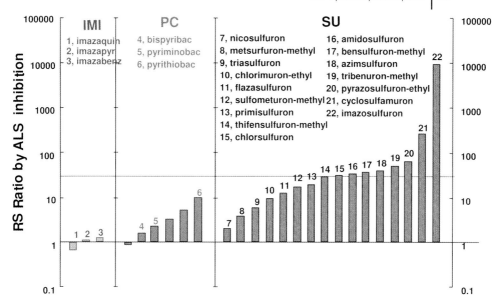

Fig. 2.5.18. Sensitivities of ALS from SU-resistant kochia to ALS-inhibiting herbicides. The data is for a kochia that has a mutation of proline to serine at position 189.

resistant trait were found in these T_2 seeds through examination of their sensitivities to bispyribac-sodium. This homozygote grew normally without bispyribac-sodium and exhibited resistance to bispyribac-sodium. These results suggested that the two-point mutated rice ALS gene functionally worked in rice and had no bad effects on rice [49].

The marker system is needed not only for the general recombinant technology but also for the gene targeting such as the homologous recombination and the mismatch repair. The mutated rice ALS gene can be used for such gene technologies. Novel binary vectors which have the S627I one point mutated ALS gene driven by the rice ALS promoter or the rice callus specific promoter have been developed and are now on the market (http://www.kumiai-chem.co.jp/).

2.5.12
Use of the Mutated ALS Genes for Resistance Management of ALS-inhibiting Herbicides

Using the accumulated knowledge concerning rice ALS genes, rice mutated ALS genes were artificially prepared [51]. Each recombinant ALS was prepared and the sensitivity of each protein to the ALS-inhibiting herbicides was examined. The mutated ALSs that have one-point mutation in proline at position 171 exhibited a high resistance to the SU herbicide, chlorsulfuron. Conversely, the resistance level of these mutated ALSs to the IMI herbicide, imazaquin, was lower than that of chlorsulfuron. In contrast, the resistance level of the proline mutated ALSs to the PSs was moderate, between those of chlorsulfuron and imazaquin [52]. These results were correlated to the cross-resistance pattern of the proline-mutated ALS of *K. scoparia* (Fig. 2.5.18). From these results, it is considered that rice mutated recombinant ALSs are useful as resistant enzyme models for the herbicide resistance management at newly developed or developing ALS-inhibiting herbicides.

Abbreviations

ALS, acetolactate synthase
SU, sulfonylurea
IMI, imidazolinone
PS, pyrimidinylsalicylate
PC, pyrimidinylcarboxy(late)
TP, triazolopyrimidine sulfonamide
DMSP, 2-methanesulfonyl-4,6-dimethoxypyrimidine

References

1 Y. Nezu, M. Miyazaki, K. Sugiyama, I. Kajiwara, *Pestic. Sci.* **1996**, 47, 103–113.

2 T. Shimizu, I. Nakayama, K. Nagayama, T. Miyazawa, Y. Nezu, *Herbicide Classes in Development*, Springer, **2002**, pp. 1–41.

3 Y. Nezu, M. Miyazaki, K. Sugiyama, N. Wada, I. Kajiwara, T. Miyazawa, *Pestic. Sci.* **1996**, 47, 115–124.

4 Y. Nezu, N. Wada, Y. Saitoh, S. Takahashi, T. Miyazawa, *J. Pesticide Sci.* **1996**, 21, 293–303.

5 T. Shimizu, I. Nakayama, T. Nakao, Y. Nezu, H. Abe, *J. Pesticide Sci.* **1994**, 19, 59–67.

6 Y. Nezu, *Development of Agrochemicals in Japan*, Pesticide Science Society of Japan, **2003**, pp. 279–290.

7 F. Takabe, K. Kaku, N. Wada, A. Takeuchi, S. Shigematsu, *Abstracts of Papers*, 19th *Annual Meeting of the Pesticide Science Society of Japan*, **1994**, p. 55.

8 Y. Nezu, T. Nagata, S. Itoh, K. Masuda, Japan Patent JP2-85262 (**1990**).

9 S. Yokota, N. Wada, R. Hanai, T. Shimizu, *Abstracts of Papers*, 20th *Annual Meeting of the Pesticide Science Society of Japan*, **1995**, p. 78.

10 K. Hirai, A. Uchida, R. Ohno, *Herbicide Classes in Development*, Springer, Berlin **2002**, pp 202–210.

11 Y. Nezu, N. Wada, F. Yoshida, T. Miyazawa, T. Shimizu, T. Fujita, *Pestic. Sci.*, **1998**, 52, 343–353.

12 (a) K. Kaku, S. Kusano, Y. Toyokawa, T. Miyazawa, R. Yoshida, JP1-301668 (**1989**). (b) K. Kaku, N. Wada, K. Shigematsu, A. Takeuchi, JP2-85262 (**1989**).

13 Y. Saitoh, N. Wada, S. Kusano, Y. Toyokawa, T. Miyazawa, Japan Patent 2561524 (**1996**).

14 N. Wada, S. Kusano, Y. Toyokawa, Japan Patent 2558516 (**1996**).

15 M. Tamaru, N. Kawamura, M. Satoh, F. Takabe, S. Tachikawa, Japan Patent 2603557 (**1997**).

16 S. Takahashi, S. Shigematsu, A. Morita, Y. Nezu, J. S. Claus, C. S. Williams, *Proc. Brighton Crop Prot. Conf. Weeds*, **1991**, 1, 57–62.

17 Y. Nezu, Y. Saitoh, S. Takahashi, Y. Tomoda, *J. Pesticide Sci.* **1999**, 24, 217–229.

18 (a) Y. Tomoda, Japan Patent 2905900 (**1999**). (b) Y. Tomoda, Japan Patent 3060111 (**2000**).

19 Y. Saito, N. Wada, S. Kusano, T. Miyazawa, S. Takahashi, Y. Toyokawa, Y. Kajiwara, US Patent 4932999 (**1990**).

20 O. Watanabe, M. Yokoyama, S. Fujita, T. Miyazaki, N. Wada, *J. Weed Sci. Technol.*, **2003**, 48, 24–30.

21 M. Yokoyama, O. Watanabe, K. Yanagisawa, Y. Ogawa, N. Wada, S. Shigematsu, *J. Weed Sci. Technol.*, **1994**, 42(Supplement), 32–33.

22 M. Yokoyama, O. Watanabe, K. Kawano, S. Shigematsu, N. Wada, *Proc. Brighton Crop Prot. Conf. Weeds*, **1993**, 1, 61–66.

23 N. Wada, S. Kusano, Y. Toyokawa, US Patent 4906285 (**1990**).

24 S. Tachikawa, T. Miyazawa, H. Sadohara, *Proc. 16th Asian-Pacific Weed Sci. Soc. Conf.*, **1997**, 2A, 114–117.

25 M. Tamaru, T. Takehi, N. Matsuzawa, R. Hanai, *Petic. Sci.*, **1996**, 47, 327–335.

26 M. Tamaru, J. Inoue, R. Hanai, S. Tachikawa, *J. Agric. Food Chem.*, **1997**, 45, 2777–2783.

27 R. Hanai, K. Kawano, S. Shigematsu, M. Tamaru, *Proc. Brighton Crop Prot. Conf. Weeds*, **1993**, 1, 47–52.

28 (a) K. Umezu, K. Isozumi, T. Miyazaki, M. Tamaru, F. Takabe, N. Masuyama, Y. Kimura, *Synlett*, **1994**, 1, 61–62. (b) M. Tamaru, K. Umezu, K. Isozumi, C. Maejima, H. Kageyama, Y. Kimura, *Synth. Commun.*, **1994**, 24, 2749–2756.

29 (a) M. Tamaru, N. Kawamura, M. Satoh, F. Takabe, S. Tachikawa, Japan Patent 2977591 (**1999**). (b) N. Kawamura, N. Masuyama, F. Takabe, M. Tamaru, K. Isozumi, K. Umezu, Japan Patent 3046401 (**2000**). (c) K. Isozumi, K. Umezu, T. Miyazaki, Y. Kimura, Japan Patent 3258043 (**2001**).

30 M. Tamaru, N. Kawamura, M. Sato, S. Tachikawa, R. Yoshida, F. Takabe, EP Patent 435170 (**1991**).

31 (a) J. H. Cho, S.-C. Ahn, S. J. Koo, K. H. Joe, H. S. Oh, *Proc. Brighton Crop Prot. Conf. Weeds*, **1997**, 1, 39–44. (b) S. J. Koo, S.-C. Ahn, J. S. Lim, S. H. Chae, J. S. Kim, J. H. Lee, J. H. Cho, *Pestic. Sci.*, **1997**, 51, 109–114.

32 C. Luthy, H. Zondler, T. Papold, G. Seifert, B. Urwyler, T. Heinis, H. C. Steinrucken, J. Allen, *Pestic. Manag. Sci.*, **2001**, 57, 205–224.

33 T. Shimizu, *J. Pestic. Sci.*, **1997**, 22, 245–256.

34 J. Durner, V. Gailus, P. Böer, *Plant Physiol.*, **1991**, 95, 1144–1149.

35 M. V. Subramanian, B. C. Gerwick, *ACS Symp. Ser.*, **1989**, 389, 277–288.

36 D. L. Shaner, P. C. Anderson, M. A. Stidham, *Plant Physiol.*, **1984**, 76, 545–546.

37 T. Shimizu, I. Nakayama, N. Wada, T. Nakao, H. Abe, *J. Pestic. Sci.*, **1994**, 19, 257–266.

38 T. Shimizu, K. Yamashita, H. Kato, N. Hashimoto, H. Abe, I. Nakayama, *Abstracts of Papers, 20th Annual Meeting of the Pesticide Science Society of Japan*, **1995**, p. 136.

39 T. Shimizu, I. Nakayama, T. Nakao, K. Yamashita, K. Nagayama, H. Abe, *Abstracts of Papers, 18th Annual Meeting of the Pesticide Science Society of Japan*, **1993**, p. 76.

40 J. V. Schloss, L. M. Ciskanik, D. E. VanDyk, *Nature*, **1988**, 331, 360–362.

41 M. J. Muhitch, D. L. Shaner, M. A. Stidham, *Plant Physiol.*, **1987**, 83, 451–456.

42 T. R. Hawkes, *BCPC Monograph*, **1989**, pp. 131–138.

43 T. R. Hawkes, S. E. Thomas, *Biosynthesis of Branched Chain Amino Acids*, Balaban Publishers, Weinheim **1990**, pp. 373–389.

44 F. Ortega, J. Bastide, T. R. Hawkes, *Pestic. Biochem. Physiol.*, **1996**, 56, 231–242.

45 S. Sunderland, J. D. Burton, H. D. Coble, E. P. Maness, *Weed Sci.*, **1995**, 43, 21–27.

46 H. Matsushita, Y. Hukai, T. Unai, K. Ishikawa, Y. Yusa, *Abstracts of Papers, 19th Annual Meeting of the Pesticide Science Society of Japan*, **1994**, p. 127.

47 I. Nakayama, T. Shimizu, T. Nakao, H. Abe, The *Abstracts of Papers, 18th Annual Meeting of the Pesticide Science Society of Japan*, **1993**, p. 77.

48 H. Mizutani, K. Shinba, Y. Asano, Y. Yusa, *Abstracts of Papers, 23th Annual Meeting of the Pesticide Science Society of Japan*, **1998**, p. 106.

49 T. Shimizu, I. Nakayama, K. Nagayama, A. Hukuda, Y. Tanaka, K. Kaku, PCT Int. Appl. WO 0244385 (**2002**).

50 T. Shimizu, K. Kaku, K. Kawai, T. Miyazawa, Y. Tanaka, *ACS Symp. Ser.*, **2005**, 899, 256–271.

51 K. Kaku, T. Shimizu, K. Kawai, K. Nagayama, A. Fukuda and Y. Tanaka, PCT Int. Appl. WO 03083118 (**2003**).

52 K. Kaku, S. Ohno, Y. Ogawa, T. Shimizu, *J. Weed Sci. Technol.*, **2006**, 51 (Supplement), 92–93.

2.6
Sulfonylaminocarbonyl-triazolinones

Klaus-Helmut Müller

2.6.1
Introduction

The first examples of the new herbicidal class of sulfonylaminocarbonyl-triazolinones (SACTs) were reported in 1989 [1]. Following intensive chemical optimization two representatives were developed and commercialized for selective weed control in cereals. In 2000, flucarbazone-sodium (1) was introduced in the Canadian market under the trade name Everest® for the control of wild oats (*Avena fatua*) and green foxtail (*Setaria viridis*) in spring wheat (*Triticum aestivum*) and durum wheat (*Triticum durum*) (Fig. 2.6.1).

Propoxycarbazone-sodium (2) was first launched in Kenya in 2000. It is now registered in the major cereal producing European countries as Attribut® and in the United States as Olympus®. It is especially effective against brome (*Bromus* spp.), loose silky-bent (*Apera spica-venti*), common couchgrass (*Elymus repens*), blackgrass (*Alopecurus myosuroides*), jointed goatgrass (*Aegilops cylindrica*) and several broadleaf weeds from the mustard family (Fig. 2.6.2).

Both compounds are inhibitors of the acetolactate synthase enzyme, also known as aceto hydroxy acid synthase (AHAS) and are classified in group B by the Herbicide Resistance Action Committee HRAC. Table 2.6.1 gives the physicochemical properties of 1 and 2.

2.6.2
Discovery of the Active Ingredients

The discovery of the herbicidal class of SACTs is outlined in detail in Ref. [2]. Starting from the concept of seeking new applications for the Nylon 6 intermedi-

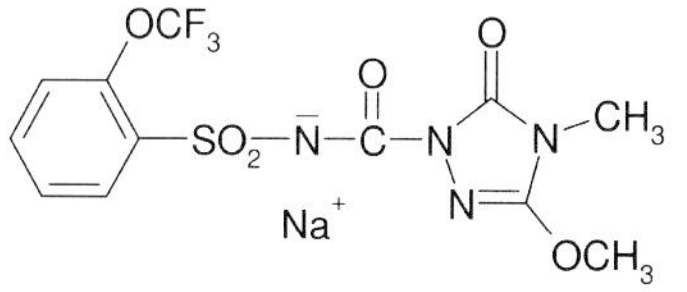

Fig. 2.6.1. Compound **1**, flucarbazone-sodium, Everest®, Vulcano®.

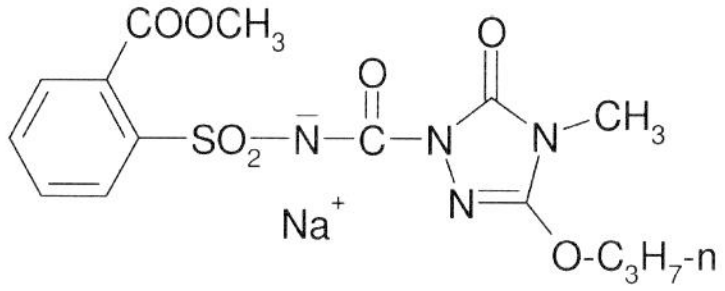

Fig. 2.6.2. Compound **2**, propoxycarbazone-sodium, Attribut®, Olympus®.

ate ε-caprolactam, the bicyclic triazolinone (**3**) was synthesized in the late 1970s as a possible intermediate for potential fungicides (Scheme 2.6.1) [3].

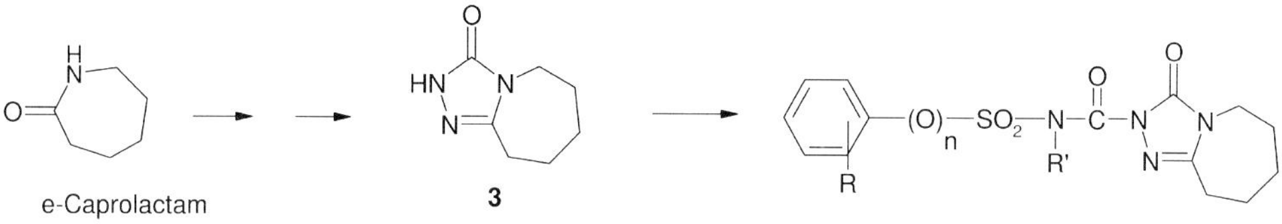

Scheme 2.6.1

Amongst many other derivatives (by NH-acylation, sulfonylation, alkylation, arylation) a sulfonylaminocarbonyl-triazolinone with the internal code no. BAY DAM 4493 was synthesized in 1985 (Fig. 2.6.3). It showed not only activity against rice blast (*Pyricularia oryzae*) but also phytotoxic symptoms at application rates of 500 g a.i. ha^{-1}.

About two years later this compound was identified in an *in vitro* assay as an unusual ALS inhibitor [4–6] and was the starting signal for a major synthesis program.

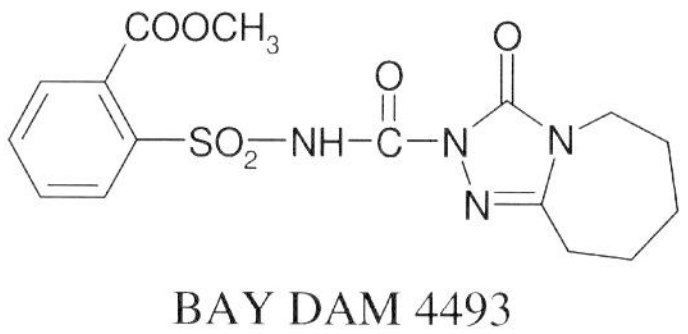

BAY DAM 4493

Fig. 2.6.3

Table 2.6.1 Physicochemical properties of flucarbazone-sodium and propoxycarbazone-sodium.

Property	Flucarbazone-sodium (1)	Propoxycarbazone-sodium (2)
CAS-No.	181274-17-9	181274-15-7
Code numbers	BAY MKH 6562	BAY MKH 6561
Melting point (°C)	200 (with decomposition)	230–240 (under decomposition)
Vapor pressure	Cannot be determined directly due to its extremely low value. From experimental results obtained for 70 °C a value of $<4 \times 10^{-8}$ Pa can be estimated as an upper limit. This limit would correspond to $<1 \times 10^{-9}$ Pa at 20 °C	Cannot be determined directly due to its extremely low value. From experimental results obtained for 70 °C a value of $<9 \times 10^{-8}$ Pa can be estimated as an upper limit. This would correspond to $<1 \times 10^{-8}$ Pa at 20 °C
Dissociation constant (at 20 °C)	The free acid produced by protonation under acidic conditions has a pK_a of 1.9.	The free acid produced by protonation under acidic conditions has a pK_a of 2.1.
Solubility in water (at 20 °C)	Water solubility is 44 g L^{-1} in unbuffered aqueous solutions in the range pH 4–9. Solubility is not influenced by pH in the range pH 4–9.	Unbuffered water and buffered between pH 7 and 9: 42 g L^{-1}. Solubility is not influenced by pH in the range pH 7–9. Solubility at pH 4.5: 2.9 g L^{-1}
Solubilities in organic solvents (g L^{-1} at 20 °C)	Dimethyl sulfoxide: 250 Poly(ethylene glycol): 48 Acetonitrile: 6.4 2-Propanol: 0.27 Xylene: <0.1	Dimethyl sulfoxide: 190 Poly(ethylene glycol): 5.2 Acetonitrile: 0.9 2-Propanol: <0.1 Xylene: <0.1
Partition coefficient log P_{OW} (octanol–water) (20 °C)	−2.85 (unbuffered) −0.89 (pH 4.0) −1.84 (pH 7.0) −1.88 (pH 9.0)	−2.60 (unbuffered) −0.30 (pH 4.0) −1.55 (pH 7.0) −1.59 (pH 9.0)

2.6.3
Optimization of the Lead Structure

All efforts to improve the herbicidal activity of BAY DAM 4493 by variation of the seven-membered ring were unsuccessful. Ring contraction to the five- and

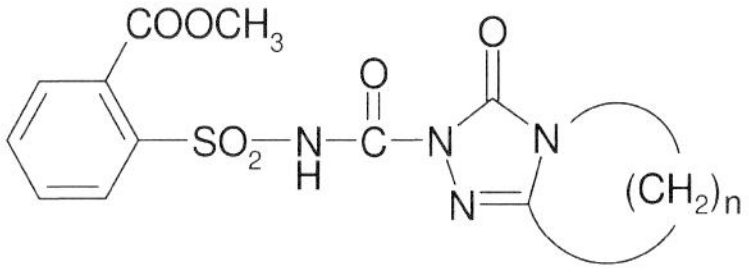

Fig. 2.6.4

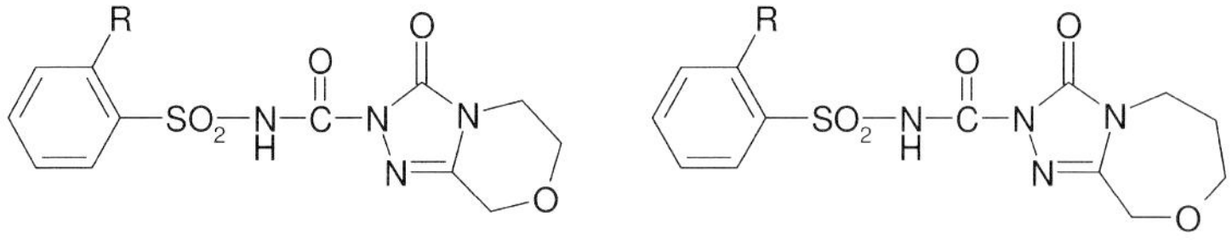

R = COOCH$_3$, Br, CF$_3$, OC$_2$H$_5$

Fig. 2.6.5

six-membered analogues ($n = 3, 4$) and ring enlargement ($n = 5, 6, 7$) up to the thirteen-membered ring ($n = 11$) significantly reduced the herbicidal potency (Fig. 2.6.4) [1, 7, 8].

Similar results were obtained upon the introduction of oxygen (Fig. 2.6.5) [9].

In contrast to this observation the introduction of monocyclic triazolinones dramatically enhanced the herbicidal activity. In a systematic manner hundreds of previously known and new intermediates were prepared and transformed into SACTs. Many new synthetic procedures have been developed. To date, derivatives of the type shown in Table 2.6.2 have been synthesized and published.

Table 2.6.2 Synthesized derivatives of monocyclic triazolinones and the relevant references.

A	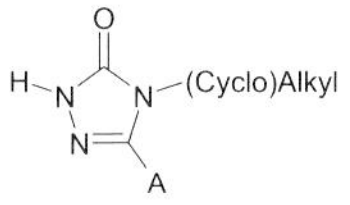	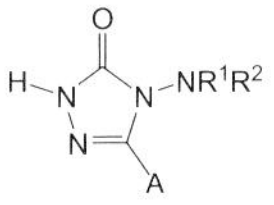
A = (cyclo)alkyl	[1, 10]	[1, 10]
S-R^3	[11, 12]	[11]
Hal	[13]	[13]
NR^{4}R^5	[10]	[10]

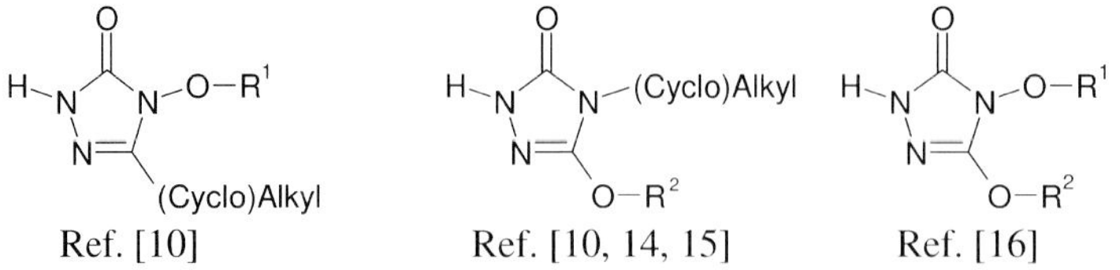

Fig. 2.6.6

The major breakthrough was the introduction of oxygen-bound residues, either on nitrogen and/or on carbon, (Fig. 2.6.6).

Special interest in the SACTs was generated by their biological profile. In the early 1990s most known ALS inhibitors like sulfonylureas had their main focus on dicotyledons. In contrast, the new class of SACTs exhibits in general high activity against grassy weeds, with sometimes rather good dicot activity and selectivity in cereals.

The first outdoor trials were undertaken in 1991 with the N-ethoxy derivative BAY MKH 4340 (internal code) (Fig. 2.6.7).

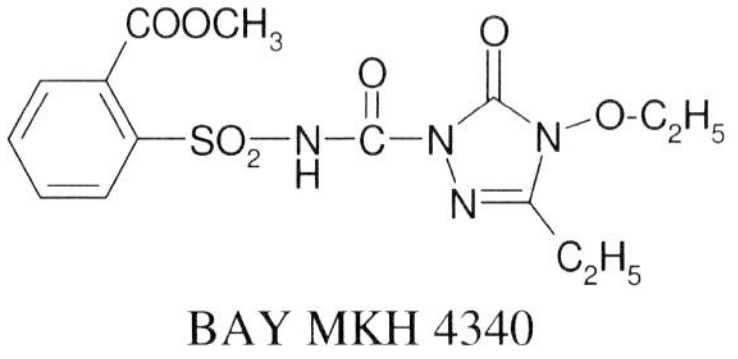

BAY MKH 4340

Fig. 2.6.7

In the following two years a further seven compounds were tested in parallel in the form of their sodium salts in Europe, the United States and Canada (Fig. 2.6.8).

The special biological spectrum – especially against difficult to control monocotyledonous weeds – and the selectivity in cereals led to development of two compounds, later known as flucarbazone-sodium (**1**) and propoxycarbazone-sodium (**2**).

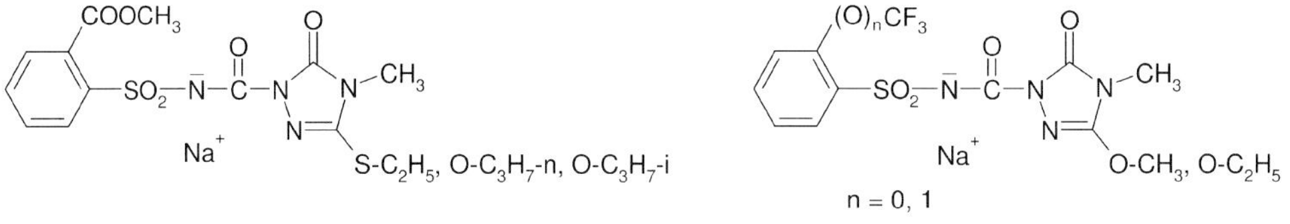

Fig. 2.6.8

2.6.4
Synthesis

Flucarbazone and propoxycarbazone can be prepared by various methods:

(a) Sulfonyl isocyanate addition (Scheme 2.6.2) [1, 17].

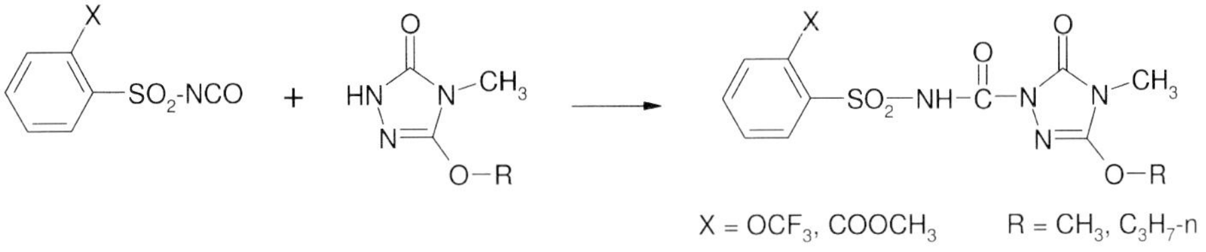

X = OCF$_3$, COOCH$_3$ R = CH$_3$, C$_3$H$_7$-n

Scheme 2.6.2

(b) The phenylurethane route (Scheme 2.6.3) [10].

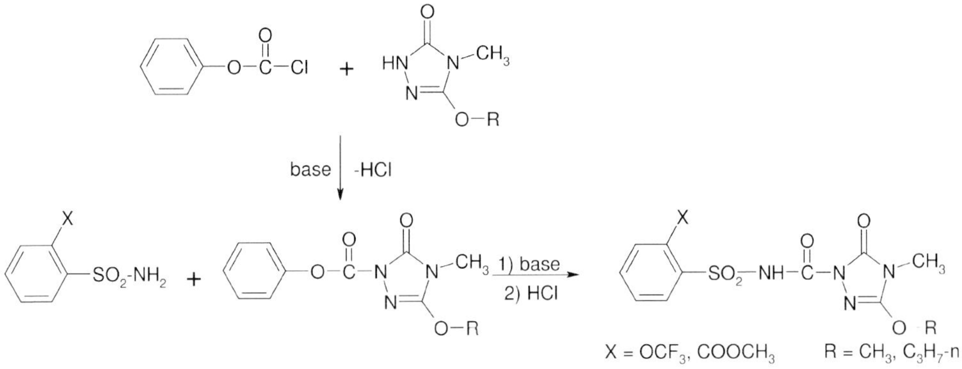

base | -HCl

1) base
2) HCl

X = OCF$_3$, COOCH$_3$ R = CH$_3$, C$_3$H$_7$-n

Scheme 2.6.3

(c) The cyanate route (Scheme 2.6.4) [18].

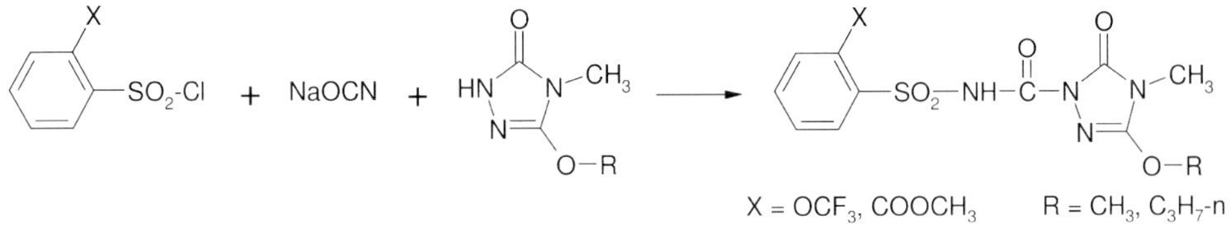

X = OCF$_3$, COOCH$_3$ R = CH$_3$, C$_3$H$_7$-n

Scheme 2.6.4

For stability reasons both compounds are formulated as sodium salts. Several synthetic procedures describe the salt formation [19, 20].

2.6.4.1 Sulfonyl Components

The sulfonyl component of propoxycarbazone-sodium (**2**) is an integral part of several commercial sulfonylureas [21]. Technical procedures exist for sulfonyl isocyanate preparation starting from saccharin (Scheme 2.6.5) [22, 23].

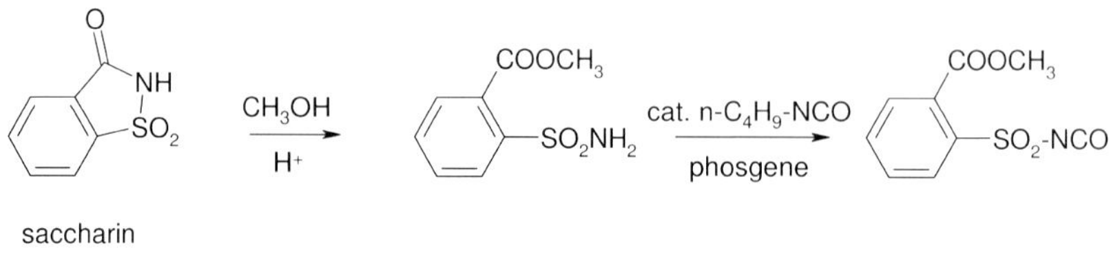

saccharin

Scheme 2.6.5

The sulfonyl part **4** of flucarbazone-sodium (**1**) can be prepared by a classical Meerwein reaction from the aniline (**5**) prepared according to Refs. [24, 25] (Scheme 2.6.6).

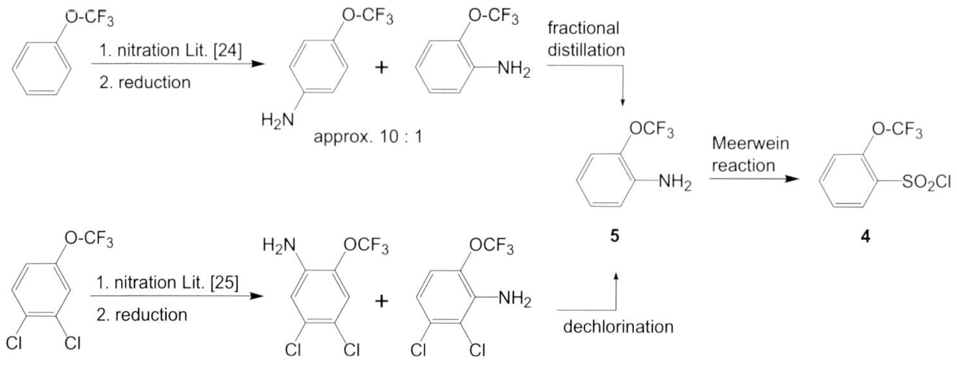

Scheme 2.6.6

Another method is based on sulfochlorination and the different solubilities of the isomeric sulfonamides **6a** and **6b** (Scheme 2.6.7) [26].

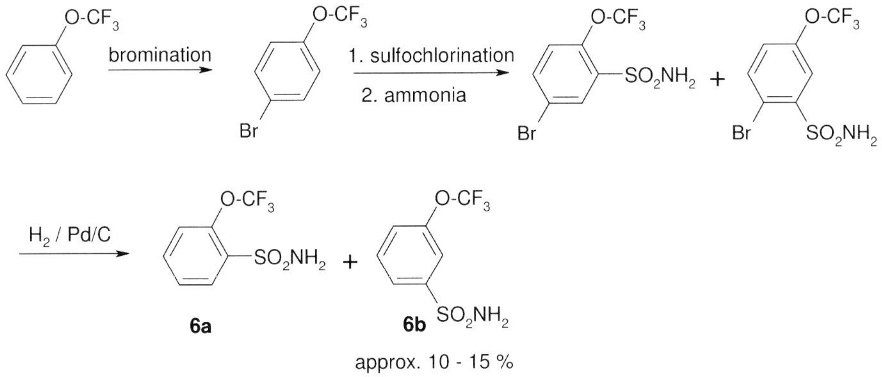

Scheme 2.6.7

2.6.4.2 Triazolinone Synthesis

The flucarbazone-sodium intermediate (**7**) has long been known [27], but for safety reasons there was a great need for alternative procedures (Scheme 2.6.8):

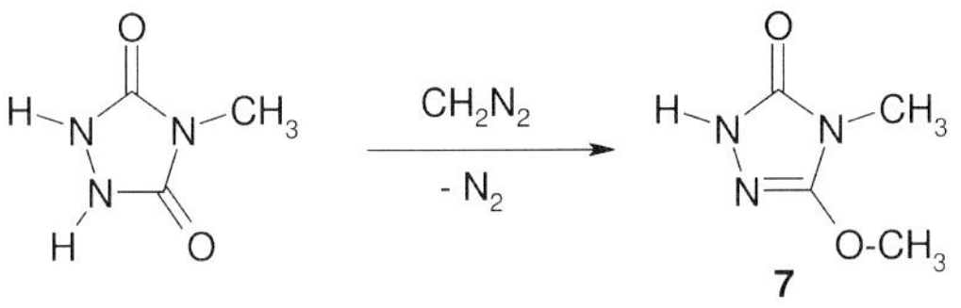

Scheme 2.6.8

(a) Iminoester route ("tin pathway") (Scheme 2.6.9) [28–31].

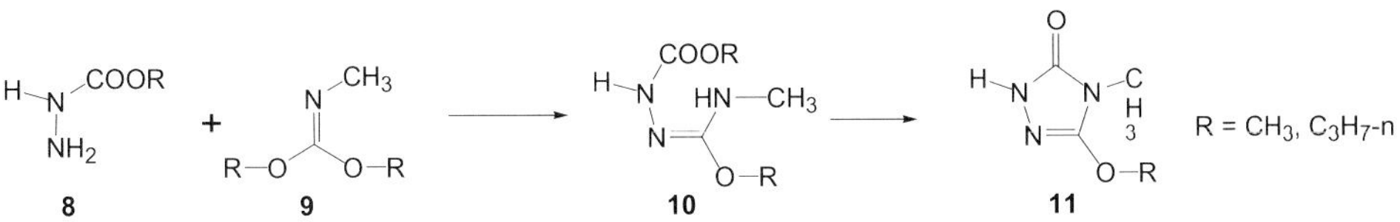

Scheme 2.6.9

Here, the iminoester (**9**) reacts with carbazate (**8**) to give **10**, which cyclizes under basic conditions to the triazolinones (**11**).

(b) Iminothioester route (Scheme 2.6.10) [32, 33].

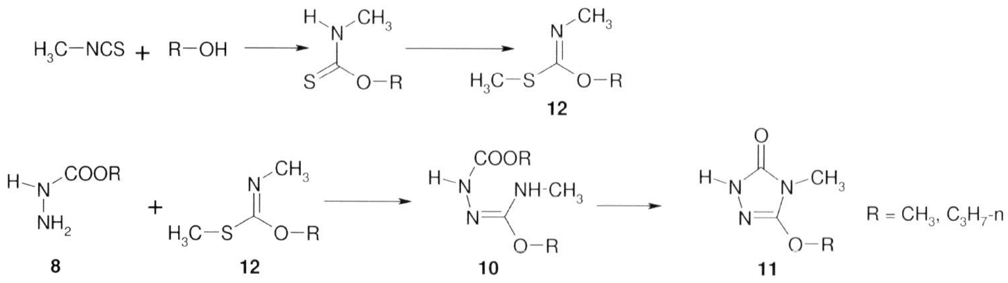

Scheme 2.6.10

The iminothioester (**12**) is much more easily prepared than the iminoester (**9**).

(c) Methylation of NH-triazolinones (**14**) (Scheme 2.6.11) [34, 35].

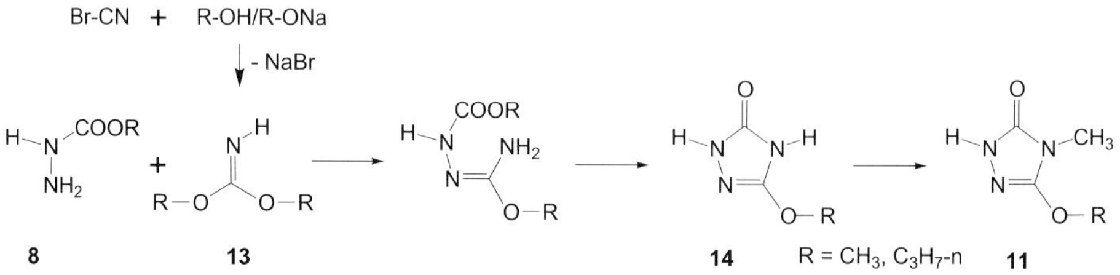

Scheme 2.6.11

The methylation of **14**, prepared via imidoester **13** [36], takes place exclusively at the amidic nitrogen.

Various patents describe alternative methods for the formation of the NH-triazolinones (**14**) (Scheme 2.6.12) [35, 37–39].

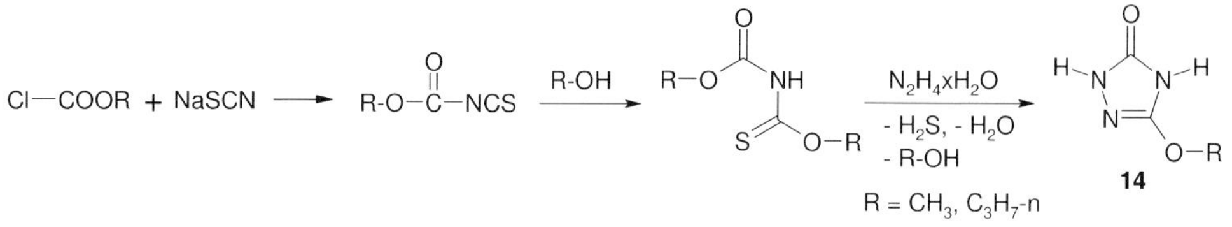

Scheme 2.6.12

2.6.5
Biology

Flucarbazone-sodium (**1**) was discovered and developed by the former Plant Protection Division of Bayer AG (now Bayer CropScience) [40–42]. Since 2002 it is commercialized by Arysta LifeScience in Canada and the USA under the trade name Everest® [43] and in Chile as Vulcano®.

It is a selective herbicide for the control of wild oats (*Avena fatua*) [44], green foxtails (*Setaria viridis*), Italian ryegrass (*Lolium multiflorum*), windgrass (*Apera spica-venti* and *A. interrupta*) and *Bromus* species like cheatgrass (*Bromus secalinus*) and Japanese brome (*Bromus japonicus*) in spring, durum and winter wheat [40–43].

In addition to these grasses, numerous broadleaf weeds are controlled, such as redroot pigweed (*Amaranthus retroflexus*), wild mustard (*Brassica kaber*), stinkweed (*Thlaspi arvense*), shepherd's purse (*Capsella bursa-pastoris*), green smartweed (*Polygonum scabrum* Moench.) and volunteer canola (*Brassica napus*). For broader spectrum control of broadleaf weeds Everest® may be mixed with broadleaf herbicides like 2,4-D (MCPA) amine or ester, bromoxynil, dicamba, fluroxypyr or sulfonylureas like metsulfuron-methyl, triasulfuron, tribenuron-methyl, chlorsulfuron, thifensulfuron-methyl or prosulfuron [43]. Currently application rates of 15–20 g a.i. ha^{-1} are registered in Canada and 21–42 g a.i. ha^{-1} in the USA.

Propoxycarbazone-sodium (**2**) is a new selective herbicide for grass control in wheat, rye and triticale [45]. In Europe it is registered as Attribut® at rates of 28 to 70 g a.i. ha^{-1}. It acts predominantly against important grasses such as *Bromus* species [46, 47], blackgrass (*Alopecurus myosuroides*) [48, 49] and loose silkybent (*Apera spica-venti*) [49]. It is applied post-emergence in spring at a core use rate of 42 g a.i. ha^{-1}. Compared with single applications a split application or sequences following an autumn-standard treatment is more favorable, giving good to very good grass control. The following *Bromus* species can be controlled effectively [46, 47]: field brome (*B. arvensis*), meadow brome (*B. commutatus*), Japanese brome (*B. japonicus*), soft brome (*B. mollis*), rye brome (*B. secalinus*), barren brome (*B. sterilis*) and drooping brome (*B. tectorum*). Propoxycarbazone-sodium is taken up via leaves and, particularly, via roots. Especially on light soils, and under moist conditions, it controls couch grass (*Elymus repens*) at a commercially acceptable level. On heavy soils higher rates are recommended. Owing to its systemic mobility it also kills the rhizomes of *E. repens* [50].

In the USA propoxycarbazone-sodium is registered in spring, winter and durum wheat with application rates of 30 to 45 g a.i. ha^{-1} and sold under the trade name Olympus®. Bromus control is the primary target and, as in Europe, all important species, including cheatgrass (*B. secalinus*) and downy brome (*B. tectorum*), are well controlled [51]. Besides brome the following grasses can be economically reduced or suppressed [51]: loose silky-bent (*Apera spica-venti*), wild oats (*Avena fatua*), littleseed canarygrass (*Phalaris minor*) [52] and paradoxagrass (*P. paradoxa*). Suppression of jointed goatgrass (*Aegilops cylindrica*) can be achieved by two sequential treatments of 30 g a.i. ha^{-1} in autumn and spring [53].

Besides grasses, broadleaf species belonging to the mustard complex like *Sisymbrium*, *Brassica*, *Descurainia*, *Chorispora*, *Camelina*, *Capsella* and *Thlaspi* [51] represent further target weeds for this herbicide.

Propoxycarbazone-sodium (**2**) can be applied straight or in tank mixtures with other herbicides such as triasulfuron, metsulfuron-methyl, chlorsulfuron, thifensulfuron-methyl, prosulfuron, carfentrazone, dicamba, bromoxynil, clopyralid, MCPA amine or ester, 2,4-D amine or ester, metribuzin or fluroxypyr. Olympus™ flex is a ready to use formulation with mesosulfuron.

2.6.6
Conclusion

Sulfonylaminocarbonyl-triazolinones are a new ALS inhibitor class discovered in 1987 by the former Plant Protection Division of Bayer AG. In the 1990s two compounds were developed for selective grass control in cereals. Flucarbazone-sodium (**1**) acts predominantly against green foxtail and wild oats and is registered in US and Canada as Everest® and in Chile as Vulcano®. Propoxycarbazone-sodium (**2**) is a brome specialist both in Europe (Attribut®) and the US (Olympus®). Besides other grasses, such as loose silky-bent and blackgrass, the rhizomes and the green part of couchgrass can be controlled. Additionally, there is a good activity against several broadleaf weeds from the mustard family.

References

1 W. Daum, K.-H. Müller, M. Schwamborn, P. Babczinski, H.-J. Santel, R. R. Schmidt, H. Strang, **1989**, Sulfonylaminocarbonyltriazolinone, EP 341489 (Prio: 09. 05. 1988), Bayer AG, Leverkusen, Germany.

2 K.-H. Müller, *Pflanz.-Nachrichten*, **2002**, 55(1), 15–28.

3 W. Daum, **1984**, Derivatisierung von 2,5,6,7,8,9-Hexahydro-3H-triazolo[4,3a]azepin-3-on, Internal Report, Bayer AG, Uerdingen, Germany.

4 P. Babczinski, **1987**, Unusual ALS Inhibitors, Internal Report, Bayer AG, Leverkusen, Germany.

5 P. Babczinski, **1988**, Annual Report 1988, Internal Report, Bayer AG, Leverkusen, Germany.

6 P. Babczinski, *Pflanz.-Nachrichten*, **2002**, 55(1), 5–14.

7 W. Daum, retired, partly unpublished results from 1978–1989, formerly Central Research, Bayer AG, Uerdingen, Germany.

8 W. Daum, **1989**, 4,5-Alkanylen-1,2,4-triazol-5-in-3-on – potentielle Fungicide und Herbicide, Internal Report, Bayer AG, Uerdingen, Germany.

9 K. König, K.-H. Müller, unpublished results, Bayer AG, Leverkusen, Germany.

10 K.-H. Müller, P. Babczinski, H.-J. Santel, R. R. Schmidt, **1991**, Sulfonylaminocarbonyltriazolinone, EP 422469 (Prio. 12. 10. 1989), Bayer AG, Leverkusen, Germany.

11 K.-H. Müller, P. Babczinski, H.-J. Santel, R. R. Schmidt, **1991**, Sulfonylaminocarbonyl-triazolinones having substituents attached through sulfur, EP 431291 (Prio: 03. 11. 1989), Bayer AG, Leverkusen, Germany.

12 K.-H. Müller, R. Kirsten, E. R. F. Gesing, J. Kluth, K. Findeisen, J. R. Jansen, K. König, H.-J. Riebel, D. Bielefeldt, M. Dollinger, H.-J. Santel, K. Stenzel, **1996**, Herbicidal or fungicidal sulfonylaminocarbonyl-triazolinones with halogenated alk(en)ylthio substituents, WO 1996/27591 (Prio: 08. 03. 1995), Bayer AG, Leverkusen, Germany.

13 K.-H. Müller, P. Babczinski, H.-J. Santel, R. R. Schmidt, **1991**, Halogenierte Sulfonylaminocarbonyltriazolinone, EP 425948 (Prio: 03. 11. 1998), Bayer AG, Leverkusen, Germany.

14 K.-H. Müller, K. König, J. Kluth, K. Lürssen, H.-J. Santel, R. R. Schmidt, **1992**, Sulfonylaminocarbonyltriazolinones with oxygen-bound substituents, EP 507171 (Prio: 04. 04. 1991), Bayer AG, Leverkusen, Germany.

15 K.-H. Müller, R. Kirsten, E. R. F. Gesing, J. Kluth, K. Findeisen, J. R. Jansen, K. König, H.-J. Riebel, O. Schallner, H.-J. Wroblowsky, M. Dollinger, H.-J. Santel, K. Stenzel, **1996**, Herbicidal or fungicidal sulfonylaminocarbonyltriazolinones with halogenated alk(en)oxy substituents, WO 1996/27590 (Prio: 08. 03. 1995), Bayer AG, Leverkusen, Germany.

16 W. Haas, K.-H. Müller, K. König, H.-J. Santel, K. Lürssen, R. R. Schmidt, **1993**, Sulfonylaminocarbonyltriazolinones with two through oxygen bonded substituents, EP 534266 (Prio: 25. 09. 1991), Bayer AG, Leverkusen, Germany.

17 V. A. Prasad, K. Jelich, **2000**, Process for the manufacture of sulfonylaminocarbonyltriazolinones in the presence of xylene as solvent, US 6,160,125 (Prio: 27. 12. 1999), Bayer Corporation, Pittsburgh, PA, USA.

18 J. Kluth, K.-H. Müller, **1995**, Process for the preparation of sulfonylaminocarbonyltriazolinones, EP 659746 (Prio: 21. 12. 1993), Bayer AG, Leverkusen, Germany.

19 V. A. Prasad, S. V. Kulkarny, E. Rivadeneira, V. C. Desai, **2000**, Process for the manufacture of sulfonylaminocarbonyltriazolinones and salts thereof, US 6,147,221 (Prio: 27. 12. 1999), Bayer Corporation, Pittsburgh, PA, USA.

20 V. A. Prasad, K. Jelich, **2000**, Process for the manufacture of sulfonyl-aminocarbonyltriazolinones and salts thereof under pH controlled conditions, US 6,147,222 (Prio: 27. 12. 1999), Bayer Corporation, Pittsburgh, PA, USA.

21 Sulfometuron-methyl, Metsulfuron-methyl, Tribenuron-methyl, Ethametsulfuron-methyl (all from E. I. Du Pont de Nemour and Company), Primisulfuron-methyl (Syngenta Corporation), and the experimental herbicide HNPC-C9908 (Hunan Branch of the National Pesticide R&D South Center, Changsha, China).

22 G. Levitt, **1981**, 2-Isocyanatosulfonyl-benzoic acid esters and preparation thereof, EP 34431 (Prio: 06. 02. 1980), E. I. Du Pont de Nemour and Company, Wilmington, Del, USA.

23 G. Levitt, **1982**, Substituted benzenesulfonyl isocyanates and preparation thereof, EP 46626 (Prio: 30. 05. 1978), E. I. Du Pont de Nemour and Company, Wilmington, Del, USA.

24 G. A. Olah, T. Yamato, T. Hashimoto, J. G. Shih, N. Trivedi, B. P. Singh, M. Piteau, J. A. Olah, *J. Am. Chem. Soc.* **1987**, 109, 3708–3713.

25 R. Lantzsch, A. Marhold, **1998**, Process for the preparation of 2-trifluoromethoxy-aniline, EP 820981 (Prio.: 26. 07. 1996), Bayer AG, Leverkusen, Germany.

26 R. Lantzsch, A. Marhold, E. Kysela, **1997**, Method of preparing 2-trifluoromethoxy benzene sulfonamide, WO 1997/19056 (Prio.: 21. 11. 1995), Bayer AG, Leverkusen, Germany.

27 F. Arndt, L. Loewe, A. Tarlan-Akön, *Rev. Faculté Sci. Univ. Istanbul*, **1948**, 13A, 127–146.

28 D. L. Alleston, A. G. Davies, *J. Chem. Soc.* **1962**, 2050–2054.

29 S. Sakai, H. Niimi, Y. Kobayashi, Y. Ishii, *Bull. Chem. Soc. Jpn.*, **1977**, 50, 3271–3275.

30 H.-J. Wroblowsky, K. König, **1996**, Process for the preparation of alkoxytriazolinones, EP 703225 (Prio: 23. 09. 1994), Bayer AG, Leverkusen, Germany.

31 H.-J. Wroblowsky, K. König, **1996**, Process for the preparation of alcoxytriazolinones, EP 703226 (Prio: 23. 09. 1994), Bayer AG, Leverkusen, Germany.

32 H.-J. Wroblowsky, K. König, J. Kluth, K.-H. Müller, **1996**, Process for the preparation of alkoxytriazolinones, EP 703224 (Prio: 23. 09. 1994), Bayer AG, Leverkusen, Germany.

33 D. E. Jackman, **2001**, Process for manufacturing substituted triazo-linones, US 6,222,045 (Prio: 20. 09. 2000), Bayer Corporation, Pittsburgh, PA, USA.

34 H.-J. Wroblowsky, K. König, **1996**, Process for the preparation of alkoxytriazoles, EP 708095 (Prio: 23. 09. 1994), Bayer AG, Leverkusen, Germany.

35 S. V. Kulkarni, V. A. Prasad, V. C. Desai, E. Rivadeneira, K. Jelich, **2001**, Process for the manufacture of substituted triazolinones, US 6,197,971 (Prio: 27. 12. 1999), Bayer Corporation, Pittsburgh, PA, USA.

36 V. A. Lopyrev, V. N. Kurochkina, I. A. Titova, M. G. Voronkov, *Bull.-Acad. Sci. USSR, Div. of Chem. Sci.*, **1989**, 10, 2174–2175.

37 M. Conrad, R. Lantzsch, V. C. Desai, S. V. Kulkarni, **1999**, Process for preparing alkoxytriazolinones, US 5,917,050 (Prio: 11. 02. 1998), Bayer Corporation, Pittsburgh, PA, USA, Bayer AG, Leverkusen, Germany.

38 S. Kulkarni, **2000**, Process for manu-facturing N-alkoxy (or aryloxy)-carbonyl isothiocyanate derivatives using N,N-dialkylarylamine as cata-

lyst, US 6,066,754 (Prio: 10. 06. 1999), Bayer Corporation, Pittsburgh, PA, USA.

39 S. Kulkarni, V. C. Desai, **2001**, Process for manufacture of N-alkoxy (or aryloxy)carbonyl isothiocyanate derivatives in the presence of N,N-dialkylarylamine catalyst and aqueous solvent, US 6,184,412 (Prio: 10. 06. 1999), Bayer Corporation, Pittsburgh, PA, USA.

40 H. J. Santel, B. A. Bowden, V. M. Sorensen, K. H. Müller, *Proc. Brighton Conference – Weeds*, **1999**, Vol. 1, 23–28.

41 H. J. Santel, B. A. Bowden, V. M. Sorensen, K. H. Mueller, J. Reynolds, *Proc. West. Soc. Weed Sci.*, **1999**, 52, 124.

42 R. Brenchley, D. E. Rasmussen, H. J. Santel, V. M. Sorensen, B. A. Bowden, P. G. M. Bulman, D. E. Feindel, B. Gibb, B. M. Tomolak, *Proc. West. Soc. Weed Sci.*, **1999**, 52, 125.

43 Everest 70 WDG Label, Arysta LifeScience North America Corp., San Francisco, EPA Registration No. 66330-49, EPA Est. No. 554-ND-002, Label Number 20444-D, http://www.cdms.net/ldat/ld48U014.pdf.

44 K. J. Kirkland, E. N. Johnson, F. C. Stevenson, *Weed Techno.*, **2001**, 15, 48–55.

45 D. Feucht, K.-H. Müller, A. Well-mann, H. J. Santel, *Proc. Brighton Conference – Weeds*, **1999**, Vol. 1, 53–58.

46 A. Amann, A. Wellmann, *Proc. Brighton Conference – Weeds*, **2001**, Vol. 2, 469–474.

47 A. Amann, *Pflanz.-Nachrichten*, **2002**, 55(1), 87–100.

48 N. P. Godley, G. W. Bubb, *Proc. Brighton Conference – Weeds*, **2001**, Vol. 2, 633–638.

49 A. Wellmann, D. Feucht, *Pflanz.-Nachrichten*, **2002**, 55(1), 67–86.

50 A. Amann, A. Wellmann, Wageningen 2002, *Proceedings 12th EWRS (European Weed Research Society) Symposium*, Papendal, Netherlands, June 24–27, **2002**, 188–189.

51 A. C. Scoggan, H. J. Santel, J. W. Wollam, R. D. Rudolph, *Proc. Brighton Conference – Weeds*, **1999**, Vol. 1, 93–98.

52 C. E. Bell, *Proc. Brighton Conference – Weeds*, **1999**, Vol. 1, 211–216.

53 H. J. Santel, J. E. Anderson, R. G. Brenchley, J. E. Cagle, A. C. Scoggan, A. Wellmann, P. Stahlman, P. Geier, *Proc. Brighton Conference – Weeds*, **2001**, Vol. 2, 487–492.

3
Protoporphyrinogen-IX-oxidase Inhibitors

George Theodoridis

3.1
Introduction

Rarely do we encounter an area of agrochemical research with both the chemical diversity and the very specific and often conflicting structure–activity relationship (SAR) rules as is the case with the herbicidal protoporphyrinogen oxidase (Protox) inhibitors. It was this incredible array of possibilities that lured every single agrochemical organization during the 1980s and 1990s in the United States, Europe, and Asia into initiating a research effort, in the hope of finding the next blockbuster herbicide. Soon, many Protox areas that were initially seen as having unlimited potential turned into dead ends, with only a handful of commercial products achieving significant market share. Part of the difficulty in exploiting the Protox area of herbicide chemistry was the fact that even though it was relatively easy to find chemistries with good biological activity, it was much harder to find clear crop selectivity, either on pre-emergently or post-emergently applied materials.

The lack of clear selectivity of several commercially significant row crops was overcome following the discovery of several highly active and selective Protox herbicides such as the post-emergence soybean selective herbicide fomesafen **7** (Flex[R], Flexstar[R], Reflex[R]) [1, 2], introduced in 1983 by ICI Plant Protection Division, and the soybean selective pre-emergence herbicides F5231, compound **14** [3–5], and sulfentrazone (**15**) (Authority[R], Boral[R], Capaz[R]) [6–8] introduced by FMC in 1991.

Research in Protox herbicides peaked in the early 1990s [9] and diminished soon after as the use of glyphosate-resistant crops gained increased market share. Glyphosate, N-(phosphonomethyl)glycine, is a broad-spectrum, post-emergence, systemic herbicide that has been used extensively over the past 30 years. This intense and prolonged use of glyphosate has resulted in documented resistance to glyphosate in several weed populations [10], which, in turn, has stimulated new interest in Protox-inhibiting herbicides.

Modern Crop Protection Compounds. Edited by W. Krämer and U. Schirmer
Copyright © 2007 WILEY-VCH Verlag GmbH & Co. KGaA, Weinheim
ISBN: 978-3-527-31496-6

The mode of action of Protox herbicides has been extensively reviewed [11–17]. Protox herbicides act by inhibition of the enzyme protoporphyrinogen oxidase, the last common enzyme to both heme and chlorophyll biosynthesis [18–23]. The protoporphyrinogen oxidase enzyme catalyzes the oxidation of protoporphyrinogen IX to protoporphyrin IX by molecular oxygen. Inhibition of the Protox enzyme results in the accumulation of the enzyme product protoporphyrin IX, but not the substrate, via a complex process that has not been entirely elucidated. In the presence of light, protoporphyrin IX generates large amounts of singlet oxygen, which results in the peroxidation of the unsaturated bonds of fatty acids found in cell membranes (Fig. 3.1). The end result of this peroxidation process is the loss of membrane integrity and leakage, pigment breakdown, and necrosis of the leaf that results in the death of the plant. This is a relatively fast process, with leaf symptoms such as a flaccid wet appearance observed within hours of plant exposure to the Protox herbicide under sunlight.

In this chapter, we discuss recent developments and challenges in the field of Protox-inhibiting agrochemicals and place those agrochemicals in the context of research done in this area of chemistry in the past four to five decades.

3.2
Historical Development

The diphenyl ether nitrofen (**1**) [24], introduced in 1963 by Rohm and Haas, now Dow AgroSciences; the oxadiazolinone oxadiazon (**2**) [25, 26] (Explorer®, Herbstar®, Romax®, Ronstar®), introduced in 1968 by Rhone-Poulenc; and the tetrahydrophthalimide chlorophthalim **3** [27], introduced in 1972 by Mitsubishi, represent the earliest examples of Protox herbicides (Fig. 3.2). Though all three classes are chemically quite different, they share a common mode of action, inhibition of the protoporphyrinogen oxidase enzyme, though this was not known until the late 1980s.

Each of these chemistries generated intensive work in the 1960s–1980s, which resulted in numerous diverse chemistries, from which several useful commercial products were obtained.

3.2.1
Diphenyl Ether

Following the discovery of the herbicidal activity of nitrofen (**1**) in 1963, intense research by several agrochemical companies resulted in a vast number of highly active and diverse chemistries [28, 29]. As mentioned earlier, the diphenyl ether chemistry represents the first class of Protox herbicides. Replacement of the aromatic 4-chloro group with a trifluoromethyl, as is the case with oxyfluorfen (**5**) (Goal®) [30], resulted in a significant improvement in biological activity, and 2-

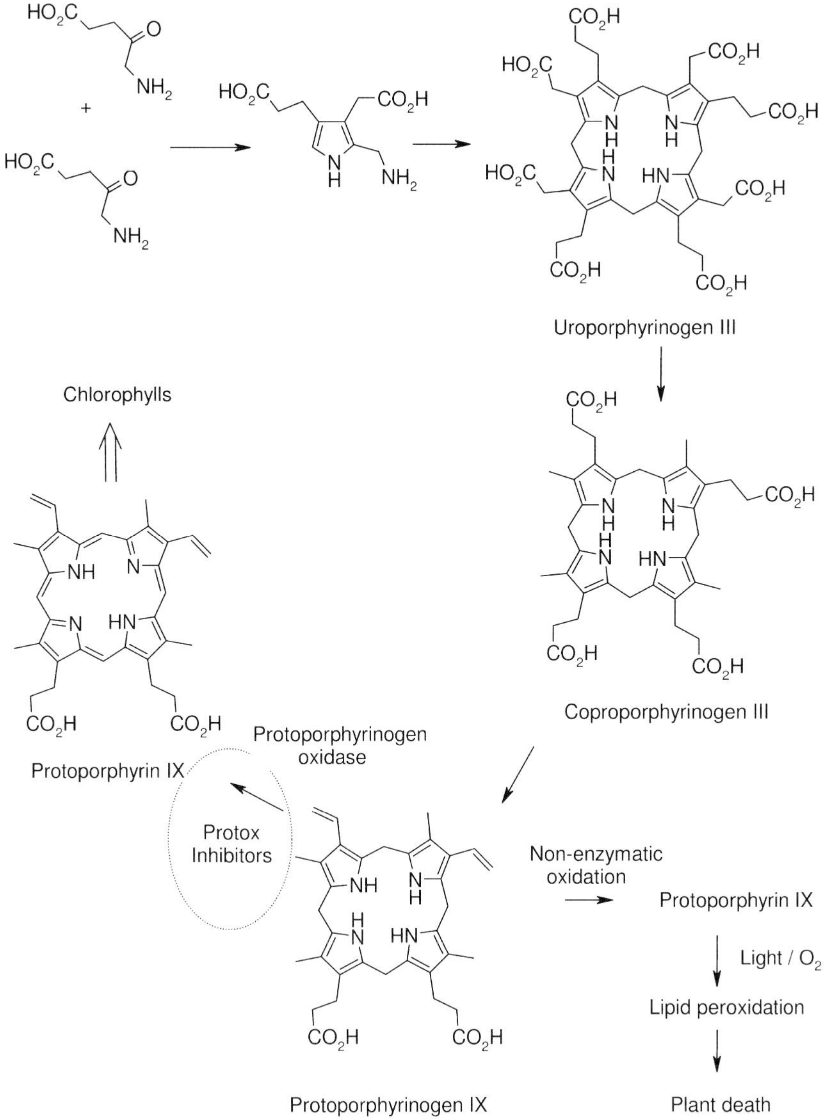

Fig. 3.1. Chlorophyll biosynthetic pathway.

chloro-4-(trifluoromethyl)benzene became the dominant substitution pattern for the second generation of diphenyl ethers (Fig. 3.3), eventually replacing products such as nitrofen (**1**) and bifenox (**4**) (Foxpro[R], Modown[R]) [31]. As can be seen from Fig. 3.3, the 2-chloro-1-(3-substituted-4-nitrophenoxy)-4-(trifluoromethyl)-benzene became the most successful diphenyl ether chemistry scaffold, with four

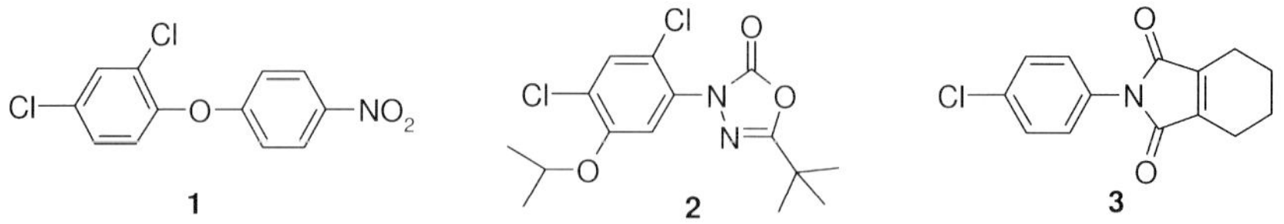

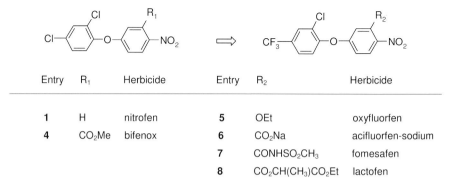

Fig. 3.2. Chemical structures of three early examples of Protox inhibitors.

significant products launched in fewer than ten years. In general, diphenyl ether herbicides such as oxyfluorfen (**5**) (Goal[R]), acifluorfen-sodium (**6**) (Blazer[R]) [32, 33], fomesafen (**7**) (Flex[R], Flexstar[R], Reflex[R]) [1, 2], and lactofen (**8**) (Cobra[R]) [34] are more effective when applied post-emergently, and are more effective for the control of broadleaf than of grass weeds.

Though the 1980s and early 1990s were a period of intense research in diphenyl ether chemistry, the main products described above were all introduced by 1987. By 1996, sales of diphenyl ethers had peaked at $381 million, steadily declining to $200 million by 2001 [35]. This decline was due to the introduction of more effective herbicides, as well as the increasing adoption of herbicide-tolerant crops. Despite this decline, research in diphenyl ether chemistry continued, resulting in the third generation of diphenyl ethers. This newer group of diphenyl ether herbicides consisted of compounds in which either the nitrophenyl ring was replaced by various fused benzoheterocycles, such as benzotriazole [36], benzoisoxazole [37], and indolin-2(3H)-ones [38], or the 2-chloro-4-(trifluoromethyl)benzene group was replaced by a heterocyclic ring such as pyrazole [39].

The extensive research invested by many companies in this third generation of diphenyl ether chemistry resulted in many active molecules, but no successful commercial product.

Entry	R_1	Herbicide	Entry	R_2	Herbicide
1	H	nitrofen	5	OEt	oxyfluorfen
4	CO_2Me	bifenox	6	CO_2Na	acifluorfen-sodium
			7	$CONHSO_2CH_3$	fomesafen
			8	$CO_2CH(CH_3)CO_2Et$	lactofen

Fig. 3.3. Evolution of diphenyl ether herbicides.

3.2.2
Phenyl Ring Attached to Heterocycle

Several discoveries made in the 1960s had a significant impact on our under-standing of the structure–activity of Protox herbicides. The first breakthrough was the discovery of the importance of the 2,4-dihalo-5-substituted-phenyl sub-stitution pattern. Rhone-Poulenc first introduced 3-(2,4-dichlorophenyl)-1,3,4-oxadiazol-2(3*H*)-one (**9**) in 1965 [40]. Further structure–activity optimization at the phenyl ring soon led to the discovery in 1968 of the 2,4-dichloro-5-isopropoxyphenyl substitution pattern of the herbicide oxadiazon (**2**) [41, 42]. The 2,4-dihalo-5-substituted pattern at the aromatic ring would become the basis for much of the 2,4,5-trisubstituted phenyl tetrahydrophthalimide **10** [43] re-search that followed in the Protox area of chemistry.

Another breakthrough discovery was the boost in biological activity caused by the replacement of chlorine by fluorine at the 2-phenyl position. In 1976, DuPont introduced the first example of a 2-fluoro-4-chlorophenyl tetrahydrophthalimide Protox inhibitor (**11**) [44] (Fig. 3.4). The dramatic increase in biological activity caused by the fluorine in the 2 position of the phenyl ring would, in the next de-cade, the 1980s, influence work in the Protox area, such as the discovery of the 4-chloro-2-fluorophenyltetrahydrophthalimide herbicide S-23142 (**12**) [45].

The herbicide oxadiazon (**2**) is used for the pre-emergence control of annual broadleaf weeds and grasses and bindweed, and for the post-emergence control of annual broadleaf weeds in ornamentals such as carnations and roses, as well as in fruit trees, vines, cotton, rice, and turf. It requires high application rates of 1 kg-a.i. ha^{-1} for weed control in rice, and up to 4 kg-a.i. ha^{-1} for pre-emergence weed control in vines and orchards [25, 26]. The high rates of pre-emergence ap-plication, the limited selectivity in several row crops, and the introduction of

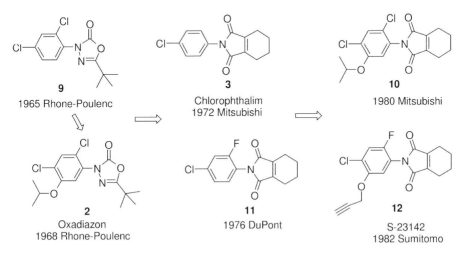

Fig. 3.4. Incorporation of the 2,4-dihalo-5-alkoxy aromatic pattern of oxadiazon into new phenyl tetrahydrophthalimide ring systems.

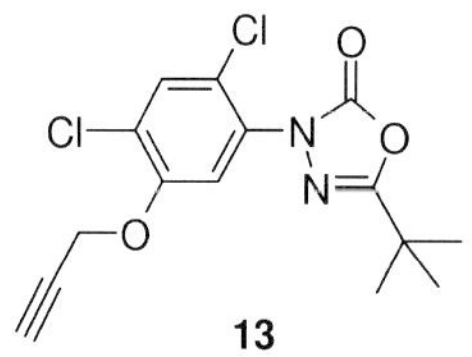

13

Oxadiargyl

Fig. 3.5. Chemical structure of oxadiargyl (**13**).

newer, more effective herbicides all have served to limit the commercial use of oxadiargyl (**2**). Later, Rhone-Poulenc introduced oxadiargyl (**13**) (Raft[R], Topstar[R]) [46, 47] (Fig. 3.5), a compound related to oxadiazon, for the control of broadleaf weeds, grass, and annual sedge in transplanted rice.

In the 1980s, several chemical changes on the 1,3,4-oxadiazol-2(3*H*)-one heterocyclic system resulted in several significant improvements in the pre- and post-emergence biological efficacy and crop selectivity of Protox herbicides. Detailed discussion of the various classes of phenyl heterocycles introduced several decades ago is beyond the scope of this chapter; they have been previously reviewed [28]. The introduction in 1985 of the 5-aminosulfonyl group in the phenyl ring of 2,4,5-trisubstituted-phenyl tetrazolinones was one such significant change. F5231 (**14**) [5], a molecule under development consideration by FMC in the late 1980s, was the first Protox inhibitor to provide excellent pre-emergence broadleaf control and clear selectivity at low application rates on several crops such as soybean, rice, corn, and wheat. FMC later replaced F5231 (**14**) with the phenyl triazolinone sulfentrazone (**15**) for soybean, sugarcane, and other crops [6–8]. Sulfentrazone (**15**) provides pre-emergence control of several broadleaf weeds – as well as several selected grass weeds – for the soybean market. A few years later, FMC introduced a second commercial phenyl triazolinone, the post-emergence cereal and corn herbicide carfentrazone-ethyl (**16**) (Aim[R], Affinity[R], Aurora[R]) [48, 49]. At low rates of 20–35 g-a.i. ha^{-1} carfentrazone-ethyl (**16**) provides excellent control of weeds in commercially important cereal crops – weeds such as bedstraw, speedwell, morning-glory, kochia, spurge, and deadnettle [50] (Fig. 3.6).

14	**15**	**16**
F5231	Sulfentrazone	Carfentrazone-ethyl

Fig. 3.6. Chemical structure of F5231 (**14**), sulfentrazone (**15**), and carfentrazone-ethyl (**16**).

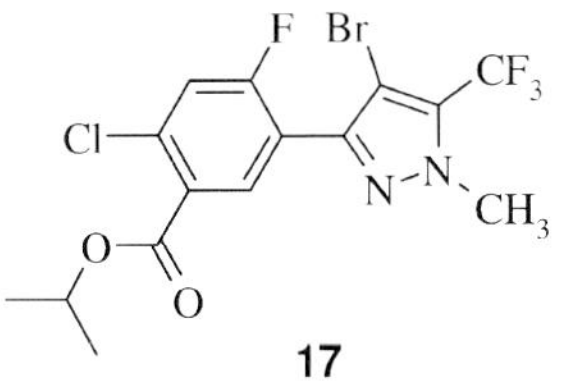

17

Fluazolate

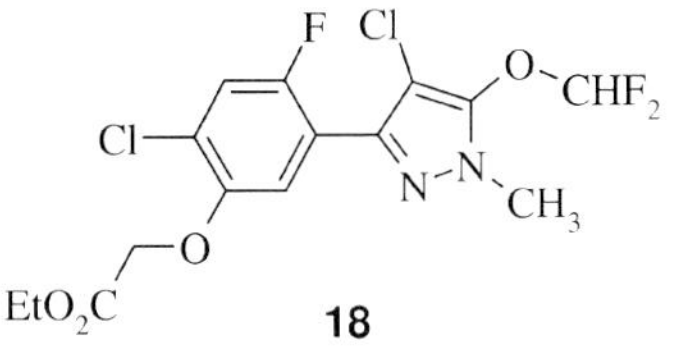

18

Pyraflufen-ethyl

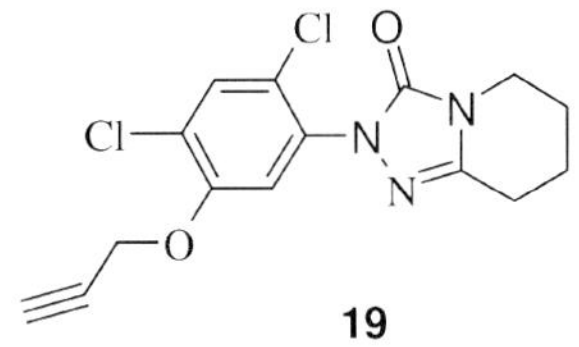

19

Azafenidin

Fig. 3.7. Chemical structure of fluazolate (**17**), pyraflufen-ethyl (**18**), and azafenidin (**19**).

In addition to the oxadiazolinone, tetrazolinone, and triazolinone heterocyclic ring systems, other five-membered ring systems investigated in the 1980s included pyrazoles, such as fluazolate (**17**) [51] from Monsanto; pyraflufen-ethyl (**18**) (Ecopart®) [52, 53], introduced in 1993 by Nihon Noyaku as a post-emergence broadleaf herbicide in cereals; and fused triazolinone rings such as azafenidin (**19**) [54, 55] from DuPont (Fig. 3.7).

3.2.3
Phenyl Tetrahydrophthalimide

Phenyl tetrahydrophthalimides represent the third class of early Protox herbicides. They were introduced in the early 1970s, after the diphenyl ether and 1,3,4-oxadiazol-2(3H)-one chemistries. Following the introduction by Mitsubishi of chlorophthalim (**3**) [27] in 1972, incorporation of the 2,4,5-trisubstituted-phenyl pattern in the 1980s resulted in the synthesis of highly active molecules such as S-23142 (**12**) [45], and S-23121 (**20**) [45, 56] (Fig. 3.8).

The tetrahydrophthalimide area of chemistry generated a great deal of interest between 1980 and 2000, with hundreds of patents issued by a wide range of agrochemical companies [29]. Once the fluorine group at the 2-phenyl position and the chlorine group at the 4-phenyl position were established as the optimum substituents for the aromatic ring, the 5 position of the phenyl ring and the tetrahydrophthalimide heterocycle became the target of intense research. A wide variety of oxygen (**21**), sulfur (**22**), amino (**23**), and carbonyl (**24**) derivatives at the 5 posi-

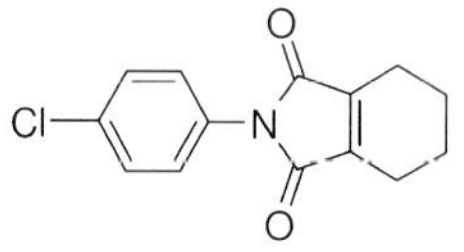

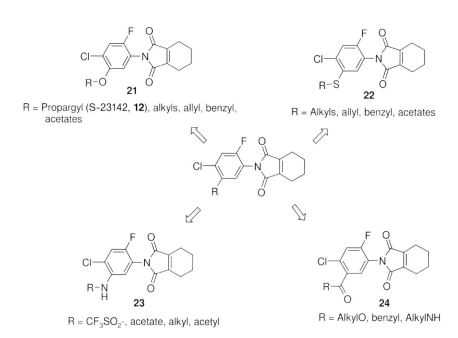

3

Chlorophthalim

1972 Mitsubishi

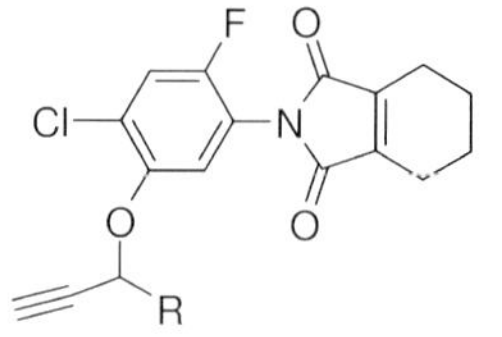

12 R = H   S-23142 1982 Sumitomo

20 R = CH₃ S-23121 1982 Sumitomo

Fig. 3.8. Chemical structures of chlorophthalim (**3**), S-23142 (**12**), and S-23121 (**20**).

tion of the aromatic ring were introduced [29] (Fig. 3.9). Some of these phenyl tetrahydrophthalimides reached the market, such as flumiclorac-pentyl (**25**) (Resource™) [57] and cinidon-ethyl (**26**) (Lotus™) [58] (Fig. 3.10).

The phenyl tetrahydrophthalimides are primarily post-emergence herbicides for the control of broadleaf weeds, though they will show pre-emergence activity at higher rates of application. Flumiclorac-pentyl is a post-emergence herbicide for the control of broadleaf weeds such as cocklebur, common lambsquarters, jimsonweed, amaranthus, prickly sida, and velvetleaf in soybean and corn at 25–

21

R = Propargyl (S-23142, **12**), alkyls, allyl, benzyl, acetates

22

R = Alkyls, allyl, benzyl, acetates

23

R = CF₃SO₂⁻, acetate, alkyl, acetyl

24

R = AlkylO, benzyl, AlkylNH

Fig. 3.9. Derivatization of aromatic position 5 of phenyl tetrahydrophthalimide.

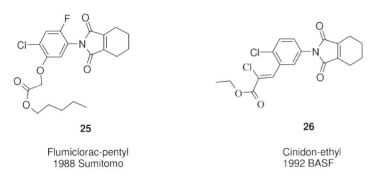

25

Flumiclorac-pentyl
1988 Sumitomo

26

Cinidon-ethyl
1992 BASF

Fig. 3.10. Chemical structures of flumiclorac-pentyl (**25**) and cinidon-ethyl (**26**).

50 g-a.i. ha^{-1}. Cinidon-ethyl is used for the post-emergence control of annual broadleaf weeds, particularly bedstraw, deadnettle, and speedwell, in winter and spring small grain cereals at 50 g-a.i. ha^{-1}.

In addition to the phenyl tetrahydrophthalimides discussed, several significant Protox herbicides with a phenyl ring attached to a fused thiadiazole[3,4-a]pyridazine or an oxazolidinedione ring were reported. Two examples of phenyl thiadiazole[3,4-a]pyridazine systems are fluthiacet-methyl (**27**) (Appeal®) [59] and NCI-876-648 (**29**) [60]. These compounds are said to act as pro-herbicides, converted in the plant into the corresponding phenyl triazolo[3,4-a]pyridazines [61] (Fig. 3.11).

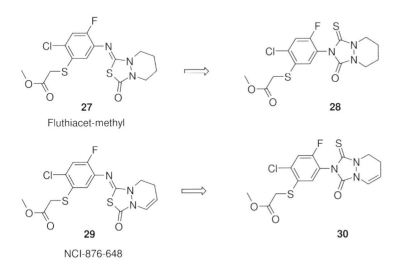

27

Fluthiacet-methyl

28

29

NCI-876-648

30

Fig. 3.11. Rearrangement of phenyl thiadiazolo[3,4-a]pyridazine to phenyl triazolo[3,4-a]pyridazines.

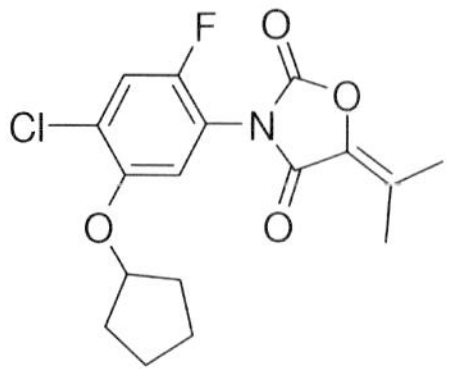

31

Pentoxazone

Fig. 3.12. Chemical structure of pentoxazone (**31**).

Phenyl oxazolidinedione chemistry is best exemplified by pentoxazone (**31**) (Wechser[®]) [62, 63] (Fig. 3.12) – discovered by Sagami for the pre-emergence control of weeds such as barnyard grass in rice at application rates of 200–450 g-a.i. ha^{-1}.

3.3
Non-classical Protox Chemistries

Several chemistries introduced in the 1990s did not conform to the established structure–activity relationships (SARs) of previous chemistries like the diphenyl ethers and the 2,4-dihalo-5-substituted-phenyl heterocycles. Figure 3.13 shows the SARs of 2-fluoro-4-chloro-5-substituted-phenyl heterocycles [5, 64]. These newer developments impacted both the aromatic and the heterocyclic portion of N-phenyl heterocycles.

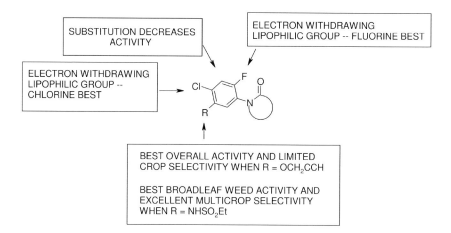

Fig. 3.13. Structure–activity relationships of the two aromatic rings of 2,4,5-trisubstituted-phenyl heterocyclic systems.

3.3.1
N-Phenyl Heterocycle – New Heterocyclic Systems

Many heterocyclic systems, usually attached to aromatic rings via a nitrogen or carbon atom, have been introduced in the past fifteen years. We have already discussed some of these heterocyclic rings, such as oxadiazolinone [25, 26], tetrahydrophthalimide [27], tetrazolinone [5], triazolinone [6–8, 48], pyrazole [51–

Table 3.1 Introduction of the uracil and pyridazinyl heterocyclic ring systems in the 1990s.

Ring system	Heterocycle		Heterocycle	Ring system
Oxadiazolinones				
Tetrahydrophthalimides				
				Uracil
Triazolinones		1990		
Tetrazolinones				Pyridazinone
Pyrazoles				
Oxazolidinediones				

53], and oxazolidinedione [62]. Extensive reviews of these heterocyclic systems, their properties, and their synthesis have been published [28, 29].

Of the several dozen new heterocyclic systems introduced during the period 1990–2005, the one that stands out as having the greatest impact in terms of significant increase in biological activity is 6-trifluoromethyl-2,4(1H,3H)-pyrimidinedione ring – commonly referred to as uracil – initially introduced by Hoffman-La Roche and Uniroyal [65, 66] (Table 3.1).

Replacement of the tetrahydrophthalimide and other heterocyclic rings such as the triazolinone **32** with the uracil **33** ring resulted in a significant improvement in biological activity, particularly when applied pre-emergently [67] (Table 3.2).

Because of this significant improvement in herbicidal activity, the uracil ring has become a standard ring in any N-aryl heterocycle Protox-related patent application. Initial SAR studies at the uracil nitrogen position showed that methyl and amino resulted in optimum activity [67]. Increasing the size and length of the R group in compound **34** resulted in a significant reduction in biological activity. Interestingly, both the lipophilic methyl group (**35**) and the hydrophilic amino group (**36**) are equally active (Table 3.3).

Three examples of molecules that contained the uracil ring and reached an advanced stage of development are benzfendizone (**41**) [68, 69], butafenacil (**42**) (Inspire®, Rebin®) [70], and flufenpyr-ethyl (**43**) [71] (Fig. 3.14). Benzfendizone is a post-emergence herbicide that provides good control of grass and broadleaf weeds in tree fruits and vines, acts as a cotton defoliant, and has applications in total vegetation control. The 6-trifluoromethyl group in the uracil ring is essential for biological activity; replacing it with methyl results in complete loss of activity.

Table 3.2 Comparison of biological activity of triazolinone and uracil heterocycles.

Weed species	Pre-emergent biological activity ED$_{85}$ (g-a.i. ha^{-1})	
	32	33
Morning-glory	395	22
Johnson grass	300	10

Table 3.3 Effects of uracil N-substituents on herbicidal activity of analogs of compound **34**.

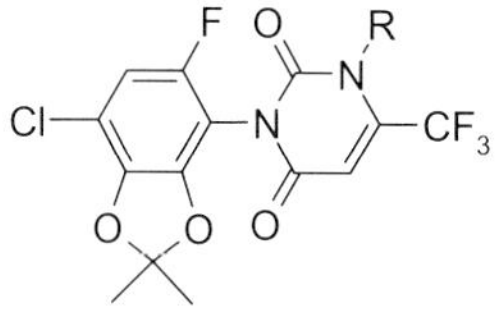

34

Compound	R	Pre-emergent biological activity ED$_{85}$ (g-a.i. ha^{-1})		
		Velvetleaf	Morning-glory	Green foxtail
35	CH$_3$	3	3	3
36	NH$_2$	3	3	3
37	CH$_2$CH$_3$	8	17	3
38	CH$_2$OCH$_3$	18	52	44
39	CH$_2$C$_6$H$_5$	958	>1000	307
40	CH$_2$CH$_2$CH$_3$	>1000	>10000	>1000

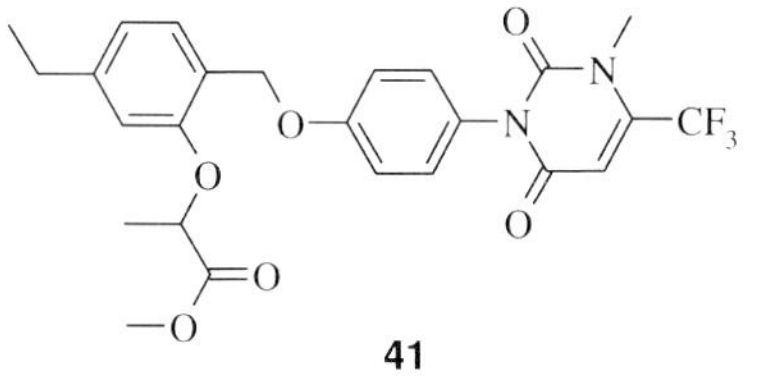

41

Benzfendizone

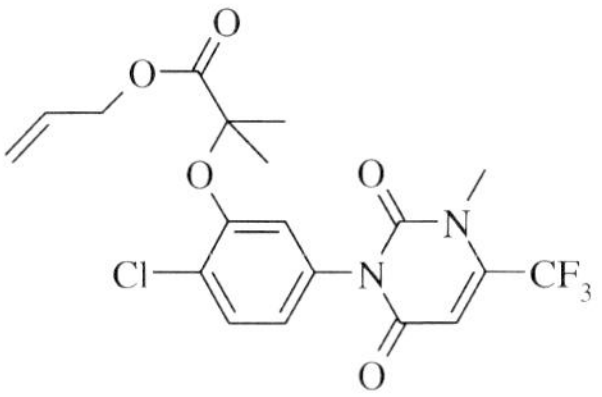

42

Butafenacil

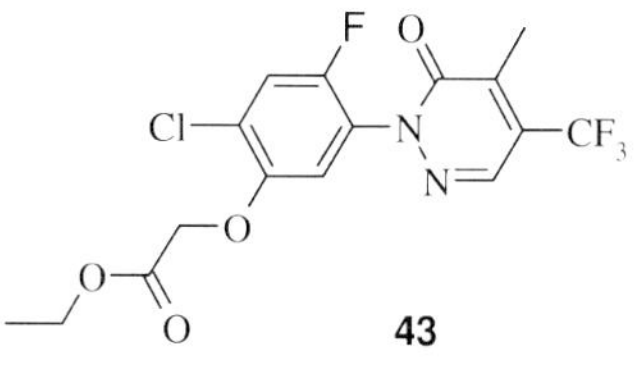

43

Flufenpyr-ethyl

Fig. 3.14. Chemical structures of benzfendizone (**41**), butafenacil (**42**), and flufenpyr-ethyl (**43**).

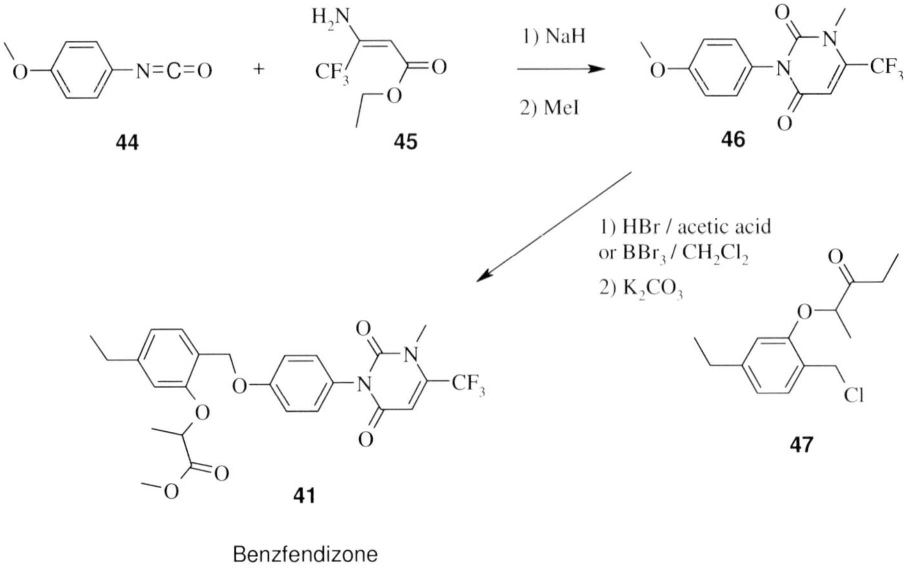

Scheme 3.1. Synthesis of the uracil heterocyclic ring of benzfendizone (41).

The uracil heterocycles are readily prepared in high yields from the corresponding aryl isocyanates 44 and from ethyl trifluoromethylaminocrotonate (45) in the presence of a base [72]. The uracil heterocycle is then directly N-methylated with methyl iodide in a one-pot reaction (Scheme 3.1). The uracil ring is stable to treatment with strong acids such as HBr and weak bases such as potassium carbonate, both reagents used in the further derivatization of the intermediate 46 with the benzyl chloride 47 to produce benzfendizone (41).

A less familiar ring system, but one that was part of a molecule selected for advanced testing, was the 5-methyl-6-oxo-4-(trifluoromethyl)-1-(6H)-pyridazinyl ring system of flufenpyr-ethyl (43). The pyridazinyl heterocycle can be prepared from the reaction of 4-chloro-2-fluoro-5-hydroxyphenyl hydrazine (48) and 1,1-dibromo-3,3,3-trifluoroacetone (49) to give the corresponding hydrazone 50, which when reacted with methyl malonic acid (51), in the presence of a base, provides the intermediate 52. Acid-catalyzed ring closure of 52, followed by O-alkylation of 53 with ethyl chloroacetate, results in the synthesis of flufenpyr-ethyl (43) [73] (Scheme 3.2).

3.3.2
Phenoxyphenyl and Benzyloxyphenyl Attached to Heterocycle

Extensive modeling and quantitative structure–activity relationship (QSAR) studies of Protox herbicides have been reported [12, 74, 75]. Earlier, it was postulated that Protox herbicides act by mimicking the protoporphyrinogen oxidase

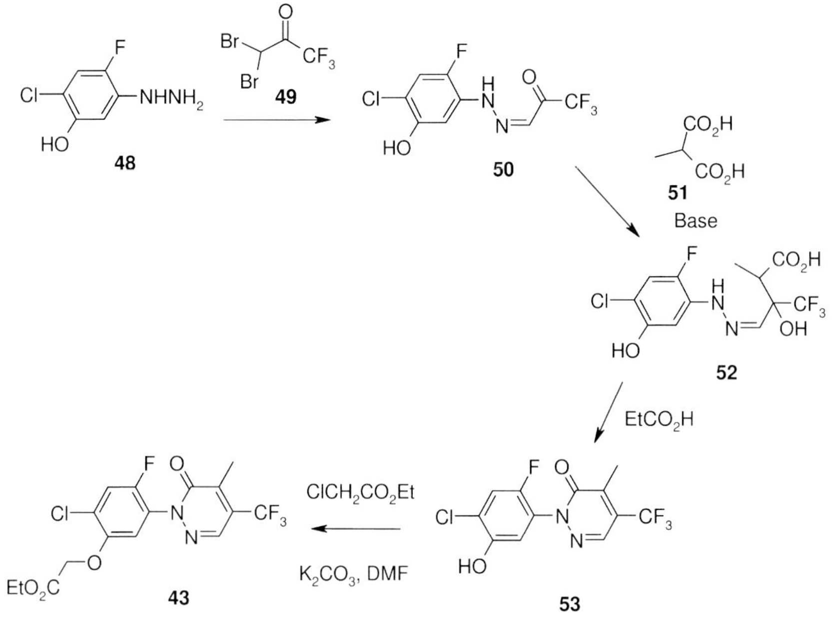

Scheme 3.2. Synthesis of the pyridazinyl heterocyclic ring of flufenpyr-ethyl (**43**).

substrate, protoporphyrinogen IX [76] (Fig. 3.15). This observation resulted in the discovery of the three-ring 2,4-dihalo-5-phenoxyphenyl propionate heteroaryl herbicide **54**, and later the heterocyclic phenoxymethylphenoxy propionate chemistry **56**, a significant improvement over the 4-chlorobenzyloxyphenyl heterocycles **55**. Both **54** and **56** are highly potent classes of Protox herbicides.

Subsequently, molecular modeling studies found good overlap between the diphenyl ether aromatic rings and protogen [77], as well as between a set of imide-type Protox inhibitors and protogen [78]. These studies were important in advancing the hypothesis that the diphenyl ethers mimicked protogen, though they were of limited practical value, having failed to reveal any new chemical structures.

A class of Protox inhibitors that redefined the accepted SARs and QSARs of the aromatic 4 position was the substituted benzyloxyphenyl heteroaryl area. As discussed earlier, SAR and QSAR studies of the phenyl ring of Protox herbicides demonstrated the need for halogens in the 2- and 4 positions of the phenyl ring, with the exception of the 4-chlorobenzyloxy group such as that of 4-chlorobenzyloxyphenyl tetrahydrophthalimide outlier **55** (Fig. 3.15) and reported by Ohta and coworkers in 1980 [79]. Chlorine at the para position of the benzyloxy was reported to provide optimum biological activity.

Further QSAR studies by Fujita [80] rationalized the high activity of this outlier 4-chlorobenzyloxy group by stating that the unexpected activity of the 4-benzyloxy ring was due to additional interactions of this group with the target enzyme.

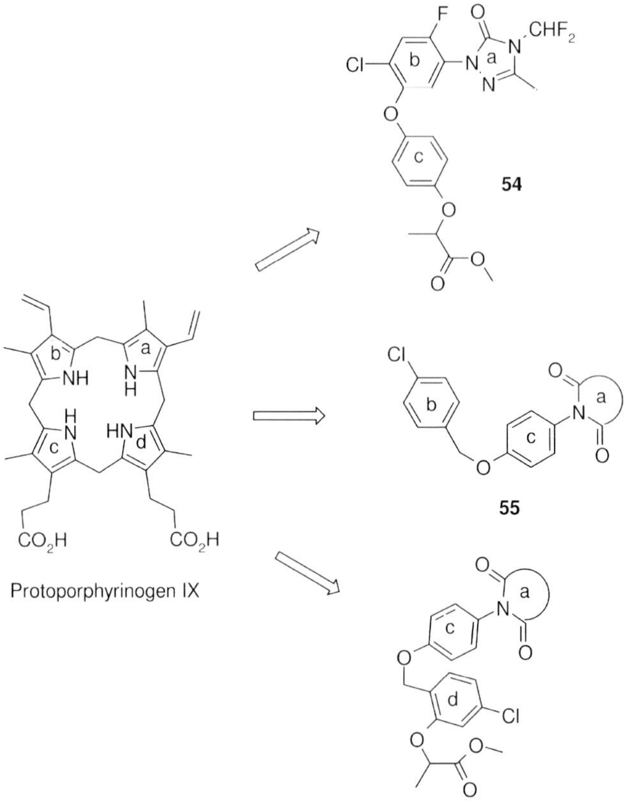

Fig. 3.15. Phenoxyphenoxy and benzyloxyphenyl as three-ring mimics of tetrapyrroles.

Given the strict steric and electronic requirements of groups at the 4 position of the phenyl ring, with chlorine as the optimum group, Fujita's explanation for the unexpected activity of a bulky electron-donating group such as the 4-benzyloxyphenyl is highly unlikely. In general, the presence of outliers in SAR or QSAR analysis indicates an unusual property of that group – such as a switch in the nature of the binding of the outlier in the enzyme site – and an opportunity for a major breakthrough. It was speculated that the activity of the 4-benzyloxy outlier was potentially due to a shift in binding mode for phenyl rings attached to heterocycles containing two flanking carbonyl rings, such as tetrahydrophthalimides and the newer uracil rings [64, 69]. Based on this new binding mode, the benzyloxy group could mimic the lipophilic portion of protoporphyrinogen IX, ring b, or the hydrophilic portion of protoporphyrinogen IX, ring d. Based on this working hypothesis, a series of compounds were prepared that contained an oxypropionate side chain, in addition to chlorine, in the benzyloxy group [69]. This work resulted in benzfendizone, a highly active broad-spectrum post-emergence herbicide.

3.3.3
Benzoheterocyclic Attached to Heterocycle

As discussed earlier, extensive studies of the 2, 4, and 5 positions of the phenyl ring of Protox inhibitors revealed very specific electronic, lipophilic, and steric requirements for chemical groups at these positions. Thus, it was rather surprising when it was discovered that it was possible to obtain highly active molecules by linking the 4 and 5 or the 5 and 6 positions of the phenyl ring to yield a wide variety of benzoheterocycles, such as those in Figs. 3.16 and 3.17.

Linking the 4 and 5 or 5 and 6 positions of the phenyl ring of Protox inhibitors to give a new heterocyclic ring resulted in two new classes of Protox herbicides, both with increased biological efficacy and new SARs. In the first instance, previous SAR studies of 2,4,5-trisubstituted-phenyl heterocycles have shown that position 2 of the phenyl ring required a halogen group for optimum biological activity, with chlorine and fluorine generating the highest overall activity. Position 4 of

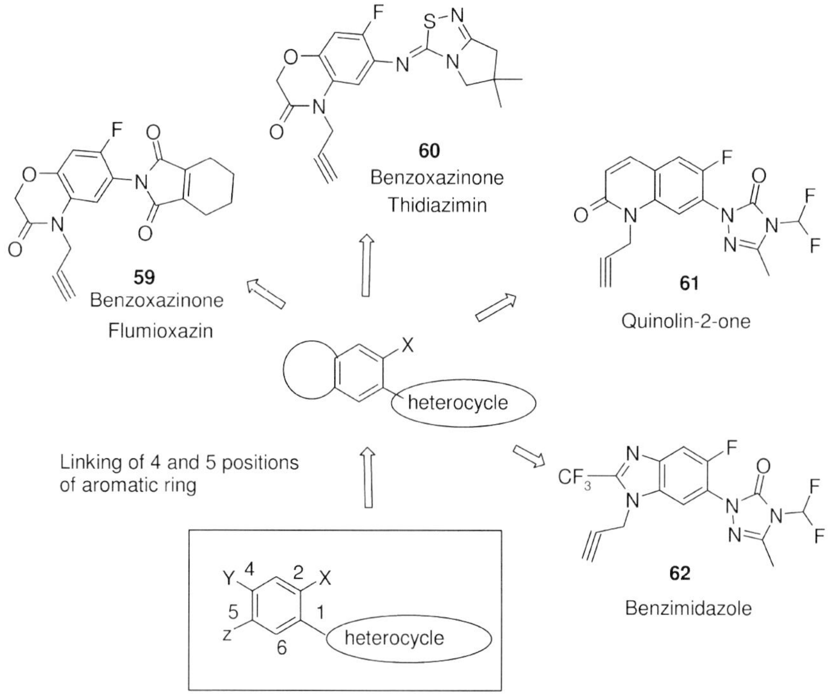

Fig. 3.16. Benzoheterocycles resulting from linking aromatic positions 4 and 5 of phenyl heterocyclic Protox inhibitors.

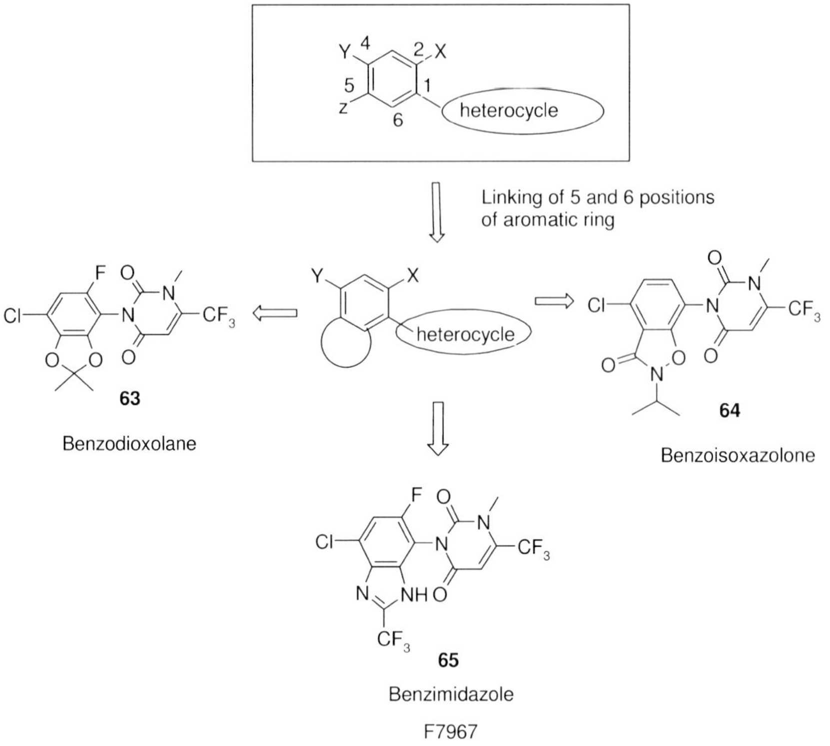

Fig. 3.17. Benzoheterocycles resulting from linking aromatic positions 5 and 6 of phenyl heterocyclic Protox inhibitors.

the phenyl ring required a hydrophobic, electronegative group such as halogen for optimum activity, with chlorine and bromine resulting in the best activity. Electron-donating groups such as methoxy resulted in significant loss of biological activity. The benzoxazinone SAR does not fit these rules, with compound **58** being far more active than its open chain analog **57** [81, 82] (Table 3.4).

Incorporating the benzoxazinone ring in Protox herbicides resulted in several commercial molecules, such as flumioxazin (**59**, Sumisoya®) [83] and thidiazimin (**60**) [84] (Fig. 3.16). Other heteroaryl rings include the quinolin-2-one **61** [82] and benzimidazole **62** [85].

The second class of benzoheteroaryl Protox herbicides are obtained when aromatic positions 5 and 6 are linked together to form various 3-(4,6-substituted benzoheterocyclyl) rings, which can be attached to a wide range of heterocycles (Fig. 3.17). The 3-(4,6-substituted benzoheterocyclyl) ring system represents a highly active area of Protox inhibitors, particularly when applied pre-emergently. The benzodioxolane uracil **63** provides complete control of pigweed, wild mustard, velvetleaf, green foxtail, and johnson grass at rates as low as 10 g-a.i. ha^{-1} when applied pre-emergently [72]. Other rings include benzoisoxazolone **64** [72]

Table 3.4 Comparison of biological activity of open vs. fused ring analogs.

57   58

Compound	Pre-emergent biological activity ED_{85} (g-a.i. ha^{-1})	
	Velvetleaf	Morning-glory
57	2000	>4000
58	62.5	125

and the corn and rice benzimidazole F7967, compound **65**, a new pre-emergence herbicide from FMC [86]. In pre-emergence applications, under greenhouse conditions, **65** controlled at rates as low as 10–30 g-a.i. ha^{-1} several broadleaf weeds – velvetleaf, morning-glory, pigweed, bindweed, nightshade, kochia, and chickweed – and grass weeds such as crabgrass, foxtails, johnson grass, and shattercane.

The SARs of these 3-(4,6-substituted benzoheterocyclyl) heterocycle herbicides differ from the more traditional 2,4-dihalo-5-substituted-phenyl heterocycles discussed earlier. As shown in Table 3.5, introducing a methoxy group at position 6 of compound **66** results in a dramatic loss of biological activity, the resulting com-

Table 3.5 Comparison of biological activity of open vs. fused ring analogs.

63   66   67

Compound	Pre-emergent biological activity ED_{85} (g-a.i. ha^{-1})	
	Velvetleaf	Green foxtail
63	3	4
66	6	9
67	32	143

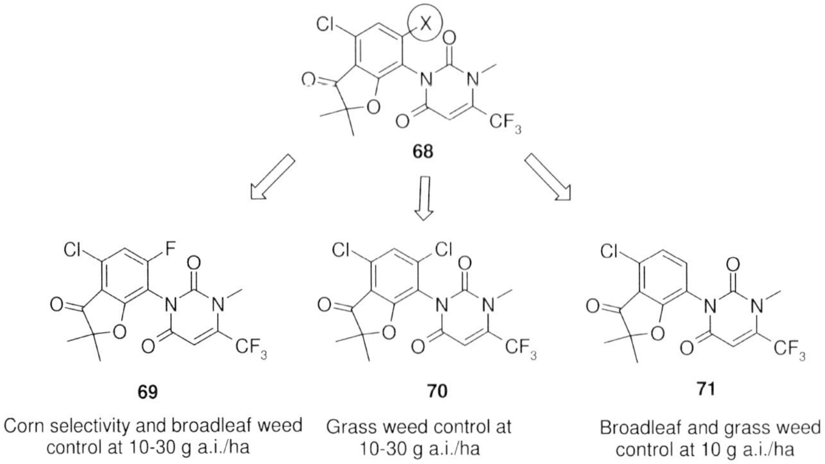

Fig. 3.18. Effect on biological activity of substituents at position 6 of the 2,3-dihydrobenzofuran ring.

pound **67** being more than five-fold less active than **66**. Linking together the aromatic positions 5 and 6 into a benzodioxolane ring resulted in compound **63**, which was more active than either compound **66** or **67** [72].

Substituents at the 6 position of the benzoheterocyclic ring had a dramatic effect on the weed spectrum and crop selectivity of these compounds when applied pre-emergently. First, as Fig. 3.18 shows, in the case of compound **68**, a fluorine at position 6 of the 2,3-dihydrobenzofuran ring gives compound **69**, which has excellent corn selectivity and control of broadleaf weeds (velvetleaf, wild mustard, and pigweed) at $10–30$ g-a.i. ha^{-1}. Next, replacing the 6-fluoro group with a chlorine resulted in a compound **70**, which has good grass weed control (barnyardgrass, green foxtail, and johnson grass) at $10–30$ g-a.i. ha^{-1}. Finally, compound **71**, with a hydrogen substituent at the 6 position, resulted in broad-spectrum control of both grass and broadleaf weeds at 10 g-a.i. ha^{-1}. Compounds **70** and **71** did not provide the same degree of corn selectivity as compound **69** [72].

3.3.4
Benzyl Attached to Heterocycle

This very interesting chemistry class of Protox inhibiting herbicides has received less attention than other Protox herbicides. It is the only class that introduces the use of a benzyl ring instead of a phenyl ring, and in so doing it has redefined the structure–activity of the aromatic ring. SAR studies of the benzyl uracil series resulted in compound **72**, with a 2,3,5-trisubstitution pattern of the phenyl ring, a significant difference from that of 3-phenyl-6-trifluoromethyluracils, where the optimum substitution pattern is that with substituents at the 2, 4, and 5 positions of the phenyl ring, as in **66** [87] (Fig. 3.19).

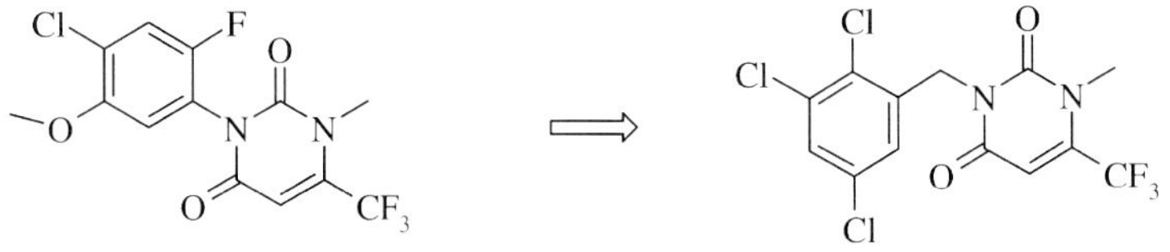

Fig. 3.19. Substitution patterns of phenyl and benzyl uracils.

3.3.5
Replacement of Phenyl Ring with Pyrazole

In this section we discuss the unusual replacement of the phenyl portion of Protox herbicides with a pyrazole ring to give pyrazogyl (**78**) [88], a rice herbicide initially discovered by Aventis, now Bayer CropScience.

There are several examples of Protox inhibitors in which the phenyl ring has been replaced with pyridine [89, 90] – a fairly common bioisosteric move – while preserving the 2,4-dihalo-5-substituted pattern in the heteroaromatic. Pyrazogyl (**78**) is unusual in that it involves several changes, such as the nature and placement of substituents, the size of the ring, and the replacement of phenyl with a heterocycle. It can be prepared in several steps from 2-hydrazino-4,5,6,7-tetrahydropyrazo[4,5-*a*]pyridine (**73**) and ethoxymethylenemalononitrile (**74**), followed by bis-chlorination of the pyrazole rings in **75**, N-methylation of **76**, and, finally, N-propargylation of **77** [88] (Scheme 3.3).

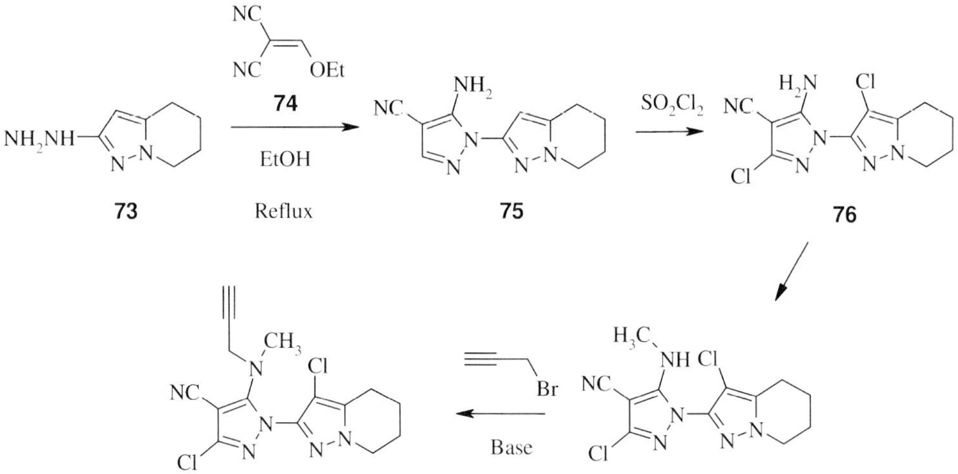

Scheme 3.3. Synthesis of pyrazogyl (**78**).

3.4
Recent Developments

Several reviews of Protox herbicides cover the period from the 1960s to 2002 [11–13, 28, 73, 91]. In this section we discuss Protox-related work conducted between 2003 and 2006. Following the momentous volume of research in all aspects of Protox herbicides – their chemistry, biology, biochemistry – in the decades between 1970 and 1990, work in this area of herbicide chemistry has significantly slowed in more recent years. Although corporations have continued to invest in Protox research, with several new structures introduced recently, none of these new chemical structures differ significantly from those already discussed.

The crystal structure of the mitochondrial protoporphyrinogen IX oxidase enzyme obtained from tobacco, and complexed with phenyl pyrazole Protox inhibitors, was published in 2004 [92]. As discussed in the introduction, the membrane-embedded protoporphyrinogen oxidase enzyme is the target of the Protox herbicides. It was also mentioned in Section 3.3.2 that molecular modeling studies of Protox inhibitors found good overlap between the diphenyl ether aromatic rings and protoporphyrinogen IX (protogen) [77], and between a set of imide-type Protox inhibitors and protogen [78]. The paper on the protoporphyrinogen IX oxidase crystal structure, a collaboration between the Max-Plank Institute, Bayer CropScience, and Proteros, discusses how the active site architecture suggests a specific substrate-binding mode that is compatible with the rare six-electron oxidation. It also proposes that the pyrazole ring of 4-bromo-3-(5-carboxy-4-chloro-2-fluorophenyl)-1-methyl-5-trifluoromethylpyrazole (**79**) matches ring A, and the phenyl ring matches ring B of protoporphyrinogen IX (Fig. 3.20).

In terms of recent patent activity related to Protox inhibitors, a series of N-substituted phenyl isothiazolone Protox herbicides were prepared to investigate the potential of the isothiazolone heterocycle ring to act as a bioisostere for comparable tetrahydrophthalimides such as compound **80** [93] (Fig. 3.21). The 2-(4-chloro-3-isopropoxycarbonyl)phenyl isothiazole-1,1-dioxide **83** was the most active

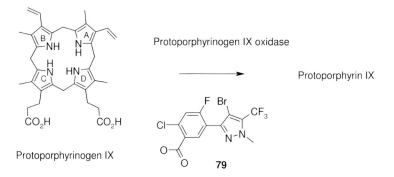

Fig. 3.20. Protox inhibitor 4-bromo-3-(5-carboxy-4-chloro-2-fluorophenyl)-1-methyl-5-trifluoromethylpyrazole (**79**) used in protoporphyrinogen IX oxidase binding studies.

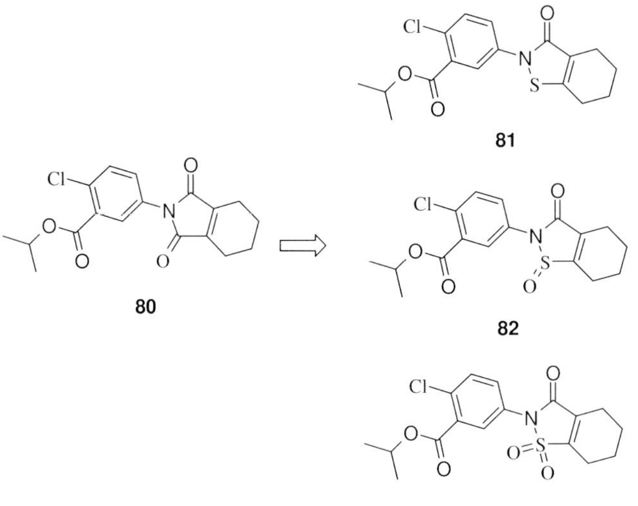

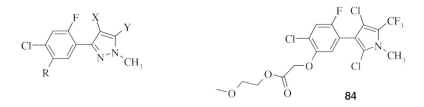

Fig. 3.21. Isothiazolone bioisosteres for tetrahydrophthalimide Protox inhibitors.

in the isothiazolone series as measured by inhibition of protoporphyrinogen IX oxidase isolated from corn, as well as by growth inhibition, chlorophyll decrease, and peroxidative destruction of cell membranes of green microalga *Scenedesmus acutus* [94]. Compound **83** was more active than either compound **81** or **82**, but about 100× weaker than the reference compound **80**.

Also published in 2004 were the synthesis and structure–activity of several 2-fluoro-4-chloro-5-substituted-phenyl pyrrole Protox herbicides, such as compound **84** [95]. This interesting pyrrole class of chemistry further extends the structure–activity of the 2-fluoro-4-chloro-5-substituted-phenylpyrazoles fluazolate (**17**) and pyraflufen-ethyl (**18**) discussed in Section 3.2.2 (Fig. 3.22).

R	X	Y	Compound	
CO₂CH(CH₃)₂	Br	CF₃	Fluazolate	17
OCH₂CO₂Et	Cl	OCHF₂	Pyraflufen-ethyl	18

R	X	Y	Compound	
$CO_2CH(CH_3)_2$	Br	CF_3	Fluazolate	**17**
OCH_2CO_2Et	Cl	$OCHF_2$	Pyraflufen-ethyl	**18**

Fig. 3.22. Phenyl pyrazole and phenyl pyrrole Protox inhibitors.

Compound **84** was extensively field tested in cereals and soybeans between 1999 and 2002 in France, Italy, and the United States. Post-emergence field application of **84** at 50 g-a.i. ha^{-1} demonstrated broadleaf weed control, with soybean tolerance, of morning-glory, redroot pigweed, and prickly sida. Soybean plants eventually outgrew initial injury at seven days after application. Field testing on winter wheat provided >80% control of several broadleaf weeds, including cleavers, at application rates of 50–60 g-a.i. ha^{-1}. The pyrazole **84** can be prepared in several steps starting from the 1,3-dipolar cycloaddition of 2-trifluoromethyl-3-methyl-1,3-oxazolium-5-olate **86** to 2-chloro-4-fluoro-5-ethynylphenol (**85**), followed by chlorination of the resulting pyrrole **87**, and reaction of **88** with the corresponding bromo acetate [95] (Scheme 3.4).

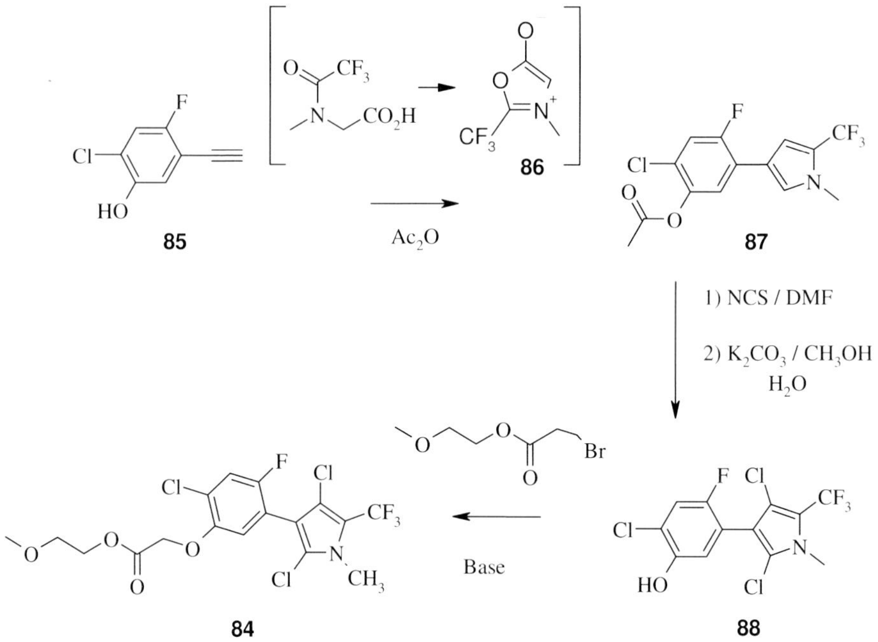

Scheme 3.4. Synthesis of phenyl pyrrole Protox inhibitors.

Another area related to fluazolate (**17**) and pyraflufen-ethyl (**18**) chemistry is a series of 2,4,5,6-tetrasubstituted-phenyl pyrazoles **89** (Fig. 3.23) from Ishihara Sangyo Kaisha [96]. These compounds differ from previous phenyl pyrazoles in that they have substituents at the 6 position of the phenyl ring. Pre-emergence application of **89** provided 100% control at 63 g-a.i. ha^{-1} of barnyardgrass, crabgrass, green foxtail, redroot pigweed, prickly sida, and velvetleaf. Soybean was reported to have 20% injury for compound **89** at this rate of application.

Several 2-phenyl-4,5,6,7-tetrahydro-2*H*-indazoles with several isoxazolinylmethoxy groups at the 5 position of the aromatic ring, such as compound **91**, were

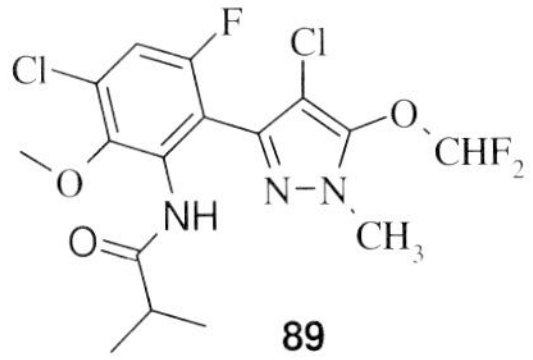

89

Fig. 3.23. Chemical structure of a tetrasubstituted-phenyl pyrazole.

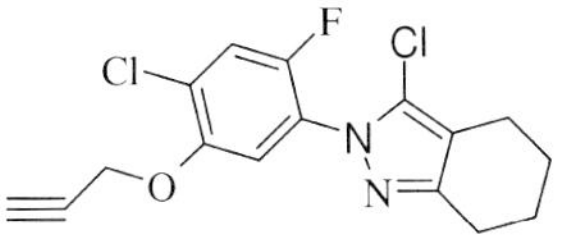

90

S-275

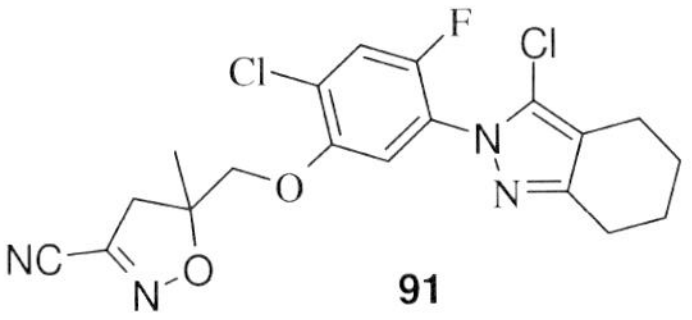

91

Fig. 3.24. Chemical structure of 2-fluoro-4-chloro-5-isoxazolinylmethoxy tetrahydroindazole **91**. Such compounds have the general structure of S-275 (**90**).

introduced by the Korea Research Institute of Chemical Technology [97, 98] as paddy rice herbicides (Fig. 3.24). These compounds have the general structure of S-275 (**90**), from Sumitomo [99]. Introduction of the isoxazolinylmethoxy groups at the 5 position of the aromatic ring is said to provide good broadleaf control with good tolerance by transplanted rice seedlings.

Herbicidal activity on several weeds, such as hairy beggarsticks, black nightshade, and knotweed, was reported in 2004 for a series of 2,4,5-imidazolidine triketones, such as compound **92** [100] (Fig. 3.25).

A series of four- and five-membered benzoheterocycle uracils derived from tying back the 4 and 5, as well as the 2 and 3, aromatic positions were disclosed. The benzoheterocycles obtained from linking aromatic positions 4 and 5 were developed by Bayer in 2003 [101]. The differentiating feature between these benzoheterocyclic uracils and earlier ones discussed in Section 3.3.1 is the replacement of the N-methyl group with an amino group in the uracil heterocycle, as exemplified by **93** and **94** (Fig. 3.26).

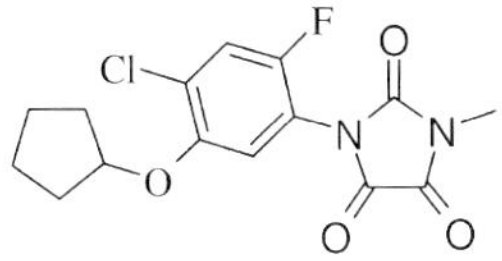

92

Fig. 3.25. Chemical structure of a 2-fluoro-4-chloro-5-alkoxy phenyl imidazolidine triketone Protox inhibitor.

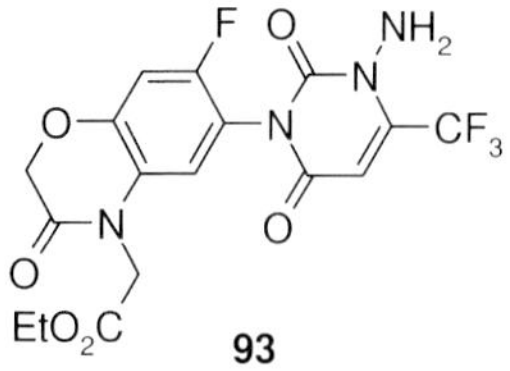

Fig. 3.26. Benzoheteroaryl N-amino uracil derivatives.

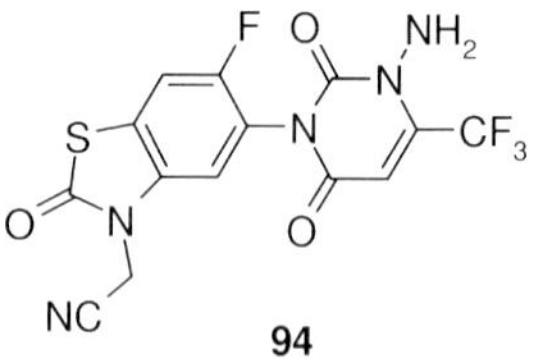

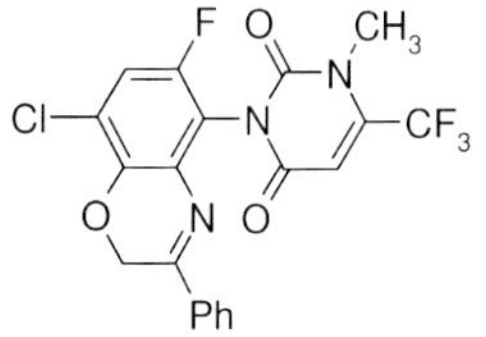

Fig. 3.27. Benzoheterocycle uracils.

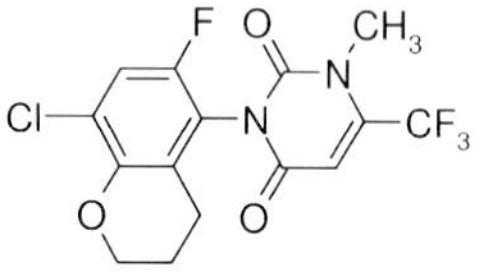

Ishihara Sangyo Kaisha disclosed a series of benzoheterocycles derived from linking the 2 and 3 aromatic positions, such as compounds **95** [102] and **96** [103] (Fig. 3.27).

Further derivatization at the 5 position of the phenyl ring of 2,4,5-trisubsitutedphenyl heterocycles has resulted in several new Protox herbicide patents. Ishihara Sangyo Kaisha introduced several benzohydrazide derivatives such as **97** [104] for use as herbicides, desiccants, and defoliants. Pre- and post-emergence control of several weeds, such as redroot pigweed, velvetleaf, sicklepod, ivyleaf morning-glory, and cocklebur, was demonstrated at application rates as low as 63 g-a.i. ha^{-1}. BASF reported the following new chemistries: the benzoic acid derivatives **98**, with good post-emergence activity in redroot pigweed and common lambsquarter, as well as potential use as cotton desiccants or defoliants [105], and **99** [106]; aminosulfonylamino phenyl uracil derivatives (**100**) [107]; and benzosulfonamides (**101**) [108]. Figure 3.28 shows these and other Protox inhibitors with diverse groups at the aromatic meta position, which are discussed below.

Bayer introduced 2-aryl-1,2,4-triazine-3,5-diones with the 2,4-dihalo-5-aminoalkylsulfonylphenyl, such as **102** [109] and **103** [110]; the aromatic substitution pattern is reminiscent of sulfentrazone (**15**).

In addition, in 2003 Bayer introduced phenyluracil derivatives with heteroaryl-methyleneoxy groups at the 5 position of the phenyl ring, as in **104** [111], and N-(thiocarbonylaminophenyl)uracils such as **105** and **106** [112].

Isagro Ricerca claimed good pre-emergence and post-emergence weed control at rates as low as 15 g-a.i. ha^{-1} for several Protox inhibitors with a wide variety of groups in the 5 aromatic position of 2,4-dihalo-5-substituted uracils, such as

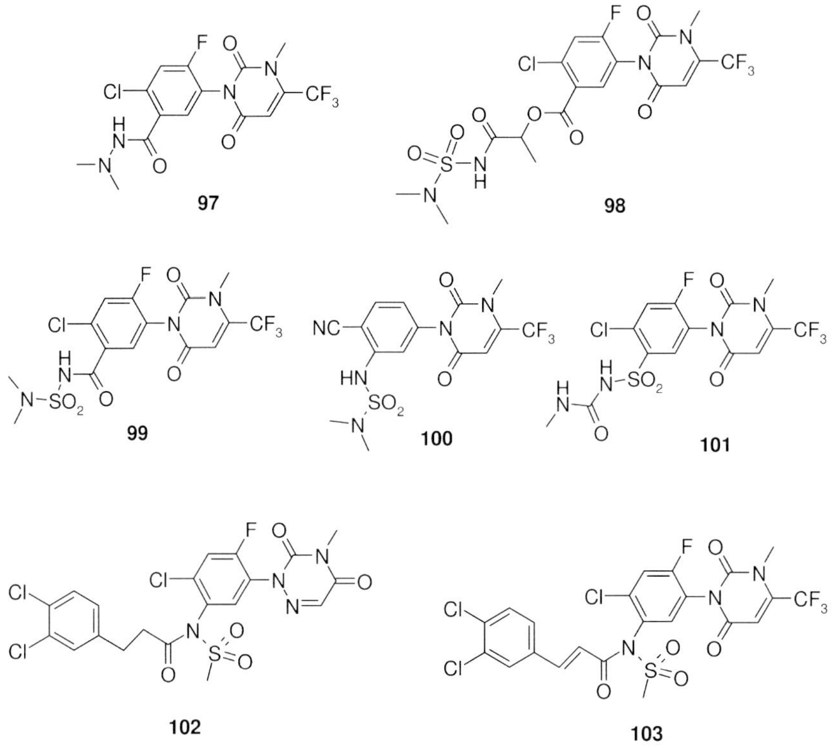

Fig. 3.28. Protox inhibitors with diverse groups at the aromatic meta position (see text for details).

107 and **108** (Fig. 3.28) [113]. Among the weeds controlled were bedstraw, barnyardgrass, redroot pigwccd, prickly sida, and velvetleaf, with crop selectivity in rice, wheat, barley, corn, and soybean.

Researchers at Central South University and Hunan Research Institute of Chemical Industry in Changsha, Hunan, China have reported the herbicidal activity of several isoindoline-1,3-diones molecules such as **109**, and compared their biological activity to flumioxazin (**59**) (Fig. 3.29). Compound **109** provided >80% control at 75 g-a.i. ha^{-1} in both pre- and post-emergence treatments against broadleaf weeds such as velvetleaf, common lambsquarter, and redroot pigweed, and against grass weeds such as large crabgrass, barnyardgrass, and green foxtail. Compound **109** was reported to be safe on cotton and corn at an application rate of 150 g-a.i. ha^{-1} when applied pre-emergently, and it also provided good wheat safety when applied post-emergently at 7.5–30 g-a.i. ha^{-1}. The IC$_{50}$ (inhibitive concentration, in g-a.i. ha^{-1}, to obtain 50% growth inhibition) values for the post-emergence control of velvetleaf and crabgrass were given for **109**, IC$_{50}$ = 3.6 and 4.8, respectively, and compared to those of flumioxazin, IC$_{50}$ = 1.0 and 2.5 [114].

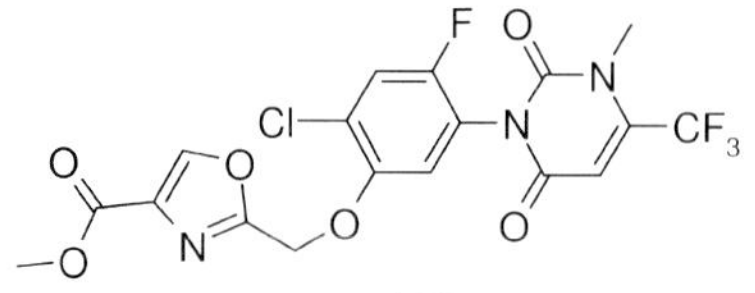

104

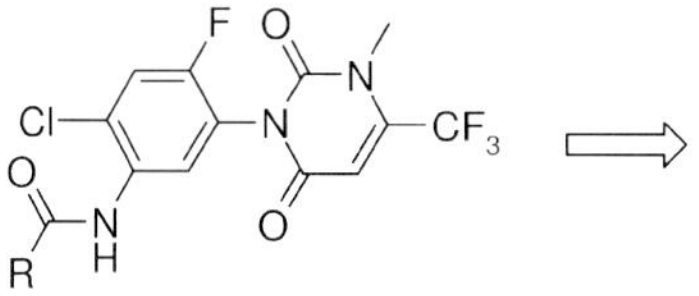

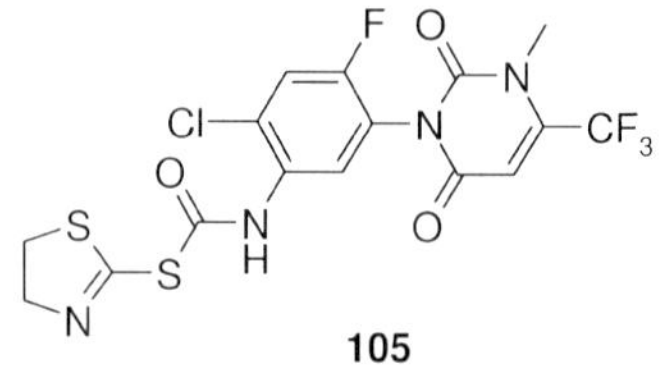

105

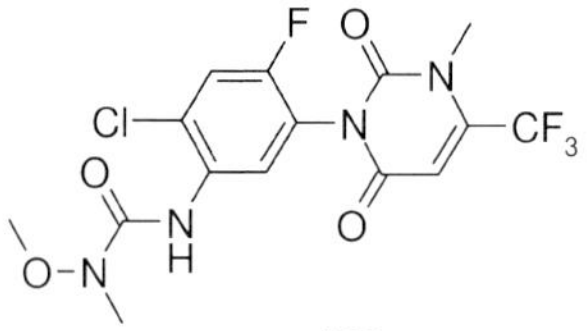

106

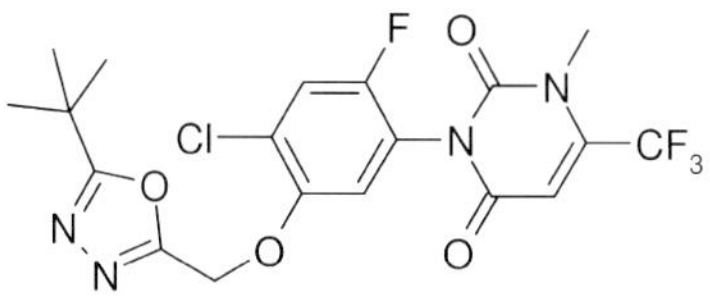

107

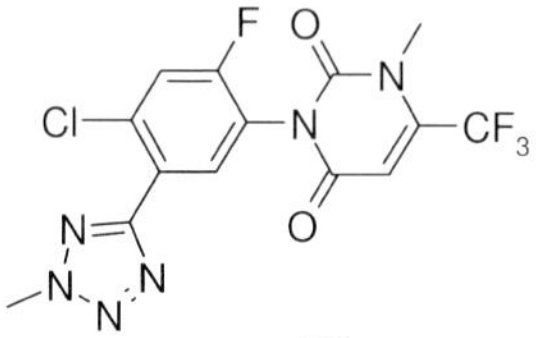

108

Fig. 3.28. (*continued*)

Bencarbazone (**114**) [115] is a recent Protox inhibitor triazolinone herbicide from Arvesta for the post-emergence control of broadleaf weeds in cereals and corn. It provides good control of bedstraw, velvetleaf, redroot pigweed, common lambsquarter, and speedwell at rates of application of 20–30 g-a.i. ha^{-1}. Bencarba-

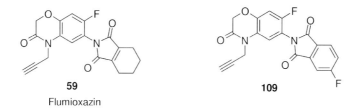

59
Flumioxazin

109

Fig. 3.29. Chemical structure of benzoxazine isoindoline-1,3-diones.

zone (**114**) has many of the features associated with Protox herbicides, particularly those of the Protox herbicide sulfentrazone (**15**). The most striking chemical feature of bencarbazone (**114**) is the replacement of the phenyl 4-chloro group with a thioamide group.

Bencarbazone **114** can be prepared in several steps from the nucleophilic displacement reaction of 2,4,5-trifluorobenzonitrile (**110**) with 4-methyl-3-trifluoromethyl-1,2,4-triazolin-5-one (**111**) to give 1-(4-cyano-2,5-difluorophenyl)-4-methyl-3-trifluoromethyl-1,2,4-triazolin-5-one (**112**). Reaction of **112** with ethanesulfonamide in the presence of a base such as potassium carbonate gives **113**, which on reaction with hydrogen sulfide gives bencarbazone (**114**) [116] (Scheme 3.5).

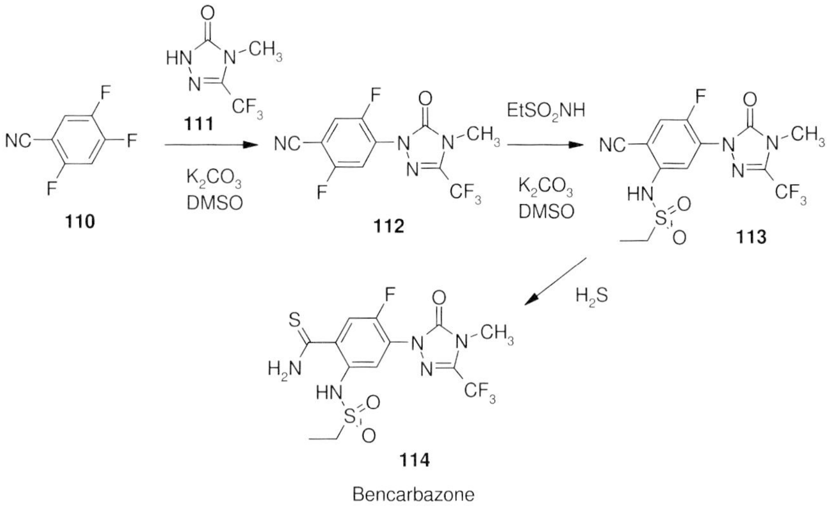

Scheme 3.5. Chemical structure and synthesis of bencarbazone (**114**).

3.5
Toxicology

The toxicology of Protox inhibitors has been discussed previously [13, 117]. It was shown that the addition of high doses of the Protox-inhibiting herbicides fomesafen, oxyfluorfen, and oxadiazon to the diet of mice increased the porphyrin content of liver, bile, and feces. The porphyrin accumulation induced by high-dose, short-term herbicide treatment is reversible. Within days after withdrawal of herbicide treatment, porphyrin levels returned to normal. Based on these findings – the high dose required to elicit an effect and the reversible nature of that effect – the authors, Krijt et al., concluded that the toxicological risk resulting from exposure to Protox-inhibiting herbicides is small [117].

3.6
Summary

Protox-inhibiting herbicides continue to be an area of interest to agrochemical companies, with most effort focused on fine tuning the 5 position of the aromatic ring of N-phenyl uracil to gain both a particular crop/weed/application method as well as a proprietary position.

In addition to Protox herbicide activity reported in the patent literature, there is continued interest in understanding the structure–activity relationships of Protox inhibitors [118–120]. Research efforts continue to be devoted to the development of Protox inhibitor-resistant crops [121]. In 1999, Syngenta announced its discovery of a novel gene technology, under the trademark Acuron™, that provides crops with tolerance to Protox inhibitors.

Finally, weed shifts observed in genetically modified crops, caused by the development of weed resistance to the widely used glyphosate herbicide, will offer market opportunities for herbicides with other modes of action, such as Protox-inhibiting herbicides.

References

1 D. Cartwright, D. J. Collins, US Patent 4,285,723 (**1981**).

2 S. R. Colby, J. W. Barnes, T. A. Sampson, J. L. Shoham, D. J. Osborn, 10th International Congress of Plant Protection, BCPC, Croydon, England, **1983**, 1, 295–302.

3 G. Theodoridis, US Patent 4,868,321 (**1989**).

4 G. Theodoridis, F. W. Hotzman, L. W. Scherer, B. A. Smith, J. M. Tymonko, M. J. Wyle, *Pestic. Sci.* **1990**, 30(3), 259–274.

5 G. Theodoridis, F. W. Hotzman, L. W. Scherer, B. A. Smith, J. M. Tymonko, M. J. Wyle, in *Synthesis and Chemistry of Agrochemicals III*, ed. D. R. Baker, J. G. Fenyes, J. J. Steffens, ACS Symposium series 504, ACS, Washington, D.C., **1992**, 122–133.

6 G. Theodoridis, US Patent 4,818,275 (**1989**).

7 G. Theodoridis, J. S. Baum, F. W. Hotzman, M. C. Manfredi, L. L. Maravetz, J. W. Lyga, J. M. Tymonko, K. R. Wilson, K. M. Poss, M. J. Wyle, in *Synthesis and Chemistry of Agrochemicals III*, ed. D. R. Baker, J. G. Fenyes, J. J. Steffens, ACS

Symposium series 504, ACS, Washington, D.C., **1992**, 134–146.

8 W. A. Van Saun, J. T. Bahr, G. A. Crosby, Z. A. Fore, H. L. Guscar, W. N. Harnish, R. S. Hooten, M. S. Marquez, D. S. Parrish, G. Theodoridis, J. M. Tymonko, K. R. Wilson, M. J. Wyle, *Proc. Br. Crop Prot. Conf. Weeds* **1991**, 1, 77.

9 I. Iwataki, in *Peroxidizing Herbicides*, ed. P. Böger, K. Wakabayashi, Springer-Verlag, Berlin, **1999**, 73–89.

10 V. K. Nandula, K. N. Reddy, S. O. Duke, D. H. Poston, *Outlooks Pest Manage.* **2005**, 16(4), 183–187.

11 H. Matsumoto, in *Herbicide Classes in Development*, ed. P. Böger, K. Wakabayashi, K. Hirai, Springer-Verlag, Berlin, **2002**, 151–161.

12 K. Wakabayashi, P. Böger, in *Peroxidizing Herbicides*, ed. P. Böger, K. Wakabayashi, Springer-Verlag, Berlin, **1999**, 163–190.

13 F. E. Dayan, S. O. Duke, Phytotoxicity of Protoporphyrinogen Oxidase Inhibitors: Phenomenology, Mode of Action and Mechanisms of Resistance, in *Herbicide Activity: Toxicology, Biochemistry and Molecular*

Biology, ed. R. M. Roe, J. D. Burton, R. J. Kuhr, IOS Press, Amsterdam, **1997**, 11–35.

14 R. Scalla, M. Matringe, *Rev. Weed Sci.* **1994**, 6, 103–132.

15 J. M. Jacobs, N. J. Jacobs, *Am. Chem. Soc. Sym. Ser.* **1994**, 559, 105–119.

16 R. Scalla, M. Matringe, J. M. Camadro, P. Labbe, *Z. Naturforsch.* **1990**, 45(5), 503–511.

17 S. O. Duke, J. M. Becerril, T. D. Sherman, J. Lydon, H. Matsumoto, *Pestic. Sci.* **1990**, 30(4), 367–378.

18 M. Matringe, R. Scalla, *Pest. Biochem. Physiol.* **1988**, 32(2), 164–172.

19 M. Matringe, R. Scalla, *Plant Physiol.* **1988**, 86(2), 619–622.

20 D. A. Witkowski, B. P. Halling, *Plant Physiol.* **1988**, 87(3), 632–637.

21 J. Lydon, S. O. Duke, *Pestic. Biochem. Physiol.* **1988**, 31(1), 74–83.

22 M. Matringe, J. M. Camadro, P. Labbe, R. Scalla, *Biochem. J.* **1989**, 260(1), 231–235.

23 M. Matringe, J. M. Camadro, P. Labbe, R. Scalla, *FEBS Lett.* **1989**, 245(1–2), 35–38.

24 H. F. Wilson, US Patent 3,080,225 (**1963**).

25 J. Metivier, R. Boesch, US Patent 3,385,862 (**1968**).

26 L. Burgaud, J. Deloraine, M. Guillot, M. Riottot, *Proc. 10ᵗʰ Brit. Weed Control Conf.* **1970**, 2, 745–751.

27 K. Matsui, H. Kasugai, K. Matsuya, H. Aizawa, FR 2119703 (**1972**).

28 R. J. Anderson, A. E. Norris, F. D. Hess, in *Porphyric Pesticides: Chemistry, Toxicology, and Pharmaceutical Applications*, ed. S. O. Duke, C. A. Rebeiz, ACS Symposium Series 559, ACS, Washington, D.C., **1994**, 18–33.

29 K. Hirai, in *Peroxidizing Herbicides*, ed. P. Böger, K. Wakabayashi, Springer-Verlag, Berlin, **1999**, 15–70.

30 R. Y. Yih, C. Swithenbank, *J. Agric. Food Chem.* **1975**, 23(3), 592–593.

31 R. J. Theissen, US Patent 3,652,645 (**1972**).

32 H. O. Bayer, C. Swithenbank, R. Y. Yih, US Patent 3,928,416 (**1975**).

33 W. O. Johnson, G. E. Kollman, C. Swithenbank, R. Y. Yih, *Agric. Food Chem.* **1978**, 26(1), 285–286.

34 W. O. Johnson, EP 20052 (**1980**).

35 Phillips McDougall AgriService – 2002 Market Report.

36 M. E. Condon, S. I. Alvarado, F. J. Arthen, J. H. Birk, T. E. Brady, A. D. Crews, P. A. Marc, G. M. Karp, J. M. Lavanish, D. R. Nielsen, T. A. Lies, in *Synthesis and Chemistry of Agrochemicals IV*, ed. D. R. Baker, J. G. Fenyes, G. S. Basarab, ACS Symposium Series 584, ACS, Washington, D.C., **1995**, 122–135.

37 P. Wepplo, J. H. Birk, J. M. Lavanish, M. Manfredi, D. R. Nielsen, in *Synthesis and Chemistry of Agrochemicals IV*, ed. D. R. Baker, J. G. Fenyes, G. S. Basarab, ACS Symposium Series 584, ACS, Washington, D.C., **1995**, 149–160.

38 G. M. Karp, M. E. Condon, F. J. Arthen, J. H. Birk, P. A. Marc, D. A. Hunt, J. M. Lavanish, J. A. Schwindeman, in *Synthesis and Chemistry of Agrochemicals IV*, ed. D. R. Baker, J. G. Fenyes, G. S. Basarab, ACS Symposium Series 584, ACS, Washington, D.C., **1995**, 136–148.

39 K. Moedritzer, S. G. Allgood, P. Charumilind, R. D. Clark, B. J. Gaede, M. L. Kurtzweil, D. A. Mischke, J. J. Parlow, M. D. Rogers, R. K. Singh, G. L. Stikes, R. K. Webber, in *Synthesis and Chemistry of Agrochemicals III*, ed. D. R. Baker, J. G. Fenyes, J. J. Steffens, ACS Symposium Series 504, ACS, Washington, D.C., **1992**, 147–160.

40 R. Boesch, J. Metivier, FR 1394774 (**1965**).

41 R. Boesch, J. Metivier, GB 1110500 (**1968**).

42 A. Blind, J. M. Cassal, R. Boesch, DE 2039397 (**1971**).

43 O. Wakabayashi, K. Matsuya, H. Ota, T. Jikihara, S. Susuki, DE 3013162 (**1980**).

44 S. J. Goddard, US Patent 536,322 (**1976**).

45 E. Nagano, S. Hashimoto, R. Yoshida, H. Matsumoto, H. Oshio, K. Kamoshita, EP 61741 (**1982**).

46 R. Dickmann, J. Melgarejo, P. Loubiere, M. Montagnon, *Proc. Br. Crop Prot. Conf. Weeds* **1997**, 1, 51–57.

47 R. Boesch, DE 2227012 (**1972**).

48 K. M. Poss, US Patent 5,125,958 (**1992**).

49 G. Theodoridis, J. T. Bahr, B. L. Davidson, S. E. Hart, F. W. Hotzman, K. M. Poss, S. F. Tutt, in *Synthesis and Chemistry of Agrochemicals IV*, ed. D. R. Baker, J. G. Fenyes, G. S. Basarab, ACS Symposium Series 584, ACS, Washington, D.C., **1995**, 90–99.

50 W. A. Van Saun, J. T. Bahr, L. J. Bourdouxhe, F. J. Gargantiel, F. W. Hotzman, S. W. Shires, N. A. Sladen, S. F. Tutt, K. R. Wilson, *Proc. Br. Crop Prot. Conf. Weeds* **1993**, 1, 19–28.

51 S. D. Prosch, A. J. Ciha, R. Grogna, B. C. Hamper, D. Feucht, M. Dreist, *Proc. Br. Crop Prot. Conf. Weeds* **1997**, 1, 45–50.

52 Y. Miura, M. Ohnishi, T. Mabuchi, I. Yanai, *Proc. Br. Crop Prot. Conf. Weeds* **1993**, 1, 35–40.

53 Y. Miura, H. Takaishi, M. Ohnishi, K. Tsubata, *Yuki Gosei Kagaku Kyokaishi* **2003**, 61(1) 4–15.

54 A. D. Wolf, US Patent 4,139,364 (**1979**).

55 R. Shapiro, R. DiCosimo, S. M. Hennessey, B. Stieglitz, O. Campopiano, G. C. Chiang, *Org. Process Res. Develop.* **2001**, 5(6), 593–598.

56 E. Nagano, S. Hashimoto, R. Yoshida, H. Matsumoto, K. Kamoshita, US Patent 4,484,941 (**1984**).

57 E. Nagano, S. Hashimoto, R. Yoshida, H. Matsumoto, K. Kamoshita, US Patent 4,770,695 (**1988**).

58 K. Grossmann, H. Schiffer, *Pestic. Sci.* **1999**, 55(7), 687–695.

59 T. Miyazawa, K. Kawano, S. Shigematsu, M. Yamaguchi, K. Matsunari, P. Porpiglia, K. G. Gutbrod, *Proc. Br. Crop Prot. Conf. Weeds* **1993**, 1, 23–28.

60 J. Satow, K. Fukuda, K. Itoh, T. Nawamaki, in *Synthesis and Chemistry of Agrochemicals IV*, ed. D. R. Baker, J. G. Fenyes, G. S. Basarab, ACS Symposium Series 584, ACS, Washington, D.C., **1995**, 100–113.

61 T. Shimizu, N. Hashimoto, I. Nakayama, T. Nakao, H. Mizutani, T. Unai, M. Yamaguchi, H. Abe, *Plant Cell Physiol.* **1995**, 36, 625–632.

62 K. Hirai, T. Futikami, A. Murata, H. Hirose, M. Yokota, US Patent 4,818,272 (**1989**).

63 K. Hirai, T. Yano, S. Ugal, T. Yoshimura, M. Hori, *J. Pestic. Sci.* **2001**, 26(2) 194–202.

64 G. Theodoridis, *Pestic. Sci.* **1997**, 50, 283–290.

65 J. Wenger, P. Winternitz, M. Zeller, WO 8810254 (**1988**).

66 A. Bell, US Patent 4,943,309 (**1990**).

67 G. Theodoridis, J. T. Bahr, S. Crawford, B. Dugan, F. W. Hotzman, L. L. Maravetz, S. Sehgel, D. P. Suarez, in *Synthesis and Chemistry of Agrochemicals VI*, ed. D. R. Baker, J. G. Fenyes, G. P. Lahm, T. P. Selby, T. M. Stevenson, ACS Symposium Series 800, ACS, Washington, D.C., **2002**, 96–107.

68 G. Theodoridis, US Patent 5,344,812 (**1994**).

69 G. Theodoridis, J. T. Bahr, F. W. Hotzman, S. Sehgel, D. P. Suarez, *Crop Protection* **2000**, 19, 533–535.

70 W. Kunz, U. Siegrist, P. Baumeister, WO 9532952 (**1995**).

71 T. Katayama, S. Kawamura, Y. Sanemitsu, Y. Mine, WO 9707104 (**1997**).

72 G. Theodoridis, US Patent 5,798,316 (**1998**).

73 T. Furukawa, EP 943610 (**1999**).

74 U. Nandihalli, S. O. Duke, in *Porphyric Pesticides: Chemistry, Toxicology, and Pharmaceutical Applications*, ed. S. O. Duke, C. A. Rebeiz, ACS Symposium Series 559, ACS, Washington, D.C., **1994**, 133–146.

75 F. E. Dayan, K. N. Reddy, S. O. Duke, in *Peroxidizing Herbicides*, ed. P. Böger, K. Wakabayashi, Springer-Verlag, Berlin, **1999**, 141–161.

76 G. Theodoridis, K. M. Poss, F. W. Hotzman, in *Synthesis and Chemistry of Agrochemicals IV*, ed. D. R. Baker, J. G. Fenyes, G. S. Basarab, ACS Symposium Series 584, ACS, Washington, D.C., **1995**, 78–89.

77 U. Nandihalli, M. V. Duke, S. O. Duke, *Pestic. Biochem. Physiol.* **1992**, 43(3), 193–211.

78 R. Uraguchi, Y. Sato, A. Nakayama, M. Sukekawa, I. Iwataki, P. Böger,

K. Wakabayashi, *J. Pestic. Sci.* **1997**, 22(4), 314–320.

79 H. Ohta, T. Jikihara, K. Wakabayashi, T. Fujita, *Pestic. Biochem. Physiol.* **1980**, 14(2), 153–160.

80 T. Fujita, A. Nakayama, in *Peroxidizing Herbicides*, ed. P. Böger, K. Wakabayashi, Springer-Verlag, Berlin, **1999**, 92–139.

81 G. Theodoridis, J. S. Baum, J. H. Chang, S. D. Crawford, F. W. Hotzman, J. W. Lyga, L. L. Maravetz, D. P. Suarez, H. Hatterman-Valenti, in *Synthesis and Chemistry of Agrochemicals V*, ed. D. R. Baker, J. G. Fenyes, G. S. Basarab, D. A. Hunt, ACS Symposium Series 686, ACS, Washington, D.C., **1998**, 55–66.

82 J. W. Lyga, J. H. Chang, G. Theodoridis, J. S. Baum, *Pestic. Sci.* **1999**, 55, 281–287.

83 E. Nagano, T. Haga, R. Sato, K. Morita, US Patent 4,640,707 (**1987**).

84 M. Ganzer, W. Franke, G. Dorfmeister, G. Johann, F. Arndt, R. Rees, EP 311135 (**1989**).

85 S. D. Crawford, L. L. Maravetz, G. Theodoridis, US Patent 5,661,108 (**1997**).

86 S. D. Crawford, L. L. Maravetz, G. Theodoridis, B. Dugan, US Patent 6,077,812 (**2000**).

87 M. J. Konz, H. R. Wendt, T. G. Cullen, K. L. Tenhuisen, O. M. Fryszman, in *Synthesis and Chemistry of Agrochemicals V*, ed. D. R. Baker, J. G. Fenyes, G. S. Basarab, D. A. Hunt, ACS Symposium Series 686, ACS, Washington, D.C., **1998**, 67–78.

88 G. Dorfmeister, H. Franke, J. Geisler, U. Hartfiel, J. Bohner, R. Rees, WO 09408999 (**1994**).

89 W. Kunz, K. Nebel, J. Wenger, WO 9952892 (**1999**).

90 K. Nebel, W. Kunz, J. Wenger, WO 9952893 (**1999**).

91 H. Matsumoto, in *Herbicide Classes in Development*, ed. P. Böger, K. Wakabayashi, K. Hirai, Springer-Verlag, Berlin, **2002**, 255–289.

92 M. Koch, C. Breithaupt, R. Kiefersauer, J. Freigang, R. Huber, A.

Messerschmidt, *EMBO J.* **2004**, 23, 1720–1728.

93 O. Yamada, M. Yanagi, F. Futatsuya, K. Kobayashi, GB 2071100 (**1981**).

94 Y. Miyamoto, Y. Ikeda, K. Wakabayashi, *J. Pestic. Sci.* **2003**, 28, 293–300.

95 G. Meazza, F. Bettarini, P. La Porta, P. Piccardi, E. Signorini, D. Portoso, L. Fornara, *Pest Manag. Sci.* **2004**, 60(12), 1178–1188.

96 H. Shimoharada, M. Tsukamoto, H. Kikugawa, Y. Kitahara, US Patent Application Publication 2005/0245399 (**2005**).

97 I. T. Hwang, H. R. Kim, D. J. Jeon, K. S. Hong, J. H. Song, C. K. Chung, K. Y. Cho, *Pest. Manag. Sci.* **2005**, 61(5), 483–490.

98 I. T. Hwang, K. S. Hong, J. S. Choi, H. R. Kim, D. J. Jeon, K. Y. Cho, *Pestic. Biochem. Physiol.* **2004**, 80(2), 123–130.

99 E. Nagano, I. Takemoto, M. Fukushima, R. Yoshida, H. Matsumoto, GB 2127410 (**1984**).

100 B. Li, J. Xu, Y. Man, CN 1515560 (**2004**).

101 O. Schallner, D. Hoischen, M. W. Drewes, P. Dahmen, D. Feucht, R. Pontzen, WO 2003006461 (**2003**).

102 M. Tsukamoto, S. Gupta, S.-Y. Wu, B.-P. Ying, D. A. Pulman, US Patent 6,573,218 (**2003**).

103 M. Tsukamoto, H. Kikugawa, M. Sano, US Patent Application 2004157738 (**2004**).

104 M. Tsukamoto, M. Read, US Patent 6,770,597 (**2004**).

105 M. Puhl, G. Hamprecht, R. Reinhard, I. Sagasser, W. Seitz, C. Zagar, M. Witschel, A. Landes, WO 2004009561 (**2004**).

106 C. Zagar, M. Witschel, A. Landes, WO 2004080183 (**2004**).

107 R. Reinhard, G. Hamprecht, M. Puhl, I. Sagasser, W. Seitz, C. Zagar, M. Witschel, A. Landes, WO 2004007467 (**2004**).

108 G. Hamprecht, M. Puhl, R. Reinhard, W. Seitz, C. Zagar, M. Witschel, A. Landes, WO 2004089914 (**2004**).

109 K.-H. Linker, R. Andree, D. Hoischen, H.-G. Schwarz, J. Kluth,

M. W. Drewes, D. Feucht, R. Pontzen, DE 10255416 (**2004**).

110 R. Andree, M. W. Drewes, P. Dahmen, D. Feucht, R. Pontzen, P. Loesel, WO 2003043994 (**2003**).

111 H.-G. Schwarz, R. Andree, D. Hoischen, K.-H. Linker, M. W. Drewes, P. Dahmen, D. Feucht, R. Pontzen, WO 2003099009 (**2003**).

112 H.-G. Schwarz, R. Andree, D. Hoischen, J. Kluth, K.-H. Linker, A. Vidal-Ferran, M. W. Drewes, P. Dahmen, D. Feucht, R. Pontzen, WO 2003093244 (**2003**).

113 G. Meazza, P. Paravidino, F. Bettarini, L. Fornara, WO 2004056785 (**2004**).

114 M.-Z. Huang, K.-L. Huang, Y.-G. Ren, M.-X. Lei, L. Huang, Z.-K. Hou, A.-P. Liu, X.-M. Ou, *J. Agric. Food Chem.* **2005**, 53, 7908–7914.

115 H.-J. Wroblowsky, R. Thomas, WO 9733876 (**1997**).

116 K.-H. Linker, K. Findeisen, R. Andree, M.-W. Drewes, A. Lender, O. Schallner, W. Haas, H.-J. Santel, M. Dollinger, US Patent 6,451,736 (**2002**).

117 J. Krijt, M. Vokurka, J. Sanitrák, V. Janousek, in *Porphyric Pesticides: Chemistry, Toxicology, and Pharmaceutical Applications*, ed. S. O. Duke, C. A. Rebeiz, ACS Symposium Series 559, ACS, Washington, D.C., **1994**, 247–254.

118 N.-D. Sung, J.-H. Song, K.-Y. Park, *Han'guk Eungyong, Sangmyong Hwahakhoeji* **2004**, 47(4), 414–421.

119 J. Wan, L. Zhang, G. Yang, C.-G. Zhan, *J. Chem. Inf. Comput. Sci.* **2004**, 44, 2099–2105.

120 L. Zhang, J. Wan, G. Yang, *Bioorg. Med. Chem.* **2004**, 12, 6183–6191.

121 X. Li, D. Nicholl, *Pest. Manag. Sci.* **2005**, 61(3), 277–285.

4
Herbicides with Bleaching Properties

4.1
Phytoene Desaturase Inhibitors

Gerhard Hamprecht and Matthias Witschel

4.1.1
Introduction

Herbicidal activity through inhibition of phytoene desaturase (PDS) can be easily detected by a striking whitening effect of tissues in newly grown plant leaves in the light. These symptoms led to their classification as "bleaching herbicides", i.e., herbicides interfering with the biosynthesis of photosynthetic pigments, chlorophylls or carotenoids [1, 2]. While norflurazon, as the oldest representative, was introduced by Sandoz as a spin-off of phenylpyridazinone chemistry (see Section 4.1.4.6) as early as 1968, it took almost two decades for the Mode of Action (MoA) – inhibition of PDS and consequently carotenoid biosynthesis – to become fully known. Since then, due to their low application rates, lack of resistance in the field – which could only be introduced genetically [3, 4] – and favorable mammalian toxicity, industrial research concentrated on this new MoA, leading to several potent herbicides for modern agriculture.

4.1.2
Carotenoid Biosynthesis and Phytotoxic Effects of Bleaching Herbicides

4.1.2.1 Targets for Bleaching Herbicides

Bleaching may be a result of photooxidative events generated within the plant cell or chloroplast, leading to the destruction of the plant pigments or direct inhibition of pigment biosynthesis, whereby carotenoid and chlorophyll formation is prevented [5].

With carotenoid biosynthesis, plastoquinone is involved as an electron acceptor, which we encounter further in photosynthetic electron transport [2]. An important precursor in the synthesis of plastoquinone, which also serves as a cofactor for the PDS enzyme, is homogentisic acid, which is formed from 4-hydroxyphenyl-pyruvate by 4-hydroxyphenylpyruvate dioxygenase (HPPD)

Modern Crop Protection Compounds. Edited by W. Krämer and U. Schirmer
Copyright © 2007 WILEY-VCH Verlag GmbH & Co. KGaA, Weinheim
ISBN: 978-3-527-31496-6

[6, 7]. Inhibition of plastoquinone biosynthesis through HPPD blockade, therefore, causes herbicidal and bleaching phytotoxicity symptoms similar to those of PDS inhibition [6, 8]. However, HPPD inhibition induces reduced growth and chlorosis, which can be antagonized by homogentisic acid. Additionally α-tocopherol synthesis – a scavenger of activated singlet oxygen – is blocked, leading ultimately to oxidation of the D1 protein chain with the nonheme iron of the photosystem II reaction center, oxidative tissue damage, and bleaching [6]. The story of the discovery of HPPD herbicides and the structural requirements for herbicidal diketones have been described [9, 10]; see also Chapter 4.2 [Hydroxyphenylpyruvate Dioxygenase (HPPD) the Herbicide Target] of this book. In addition to PDS and HPPD inhibitors, other herbicides became known for their bleaching properties: amitrole – an oldtimer herbicide, applied in the 1950s – and clomazone, both inhibiting an early step in carotenoid biosynthesis [6, 11, 12].

4.1.2.2 Carotenoids – Properties and Function

Carotenoids are constituents of the photosynthetic reaction centers and the light-harvesting complexes of the antennae [13], playing their role as redox intermediates in electron transfer processes of photosystem II [14] and as accessory pigments in light harvesting [5, 15].

The reaction centers are rich in β-carotene and in some plant species may also contain α-carotene. In contrast, the peripheral light-harvesting complex contains several xanthophylls, including lutein, violaxanthin and neoxanthin [5].

Carotenes play a vital role in the protection of the chloroplast against photooxidative damage. At high light intensities, the chlorophyll molecules are exposed to more light than they can direct into electron transport, leading to chlorophyll fluorescence as one way of energy offtake [15] and intersystem crossing of the excited singlet chlorophyll to the longer-lived triplet state as a second way of energy offtake [5]. This triplet-state chlorophyll can use its energy to convert molecular oxygen into the highly active and destructive singlet oxygen ($^1O_2$). The latter will lead to destruction of lipids, membranes, nucleic acids and whole tissues. As a result, the degradation of chlorophyll, depending on the intensity of illumination, leads to the typical bleaching symptoms in plants and decline of photosynthetic activity. Typically, only the newly formed green leaves are affected and fade away by bleaching. Carotenoids protect against this photosensitized damage by direct quenching of the excitation energy of triplet-state chlorophyll. Secondly, the carotenoid molecule can also quench any $^1O_2$ build up, producing carotenoid triplets, which then decay harmlessly, developing heat rather than toxic products [5, 15].

Besides their function as light collectors and photoprotectors, carotenoids also have important effects as membrane stabilizers in chloroplasts. The xanthophyll violaxanthin and its enzymatic de-epoxidation products antheraxanthin and zeaxanthin partition between the light-harvesting-complexes (LHCs) of PS I and PS II and the lipid phase of the thylakoid membranes, bringing about a decrease in membrane fluidity, an increase in membrane thermostability and a lowered susceptibility to lipid peroxidation [16].

4.1.2.3 Carotenoid Biosynthesis in Higher Plants

4.1.2.3.1 The Biosynthetic Pathway

Carotenoids of higher plants, algae, and fungi are C_{40} tetraterpenes biosynthesized by the well-known isoprenoid pathway [1, 5, 6, 8, 17, 18]. The early steps, involving the formation of the C_5 isoprenoid units and the subsequent synthesis of prenyldiphosphate intermediates, are common to all classes of terpenoids.

4.1.2.3.2 Early Steps and Formation of Phytoene

The first specific precursor for terpenoids in the cytoplasma is the C_6 molecule mevalonic acid (MVA), which is built via the classical acetate/mevalonate pathway and converted by a series of phosphorylating and decarboxylation reactions into C_5 isopentenyldiphosphate (IPP), the universal building block for chain elongation up to C_{20}. In the chloroplasts, the biosynthesis of IPP starts from glyceraldehyde-3-phosphate and pyruvate to give 1-deoxy-D-xylulose-5-phosphate (DOXP) via the non-mevalonate pathway as a recently detected alternative IPP route [19]. The reaction is catalyzed by the enzyme DOXP synthase and can be inhibited by a breakdown product of the herbicide clomazone [12].

   After 1,3-allylic isomerization of IPP to dimethylallyl pyrophosphate (DMAPP) by the enzyme IPP isomerase, another IPP unit is added to yield C_{10} geranylpyrophosphate (GPP).

   Subsequent addition of a second or third molecule of IPP leads to the formation of C_{15} farnesyl pyrophosphate (FPP) and the C_{20} geranylgeranyl pyrophosphate (GGPP). The chain elongation is a head-to-tail condensation process, which forms carbon–carbon bonds between C-4 of IPP and C-1 of the allylic substrate.

4.1.2.3.3 The Specific Carotene Pathway

The stages unique to carotenoid biosynthesis start with the formation of the C_{40} phytoene (7,8,11,12,7′,8′,11′,12′-octahydro-ψ,ψ-carotene) from two molecules of GGPP via the C_{40} intermediate prephytoene pyrophosphate (PPPP), from which phytoene with its central double bond is directly derived (Fig. 4.1.1). It is colorless, being formed by head-to-head condensation of two molecules of GGPP (all-trans) and obtained in all photosynthetic organisms as the 15-*cis*-phytoene [20]. The condensation is catalyzed by the enzymes PPPP synthase and phytoene synthase. Desaturation starts from the symmetrical phytoene on both of its identical halves to give, in a first step, phytofluene as an intermediate and then ζ-carotene, catalyzed by the enzyme phytoene desaturase (PDS). Further desaturation of the latter occurs by a stepwise sequence of reactions to form neurosporene and the maximally desaturated lycopene. At each stage two *anti*-hydrogen atoms from adjacent functions are lost by oxidation to extend the chromophore by two double bonds. Starting with three conjugated double bonds in phytoene, one ends up with 11 in lycopene. The other enzyme involved is ζ-carotene desaturase (ZDS), which catalyzes a closely similar desaturation to PDS (Fig. 4.1.1) [21].

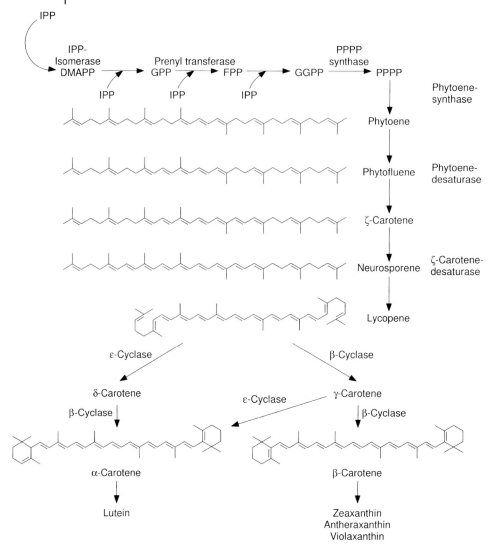

DMAPP dimethylallyl pyrophosphate; FPP farnesyl pyrophosphate; GPP geranyl pyrophosphate;
GGPP geranylgeranyl pyrophosphate; IPP isopentenyl pyrophosphate; PPPP prephytoene pyrophosphate.

Fig. 4.1.1. Pathway of carotene biosynthesis from IPP to α- and β-carotene.

4.1.2.3.4 **Cyclization**

Lycopene is the starting building block for the cyclization reactions to the final α-
and β-carotenes via their intermediates δ-carotene (with one ε-ionone ring) and
γ-carotene (with one β-ionone ring) respectively. Two different enzymes are responsible for the β- and ε-cyclization, called lycopene β- and ε-cyclase respectively
[22, 23].

In contrast to the 15-*cis* phytoene, the colored, fully desaturated carotenoids present in photosynthetic tissues are usually in the (all-ε) all-*trans* form, e.g., β-carotene or lutein. By hydroxylation of carotenes with molecular oxygen and in the presence of NADPH-dependent mixed-function oxygenase, hydroxy groups are introduced and epoxidation is another path for further derivatization, though little is known about the epoxidase involved [5]. Typical representatives are xanthophylls containing a hydroxy group at C-3 in the β- or ε-ring, violaxanthin (5,6,5′,6′-diepoxy-5,6,5′,6′-tetrahydro-β, β-carotene-3,3′-diol) or zeaxanthin (β,β-carotene-3,3′-diol). The importance of the violaxanthin–zeaxanthin cycle for both high rates of photosynthesis and energy dissipation has been described [5, 24].

4.1.2.3.5 Isolated Enzymes

Carotenoid biosynthesis takes place in a membrane-bound multienzyme complex, making it difficult to isolate and purify the enzymes involved. Owing to their sensitivity to detergents and low abundance, only a few have been purified from plant tissue. Many others had to be heterologously expressed in a way that high-pressure cell breaking resulted in a soluble and enzymatically active form [18, 25]. As an example, phytoene desaturase was cloned and expressed in recombinant *Escherichia coli*. To prepare the enzyme, the *E. coli* cells were disrupted by pressing them through a French Press. After centrifugation, the soluble supernatant fraction was used for enzymatic assays with HPLC recording or recording by optical absorption spectra [26].

4.1.3
Primary Targets

4.1.3.1 Inhibition of Phytoene Desaturase and ζ-Carotene Desaturase

Owing to the similarity of desaturation reactions catalyzed by PDS or ZDS, differentiation in the plant is not easy to detect. Most of the herbicidal inhibitors probably inhibit both, although to a different extent [6]. If strong inhibition of PDS has taken place with accumulation of phytoene, then the compound's ability to inhibit ZDS cannot be seen. Figure 4.1.2 shows that the commercial products primarily inhibit PDS [6, 8, 27–29]. Cell-free studies exemplified by norflurazon and fluridone have shown them to act as reversible noncompetitive inhibitors of PDS [27]. Other PDS active structures are shown below in Table 4.1.2 and in Section 4.1.4.10.

Direct interaction with the enzyme ζ-carotene desaturase was shown for the dihydropyrone LS-80707 and the pyrimidine SAN 380H [8]. Later, the compound RH 1965 and substituted 4-phenyl-3-benzylthio-4*H*-1,2,4-triazoles were reported to also inhibit ζ-carotene desaturation [30, 31].

4.1.3.2 Inhibition of Lycopene Cyclase (LCC)

Amitrole (3-amino-1*H*-1,2,4-triazole) has been known to lead to some lycopene accumulation *in vivo* at a temperature-dependent rate but it is not considered to

Inhibitors of phytoene desaturase (PDS)

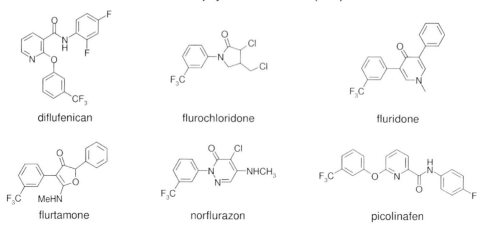

diflufenican   flurochloridone   fluridone

flurtamone   norflurazon   picolinafen

Inhibitors of ζ-carotene desaturase (ZDS)

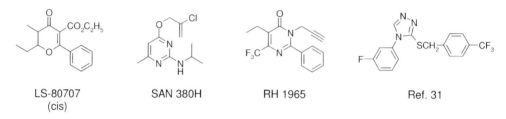

LS-80707   SAN 380H   RH 1965   Ref. 31
(cis)

Inhibitors of lycopene cyclase (LCC)

CPTA   MPTA

Fig. 4.1.2. Structure of commercial herbicides and some herbicidally active compounds that inhibit different enzymes in the biosynthetic pathway leading to the carotenoids.

be primarily an LCC inhibitor and may indirectly inhibit an early step in carotenoid biosynthesis [6, 11, 33]. The only specific more recent inhibitors are substituted diethylamines like CPTA and MPTA, which appear to inhibit both β- and ε-cyclase (Fig. 4.1.1) [6]. Their MoA is noncompetitive inhibition of lycopene cyclase versus lycopene [32].

In 2001, potent diethylamines were found as a new LCC inhibitor structural type [33]. Although very effective in seedling tests, the LCC inhibitors known so far have not shown sufficient activity for herbicide development.

4.1.3.3 Genetic Engineering of Herbicide Resistance by Modification of the Carotenogenic Pathway

The availability of numerous carotenogenic genes makes it possible to modify and engineer the carotenoid biosynthetic pathways in microorganisms and plants. Convenient tools for generation of mutants with a herbicide resistant PDS are unicellular cyanobacteria [3, 4, 7]. Various lines of resistant mutants of *Synechococcus* have been selected against norflurazon, showing not only a resistance factor of up to 70 but in most cases also cross-resistance to other PDS herbicides [7].

4.1.4
Chemical Structure and Activities of PDS Inhibitors

4.1.4.1 Enzyme Activity, Physical Data and Acute Oral Toxicity of Commercial PDS Herbicides

In recent years, structural evolution, detailed quantitative and qualitative structure–activity studies have been performed with a range of chemically different PDS inhibitors. Reference [27] reviews the early literature until about 1990. Subsequent years, to the late 1990s, is the topic of another review [34].

Table 4.1.1 presents IC_{50}s of commercial herbicides for inhibition of carotenoid biosynthesis obtained *in vivo* according to Ref. [26]. The assay is easier to run than the early radioactive approach with unicellular cyanobacteria [35], giving in three cases differing results. This may be caused by differences in target site sensitivity, uptake and translocation effects or metabolism of the herbicides in the treated bacteria cells of the early test assays. Table 4.1.1 also presents physical and acute oral toxicity data [36, 37].

4.1.4.2 Phenoxybenzamides

Removal of a p-nitro group from peroxidative diphenyl ethers drastically reduced their peroxidative activity while increasing the inhibition of carotenoid biosynthesis, provided a substituted formamide substituent is present in the meta-position (**1**, Fig. 4.1.3). Both the o- and p-derivatives are inactive (reviewed in Ref. [27]). Lipophilicity of the phenoxy ring and chain length of the alkyl group up to five carbon atoms increases activity, while branching results in a loss of activity. QSAR equations of the effect of the carbonamide substituent have been calculated [38]. No commercial product has been developed.

4.1.4.3 Phenoxypyridincarbonamides

Phenoxypyridinecarbonamides are surprisingly flexible, when the pyridine ring is substituted (review in Refs. [27, 34]). The first active pattern consisted of nicotinamides with a 2-phenoxy substitution (**3**, Fig. 4.1.4). For the latter, the m-position (R^1) was important with 3-CF_3 and 3-Cl being most active, while double substitution led to a decrease of activity. While small substituents R^2 such as H or CH_3 gave good herbicidal activity, Br or Cl were weaker. In the amide part, N-phenyl and N-benzyl derivatives showed comparable activity; ethylene as a spacer

Table 4.1.1 IC$_{50}$ values and physicochemical and oral toxicity data for commercial herbicides for carotenoid biosynthesis inhibition.

Structure	IC$_{50}$ (mol L^{-1})	log P (pH 7.5)	Vapor pressure (mbar)	M (g mol^{-1})	mp (°C)	LD$_{50}$ rats (mg kg^{-1})
diflufenican	3.40×10^{-8}	4.90	4.25×10^{-8}	394.3	159–161	>2000
flurochloridone	2.02×10^{-6}	3.36	4.40×10^{-6}	312.1	41 (eutectic)	4000
fluridone	2.93×10^{-7}	1.87	1.30×10^{-7}	329.3	154–155	>10000
flurtamone	8.21×10^{-7}	3.22	4.20×10^{-7}	333.3	152–155	500
norflurazon	5.18×10^{-7}	2.45	3.86×10^{-8}	303.7	174–180	>5000
picolinafen	8.98×10^{-8}	5.37	1.66×10^{-12}	376.3	107	>5000
beflubutamid	1.75×10^{-6}	4.28	1.10×10^{-7}	355.3	75	>5000

Enzyme values obtained from BASF Agricultural Research, other values taken from *The Pesticide Manual* [36] and SRC PhysProp Database [37].

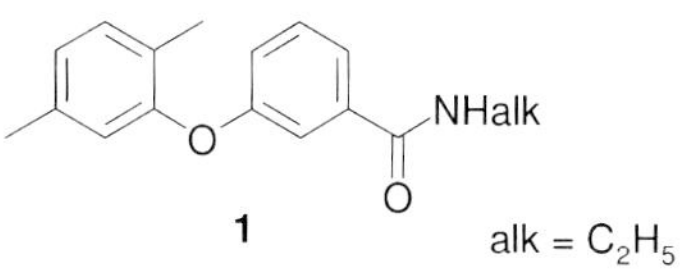

1
S 3422

Fig. 4.1.3. Phenoxybenzamide S 3422 (**1**).

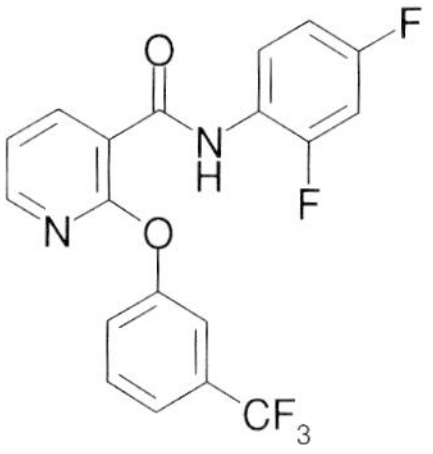

2
Diflufenican

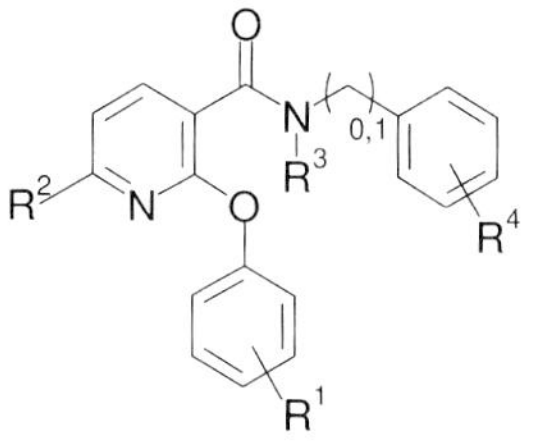

3
Substitution pattern of phenoxynicotinamides

Fig. 4.1.4. Diflufenican (**2**).

strongly decreased *in vitro* activity. Thioamides were slightly less active. Substitution of the amide hydrogen (R^3) by alkyl led to a decrease of activity parallel to their length. Single substitution of the N-phenyl ring (R^4) resulted in loss of activity, with the exception of the 4-F-moiety; the 2,4-difluoro derivative showed comparable activity to the unsubstituted compound. Most SAR contributions came from the laboratory of May & Baker, where diflufenican (**2**, Fig. 4.1.4) was found and later developed by Rhône-Poulenc [39, 40].

Researchers in the Shell laboratories later discovered a gap in the diflufenican patent, the 2,6-isomer **5**, which soon became very promising and led, after the acquisition by American Cyanamid and later BASF, to the marketing of picolinafen (**4**) in 2001 (Fig. 4.1.5) [41]. The discovery of the 2,6-pyridine cluster marked the beginning of a considerable number of follow-up patents to secure the new lead [34]. It could be shown that lower alkyl amino groups ($R^2 = CH_3$) may substitute

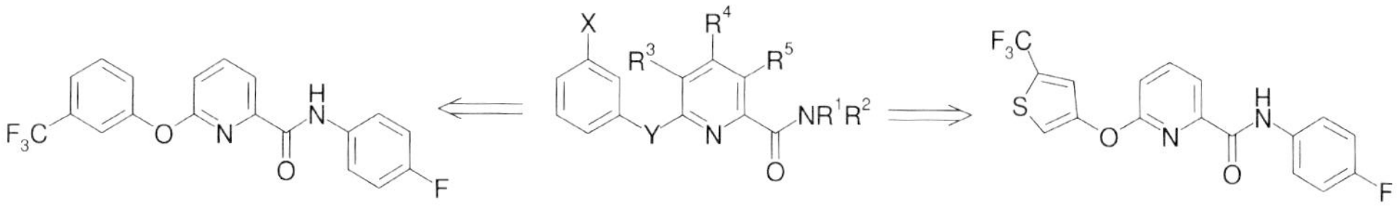

4 Picolinafen   **5**   **6**

Fig. 4.1.5. Picolinafen (**4**).

the 4-F- or 2,4-difluoroanilide moiety, while R^1 has to be H or CH_3. In addition R^3–R^5 is best with H or F. The 6-phenoxy unit could be replaced by 4-oxypyridine, 5-oxypyrazole [34] and 3-oxythiophene 6 (Fig. 4.1.5) [42]. The latter all need, again, a substituent (X) meta to the ether bridge (Y) and while the pyridine ether gives similar activity with Cl, a CF_3 group will be necessary for the phenoxy, 5-oxy-pyrazole and 3-oxy-thiophene unit.

Since the amide moiety could be totally replaced by aryl or hetaryl ethers and also directly substituted by pyrazol, several new combinations became possible, which led to the discovery of the phenoxypyridine ethers (Section 4.1.4.4).

4.1.4.4 Phenoxypyridine Ethers

In the 6-phenoxy moiety of this lead (8) a CF_3 substituent X proved best for herbicidal activity, which is also the case for 5-pyrazole-oxy and 3-thiophene-oxy substitution; see Ref. [34] and Table 4.1.2. In the 4-pyridyl-oxy group m-Cl and m-difluoromethoxy were another good choice of substituent. When 2,6-bisaryloxy-pyridines were synthesized, one phenyl group could be replaced by benzyl. Only a highly lipophilic aliphatic substituent R^4 such as trifluoromethylthiopropyl 14 (n = 1) could compete with the (hetero)-aryloxy compounds. In general, the best substituents for R^1 and R^3 were H and F. The substituent R^2 may be H, CH_3 and CH_3O and with 16 ($R^1 = R^3 = F$) brought a rise in activity. Interestingly, activity in structures 13, 15 and 16 is retained, even when replacing oxygen by a bond.

4.1.4.5 Phenylfuranones

The oldest phenylfuranone is difunon (17, Fig. 4.1.6), which turned out to be a rigid structure. Replacement of the 4-phenyl group by n-butyl, cyclohexyl and of the 3-CN group by carbonamide or an ester resulted in loss of activity (reviewed in Ref. [27]). Only the 3-position of the phenyl ring was tolerant of substitution such as $-SCH_3$, $-OCH_3$, $-C_6H_5$ and $-CF_3$, leading in some weeds to a rise of activity. The compound never became commercialized.

The other representative, flurtamone (18, Fig. 4.1.6) only has the 4-phenyl group and a basic side chain in common with difunon while the remaining substituents vary considerably. The best substituent of position 2 is phenyl; surprisingly it could be replaced by C_1–C_3-alkyl, but branching is unfavorable.

The 4-phenyl ring needs m-substitution by CF_3 whereas decreasing lipophilicity shows lower inhibition of PDS (reviewed in Ref. [27]).

4.1.4.6 Phenylpyridazinones

Pyridazinones turned out to be very flexible and, depending on position and substitution, show different MoA. While the cluster of the early chloridazon (19) is responsible for photosynthesis inhibition [51], BAS 10501W (20) inhibits fatty acid desaturation, changing the ratio of 18:2/18:3 fatty acids in plant membranes [27, 52], and norflurazon (21), finally, inhibits PDS (Fig. 4.1.7). Its inhibition went along with a CF_3-substituent R^3 in the phenyl moiety and small alkylamino groups R^2 in structure 22. Longer chains or branching lowered activity. While position 4 of the heterocycle (R^1) needs electron-withdrawing substituents, R^2 at C-5

Table 4.1.2 Structural evolution of phenoxypyridine ethers since 1994.

No.		Ref.	Year
9		43	1994
10		44	1996
11		45	1998
12		46	1999
13		47	2001
14		48	2001
15		49	2003
16		50	2003

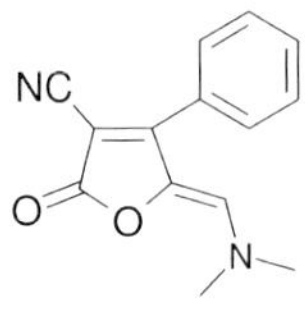

17
Difunon

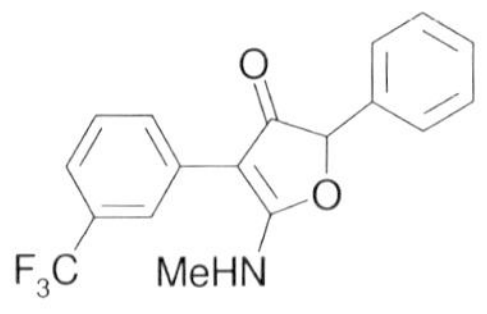

18
Flurtamone

Fig. 4.1.6. Difunon (**17**) and flurtamone (**18**).

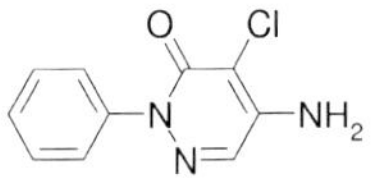

19
Chloridazon

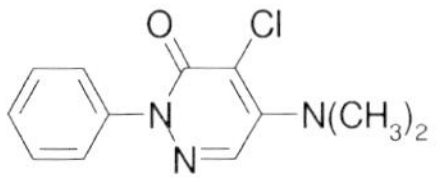

20
BAS 10501W

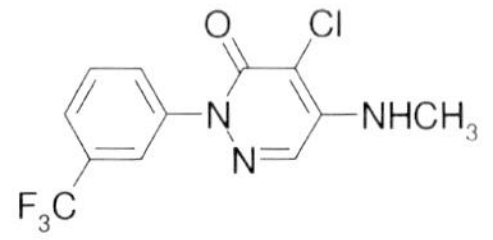

21
Norflurazon

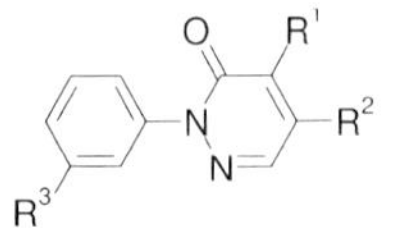

22

Fig. 4.1.7. Phenylpyridazinones.

has to be connected with electron-donating substituents, shifting electrons to-wards the heterocycle to increase activity (reviewed in Ref. [27]).

QSAR studies were performed with 2-phenylpyridazinones substituted at posi-tion 3 of the phenyl ring (R^3), where lipophilicity exerted a very strong effect on activity, counteracted by electronic properties. Steric factors did not show an influ-ence [27, 53]. The results were subsequently confirmed by new m-substituted de-rivatives [54]. Replacing the CF_3-group by fluorophenoxy or a fluorophenylalkyl side chain led to superior activity in spite of their much larger size. When Cl in R^1 was substituted later by a m-CF_3-phenyl group while R^2 was retained as CH_3NH, an early member of the diaryl heterocycle PDS inhibitor type with strong herbicidal activity was found (Section 4.1.4.10).

4.1.4.7 Phenylpyridinones

The pyridinone structure of fluridone (**23**, Fig. 4.1.8) is biologically rather inflexi-ble, so that the thiopyridinone and higher N-alkyl derivatives showed only little

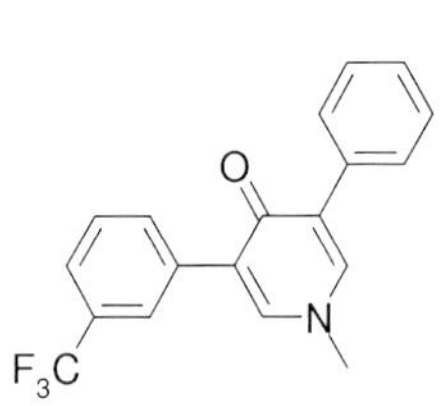

23
Fluridone

Fig. 4.1.8. Fluridone (**23**).

activity (reviewed in Ref. [27]). A m-substitution of one phenyl ring by the highly lipophilic CF_3 group is necessary, while exchange by Cl or CO_2H decreased activity. QSAR equations with whole cell data confirmed lipophilic and inductive effects; however, in vitro results only correlated to π [55]. To leave the pyridinone cluster, other contributors omitted the C=O group and continued with a series of 2,4-diphenylpyrimidines.

Although the hetero ring system of fluridone and flurtamone are completely different, their three-dimensional structures and projection to a common overlay gave rise to the concept of the "diaryl heterocyclic PDS inhibitors" of Section 4.1.4.10 [34].

4.1.4.8 Phenylpyrrolidinones

The most prominent representative of the phenylpyrrolidinones **25** is flurochloridone (**24**, Fig. 4.1.9). Again it needs an electron-withdrawing lipophilic substituent R^1 in the 3-phenyl position, such as 3-CF_3 or SCF_3, while CN or $SO_{1-2}CF_3$ were somewhat weaker (reviewed in Refs. [27, 34]). Activity ends with NO_2, NH_2 or C=O as substituents, which are no longer lipophilic and instead more prone to hydrogen bridging, with the exception of the NO_2 group. Surprisingly, high activity could be conserved by replacing Cl in R^2 with methyl and ethyl carbonamide. For reasons of activity, the chain length of R^3 is restricted to 2 and the 5-position (R^4) must be unsubstituted. Fluridone has two asymmetric carbons in the pyrrolidinone ring and 3,4-trans stereochemistry gives better herbicidal activity than the cis form. In the early 1990s, the 3-Cl was replaced by phenyl carrying an m-CF_3 group or halides in the 3–5-positions while varying R^1 and R^3. Among them 1-(3-isopropylphenyl)-3-phenyl-4-ethyl-2-pyrrolidinone was one of the herbicidally most active. When $R^2 = Cl$ was omitted, R^3 had to be CF_3. The lipophilic CF_3 in R^1 was also replaced by phenoxy units with different substituents or ring-anellated with the adjacent o-position into a 2',2'-difluordioxol-2,3-benzo ring.

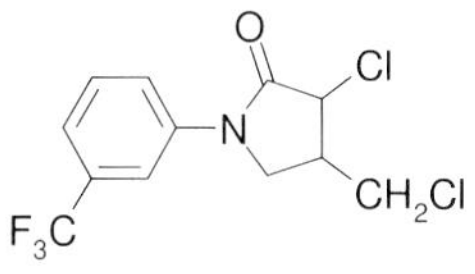

24
Flurochloridone

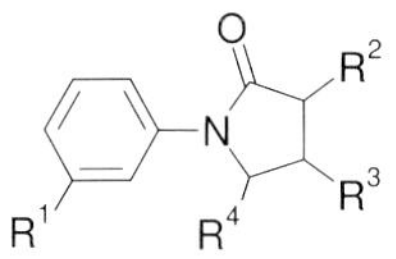

25

Fig. 4.1.9. Flurochloridone (**24**).

4.1.4.9 Phenyltetrahydropyrimidinones

Prerequisite for high herbicidal activity in phenyltetrahydropyrimidinones (**27**) is also the m-CF_3-substitution in one phenyl group (Fig. 4.1.10). Substitution of R^1 by an electron-withdrawing group shows the same biological ranking as in the other compound classes discussed before (reviewed in Ref. [27]). From the ring size of the heterocycle, a six-membered ring with X = $CHCH_3$ as optimum

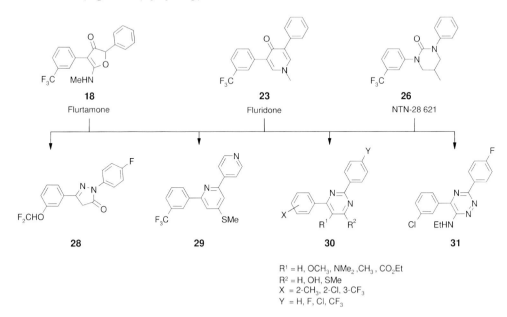

Fig. 4.1.10. Phenyltetrahydropyrimidinones.

shows the best results, while five- or seven-membered cyclic ureas are less effective. From their three-dimensional structure, there is a similarity between the phenyl-pyridinones (Section 4.1.4.7) and the saturated NTN-28621 (**26**, Fig. 4.1.10), which with its CH-CH$_3$ group indeed imitates the N-CH$_3$ group of fluridone and thus became a precursor for the compounds of Section 4.1.4.10 [34].

4.1.4.10 Structural Overlay for Diaryl Heterocycle PDS Inhibitors and Newer Developments

Structural overlay of flurtamone, fluridone and NTN-28621 led to a new pyrazolone **28** and pyridine **29**, many pyrimidines **30** and some 1,2,4-triazines **31** with the joint possession of 1,3-connected phenyl groups ("1,3-diaryl-heterocycle") (Fig. 4.1.11) [34]. A pyrimidine **30** with R^1 = CH$_3$O, R^2 = H, X = 3-CF$_3$ and

R^1 = H, OCH$_3$, NMe$_2$, CH$_3$, CO$_2$Et
R^2 = H, OH, SMe
X = 2-CH$_3$, 2-Cl, 3-CF$_3$
Y = H, F, Cl, CF$_3$

Fig. 4.1.11. Structural overlay of diaryl heterocycle PDS inhibitors and newer developments.

$Y = F$ has been reported with application rates as low as 63 g ha^{-1}. Other substituents for R^1 are NMe$_2$, CH$_3$ and CO$_2$Et while R^2 was H. In another series, R^1 was kept constant with hydrogen while R^2 could vary from OH to MeS and X was 2-CH$_3$, 2-Cl and 3-CF$_3$.

Along this line, Table 4.1.3 represents recent developments, from which **32–37** pursue substituted hetarylethers with **35** integrating its ether bridge into a heterocycle. Compounds **38–40** constitute classical pyrimidines, while pyrimidines **41** and **42** are purely aliphatically substituted. Notably, both also inhibit ZDS. Compounds **44** and **45** may be viewed as substituted phenylpyrrolidinones. The ketomorpholine **43**, the carbonamide **46** and the carbamate **47** are new PDS leads. The same holds for the pyrazolethers **36** and **37**.

4.1.4.11 Models of the Active Site – Structural Requirements

Few reports on models of the PDS herbicide binding site have appeared in the literature. The early QSAR equations correlated molecular properties like σ, π and steric parameters of one lead with its enzyme activity. Good results were obtained concerning the nature and position of substituents or when optimizing the chain length of a side chain, to get an approximate impression of electronic and steric prerequisites. However, activity prediction of structurally diverse molecules would not be possible. Also it should be kept in mind that QSAR does not take into account biological uptake, or stability against light and water at different pH, nor does it consider metabolization of a molecule in the plant, when activity in the field is desired.

A first hypothetical binding site model was proposed in studies of substituted 3(2H)-furanones [69]. Later, a steric model for the binding site of the PDS enzyme was developed by superposition of five commercial, structurally diverse inhibitors assumed to bind in the same way [64]. Conformational analysis was performed with the aid of three molecular mechanics programs to investigate three common regions in an orthogonal view: region **X** (phenyl ring preferentially substituted by the lipophilic CF$_3$ group), **Y** (central heterocyclic ring with an amide, vinylogous amide CO group), in which steric and electronic requirements appear to be relatively well defined and region **Z**, which appears to be sterically more tolerant. This model is similar to Fig. 4.1.12 when **A**, **B** and **C** are represented by **X**, **Y** and **Z**. It was used to predict the likely levels of activity of some analogues of the 6-ketomorpholine **43** and was able to show that the inhibitory activity resides almost exclusively with the (2R),(5S) form.

In 1999, another contribution described PDS active structure **34** and **35**, showing an overall similarity to the leads mentioned earlier [57]. In computational studies with SYBYL and a pharmacophore mapping approach putative receptor-bound conformations of benzoxazole and benzothiazoles differing from these were obtained, supporting coplanar geometry over the tilted conformation. In addition to previous findings, a functional group at the enzyme opposite to the central ring acting as either a hydrogen bond donor or an acceptor was suggested for interaction with the sp^2 hybridized nitrogen of the benzoxazole or the benzothia-

Table 4.1.3 Diaryl heterocycle PDS inhibitors and recent developments.

No.	Type	Ref. (year)	IC$_{50}$ (µM) Dose
Hetarylethers			
32		56 (1998)	17–140 g ha^{-1}
33		56 (1998)	
34		57 (1999)	0.75×10^{-9}
35		57 (1999)	0.35×10^{-8}
36		58 (2004)	9.6×10^{-7}
37	 KPP-856	59 (2004)	
Pyrimidines			
38		60 (2001)	1.4×10^{-6}
39		61 (2002)	5–10 g ha^{-1}

Table 4.1.3 (*continued*)

No.	Type	Ref. (year)	IC$_{50}$ (µM) Dose
40	 DPX-MY926	62 (2002)	10–15 g ha^{-1}
41		63 (2002)	8.6×10^{-7} ZDS 4.2×10^{-6}
42		63 (2002)	7.9×10^{-6} ZDS 6.5×10^{-5}

Saturated heterocycles

No.	Type	Ref. (year)	IC$_{50}$ (µM) Dose
43	 (2R), (5S)	64 (1995) 65 (2001)	1.7×10^{-7}
44		66 (2001)	1×10^{-9}
45		66 (2001)	4.8×10^{-7}

Aliphatic scaffolds

No.	Type	Ref. (year)	IC$_{50}$ (µM) Dose
46	 beflubutamid	67 (1999)	
47		68 (2003)	1.3×10^{-7}

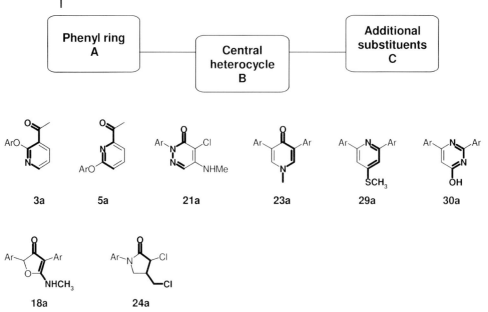

Fig. 4.1.12. Common structural elements of PDS inhibitors. (Modified from Ref. [7].)

zole heteroatoms. The herbicide binding site thus generated would be large and defined enough to fit most of the other inhibitors.

When comparing compounds with a modified central ring, it was concluded that the optimum inhibition of PDS is further reflected by a similar diagonal length from the negatively charged regions across the central heterocycle B carrying one or two substituted phenyl rings A and C (Fig. 4.1.12) [7]. A region of a total of six C- and hetero-atoms spans from one end of the substituted central heterocycle with C=O and N=C groups to the opposite site. In the nonplanar keto-morpholine 43, the optimum length is only five atoms with the 5-methyl group in the (5S)-form. The stereospecific inhibitory property of certain substituents of this heterocycle is another indication for the spatial requirement of this model. Most PDS inhibitors fit this description quite well.

However, some exceptions have to be noted. Norflurazon (21) is highly active and does not contain a ring C. In beflubutamid (46, Table 4.1.3) and in structure 47, an oxycarbonamide and oxyalkanecarbamate moiety, respectively, replace the central ring B by constituting a polar scaffold for rings A and C to interact with the binding niche, and in structures 41 and 42 only one central ring, without aromatic substituents, is left.

Finally, the question was raised whether inhibitors of PDS and ZDS can be modeled as analogues of plastoquinone because of their competitive behavior with the latter [70].

4.1.5
Biology and Use Pattern [36]

Diflufenican was introduced by May&Baker Ltd. (now Bayer CropScience). It is applied at 125–250 g ha^{-1} pre- or early post-emergence in autumn sown wheat and barley for the control of broad-leaved weeds. Degradation proceeds via the metabolites 2-(3-trifluormethylphenoxy)nicotinamide and 2-(3-trifluormethylphenoxy)nicotinic acid to bound residues and CO_2. DT$_{50}$ varies from 15 to 30 weeks, depending on soil type and water content.

Fluorochloridone was introduced by Stauffer Chemical Co. (now Syngenta AG). Rights were acquired by Agan Chemical Manufacturers Ltd. in 2002. It is applied pre-emergence at 250–750 g ha^{-1} for the control of weeds in winter wheat and winter rye, sunflowers and potatoes. It degrades in the soil under laboratory conditions, mostly forming CO_2 and a bound residue. DT$_{50}$ in the field is 9–70 days.

Fluridone was introduced by Eli Lilly&Co. (now Dow Agrosciences), and later sold to SePRO. It is used as an aquatic herbicide for control of most submerged and emerged aquatic plants in ponds, lakes, reservoirs, irrigation ditches, etc. Application rates are 45–90 ppb in ponds and 10–90 ppb in lakes. As an upland crop, only cotton has been found selective. In aquatic environments, degradation occurs mainly by photolytic processes, but microorganisms and aquatic vegetation are also factors. DT$_{50}$ in water (anaerobic) 9 months, (aerobic) about 20 days.

Flurtamone was introduced by Chevron Chemical Co. and later acquired by Rhône-Poulenc Agrochemical (now Bayer CropScience). It is incorporated preplant, pre-emergence or applied post-emergence for the control of broad-leaved and some grass weeds in small grains, peanuts, cotton, peas and sunflowers at 250–375 g ha^{-1}. The main metabolite is trifluoromethylbenzoic acid and, 10 months after application, no residues could be detected. Field DT$_{50}$ 46–65 days.

Norflurazon was introduced by Sandoz Ag (now Syngenta AG). It is used at 0.5–2 kg ha^{-1} for the pre-emergence control of grasses and sedges as well as broad-leaved weeds in cotton, Soya beans and peanuts, and at 1.5–4 kg ha^{-1} in nuts, citrus, vines, pomefruit, stone fruit, ornamentals, hops and industrial vegetation management. It dissipates in soil by photodegradation and volatilization, DT$_{50}$ about 6–9 months.

Picolinafen is the youngest commercial PDS herbicide and was discovered by Shell International Research, acquired by American Cyanamid Company (now BASF AG). It is used at 50 up to 100 g ha^{-1} as a cereal herbicide for the post-emergence control of broad-leaved weeds and marketed in mixtures with other cereal herbicides such as pendimethalin, isoproturon, and MCPA. It shows strong synergistic properties, for instance with pendimethalin, to also control grasses [71]. It is further registered in lupines. Metabolism proceeds via rapid hydrolytic cleavage of the amide bond. It is photochemically degraded in the environment; DT$_{50}$ 1 month.

Beflubutamid is under development by Ube Industries Ltd. Its intended use is alone or in mixture with isoproturon for pre- and early post-emergence control of broad-leaved weeds in wheat and barley at 170–255 g ha^{-1}. Soil DT$_{50}$ 5.4 days; the main metabolite is the corresponding butanoic acid, which itself is rapidly degraded in soil.

Table 4.1.4 summarizes an overview.

Table 4.1.4 Summary of application data of commerical PDS herbicides.

Chemical structure	Tradename (year)	ISO name Code No. Company	Dose (g ha^{-1}) Applic. method Target crops	Ref.
	Fenikan Tigrex (1987)	Diflufenican MB-38544 May & Baker Bayer CropScience	125–250 Pre, post Cereals	39
	Racer (1985)	Flurochloridone R-40244 Stauffer Chemical Syngenta	250–750 Pre Cereals, cotton, sunflowers, potatoes	72
	Sonar (1977)	Fluridone EL171 Elanco SePRO	500–2000 Pre, post Cotton aquatic herbicide 45–90 ppb	51 73
	Bacara (1997)	Flurtamone RE-40885 Chevron Bayer CropScience	250–375 Pre, post Cereals, peanuts, peas, cotton, sunflowers	74
	Solicam Zorial (1968)	Norflurazon H 9789 Sandoz Syngenta AG	500–2000 Pre Cotton, peanuts, soybeans, orchard	75 76
	Pico (2001)	Picolinafen AC 90001 ACC BASF AG	50–100 Post Cereals, lupines	41
	Herbaflex (2003)	Beflubutamid UBH-820 Ube under development	170–255 Pre, early post Wheat, barley	77

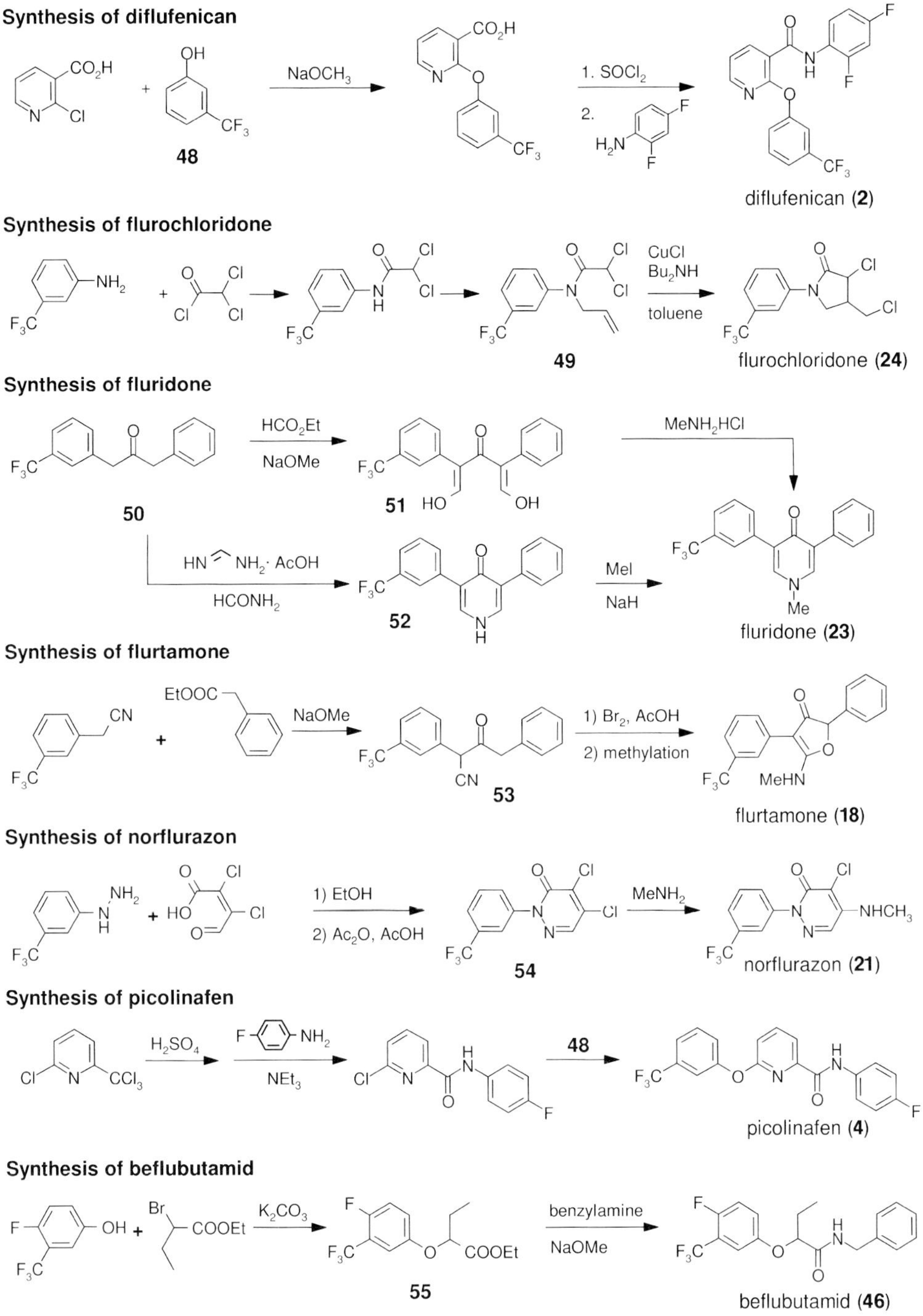

Synthesis of diflufenican

48

diflufenican (**2**)

Synthesis of flurochloridone

49

flurochloridone (**24**)

Synthesis of fluridone

50

51

52

fluridone (**23**)

Synthesis of flurtamone

53

flurtamone (**18**)

Synthesis of norflurazon

54

norflurazon (**21**)

Synthesis of picolinafen

48

picolinafen (**4**)

Synthesis of beflubutamid

55

beflubutamid (**46**)

Scheme 4.1.1. Major synthetic routes for diflufenican (**2**), flurochloridone (**24**), fluridone (**23**), flurtamone (**18**), norflurazon (**21**), picolinafen (**4**) and beflubutamid (**46**).

Engineering of resistance opens the possibility for obtaining tolerant plants to increase the crop spectrum beyond the scope described above [78]. Not only was it possible to confer resistance to tobacco but also to increase the carotenoid content in tomato fruits, rapeseed and rice [18].

4.1.6
Major Synthetic Routes for Phytoene Desaturase Inhibitors

Scheme 4.1.1 depicts the major synthetic routes, which are described below.

Diflufenican is synthesized by nucleophilic substitution of 2-chloronicotinic acid with 3-hydroxybenzotrifluoride and further reaction with thionyl chloride and 2,4-difluoroaniline to the final product [39].

Flurochloridone is made by copper chloride-catalyzed cyclocondensation of N-allyl-(3-trifluormethylphenyl)dichloroacetamide **49** [34].

Fluridone is accessible in two ways. 1-(3-Trifluormethylphenyl)-3-phenyl-2-propanone (**50**) is reacted with ethyl formate in the presence of a base to yield the diformyl derivative **51**, which is cyclized with methylamine to the final product. Alternatively, **50** is condensed with formamidine acetate in the presence of formamide to the 4(1*H*)-pyridone intermediate **52** and then methylated to give fluridone [34].

Flurtamone is prepared by cyclization of 4-phenyl-2-(3-trifluormethylphenyl)-3-oxobutyronitrile (**53**) with bromine in the presence of acetic acid and methylation to the final heterocycle [34].

Norflurazone production is based on the condensation of 3-trifluormethyl-phenylhydrazine and mucochloric acid followed by cyclization with acetic anhydride to give 4,5-dichloro-2-(3-trifluormethylphenyl)pyridazin-3-one (**54**). Nucleophilic substitution of **54** with methylamine yields norflurazon [34].

Picolinafen is built on the partial hydrolysis of 2-chloro-6-trichlormethyl-pyridine, reaction with 4-fluoroaniline and subsequent nucleophilic substitution with 3-hydroxybenzotrifluoride [79].

Beflubutamid is synthesized from ethyl 2-(4-fluoro-3-trifluormethyl-phenoxy)butanoate **55** and benzylamine in the presence of a base [34].

References

1 G. Sandmann, P. Böger, in *Target Sites of Herbicide Action*, P. Böger, G. Sandmann (Eds.), CRC Press, Boca Raton, FL, **1989**, 26–44.

2 K. Wakabayashi, P. Böger, *Pest. Manag. Sci.* **2002**, 58, 1149–1154.

3 U. Windhövel, B. Geiges, G. Sandmann, P. Böger, *Plant Physiol.* **1994**, 104, 119–125.

4 U. Windhövel, G. Sandmann, P. Böger, *Pestic. Biochem. Physiol.* **1997**, 57, 68–78.

5 A. J. Young in *Herbicides*, N. R. Baker, M. P. Percival (Eds.), Elsevier Science Publishers B.V., Amsterdam, **1991**, 132–171.

6 C. Fedtke, S. O. Duke in *Plant Toxicology*, B. Hock, E. F. Elstner

(Eds.), Marcel Dekker, New York, **2005**, 271–330.

7 G. Sandmann in *Herbicide Classes in Development*, P. Böger, K. Wakabayashi, K. Hirai (Eds.), Springer-Verlag, Berlin, Heidelberg, **2002**, 43–57.

8 P. Böger, G. Sandmann, *Pestic. Outlook* **1998**, 9, 29–35.

9 D. L. Lee, M. P. Prisbylla, T. H. Cromartie, D. P. Dagarin, S. W. Howard, W. McLean Provan, M. K. Ellis, T. Fraser, L. C. Mutter, *Weed Sci.* **1997**, 45, 601–609.

10 D. L. Lee, C. G. Knudsen, W. J. Michaely, H.-L. Chin, N. H. Nguyen, C. G. Carter, T. H. Cromartie, B. H. Lake, J. M. Shribbs, T. Fraser, *Pestic. Sci.* **1998**, 54, 377–384.

11 L. Agnolucci, F. Dalla Vecchia, R. Barbato, V. Tassani, G. Casadoro, N. Rascio, *J. Plant Physiol.* **1996**, 147, 493–502.

12 C. Mueller, J. Schwender, J. Zeidler, H. K. Lichtenthaler, *Biochem. Soc. Trans.* **2000**, 28, 792–793.

13 D. Siefermann-Harms, *Physiol. Plant* **1987**, 69, 561–568.

14 C. A. Tracewell, J. S. Vrettos, J. A. Bautista, H. A. Frank, G. W. Brudvig, *Arch. Biochem. Biophys.* **2001**, 385, 61–69.

15 R. Cogdell in *Plant Pigments*, T. W. Goodwin, T. Walworth (Eds.), Academic Press, London, **1988**, 183–230.

16 M. Havaux, *Trends Plant Sci.* **1998**, 3, 147–151.

17 P. M. Bramley, K. E. Pallett, *Proc. Br. Crop Prot. Conf. – Weeds* **1993**, 2, 713–722.

18 G. Sandmann, *Arch. Biochem. Biophys.* **2001**, 385, 4–12.

19 H. K. Lichtenthaler, M. Rohmer, J. Schwender, *Physiol. Plant.* **1997**, 101, 643–652.

20 G. Britton, *Z. Naturforsch.* **1979**, 34c, 979–985.

21 J. Breitenbach, B. Fernández-Gonzáles, A. Vioque, G. Sandmann, *Plant Mol. Biol.* **1998**, 36, 725–732.

22 F. X. Cunningham, Jr., B. Pogson, Z. Sun, K. A. McDonald, D. DellaPenna, E. Gantt, *Plant Cell* **1996**, 8, 1613–1626.

23 B. Pogson, K. A. McDonald, M. Truong, G. Britton, D. DellaPenna, *Plant Cell* **1996**, 8, 1627–1639.

24 D. R. Ort, *Plant Physiol.* **2001**, 125, 29–32.

25 G. Sandmann, *Pure Appl. Chem.* **1997**, 69, 2163–2168.

26 G. Sandmann, C. Schneider, P. Böger, *Z. Naturforsch.* **1996**, 51c, 534–538.

27 G. Sandmann, P. Böger, in *Rational Approaches to Structure, Activity and Ecotoxicology of Agrochemicals*, W. Draber, T. Fujita (Eds.), CRC Press, Inc., Boca Raton, FL, **1992**, 357–371.

28 G. Sandmann, H. Linden, P. Böger, *Z. Naturforsch.* **1989**, 44c, 787–790.

29 S. Kowalczyk-Schröder, G. Sandmann, *Pest. Biochem. Physiol.* **1992**, 42, 7–12.

30 E. L. Burdge, *Pest. Manag. Sci.* **2000**, 56, 245–248.

31 N. Yamada, D. Kusano, E. Kuwano, *Biosci. Biotechnol. Biochem.* **2002**, 66(8), 1671–1676.

32 G. Schnurr, P. Böger, G. Sandmann, *J. Pesticide. Sci. (Jpn.)* **1998**, 23, 113–116.

33 C. Fedtke, B. Depka, O. Schallner, K. Tietken, A. Trebst, D. Wollweber, H.-J. Wroblowsky, *Pest. Manag. Sci.* **2001**, 57, 278–282.

34 K. Hirai, A. Uchida, R. Ohno, in *Herbicide Classes in Development: Mode of Action, Targets, Engineering, Chemistry*, P. Böger, K. Wakabayashi, K. Hirai (Eds.), Springer–Verlag, Berlin, Heidelberg, **2002**, 213–221.

35 G. Sandmann, in *Target Assays for Modern Herbicides and Related Phytotoxic Compounds*, P. Böger, G. Sandmann (Eds.), Lewis Publ., Boca Raton, FL, **1993**, 15–20.

36 *The Pesticide Manual*, 13th edition, C. D. S. Tomlin (Ed.), BCPC, Alton, Hampshire, UK, **2003**.

37 SRC PhysProp Database in http://esc.syrres.com.

38 G. Sandmann, in *Pestic. Sci. Biotech.*, R. Greenhalgh, T. R. Roberts (Eds.), Blackwell Scientific, Oxford, **1986**, 43–48.

39 M. C. Cramp, J. Gilmour, E. W. Parnell, May & Baker Limited, EP 53011, **1982**.

40 M. C. Cramp, J. Gilmour, L. R. Hatton, R. H. Hewett, C. J. Nolan, E. W. Parnell, *Pestic. Sci.* **1987**, 18, 15–28.

41 C. J. Foster, T. Gilkersen, R. Stocker, Shell Int. Res., EP 447 004, **1991**.

42 A. Kleemann, G. M. Karp, M. E. Condon, American Cyanamid Co., US 5 869 426, **1999**.

43 A. Kleemann, Shell Int. Res., WO 94/22833, **1994**.

44 H. Kanno, Y. Kanda, S. Shimizu, Y. Kubota, T. Sato, M. Arahira, Kureha Kagaku, EP 692 474, **1996**.

45 H. S. Baltruschat, A. Kleemann, T. Maier, S. Scheiblich, J. L. Pont, T. Höllmüller, American Cyanamid Co., WO 98/04 548, **1998**.

46 H. S. Baltruschat, J. L. Pont, T. Maier, S. Scheiblich, American Cyanamid Co., US 5 922 738, **1999**.

47 T. Maier, A. Kleemann, S. Scheiblich, H. Siegfried, BASF AG, EP 1 101 764, **2001**.

48 S. Scheiblich, T. Maier, H. Baltruschat, BASF AG, WO 01/36 410, **2001**.

49 M. Hofmann, L. R. Parra, W. von Deyn, E. Baumann, M. Kordes, U. Misslitz, M. Witschel, C. Zagar, A. Landes, BASF AG, WO 03/022 843, **2003**.

50 M. Hofmann, L. R. Parra, W. von Deyn, E. Baumann, M. Kordes, U. Misslitz, M. Witschel, C. Zagar, A. Landes, BASF AG, WO 03/022 831, **2003**.

51 B. Hock, C. Fedtke, R. R. Schmidt, *Herbizide: Entwicklung, Anwendung, Wirkungen, Nebenwirkungen*, Georg Thieme Verlag, Stuttgart, New York, **1995**, 124.

52 J. B. St. John, U. Schirmer, F. R. Rittig, H. Bleiholder, in *Pesticide Synthesis through Rational Approaches*, P. S. Magee, G. K. Kohn, J. J. Menn (Eds.), ACS Symposium Series 255, American Chemical Society, Washington, D.C. **1984**, 145–162.

53 G. Sandmann, P. Böger, *Z. Naturforsch.* **1982**, 37c, 1092–1094.

54 U. Wriede, W. Freund, G. Hamprecht, A. Parg, B. Würzer, N. Meyer in Book of Abstracts, H. Frehse, E. Kesseler-Schmitz, S. Conway (Eds.), *7th Int. Congr. Pestic. Chem.*, Hamburg, **1990**, Vol. 1, 68.

55 G. Sandmann, S. Kowalczyk-Schröder, H. M. Taylor, P. Böger, *Pestic. Biochem. Physiol.* **1992**, 42, 1–6.

56 M. S. South, T. L. Yakuboski, M. J. Miller, M. Marzabadi, S. Corey, J. Molyneaux, S. A. South, J. Curtis, D. Dukesherer, S. Massey, F.-A. Kunng, J. Chupp, R. Bryant, K. Moedritzer, S. Woodward, D. Mayonado, M. Mahoney in *Synthesis and Chemistry of Agrochemicals V*, D. R. Baker, J. G. Fenyes, G. S. Basarab, D. A. Hunt (Eds.), ACS Symposium Series 686, American Chemical Society, Washington, D.C. **1998**, 107–119.

57 B. Laber, G. Usonow, E. Wiecko, W. Franke, H. Franke, A. Köhn, *Pestic. Biochem. Physiol,* **1999**, 63, 173–184.

58 R. Ohno, A. Watanabe, T. Matsukawa, T. Ueda, H. Sakurai, M. Hori, K. Hirai, *J. Pestic. Sci.* **2004**, 29, 15–26; enzyme value obtained from BASF Agricultural Research.

59 R. Ohno, A. Watanabe, M. Nagaoka, T. Ueda, H. Sakurai, M. Hori, K. Hirai, *J. Pestic. Sci.* **2004**, 29, 96–104.

60 G. Sandmann, *Pestic. Biochem. Physiol.* **2001**, 70, 86–91.

61 T. P. Selby, J. E. Drumm, R. A. Coats, F. T. Coppo, S. K. Gee, J. V. Hay, R. J. Pasteris, T. M. Stevenson, in *Synthesis and Chemistry of Agrochemicals VI*, D. R. Baker, J. G. Fenyes, G. P. Lahm, T. P. Selby, T. M. Stevenson (Eds.), ACS Symposium Series 800, American Chemical Society, Washington, D.C. **2002**, 74–84.

62 T. M. Stevenson, T. P. Selby, G. M. Koether, J. E. Drumm, X. J. Meng, M. P. Moon, R. A. Coats, T. V. Thieu, A. E. Casalnuovo, R. Shapiro, in *Synthesis and Chemistry of Agrochemicals VI*, D. R. Baker, J. G. Fenyes, G. R. Lahm, T. P. Selby, T. M. Stevenson (Eds.), ACS Symposium Series 800, American Chemical Society, Washington, D.C. **2002**, 85–95.

63 J. Breitenbach, P. Böger, G. Sandmann, *Pestic. Biochem. Physiol.* **2002**, 73, 104–109.

64 G. Mitchel, in *Synthesis and Chemistry of Agrochemicals IV*, D. R. Baker, J. G. Fenyes, G. S. Basarab (Eds.), ACS Symposium Series 584, American Chemical Society, Washington, D.C. **1995**, 161–170.

65 G. Sandmann, G. Mitchel, *J. Agric. Food Chem.* **2001**, 49, 138–141.

66 H. Ogawa, I. Yamada, K. Arai, K. Hirase, K. Moriyasu, C. Schneider, G. Sandmann, P. Böger, K. Wakabayashi, *Pest. Manag. Sci.*, **2001**, 57, 33–40.

67 S. Takamura, T. Okada, S. Fukuda, Y. Akiyoshi, F. Hoshide, E. Funaki, S. Sakai, *Proc. Br. Crop Prot. Conf. – Weeds* **1999**, 1, 41–52.

68 S. Ohki, R. Miller-Sulger, K. Wakabayashi, W. Pfleiderer, P. Böger, *J. Agric. Food Chem.* **2003**, 51, 3049–3055.

69 C. E. Ward, W. C. Lo, P. B. Pomidor, F. E. Tisdell, A. W. W. Ho, C. L. Chiu, D. M. Tuck, C. R. Bernardo, P. J. Fong, A. Omid, K. A. Buteau, in *Synthesis and Chemistry of Agrochemicals*, D. A. Baker, J. G. Fenyes, W. K. Moberg, B. Cross, Eds., ACS

Symposium Series 355, American Chemical Society, Washington, DC, **1987**, 65–73.

70 J. Breitenbach, C. Zhu, G. Sandmann, *J. Agric. Food Chem.* **2001**, 49, 5270–5272.

71 M. Gardon, N. Gosselin, O. Grosjean, T. Grollier, *La Défense des Végétaux* **2002**, 550, 56–60.

72 E. G. Teach, Stauffer Chemical Co., DE 2 612 731, **1976**.

73 H. M. Taylor, Eli Lilly, DE 2 537 753, **1976**.

74 C. E. Ward, Chevron Research, DE 3 422 346, **1984**.

75 C. Ebner, M. Schuler, Sandoz Ltd., US 3 644 355, **1972**.

76 C. Ebner, M. Schuler, Sandoz Ltd., US 3 834 889, **1974**.

77 T. Takematsu, Y. Takeuchi, M. Takenaka, S. Takamura, A. Matsushita, Ube Industries, EP 239 414, **1987**.

78 R. Arias, M. D. Netherland, B. E. Scheffler, A. Puri, F. E. Dayan in Herbicide-resistant Crops from Biotechnology, Special Issue, S. Duke, N. Ragsdale (Eds.), *Pest. Manag. Sci.* **2005**, 61, 258–268.

79 M. Knell, M. Brink, J. H. Wevers, W. Heinz, BASF AG, EP 899 262, **1999**.

4.2
Hydroxyphenylpyruvate Dioxygenase (HPPD) – the Herbicide Target

Timothy R. Hawkes

The corn herbicide sulcotrione and destosyl pyrazolate (DTP), the active hydrolysis product of the rice herbicides pyrazolate and pyrazoxyfen (Fig. 4.2.1), were known [1, 2] as bleaching herbicides before their HPPD mode of action was recognized. Loss of chlorophyll and accumulation of phytoene suggested possible sites of action in protochlorophyllide biosynthesis or at the phytoene desaturase (PDS) step in carotenoid biosynthesis [3]. However, these polar acids neither resemble typical PDS inhibitor herbicide types [4] nor inhibit PDS *in vitro* [5]. The clue to the true mode of action came from toxicological studies indicating that rats fed with experimental benzoyl cyclohexane-1,3-dione (CHD) herbicides such as nitisinone (Fig. 4.2.1) exhibit increased levels of tyrosine in blood and

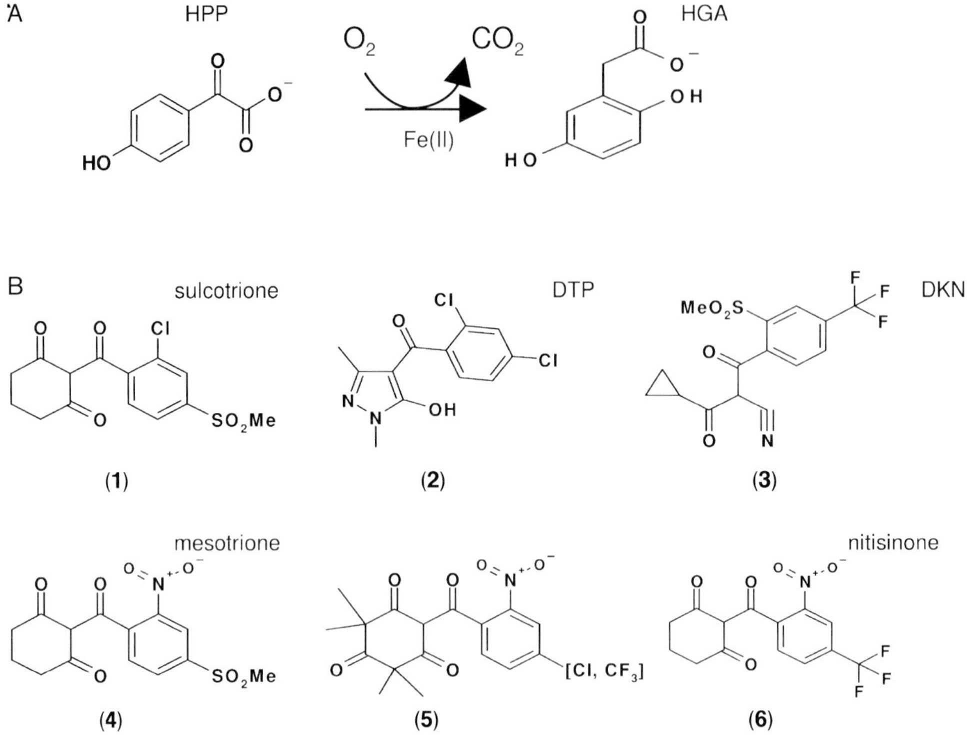

Fig. 4.2.1. (A) HPPD reaction and (B) the structures of some inhibitors.

p-hydroxyphenylpyruvate (HPP) in urine. This suggested a block in the catabolic degradation of tyrosine, and further investigative work [6–8] indicated that nitisinone is a potent inhibitor of mammalian HPPD, an enzyme that catalyzes the oxidative decarboxylation and rearrangement of HPP to homogentisate (HGA). The discovery that CHDs also inhibit HPPD in plants and the evidence firmly linking this to their herbicidal effect followed soon after [8, 9]. HGA was found to be a specific antidote for HPPD inhibitor-induced bleaching [8, 9] and a phytoene accumulating *Arabidopsis* mutant, PDS1, exhibiting a homozygous lethal bleaching phenotype rescuable by HGA mapped to a lesion in the plant HPPD gene [10]. Expression of heterologous HPPDs in transgenic plants resulted in specific tolerance to "HPPD-inhibitor" herbicides [11].

HPPD catalyzes an early step in a tyrosine degradation pathway [12] that is widely distributed in nature [13] and thus, as in animals, treatment of plants with inhibitors causes significant accumulation of tyrosine [8, 14]. HPP derived from transamination of tyrosine, is converted into HGA via HPPD, HGA is oxidized via HGA oxidase to 4-maleylacetoacetate, which is further degraded via 4-maleylacetoacetate isomerase and 4-fumarylacetoacetate lyase to fumarate and acetoacetate. In microbes the pathway provides assimilable carbon from tyrosine and phenylalanine; in higher mammals defects cause hereditary diseases [12] such as tyrosinemia type I where a lesion in 4-fumarylacetoacetate lyase causes a build up of toxic and liver-carcinogenic keto acids. Nitisisone, which blocks HPPD and thereby prevents toxin accumulation [7], is now an FDA approved treatment for tyrosinemia type I and may also find use in ameliorating other diseases arising from defects in tyrosine degradation [12].

HPPD inhibitors are only acutely toxic to photosynthetic organisms. In these HGA is not only an intermediate in tyrosine degradation but also in the biosynthesis of plastoquinone (PQ) and tocopherols [15, 16]. HPPD provides HGA from HPP, which is derived from transamination of tyrosine in the cytosol. Southern and sequence analysis of the *Arabidopsis* genome indicates the presence of only a single HPPD gene [10] for which no transit peptide leader is predicted and which is expressed only in the cytosol [17]. HGA made in the cytosol diffuses to the plastid where alternative prenyl transferase enzymes HGA solanyltranferase (HST) or HGA phytyltransferase (HPT) located in the inner plastid envelope convert it into the precursors 2-methyl-6-solanyl-1,4-benzoquinone (MSBQ) or 2-methyl-6-phytyl-1,4-benzoquinone (MPBQ), respectively. These are then further methylated (by a single enzyme, MSBQ/MPBQ methyltransferase, common to both pathways) to yield, respectively, PQ or, in the case of the tocopherol pathway, 2,3-dimethyl-6-phytyl-1,4-benzoquinone, which then gives rise to γ- and α-tocopherol (the dominant tocopherol in leaves) via further steps of cyclization and methylation.

Of the two deficiencies, lack of PQ and lack of tocopherols, which result from inhibition of HPPD, the significance of the former was more obvious [8, 9]. *In vitro* studies [18] as well as genetic mapping of phytoene accumulating mutants of arabidopsis [10] identified PQ as an essential electron acceptor in the PDS step of carotenoid biosynthesis. Herbicides such as norflurazon compete with PQ for binding to PDS [4] whereas HPPD inhibitors would appear to act indirectly by preventing PQ from being made. Consistent with this notion, PQ levels decrease in plants treated with HPPD herbicides [8] well before [14] the onset of bleaching and phytoene accumulation. Both types of herbicide ultimately cause depletion of carotenoids. Carotenoids act as accessory light-harvesting pigments and precursors of abscisic acid but their main role is in photo-protection [4]. Their extended conjugated double-bond system makes them effective quenchers of high energy triplet states of chlorophyll that would otherwise generate singlet oxygen. They mainly comprise part of the light harvesting antenna structure but β-carotene bound to the D2 protein is also a structural part of the photosystem II core. Herbicide-induced depletion of carotenoids is associated with light-dependent generation of singlet oxygen which damages lipids and proteins and causes disassembly of the photosynthetic complex and release of free chlorophyll. Free chlorophyll is photodynamically photodestructive and itself generates further singlet oxygen, eventually leading to the destruction of all leaf pigments and the characteristic white bleaching.

HPPD inhibitors might be expected to deliver the same herbicidal effects as direct inhibitors of PDS. Both are most effective in newly developing tissues that emerge bleached, presumably as a consequence of a failure to properly assemble photosynthetic units in the absence of carotenoids [19] and because those that do form bleach upon first exposure to light. Tissue damage is slower in mature tissue since it depends upon light intensity and carotenoid turnover. HPPD inhibitors are more effective applied post-emergence than are PDS inhibitors and exert greater effects on growth while PDS inhibitors cause more damage to mature leaves. These differences likely arise from differences in translocation. However, HPPD inhibitors cause some distinct phytotoxic effects. In mature cotyledons the effect of sulcotrione on both PSII quantum yield and pigment content appears intermediate between that of the PDS inhibitor fluridone and the PSII herbicide, diuron and it was thus suggested that direct inhibition of electron transport from PSII due to depletion of PQ may contribute to sulcotrione phytotoxicity [20]. Synergy between PSII and HPPD inhibitor herbicides further supports this notion. Mesotrione control of several weeds is synergistically improved with the addition of low rates of atrazine; in addition a significantly faster rate of tissue necrosis is observed in the mixture [21]. This could be rationalized purely in terms of improved competitive binding of PSII herbicides when PQ is depleted. However, there is also evidence that PSII effects might be mediated via depletion of tocopherol. Under high light and when electron transport from PSII is blocked (as by PSII herbicides) the PSII P680 reaction centre becomes over-reduced and the chlorophyll partitions toward the triplet state. Unquenched, this would generate singlet oxygen, damage the adjacent D1 protein and lead to PSII disassembly, chlorophyll release and photodynamic bleaching. Evidence from studies in *Chlamydomonas* [22] suggests that tocopherol has a key role in quenching. With careful poising of the concentration of HPPD inhibitor it was possible to partly inhibit photosynthesis in a culture of *Chlamydomonas* such that, at low light levels, PQ did not limit the photosynthetic rate and tocopherol levels were only 50% diminished. On transfer to high light the tocopherol pool diminished sharply and, within less than 3 h, PSII was inactivated and the D1 protein virtually disappeared. These effects were prevented or slowed by direct addition of short-chain cell-permeable analogues of tocopherol or diphenylamine, another direct chemical quencher of singlet oxygen. However, various tocopherol-deficient mutants of *Arabidopsis*, vte2 (HPT), vte1 (tocopherol cyclase) and vte3-1 (an MSBQ/MPBQ point mutant deficient in α- and γ-tocopherols) exhibit quite normal phenotypes [16, 23]. Only under rather drastic conditions of short exposure to high light and low temperatures could bleaching and lipid damage be induced. Under continuous high light other protective mechanisms, the xanthophyll cycle (which HPPD inhibitors would effectively block) and non-photochemical energy dissipation appear to provide compensatory protection. Overall, it seems likely that depletion of tocopherol [14] could only contribute significantly to the phytotoxic effects of HPPD herbicides under photo-inhibitory conditions. Nevertheless, it may underpin synergy between PSII and HPPD inhibitors where the two effects, of generating singlet oxygen and removing the means of protection, combine to cause a new phytotoxic symptom via rapid damage to the D1 protein.

Clearly, to be useful as herbicides in crops, HPPD inhibitors need be selective. Thus far, commercial HPPD inhibitor herbicides have been for use in corn (sulcotrione [1], mesotrione [24] and isoxaflutole [25] or rice (pyrazolate, pyrazoxyfen [2]). Development compounds, Bayer AE 0172747 (proposed name tembotriazone) and BAS670 (BASF, proposed name topramezone), are also mainly for use in corn although potential for cool-climate weed control in wheat is also indicated. Pyrazolate, pyrazoxyfen and isoxaflutole are proherbicides, with the former being detosylated [2] and the latter, (5-cyclopropylisoxazol-4-yl 2 mesyl-4-trifluoromethylphenyl ketone) quickly non-enzymically hydrolyzed to the corresponding diketonitrile (DKN) [14]. The crop safety of DKN depends on degradation to an inactive benzoic acid derivative [14]. Post-emergence safety of mesotrione in corn arises from favorable differential uptake and rapid P450-mediated hydroxylation of the cyclohexane ring [26]. In addition, mesotrione, which has a mainly broad-leaved weed spectrum, is a significantly more potent inhibitor of arabidopis HPPD (K_d 15 pM) than of HPPD from, for example, wheat (K_d 5 nM) a species to which it is much less herbicidal. Allowing for the 3–4-fold difference in the K_m for HPP this translates to about a 100-fold difference in effective potency. Accordingly, transgenic tobacco expressing wheat HPPD is highly resistant to mesotrione [26]. Where natural mechanisms of crop selectivity are inadequate such genetic engineering provides an alternative route. As yet, no transgenic HPPD-herbicide resistant crops have been commercialized although a good deal of work has been described in the patent and academic literature [11, 27]. Analogous to the engineering of resistance to glyphosate [28] mechanisms include increased expression of the target site and expression of altered target site HPPDs having enhanced resistance (e.g., by mutation or through natural tolerance). More novel are recent examples where (i) HPPD is bypassed through the expression of a three-enzyme algal pathway that provides an alternative route to provide HGA (and explains why *Synechocystis* should be insensitive to HPPD inhibitors) and (ii) resistance is considerably enhanced through co-expression of prephenate dehydrogenase with HPPD [11]. Herbicides also need to be selected for pharmacokinetic and kinetic properties that minimize impact in mammals. The pharmacokinetics of, for example, nitisinone selected to maintain a long-term block on tyrosine degradation, contrast markedly with those of the herbicide mesotrione where tyrosine accumulation effects are weak and transitory [29]. The recently solved crystal structures of rat and *Arabidopsis* HPPDs with inhibitors bound [30] promises to facilitate inherent selectivity by design.

In recent years understanding of the structure and mechanism of HPPD has progressed rapidly. A major highlight has been the X-ray crystallographic elucidation of the structure of plant, microbial and mammalian enzymes both with and without inhibitors bound [30–33]. Many aspects of HPPD structure and catalysis have been reviewed [12]. As a non-heme Fe(II)-containing dioxygenase, HPPD is a member of a wider group which couple oxidation of a substrate by dioxygen to oxidative decarboxylation of an α-keto acid (commonly α-keto-glutarate). Of the four electrons required to reduce dioxygen, two derive from oxidative decarboxylation and the other two from the substrate itself. For the HPPD reaction (Fig. 4.2.1) the decarboxylated 2-keto acid (pyruvyl) is not a separate cosubstrate but a side chain of the phenyl ring substrate that is oxidized. The pyruvyl side chain is decarboxylated to a carboxymethylene group which then migrates to the adjacent carbon of the phenyl ring while the ring is hydroxylated on the carbon at which the side chain was originally attached.

HPPD has a subunit polypeptide mass of 40–50 kDa and is typically a tetramer in bacteria or a dimer in eukaryotes (including mammals and plants) [11]. Conserved residues are found only in the C terminus and from recently solved X-ray structures (*P. fluorescens, Arabidopsis thaliana, Zea mays, S. avermitilis* and rat) the Fe and its associated inhibitor/substrate-binding site are clearly structurally well conserved and located within the C terminal part of the protein that folds as a discrete domain. At the primary sequence level, plant proteins appear somewhat distinct because they include a 15 amino acid insertion but, at a structural level, the core active site region remains similar to that in HPPDs from other phylla. As in all non-heme Fe(II) oxygenases this core consists of an active site Fe(II) coordinated by a triad of 2 histidine residues and one carboxylate [34]. In HPPD, the overall peptide fold as well as the disposition of these three residues through the primary sequence is similar to extradiol dioxygenases [12, 32] and it is suggested that HPPD exemplifies a 2-keto acid type dioxygenase that arose by convergent evolution from an extradiol type. In all HPPDs the Fe(II) is located at the centre of a cavity, between 8 and 14 Å wide, that is formed by an eight-stranded twisted half open β barrel. The three residues coordinating the Fe are located on three of these strands and the surrounding cavity environment is almost entirely conserved and dominated by hydrophobic amino acid residues within rigid secondary structural elements.

Current understanding of the catalytic mechanism derives from a combination of structural information, spectroscopy, kinetics and, since many of the proposed intermediates are too short-lived to be observed directly, also theoretical considerations. Figure 4.2.2 is based upon one current view [35].

The proposed nature of early intermediates in catalysis is consistent with the ordered mechanism observed in steady state kinetic studies [12, 36]. HPP binds first and CO_2 is the first product released. In α-ketoglutarate dioxygenases, α-ketoglutarate initially associates as a bidentate ligand of the Fe(II) and HPP appears to coordinate Fe(II) in HPPD in a similar way [12]. The initial, enzyme–Fe(II)–HPP complex (isolable under anaerobic conditions) exists as a mixture of five- and six-coordinate Fe [37] similar to the complex with inhibitor [12, 33]. In resting enzyme, the Fe(II) is relatively unreactive and, again, analogous to the other dioxygenases, HPP coordination primes it to react rapidly with dioxygen [38]. Binding of dioxygen is endergonic and the reactive Fe(III)–O_2 species thus formed predicted to be short-lived. Withdrawal of electrons into the Fe-bound dioxygen facilitates nucleophilic attack on the α carbonyl carbon resulting in decarboxylation and the generation of a theoretical short-lived Fe(II)-peracid species that heterolytically disproportionates to yield the oxo-Fe(IV) electrophile. This key species has not been detected directly but its existence is inferred from the reac-

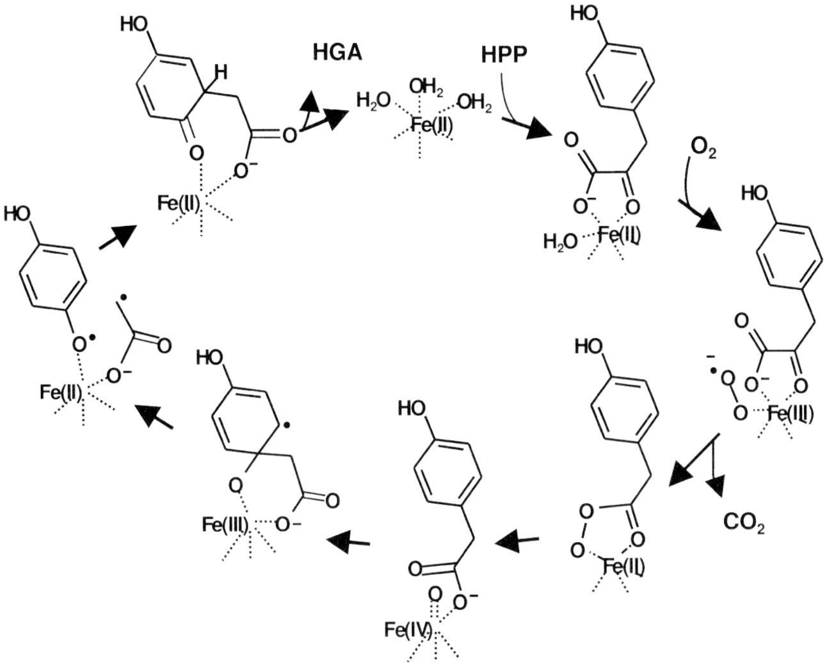

Fig. 4.2.2. Proposed intermediates in the HPPD reaction. (Based on Ref. [35].)

tion chemistry, by analogy with similar enzymes and, also, experimentally, from the observation [39] that the alternative substrate (4-hydroxyphenyl)thio-pyruvate is hydroxylated on sulphur rather than the ring. Proposals for the subsequent steps are varied. Some involve the oxo-Fe(IV) species abstracting two electrons from the aromatic ring to generate an arenium cation or benzene oxide [12]. Alternatively [35], it is argued that with two negatively charge ligands in the iron coordination shell the transfer of a second electron from the ring to the iron would be hindered and that a single electron process is more likely. This yields a radical sigma complex (that could potentially generate arene oxide via non-productive side reactions) which, in one suggested mechanism (Fig. 4.2.2) for side chain migration, undergoes homolytic C–C bond cleavage to yield a highly unstable biradical species that decays to form the new C–C bond and the ketone species. In the finals steps, re-aromatization and tautomerization to HGA could equally take place in solution as enzyme-bound.

X-Ray crystallography has provided considerable insight into how inhibitors bind. In the *S. avermitilis* HPPD/nitisinone complex, Fe is five/six-coordinate with bidentate chelation from the 5′ and 7′ oxygens of the inhibitor and a water weakly occupying the 6′th position [33]. Inhibitor binding shifts a C terminal helix to provide one of two phenylalanines that sandwich the phenyl ring of the inhibitor in a π-stacking interaction. No other energetically significant interactions between inhibitor and enzyme surfaces are evident other than exclusion of waters through space-filling Van der Waal contacts. The site around the inhibitor is highly conserved and phenylpyrazole binding into rat and plant HPPDs appears similar [30]. It is not immediately obvious how inhibitor binding should be selective between one HPPD and another. However this clearly is possible [26, 30] and 100-fold differences in K_d (12 kJ mol^{-1}) or greater may originate from the sum of structural and orientation differences too subtle to discern.

In principle inhibitors could coordinate Fe(ii) or Fe(iii) in HPPD. In free solution inhibitor complexes with Fe(iii) have the lower dissociation constants [40]. However DKN and nitisinone bind significantly only to Fe(ii) forms of the carrot and *S. avermitilis* enzyme [41, 42]. Iron(ii) enzyme inhibitor complexes are highly stable, unreactive to oxygen and, similar to the anaerobic Fe(ii) enzyme HPP complex, weakly colored due to charge transfer transitions [12, 42]. Spectroscopic studies of the Fe(ii) centre have provided quantitative insight into the relative contribution of metal coordination to overall binding. Magnetic and non-magnetic circular dichroism spectroscopy combined with calculations based on density function theory indicate that nitisinone interacts with the Fe(ii) somewhat more weakly than the substrate [43]. Thus, the π-stacking interaction with the enzyme phenylalanines makes a major contribution to inhibitor binding. Inhibitors may act as mimics of a reaction intermediate and analogous interactions may help drive catalysis. The π-stacking may electronically stabilize the putative arenium cation [12] or, equally, a phenyl radical intermediate (Fig. 4.2.2). In either case, tight binding of inhibitors as compared with substrate may be understood in terms of favorable interactions with electron deficient aromatic rings. Spectroscopic and kinetic studies [12, 43] have also provided insight into the steps involved in inhibitor binding. While di- and triketone inhibitors exist in solution as an equilibrium between several tautomers, for nitisinone at least the exocyclic enol(ate) predominates in solution at pH 7 whereas the keto form of HPP is the substrate [11]. Pre steady state spectroscopic studies of nitisinone binding to anaerobic *S. avermitilis* HPPD indicate at least three substeps before tight complex formation, a rapid weak non-chromophoric complex, a shift to a chromophoric complex (8 s^{-1}) and finally a slower chromophoric shift (0.76 s^{-1}). A solvent deuterium isotope effect of three on the latter is consistent with a proton shift being involved in final complex formation.

Whatever the sub-steps involved the orphan drugs and herbicides clearly form remarkably tight complexes with Fe(ii) HPPD. Initial studies with nitisinone and sulcotrione indicated half-times for dissociation from the rat enzyme of approximately 10 and 63 h [6]. Studies with HPPD from carrot indicated single step competitive binding of DKN to form a similar tight slow-dissociating complex with the plant enzyme [41]. The value of K_i can be evaluated from the ratio of the rate constants governing dissociation (k_{off}) and formation (k_{on}) of the enzyme inhibitor complex although relatively few measurements have been reported (Table 4.2.1).

Formation rate constants, k_{on}, have been estimated on the basis of enzyme assay [6, 26, 27, 41, 45] or quenched physical binding studies using labeled inhibitors [26, 27] and dissociation rates, k_{off}, based on rates of inhibitor exchange [26,

Table 4.2.1 Estimated inhibition constants of HPPD inhibitors.

Inhibitor (Fig. 4.2.1)	HPPD	k_{on} (M^{-1} s^{-1})	k_{off} (s^{-1})	K_d (pM)	Ref.
(3)	Carrot	1.5×10^4	9×10^{-5}	6000	41
(3)	P. fluorescens	1.6×10^4	1.8×10^{-4}	11000	27
(3)	Wheat	6.9×10^4	6.2×10^{-5}	900	27
(3)	Arabidopsis	1.8×10^5	8.3×10^{-6}	46	27
(4)	Arabidopsis	2.3×10^5	3.3×10^{-6}	14	26
(4)	Pseudomonas	1.8×10^4	2×10^{-6}	114	27
(4)	Wheat	1.8×10^5	1.0×10^{-3}	5500	26
(5) [Cl]	Arabidopsis	3.0×10^5	1.2×10^{-6}	4	27
(5) [Cl]	Pseudomonas		$>2 \times 10^{-4}$		27
(5) [Cl]	Wheat	3.0×10^5	4.2×10^{-6}	12	27
(5) [CF3]	Rat	1.5×10^4	1.9×10^{-6}	125	45
(6)	Rat	9.9×10^4	3.2×10^{-6}	32	6
(1)	Rat	3.3×10^{-4}	1.9×10^{-5}	575	6

27, 41] or activity recovery [6, 45]. Concentrations of HPPD active sites have been determined via titration with labeled inhibitors in both crude and purified preparations of enzyme. While pure preparations are preferable for kinetic studies, particularly with plant HPPDs, purification and reconstitution with Fe(II) leads to activity loss and, in principle, the possible generation of damaged species with altered binding kinetics (k_{cat}/K_m values can appear 3–10-fold greater in part- than in fully-purified preparations of E. coli-expressed recombinant plant HPPDs [26, 27, 41]). Certainly differences in assay and preparational procedures make all comparisons difficult and the absolute accuracy of, especially, the faster on and slower off rates in Table 4.2.1 is not guaranteed. Nevertheless, it is striking that many inhibitor are highly potent with K_ds in the pM range and also that there are significant, several hundred-fold, species-dependent differences in inhibitor K_d and k_{off} values with the Arabidopsis enzyme being the most sensitive of those tested. It is difficult to know which kinetic parameter to take as most predictive of biological activity. "Stickiness" (k_{off}) may be key in maintaining persistence of the pathway blockade since, once inhibited, HPPD will stay inhibited for days (until new enzyme is synthesized) and this may be important in achieving good herbicidal activity.

Interestingly, it appears that inhibition requires the binding of only a single inhibitor molecule per HPPD dimer. While only a single DKN molecule bound per dimer of carrot HPPD this "half-site" binding was nevertheless associated with complete enzyme inhibition [41]. Similar was observed in preliminary studies with part-pure Arabidopsis HPPD [27] with equilibrium binding of DKN being half that observed with CHDs (e.g., mesotrione). Mesotrione binding to Arabidopsis HPPD was biphasic with (presumptive) half-site binding and complete enzyme inhibition occurring in an initial rapid phase (1.8×10^5 M^{-1} s^{-1}) and the remaining 50% then binding much more slowly. Thus, even with CHDs that eventually bind one per monomer it appears that only the initial rapid half-site binding may be required for inhibition.

References

1 Beraud, J.M., Claument, J., Monturey, A. *Proc. Brighton Crop Protection Conference-Weeds* **1991** 51–55.

2 Matsumoto, H. *Am. Chem. Soc. Symp. Ser.* (**2005**), 892, 161–171.

3 Mayonado, D.J., Hatzios, K.K., Orcutt, D.M., Wilson, H.P. *Pest. Biochem. Physiol.* **1989** 35, 138–145.

4 Sandmann, G. in *Herbicide Classes in Development* (P. Boger, K. Wakabayashi, K. Hirai eds) **2002** pp 221–229. Springer-Verlag Berlin, Heidelberg.

5 Sandmann, G., Boger, P., Kumita, I. *Pestic. Sci.* **1990** 30, 353–355.

6 Ellis, M.K., Whitfield, A.C., Gowans, L.A., Auton, T.R., Provan, W.Mc. Lock, E.A., Smith, L.L. *Toxicol. Appl. Pharmacol.* **1995** 133(1), 12–19.

7 Lindstedt, S., Holme, E., Lock, E.A., Hjamarson, O., Strandvik, B. *Lancet* **1992** 340, 813–817.

8 Prisbylla, M.P., Onisko, B.C., Shribbs, J.M., Adams, D.O., Liu, Y., Ellis, M.K., Hawkes, T.R., Mutter, L.C. *Proc. Bright. Crop Prot. Conf. Weeds* **1993** 2, 731–738.

9 Schulz, A., Ort, O., Beyer, P., Kleinig, H. *FEBS Lett.* **1993** 318(2), 162–166.

10 Norris, S.R., Shen, X., DellaPenna, D. *Plant Physiol.* **1998** 117(4), 1317–1323.

11 Matringe, M., Sailland, A., Pelissier, B., Rolland, A., Zink, O. *Pest Manag. Sci.* **2005** 61(3), 269–276.

12 Moran, G.R. *Arch. Biochem. Biophys* **2005** 433, 117–128.

13 Fernandez-Canon, J.M., Penalva, M.A. *J. Biol. Chem.* **1995** 270, 21199–21205.

14 Pallett, K.E., Little, J.P., Sheekey, M., Veerasekaran, P. *Pest. Biochem. Physiol* **1998** 62(2), 113–124.

15 Schultz, G., Soll, J., Filder, E., Schulze-Siebert, D. *Physiol. Plant* **1985** 64, 123–129.

16 Garcia, I., Rodgers, M., Pepin, R., Hsieh, T.-F., Matringe, M. *Plant Physiol.* **1999** 119(4), 1507–1516.

17 Eckardt, N.A. *Plant Cell* **2003** 15, 2233–2235.

18 Mayer, M.P., Beyer, P., Kleinig, H. *Pestic. Biochem. Physiol.* **1990** 34, 111–117.

19 Masamoto, K., Hisatomi, S.-I., Sakurai, I., Gombos, Z., Wada, H. *Plant Cell Physiol.* **2004** 45(9), 1325–1329.

20 Kim, J.-S., Kim, T.-J., Kwon, O.K., Cho, K.Y. *Photosynthetica* **2002**, 40(4), 541–545.

21 Armel, G.R., Hall, G.J., Wilson, H.P., Cullen, N. *Weed Sci.* **2005** 53(2), 202–211.

22 Trebst, A., Depka, B., Jaeger, J., Oettmeier, W. *Pest Manag. Sci.* **2004** 60(7), 669–674.

23 Havaux, M., Eymery, F., Porfirova, S., Reay, P., Dormann, P. *Plant Cell* **2005**, 17, 3451–3469.

24 Mitchell, G., Bartlett, D.W., Fraser, T.E.M., Hawkes, T.R., Holt, D.C., Townson, J.K., Wichert, R.A. *Pest Manag. Sci.* **2001** 57(2), 120–128.

25 Luscombe, B.M., Pallett, K.E., Loubiere, P., Millet, J.C., Melgarejo, J., Vrabel, T.E. *Proc. Bright. Crop. Prot. Conf. Weeds* **1995**, 1, 35–42.

26 Hawkes, T.R., Holt, D.C., Andrews, C.J., Thomas, P.G., Langford, M.P., Hollingworth, S., Mitchell, G. in *Proc. Bright. Crop. Prot. Conf. Weeds* **2001** 2, 563–568.

27 Warner, S.A.J., Hawkes, T.R., Andrews, C.J. *PCT Patent Appln.* **2002** WO 02/46387.

28 Barry, G., Kishore, G., Padgette, S., Taylor, M., Kloacz, K., Weldon, M., Re, D., Eichholtz, D., Fincher, K., Hallas, L. in *Biosynthesis and Molecu-*

lar Regulation of Amino Acids in Plants (B.K. Singh, H.E. Floras, J.C. Shannon eds) **1992** Am. Soc. Plant Physiol. pp 139–145.

29 Hall, M.G., Wilks, M.F., Provan, W.Mc., Eksborg, S., Lumholtz, B. *Br. J. Clin. Pharmacol.* **2001** 52, 169–177.

30 Yang, C., Pflugrath, J.W., Camper, D.L., Foster, M.L., Pernich, D.J., Walsh, T.A. *Biochemistry* **2004** 43(32), 10414–10423.

31 Fritze, I.M., Linden, L., Freigang, J., Auerbach, G., Huber, R., Steinbacher, S. *Plant Physiol.* **2004** 134(4), 1388–1400.

32 Serre, L., Sailland, A., Sy, D., Boudec, P., Rolland, A., Pebay-Peyroula, E., Cohen-Addad, C. *Structure* **1999** 7(8), 977–988.

33 Brownlee, J.M., Johnson-Winters, D.H.T., Harrison, G.R., Moran, G. *Biochemistry* **2004** 43, 6370–6377.

34 Hegg, E.L., Que Jr. L. *Eur. J. Biochem.* **1997** 250, 625–629.

35 Borowski, T., Bassan, A., Siegbahn, P.E.M. *Biochemistry* **2004** 43(38), 12331–1234.

36 Rundgren, M. *J. Biol. Chem.* **1977** 252(14), 5094–5099.

37 Neidig, M.L., Kavana, M., Moran, G., Solomon, E.I. *J. Am. Chem. Soc.* **2004** 126, 4486–4487.

38 Johnson-Winters, K., Purpero, V.M., Kavan, M., Nelson, T., Moran, G.R. *Biochemistry* **2003** 42, 2072–2080.

39 Pascal, R.A. Jr., Oliver, M.A., Chen, Y.C.J. *Biochemistry* **1985** 24(13), 3158–3165.

40 Wu, C.S., Huang, J.L., Sun, Y.S., Yang, D.Y. *J. Med. Chem.* **2002** 45, 2222–2228.

41 Garcia, I., Job, D., Matringe, M. *Biochemistry* **2000** 39(25), 7501–7507.

42 Kavana, M., Moran, G. *Biochemistry* **2003** 42, 10238–10245.

43 Neidig, M.L., Decker, A., Kavana, M., Moran, G.R., Solomon, E.I. *Biochem. Biophys. Res. Commun.* **2005** 338(1), 206–214.

44 Lindstedt, S., Rundgren, M. *Biochim. Biophys. Acta* **1982** 704(1), 66–74.

45 Ellis, M.K., Whitfield, A.C., Gowans, L.A., Auton, T.R., Provan, W.Mc., Lock, E.A., Lee, D.L., Smith, L.L. *Chem. Res. Toxicol.* **1996** 9(1), 24–27.

4.3
Hydroxyphenylpyruvate Dioxygenase (HPPD) Inhibitors: Triketones

Andrew J. F. Edmunds

4.3.1
Introduction

This chapter aims to give an insight into the discovery of the triketone class of herbicides and their continuing development. A very qualitative picture of structure–activity relationships will be discussed and currently commercialized triketones, in terms of their use, weed spectrum, crop selectivity, environmental and toxicological profiles, and manufacture will be described. This chapter also contains an overview of the major companies' activities in the field in the last two decades, focusing on compounds that are likely to be brought to the market, or were putatively close to development.

4.3.2
Discovery

In 1977, at the Western Research Centre in California, scientists in Stauffer (a former legacy company of Syngenta) noticed that relatively few weeds grew under the bottle brush plant *Callistemon citrinus*. Analysis of soil samples where *C. citrinus* was growing revealed that the herbicide the plants were excreting was leptospermone (1) [1]. This natural product had previously been isolated from the volatile oils of Australian myrtacious plants [2]. Pure samples of leptospermone (1) showed unique bleaching symptoms on several weed species albeit at relatively high (5 kg ha^{-1}) rates. This herbicidal activity was patented in 1980 [3].

Independent of this discovery, in 1982 scientists from the same company were working on a project aimed at preparing novel Acetyl-CoA carboxylase inhibitors, based upon the typical cyclohexanedione structure known for this class. The first targeted compound (2), prepared as shown in Fig. 4.3.1, showed some herbicidal activity and they thus attempted preparation of a phenyl analogue in a similar manner. This led not to the expected product (3), but to the triketone (4). This compound was devoid of herbicidal activity, but (luckily!) in safener screens the compound showed antidotal effects in Soya for thiocarbamate herbicides. A further round of synthesis optimization was undertaken and it was found that the compound (5) with an ortho-chloro substituent showed reasonable herbicidal activity. Furthermore, they noticed that it exhibited the same unique bleaching symptomology observed for leptospermone (1, Fig. 4.3.1) [4]. Further optimiza-

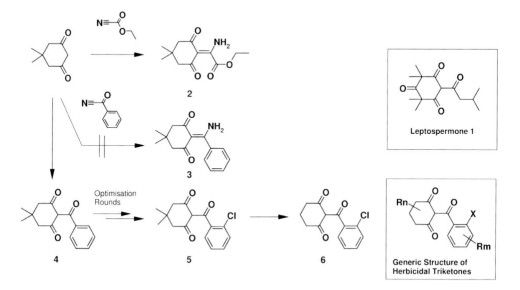

Fig. 4.3.1. Discovery of the triketone herbicides.

tion showed that removal of the methyl groups at the 5-position of the cyclohexanedione moiety (**6**) resulted in significantly enhanced herbicidal activity against a
wide range of broad-leaved weeds, with good corn tolerance, when applied pre-
and post-emergence at rates of about 2 kg ha^{-1}. The first patent was filed [5]
and the discovery of the benzoylcyclohexanedione herbicides had been made.
These events, and the generic structure of herbicidal triketones are summarized
in Fig. 4.3.1.

4.3.3
Mode of Action

As discussed in detail in Chapter 4.2, triketones exert their herbicidal mode of
action by inhibition of 4-hydroxyphenylpyruvate dioxygenase (HPPD) [6]. Triketones are not the only herbicide class that have this mode of action, and it has
retrospectively been shown that apparently structurally non-related heterocyclic
commercial herbicides such as isoxaflutole (**7**, BALANCE® and MERLIN®), and
the rice herbicides pyrazolate (**8**, SANBIRD®) and benzobicyclon (**9**, SHOW-
ACE®) also cause these bleaching symptoms by the same mode of action. However, a common feature of these herbicides, after metabolic activation to the active
metabolites (**7'**) [7], (**8'**) [8] and (**9'**) [9] is the presence of an acidic 1,3-dicarbonyl
moiety, which is also present in triketones (Fig. 4.3.2). Triketones and related her-

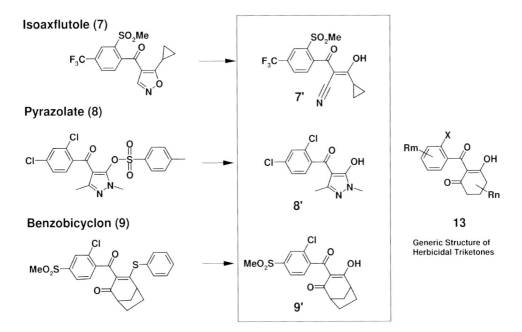

Fig. 4.3.2. Commercial herbicides with a HPPD mode of action.

bicides mimic the -keto acid group of the HPPD substrate hydroxyphenyl pyruvate, and competitively bind to the iron at the active site of the enzyme, causing its inhibition. Homology models of all structure types bound to this enzyme [10] and crystal structures of pyrazoles [11] and triketones [12] bound to the same enzyme have been published.

4.3.4
Synthesis of Triketones

Most triketone herbicides (**13**) reported in the literature are synthesized by O-acylating a cyclic 1,3-dione (**10**) with an activated aroyl acid (**11**), and then carrying out an O- to C-acyl rearrangement of the O-acyl intermediate (**12**) in the presence of a catalyst (Scheme 4.3.1). The O-acylation is generally achieved using an acid chloride in the presence of a base, but other reagents such as dicyclohexylcarbodiimide (DCC) [13], N-methyl-2-chloropyridinium iodide (Mukaiyama coupling agent) [14], 2-chloro-1,3-dimethylimidazolinium chloride [15], have been used in triketone synthesis. Typical catalysts used for the rearrangement are cyanide [16], aluminum trichloride [17], 1,2,4-triazole [18], potassium fluoride [19] and azide salts [20] whereas the cyanide source (including acetone cyanohydrin) induced O–C rearrangement has been generally the method of choice. The reaction may be carried out in a stepwise fashion (i.e., isolation of **12**) but one-pot variations have been developed in many cases by choice of the correct solvent [21]. There is also an isolated report of direct C-acylation (**10**) with an aroyl acid chloride (**11**, Z = Cl) using potassium carbonate in acetonitrile [22]. Alternatively, triketones can be obtained directly by acylation of (**10**) with the appropriate benzoyl cyanide (**11**, Z = CN) [23].

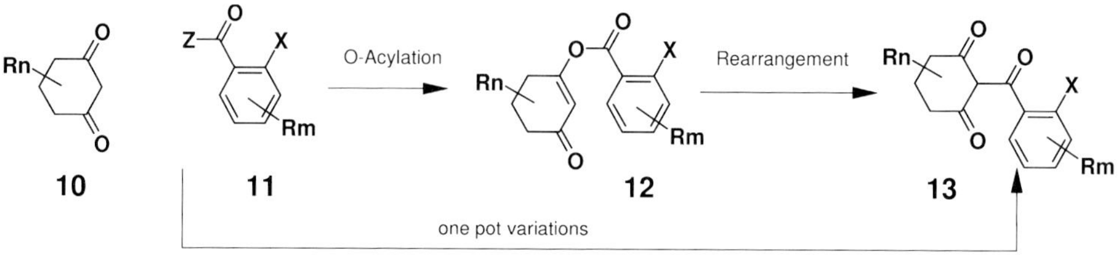

Scheme 4.3.1

Other syntheses that have been developed for preparation of triketones include activating the dione portion (to give **14**) and coupling this with an aroyl acid (**15**) in the presence of a Lewis acid catalyst [24] followed by O-acyl rearrangement, or by palladium-catalyzed carbonylation of an aroyl halide (**16**) in the presence of a dione (**10**) [25] and subsequent rearrangement of the O-acyl product (**12**) formed into the triketone (**13**) (Scheme 4.3.2).

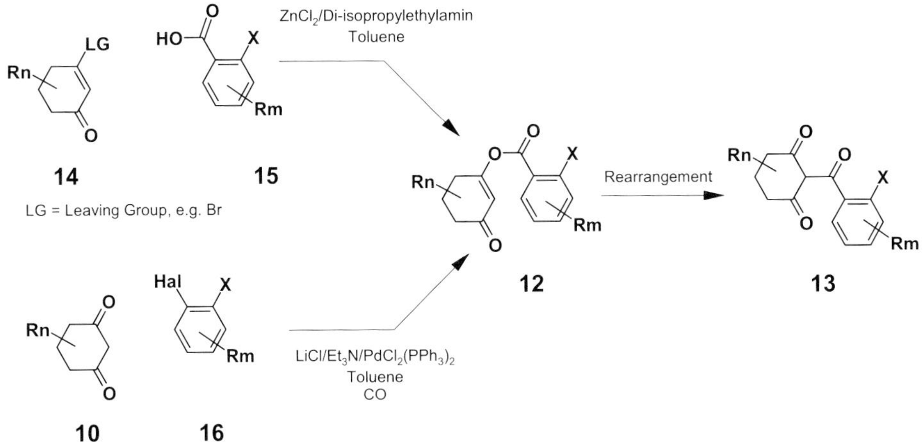

Scheme 4.3.2

Cyclohexane diones with various substitution patterns can be readily synthesized by the two general routes shown in Scheme 4.3.3 [26, 27].

The synthesis of the benzoyl portions of triketones can not be so generalized, and specific syntheses have been developed for developmental and commercialized compounds.

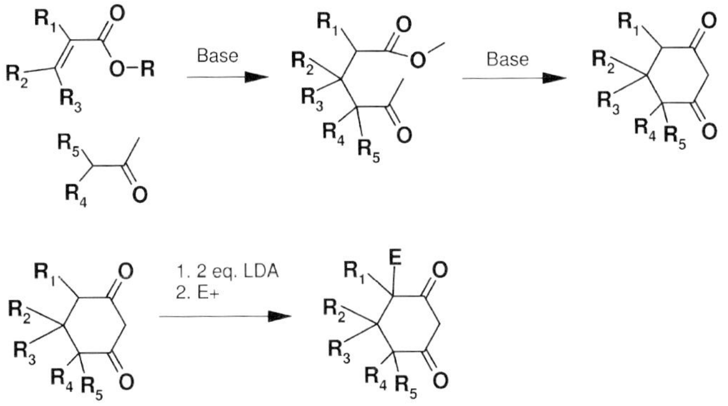

Scheme 4.3.3

4.3.5
Structure–Activity Relationships

The triketones can be separated into two parts for analysis of the structure–activity relationships, namely the benzoyl and the dione moieties. Each part can be examined independently, as they appear to play distinct and different roles in the overall expression of herbicidal activity [1]. Apart from the necessity of an ortho-substituent on the phenyl ring, it was established that 2,4,- or 2,3,4-benzoic acid substitution patterns were required for optimal activity, with the 2,5-

pattern(s) being the least effective. After more than 20 years of HPPD research, this original optimal substitution pattern still seems to be valid based on analysis of published patents. A correlation was found between the pK_a of triketones (which can be viewed as vinylogous benzoic acids) and herbicidal activity [28], with a pK_a of <6 being required for activity, as this will affect not only binding to the iron at the active site of the enzyme but also transport and translocation within the plant. As the pK_a will be affected by substituents on the aromatic ring, those that generally reduce the electron density of the aromatic ring lead to compounds with a reduced pK_a and improved herbicidal activity. A survey of the patent literature and reported SAR studies [4, 29] suggest that small ortho electron-withdrawing substituents such as Cl, NO_2 and CF_3 are particularly favored. An ortho methyl substituent is also tolerated as long as the total electron density of the aromatic ring is kept low. The para substituent is generally also an electron-withdrawing moiety, particularly halo, haloalkyl and alkylsulfonyl, with some restraints on size according to published data [4]. Zeneca (now Syngenta) arrived at several compounds with these types of substitution patterns, such as sulcotrione (**17**) and mesotrione (**18**, Scheme 4.3.4), at an early stage in triketone research. Both compounds have since reached the marketplace (Section 4.3.6).

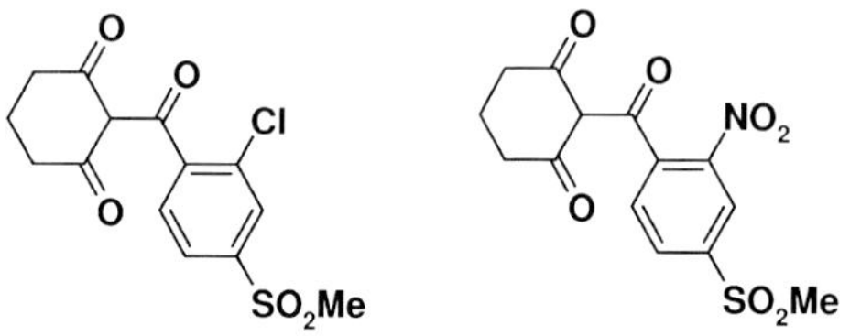

Sulcotrione (**17**)   Mesotrione (**18**)

Scheme 4.3.4

At the meta position, a multitude of functionalities have been reported to lead to herbicidally active compounds. One problem that, however, often leads to reduced potency is the presence of an electron-withdrawing substituent at the meta position combined with a potential leaving group at the 2-position (e.g., nitro or chloro) as this may give rise to dihydroxanthenones (**19**), which are known to be much less herbicidally active (Scheme 4.3.5) [4].

Two strategies have been generally used to remedy this situation: One has been to use meta substituents such as alkoxy or thioalkyl, which are reasonably electron donating at their ortho positions, thus hindering formation of **19**, but inductively electron withdrawing at the carbonyl, thus increasing overall acidity [4]. The more frequently used strategy, however, is to have a small non-leaving group at the ortho-position, such as methyl, whilst having substituents at the 3,4-positions that make the aromatic ring overall electron deficient. Many fused ring types have been reported in the patent literature and it would appear that para,meta-fused ring systems are generally more favored than ortho,meta-fused systems.

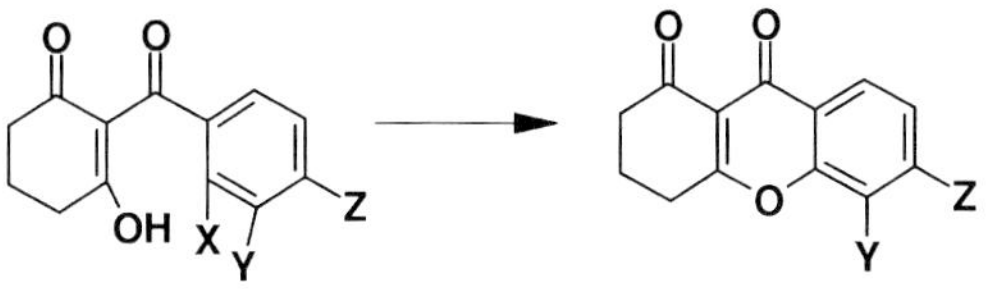

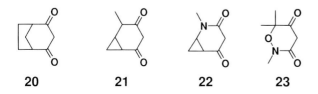

Scheme 4.3.5

The effect of adding substituents to the cyclohexanedione ring is to block site(s) of metabolism by plants [29, 30]. This results in greater herbicidal activity, as the plants have greater difficulty in detoxifying the molecule. Studies using model compounds indicate that the principal routes of metabolism of the benzoylcyclo-hexanediones in plants are hydroxylation at the 4-position of the cyclohexane-dione (if this position is blocked, then hydroxylation takes place at the chemically equivalent 6-position), and hydrolytic cleavage of the benzoyl group. It has been demonstrated that placing two methyl groups at the cyclohexanedione 4-position slows the rate of both of these metabolic processes in plants. As the sites for hydroxylation are sequentially blocked, an increase in overall activity against grasses is observed. Some of the most active triketones known contain the 2,2,4,4-tetramethyl-cyclohexane-1,3,5-trione moiety, also found in leptospermone (**1**, Fig. 4.3.1) [29]. However, reducing the potential for metabolism has other consequences, such as reduced corn selectivity and a dramatic increase in soil persistence [1]. To compensate for this effect, several important diones have regularly appeared in the patent literature, which have strained bicyclic rings and/or heterocyclic atoms (Scheme 4.3.6 compounds **20–23**), thus putatively be-ing more easily metabolized.

| 20 | 21 | 22 | 23 |

Scheme 4.3.6

4.3.6
Review of the Patent Literature

Some of the important and typical structural types patented by the various com-panies are discussed in this section. Generic structures are simplified and thus not necessarily those that appeared in the referenced patents.

Stauffer (later Zeneca, now Syngenta) patented extensively after their initial dis-covery and were able to gain good intellectual property advantage over competitor companies, particularly in terms of important 2,4- and 2,3,4-substituted benzoic

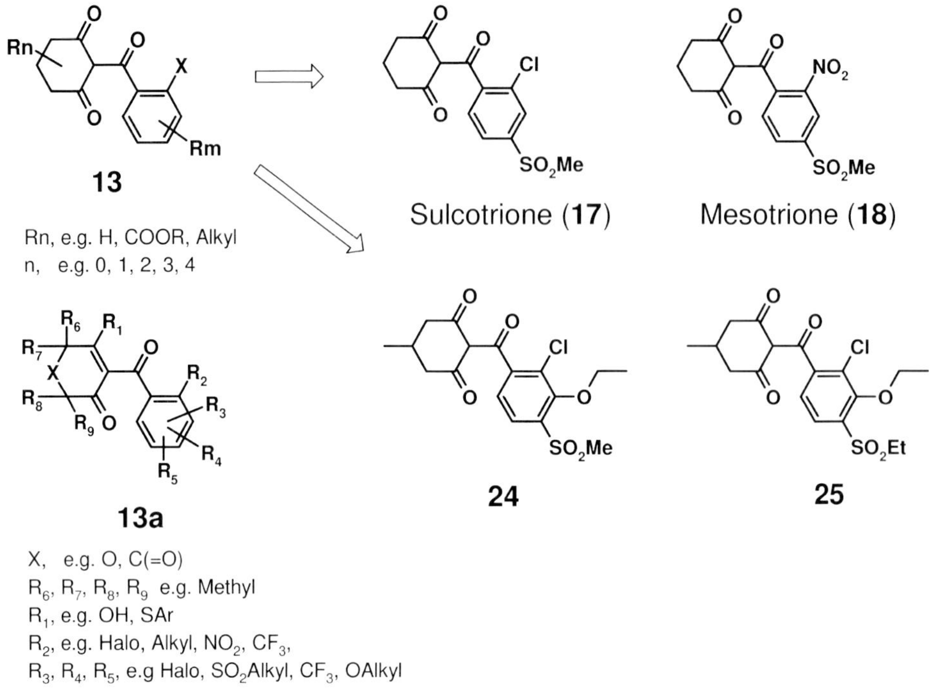
Rn
X
Rm
13
Rn, e.g. H, COOR, Alkyl
n, e.g. 0, 1, 2, 3, 4
Cl
SO2Me
Sulcotrione (17)
NO2
SO2Me
Mesotrione (18)
R6 R1
R7
X
R2
R8
R3
R9 O
R4
R5
13a
X, e.g. O, C(=O)
R6, R7, R8, R9 e.g. Methyl
R1, e.g. OH, SAr
R2, e.g. Halo, Alkyl, NO2, CF3,
R3, R4, R5, e.g Halo, SO2Alkyl, CF3, OAlkyl
Cl
SO2Me
24
Cl
SO2Et
25

Fig. 4.3.3. Typical Zeneca patents.

acid types (e.g., **17** and **18**, respectively **24** [23b] and **25** [31]), as well as important triketones with more substituted cyclohexane diones (e.g., **13a** X = C(O) [5, 31], and heteroatom containing diones (e.g., **13a** X = O, Fig. 4.3.3) [32].

They were also granted a patent that covered cyclohexane diones coupled to heteroaroyl acids (e.g., **26–28**, Fig. 4.3.4) with very broad scope [34], which made patenting rather difficult for companies following Zeneca.

Pyridines and pyrimidines were patented separately, to complete an impressive array of protection for the heterocyclic triketones [35]. Nevertheless, after the first patent appeared regarding this novel substance class, most of the major companies started programs in the field. There were basically two strategies: Some companies searched for novel diones that were at the time outside the scope of the Zeneca published patents, while other companies searched for novel aromatic acids. For example, Sandoz (now Syngenta) concentrated on the search for novel diones, and several compounds containing bicyclo[3.2.1]octane-2.4-dione, such as **29** [36, 37] and **30** [38], as well as the oxazinedione types (**31**) [39, 40], were important compounds for use in corn (Fig. 4.3.5). A collaboration between Sandoz and SDS Biotech has also led to the identification of proform triketones containing bicyclo[3.2.1]octane-2.4-diones for use in rice, such as benzobicyclon (**9**) [41].

Nippon Soda also initially investigated the dione portion of triketones, and patented extensively compounds containing bicyclo[4.1.0]heptane-2,4-diones such as **32–35** (Fig. 4.3.6) [42].

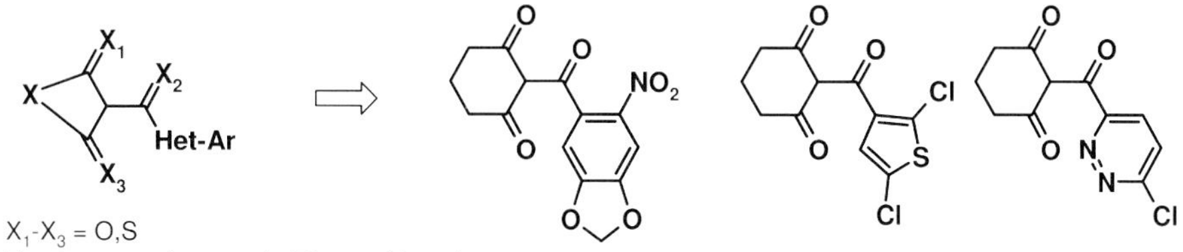

26 27 28

X_1-X_3 = O,S
Het-Ar, e.g. subs. $_5$-and $_6$-Ring and fused heterocycles
but not pyridine or pyrimidine
X e.g. $(CH_2)_3$, opt. Subst and/or interupted by C=O, O

Fig. 4.3.4. Zeneca heteroaroyl triketones.

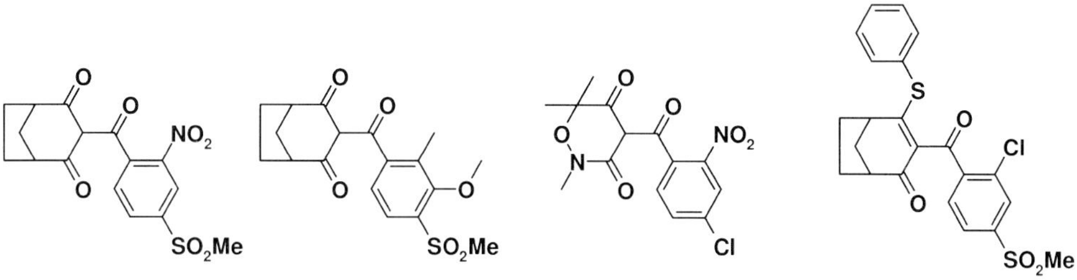

29 30 31 Benzobicyclon (**9**)

Fig. 4.3.5. Typical Sandoz compounds.

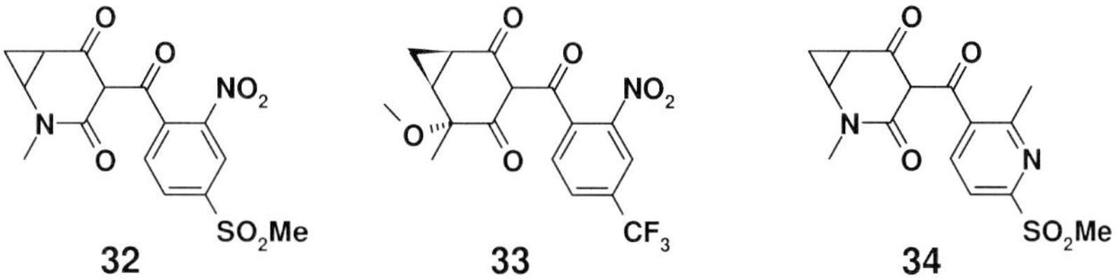

32 33 34

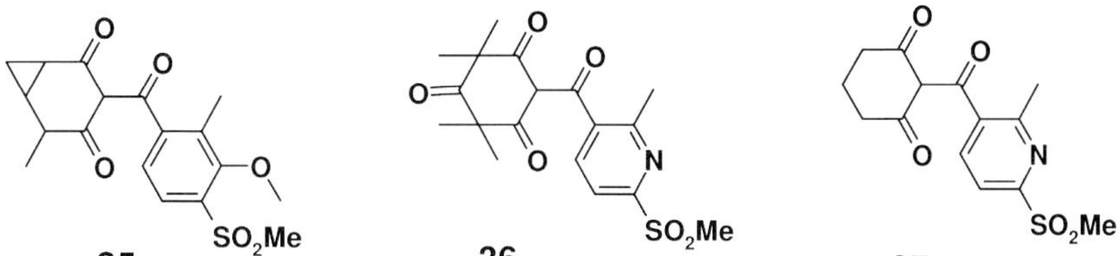

35 36 37

Fig. 4.3.6. Nippon soda types.

They also, apparently, were very interested in triketones with a nicotinoyl acid moiety, based on the number of applications filed in this area (**34**, **36**, and **37**, Fig. 4.3.6) [42, 43]. Nissan noticed the similarity of the triketones to the pyrazole type herbicides such as pyrazolate (**8**), and secured intellectual property freedom in

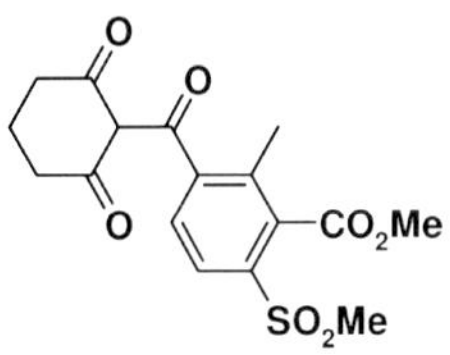

38

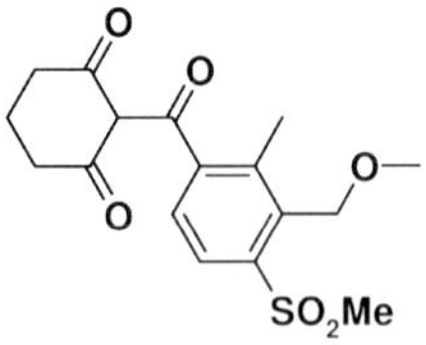

39

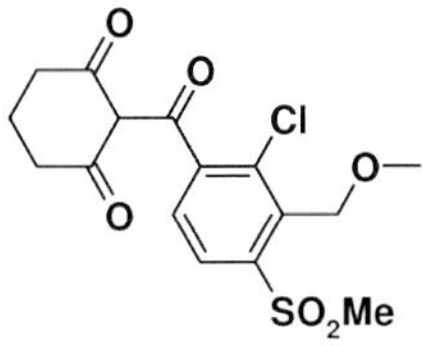

40

Fig. 4.3.7. Nissan triketones.

this area by patenting pyrazoles with the optimally substituted aroyl acids discussed previously [44]. Some important triketones containing novel trisubstituted acids were also first patented by Nissan (**38**–**40**, Fig. 4.3.7) [45].

BASF initially attempted to conquer some intellectual property by using proprietary diones from their DIMS chemistry (**41** and **42**, Fig. 4.3.8) [46]. However,

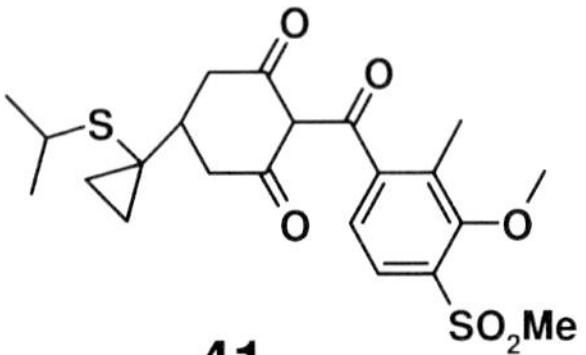

41

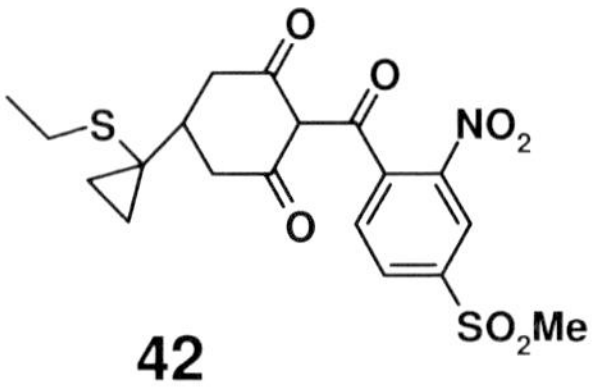

42

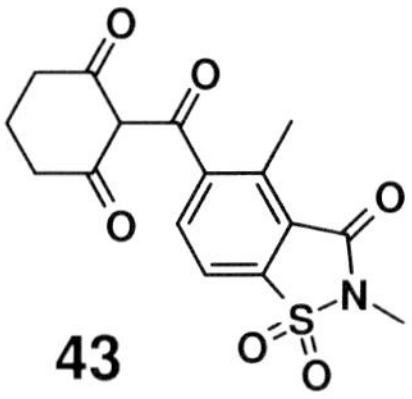

43

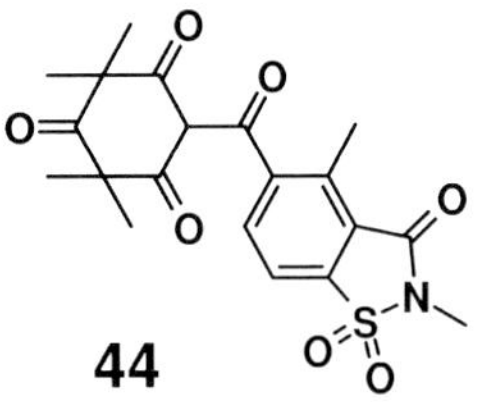

44

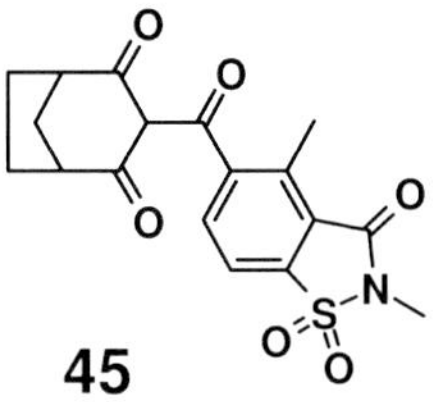

45

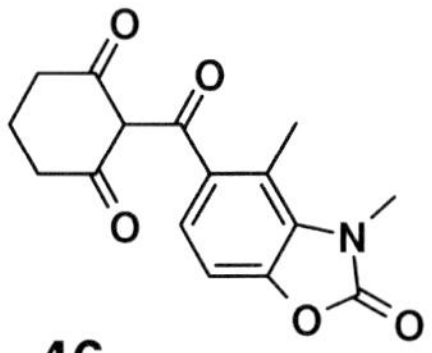

46

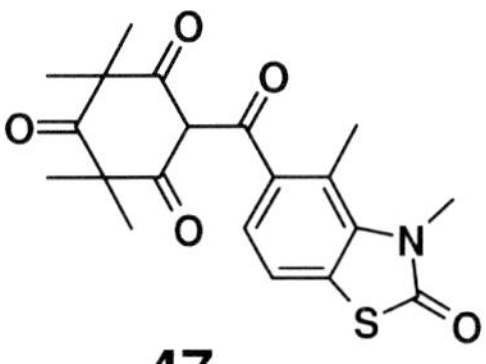

47

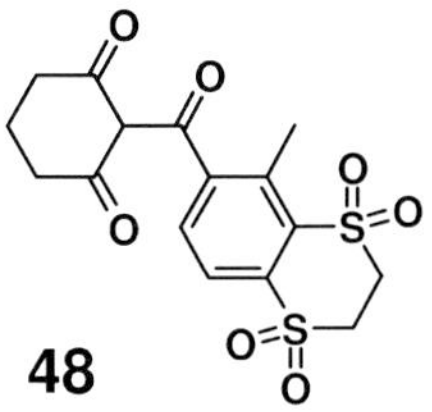

48

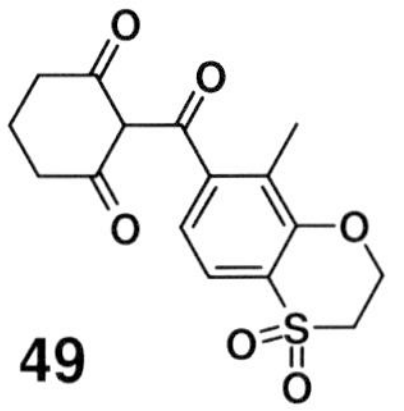

49

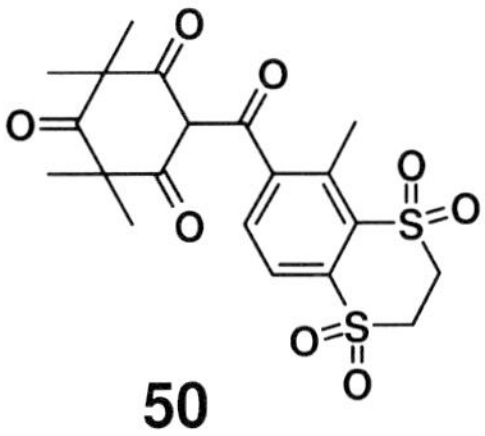

50

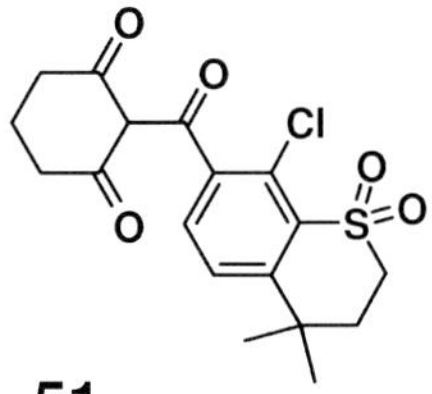

51

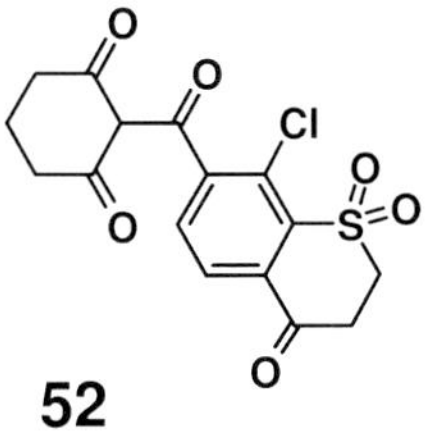

52

Fig. 4.3.8. BASF triketones.

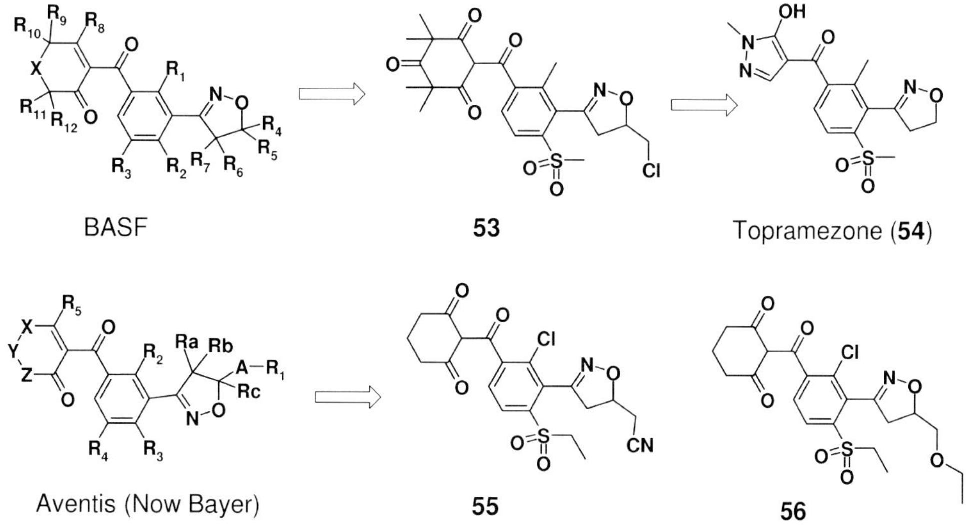

Fig. 4.3.9. BASF and Bayer meta-heterocyclic substituted triketones.

after probably realizing that large substituents at the 5-postion of cyclohexane-diones are not optimal for herbicidal activity, they switched their attention to the search for novel acids. Particularly prominent acids from BASF that have appeared in the patent literature are the saccharin's (e.g., **43–45**, Fig. 4.3.9) [47] and other fused 3,4-aroyl acids, such as **46** and **47** (Fig. 4.3.8) [48], and especially those in which the alkylsulfonyl group is incorporated into a fused ring at the 3,4-positions (**48–52**, Fig. 4.3.8) [49, 50].

BASF also explored patent free examples of triketones with novel meta-substituents, particularly acids containing heterocyclic rings at this position (e.g., **53**, Fig. 4.3.9). The 4,5-dihydro-isoxazole containing pyrazole corn herbicide top-ramezone (**54**, IMPACT®, CLIO®) [50, 51] has resulted from this work (Fig. 4.3.9).

With regard to triketones of this structure type, Aventis (now Bayer) patented substituted 4,5-dihydro-isoxazole compounds [52] prior to BASF [53], and two compounds from Bayer (**55** and **56**, Fig. 4.3.9) have frequently appeared in mixture patents with safeners and other herbicides for use in corn [54].

Idemitsu also concentrated their efforts on new acids, with emphasis on those in which the alkylsulfonyl substituent at the 4 position was joined into a ring at the 3-postion (typical Idemitsu types **57–61** are shown in Fig. 4.3.10) [55]. Although they received a patent for compounds of this type with 2-chloro substituent (e.g., **59**) they were also forced to switch to more complicated substituted het-eroaromatic systems after the publication of interfering patents from Zeneca [33], or to pyrazoles [56]. They now appear to have a pyrazole compound (generic structure **61**, Fig. 4.3.10) in development for use in corn, based on recently published mixture patents [57].

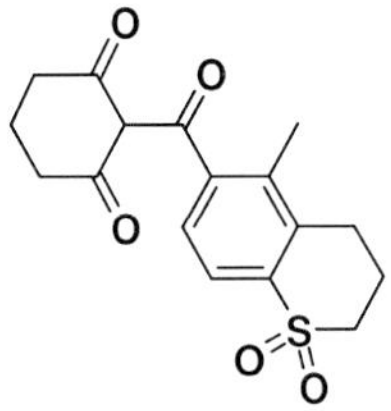

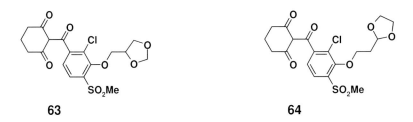

57 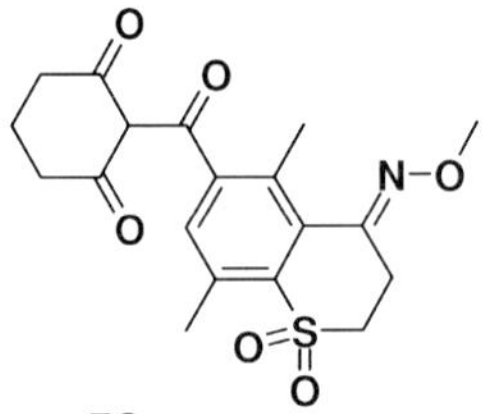   **58** 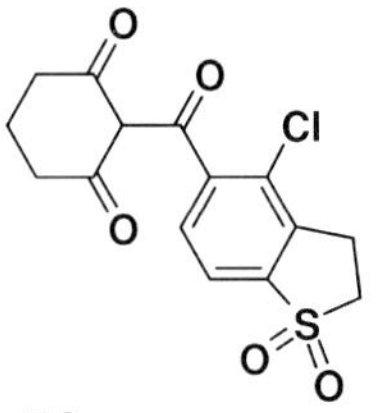   **59**

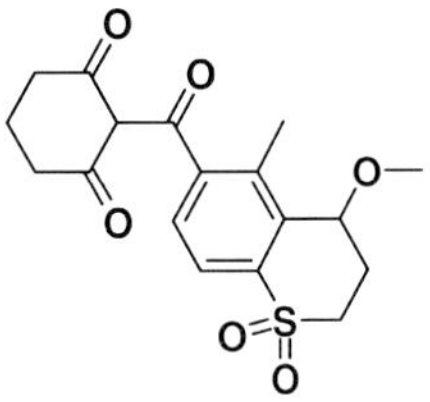

60 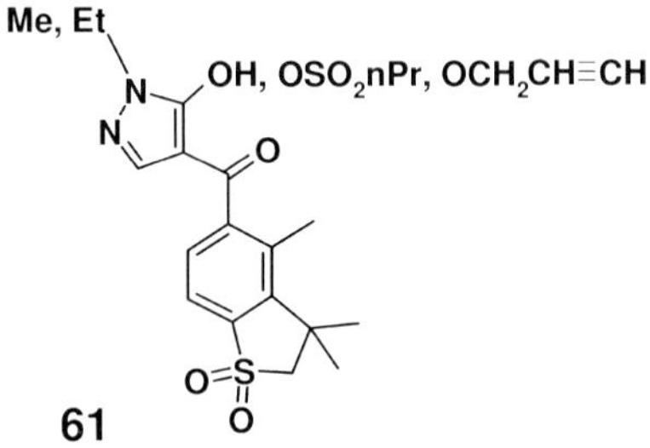   **61** 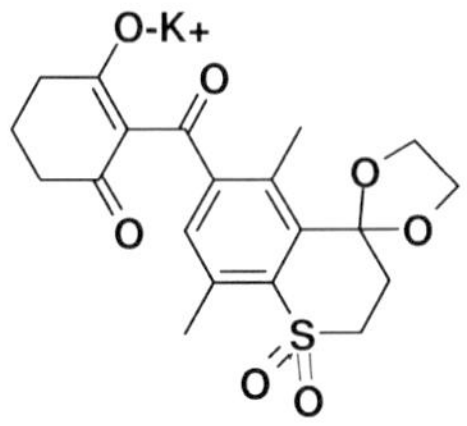   Ketospiridox (**62**)

Fig. 4.3.10. Idemitsu and Du Pont triketones.

Du Pont started relatively late in the triketone field, and directed their efforts toward novel fused bicyclic acids. As a result of their work in this area, they discovered the broad leaf weed cereal herbicide ketospiridox [58] (**62**, proposed common name, Fig. 4.3.10). Ishihara, inspired by earlier work of Hokko [59], identified some new benzoyl analogues with cyclic acetal meta-substituents (**63** and **64**, Fig. 4.3.11) that they claim have good pre-emergent activity in flooded rice paddy fields without damaging the rice seedlings [60].

63   **64**

Fig. 4.3.11. Ishihara rice triketones.

Bayer have published recently – mainly after the successful merger with Aventis – a multitude of patents [61], in which they have basically explored in more detail the effect of several novel meta-substituents on the biological activity of triketones, especially those with substituted 3-alkoxyalky-2-chloro-4-alkylsulfonyl substituents in the aromatic ring. From this work, the post-emergence corn herbicide

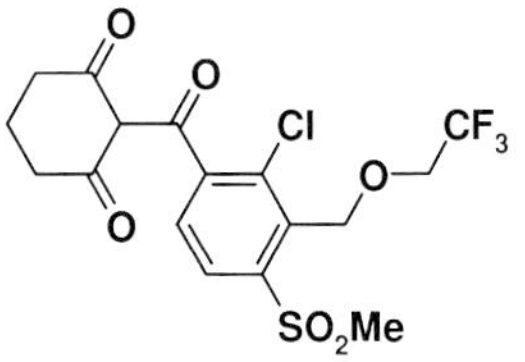

Tembotrione (**65**)

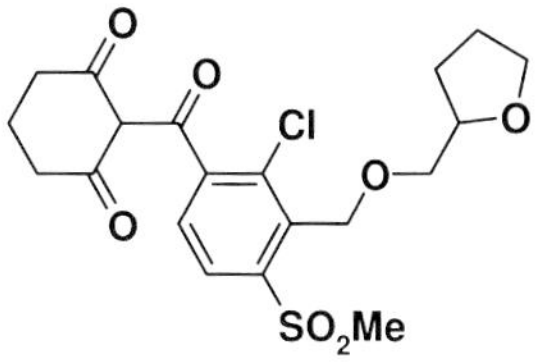

66

Fig. 4.3.12. Bayer development compounds.

tembotrione (**65**, proposed common name, Fig. 4.3.12) has resulted [62]. Tembo-trione is being promoted as a direct competitor to mesotrione (**18**) in the corn her-bicide market. Where tembotrione differentiates itself from mesotrione is im-proved grass control according to tests reported by Bayer [62]. Several mixture patents have also been disclosed, which suggest Bayer has a compound to be de-veloped in paddy rice [63]. Figure 4.3.12 shows the structure of the putative devel-opment product (**66**), and, as can be seen, it is very similar to the Ishihara's rice compounds shown in Fig. 4.3.11.

DOW invested virtually all their effort in the field of pyrazoles and has recently published several reviews in this area (Chapter 4.4) and only a few triketone pa-tents appeared. Those that did had meta-amino substituents such as **67** and **68** (Fig. 4.3.13) [64].

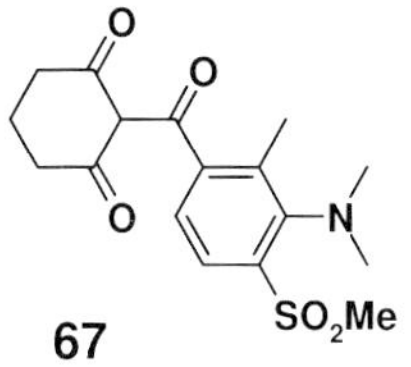

67

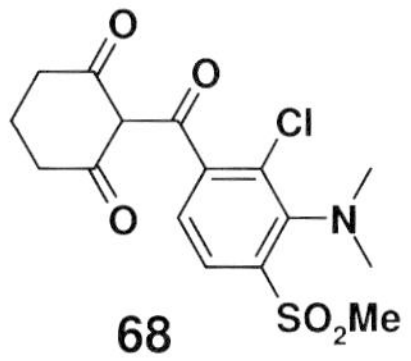

68

Fig. 4.3.13. Dow triketones.

Despite the broad granted scope of the initial Zeneca heteroaroyl triketone pa-tent [34], the pyridyl triketones were not further pursued by Zeneca, neither (be-hind the earlier Stauffer claims [36]) by Sandoz [27b] nor by Ciba-Geigy [65] (all now Syngenta). Nippon Soda too also left gaps in their patents in claiming pyrid-yls [42, 43]. All of this was exploited then once more by Novartis (now Syngenta). A series of patents around novel pyridyl acid containing triketones were pub-lished [66], and on the basis of mixture [67] and process [24, 37, 68] patents it appears that Syngenta found some very interesting new compounds for use in corn (**69** and **70**) and cereals (**71** and **72**, Fig. 4.3.14).

Particularly interesting here is that **70** and **71** containing larger ortho substitu-ents than the usual patented types (e.g., Cl or Me) are tolerated at the enzyme site, and that the picolinic acid (**72**) has no ortho-substituent, which suggests that the lone pair on the pyridyl nitrogen can act as such.

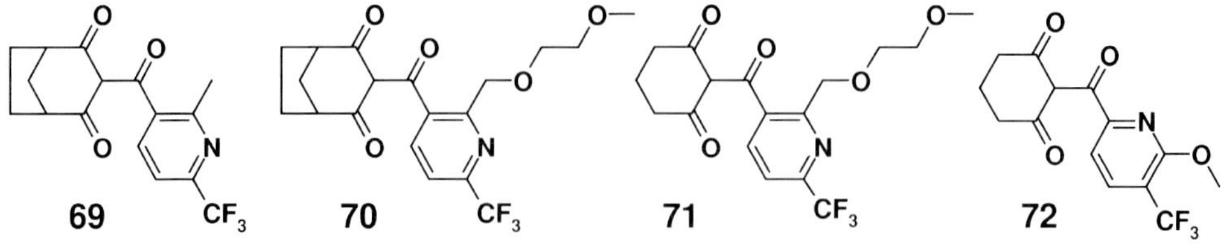

Fig. 4.3.14. Syngenta pyridyl triketones.

4.3.7
Commercialized Triketone Herbicides

The first triketone herbicide to be commercialized was sulcotrione (**17**). It was discovered by Stauffer (one of the legacy companies of Syngenta) and was registered for use in 1993 [69]. The product is sold under the trade name of MIKADO® in Europe and had sales in 2004 of $60 million (the compound is not registered for use in the USA) [70]. It was sold by Syngenta in 2001 to Bayer to satisfy conditions imposed by the European Commission in connection with the merger of Novartis Agribusiness and Zeneca Agrochemicals to form Syngenta. Bayer are apparently now seeking registration in the USA [70]. It is used for post-emergence control of (particularly) broadleaf weeds and some grass weeds (*Digitaria sanguinalis, Echinochloa crus-galli*, and *Panicum miliaceum*) in corn with application rates being 300–450 g ha^{-1}. Sulcotrione is readily absorbed through the leaves and also via the roots. The metabolism in soil is as shown in Fig. 4.3.15 [71]. Soil half-lives of 15–74 days have been measured, which causes no threat to cereals, the usual rotational crops to corn in Europe [72].

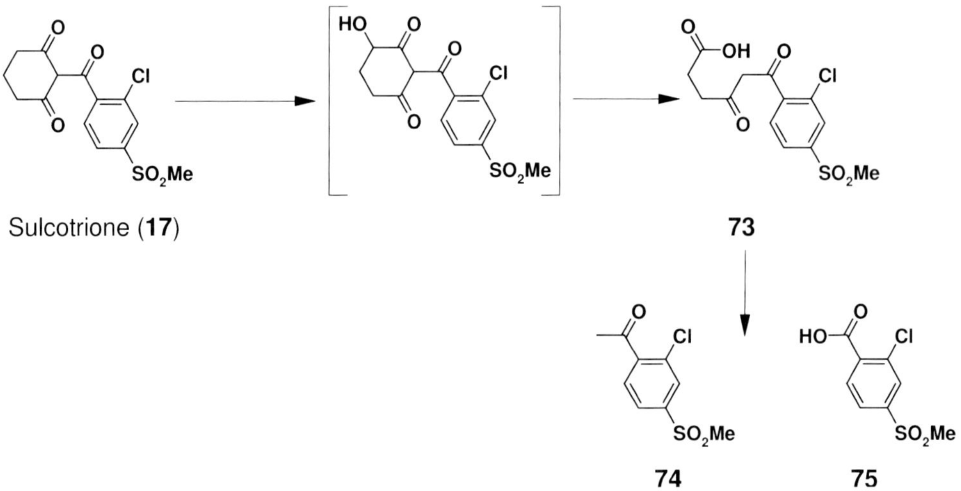

Sulcotrione (**17**)

Fig. 4.3.15. Sulcotrione soil metabolism.

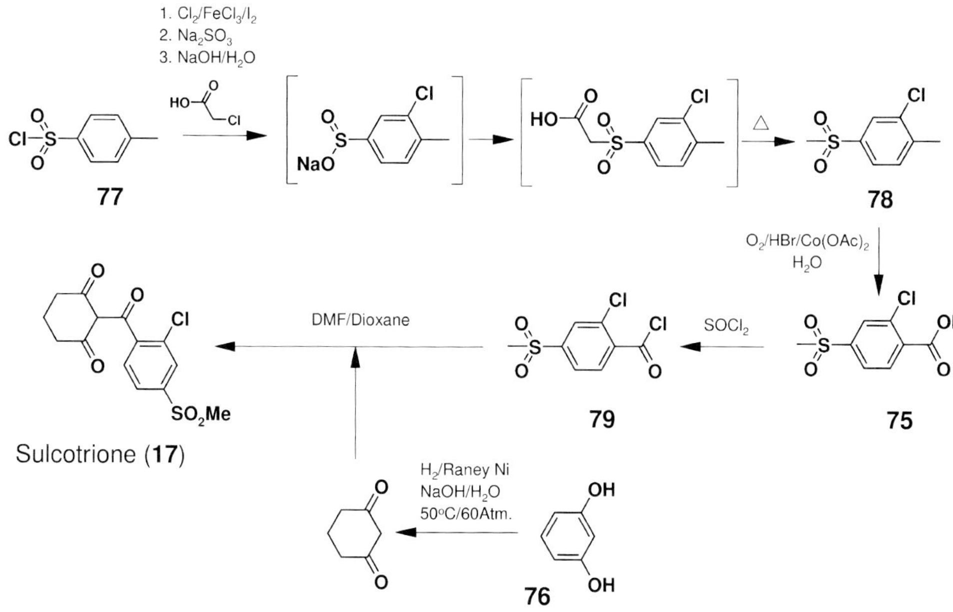

Scheme 4.3.7 Synthesis of sulcotrione.

Scheme 4.3.7 shows a recently published possible technical synthesis that gave a yield of 59–62% starting from resorcinol (**76**) and *p*-toluenesulfonyl chloride (**77**) [73].

Selected physical chemical, toxicological, and environmental properties of sulcotrione are listed in Table 4.3.1, column 1.

Mesotrione (**18**) is a second-generation triketone product developed by Zeneca (now Syngenta) as CALLISTO® for pre-emergence and post-emergence control of all the important broad-leaved weeds in corn together with some of the annual grass weeds (*Digitaria* and *Echinochloa* sp.), which are important in European corn production [1]. Typical use rates range from 100 to 225 g ha^{-1} when applied pre-emergence, and 70 to 150 g ha^{-1} for post-emergence applications. Broad-leaved weeds controlled include *Xanthium strumarium*, *Abutilon theophrasti*, *Ambrosia trifida*, together with *Chenopodium*, *Amaranthus* and *Polygonum* sp. Selectivity in corn is given by its ability to rapidly metabolize (detoxify) mesotrione into the inactive metabolites (**80**) and (**81**) (Fig. 4.3.16).

This metabolism is mediated by cytochrome P450 enzymes in both corn and weeds. In corn the detoxification process is so rapid that mesotrione is not translocated away from the directly treated area. However, the P450 enzymes in susceptible weeds can only metabolize mesotrione slowly. This allows extensive translocation throughout the weed (uptake occurs through the leaves, roots and shoots) and allows inhibition of HPPD [1]. It has also been suggested that a secondary effect contributing to corn selectivity may be the fact that foliar uptake of

Table 4.3.1 Selected physical chemical, toxicological, and environmental properties of commercial triketones.

Compound	Sulcotrione	Mesotrione	Benzobicyclon
Structure	17	18	9
Major product names	MIKADO®	CALLISTO® Other Products (mixtures): CAMIX®, CALARIS®, LEXAR®, LUMAX®	SHOW-ACE® Other products (mixtures): FOCUS SHOT®, KUSAKONTO®, SMART®
Melting point (°C)	139	165	187.3
Vapor pressure (mPa)	5×10^{-3} (25 °C)	5.69×10^{-3} (20 °C)	$<5.6 \times 10^{-2}$ (25 °C)
K_{ow} log P	<0	0.11	3.1
Solubility in water (mg L^{-1})	165 (25 °C)	15 (25 °C); 2.2 (20 °C) (pH 4.8)	0.00052 (20 °C)
Stability in water	Stable	Stable	Rapidly hydrolyzed
pK_a	3.13	3.12	–
Rat (acute oral) LD$_{50}$ (mg kg^{-1})	>5000	>5000	>5000
Birds dietary LC$_{50}$ for bobwhite quail (mg kg^{-1})	>5620	>2000	>2250
Fish LC$_{50}$ (96 h) for rainbow trout (mg L^{-1})	227	>120	LC$_{50}$ (48 h) for carp > 10 ppm
Algae	EC$_{50}$ (96 h) for *Selenastrum capricornutum* 1.2 mg L^{-1}	EC$_{50}$ (72 h) for *S. capricornutum* 4.5 mg L^{-1}	EC$_{50}$ (72 h) for *S. capricornutum* > 1 ppm
Bees LD$_{50}$ (oral and contact) (μg per bee)	>200	>9	>200
K_{oc}	44 (high pH) to 940 (low pH)	19 (pH 7.7) to 387 (soil pH 4.6)	
Soil deg DT$_{50}$ (days)	1–11	4–31.5	

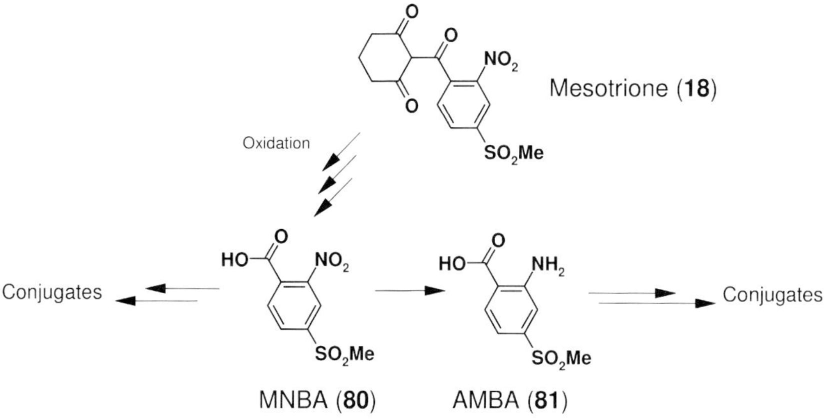

Fig. 4.3.16. Mesotrione *in planta* metabolism.

mesotrione is slower for corn than for weed species. Recent studies suggest the selectivity of sulcotrione may also be rationalized by similar arguments [74]. In extensive field tests in the USA and Europe, no corn injury has been observed with pre-emergence applications, and injury averages $\leq 3\%$ for post-emergence applications. In contrast, soybean is extremely sensitive, developing bleaching symptoms when treated with mesotrione post- emergence at application rates as low as 4 g ha^{-1}. Nevertheless, there is no risk of carry-over in rotational soybean crops due to the rapid degradation of mesotrione in soils. Mesotrione is also sold in mixtures with other herbicides to complete its spectrum (notable gaps are pre-emergent grass control and *Setaria* sp. in general). Some important brand products are LEXAR® and LUMAX® (various mixtures of mesotrione, S-metolachlor and atrazine, or alternatively terbuthylazine in countries/regions where atrazine use is prohibited). These mixtures are used as a pre-emergence broad spectrum weed control product in corn (one-shot treatment). The products have a high S-metolachlor content for areas where *Setaria* species are a major problem. Some formulations of LUMAX® also contain the S-metolachlor safener benoxacor. CAMIX® (mesotrione plus S-metolachlor) is a product that has been developed to give broad spectrum pre-emergence control of broadleaf and grass weeds in corn, where triazine herbicides are not permitted or desired, and CALARIS® (mesotrione plus terbuthylazine) is used as an early post-emergence weed control herbicide in corn (dicots and some grass weeds) for countries where atrazine use is forbidden. Mixtures with inhibitors of photosystem II such as atrazine and terbuthylazine are truly synergistic [75], which is a consequence of the complementary mode of action of triketones and PS-II inhibitors. Mesotrione has been a major success since its introduction in 2001 into the USA. Sales of mesotrione-based products have steadily increased ($270 million in 2004) [70], and it is now also a major product in Europe. Mesotrione can be synthesized similarly to the synthesis shown for sulcotrione. Scheme 4.3.8 shows a possible technical synthesis of the required benzoic acid (**80**) starting from (**82**) [76, 77].

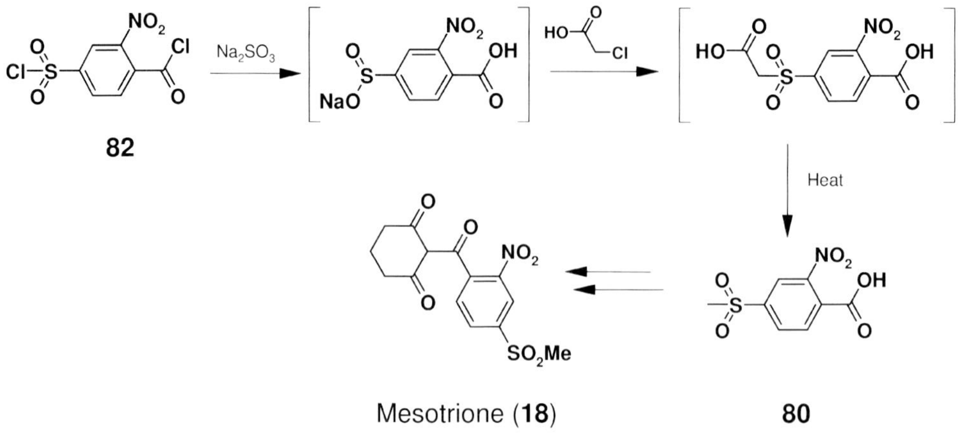

Scheme 4.3.8 Synthesis of mesotrione.

Selected physical chemical, toxicological, and environmental properties of mesotrione are listed in Table 4.3.1, column 2.

Benzobicyclon (**9**, SHOW-ACE®, Table 4.3.1, column 3) is a new triketone niche product herbicide that has been developed for control of broadleaf weeds (especially sulfonyl urea resistant weeds, e.g., *Lind* sp., *Lindernia attenuata*, *Monochoria vaginalis*) and some important grasses [e.g., *Scirpus juncoides* (sulfonyl urea resistant), *Echinochloa oryzicola* and other *Echinochloa* sp.] in paddy rice [78]. Inspection of the structure shows that it contains the sulcotrione acid moiety combined with a bicyclo[3.2.1]octane-2,4-dione, which in turn has been further elaborated to a pro-herbicide (attachment of a hydrolytically labile phenyl sulfide group to the vinylogous acid hydroxyl moiety). The latter imparts some positional selectivity to rice by decreasing the water solubility of the molecule. The phenylsulfide moiety slowly hydrolyses in water, or is metabolized in the plant to generate the active principle (**9'**, see Fig. 4.3.2). Benzobicyclon is reported to be very selective in rice and environmentally friendly due to its low water solubility ($3000\times$ less than sulcotrione) and low fish toxicity [LD_{50} (48 h), Carp > 10 ppm] [78]. Low water solubility is important for paddy rice herbicides as herbicide-containing water flowing out of the paddy field is minimized. Benzobicyclon arose out of a joint venture with SDS Biotech and Sandoz Crop Protection (now Syngenta). The presence of Sandoz's bicyclo[3.2.1]octane-2,4-dione moiety (BIOD, **20**, Scheme 4.3.9) posed a major problem for the development of this compound as the initial synthesis of benzobicyclon was cost prohibitive, particularly the synthesis of BIOD. However, after extensive process work, chemists at SDS were able to reduce the synthesis of BIOD to four steps [78, 79]. Another critical breakthrough in the technical synthesis was the finding that aluminum trichloride mediated C-acylation could be achieved in high yield directly from BIOD and the acid chloride (**79**), as the commercial use of cyanide catalyzed O-

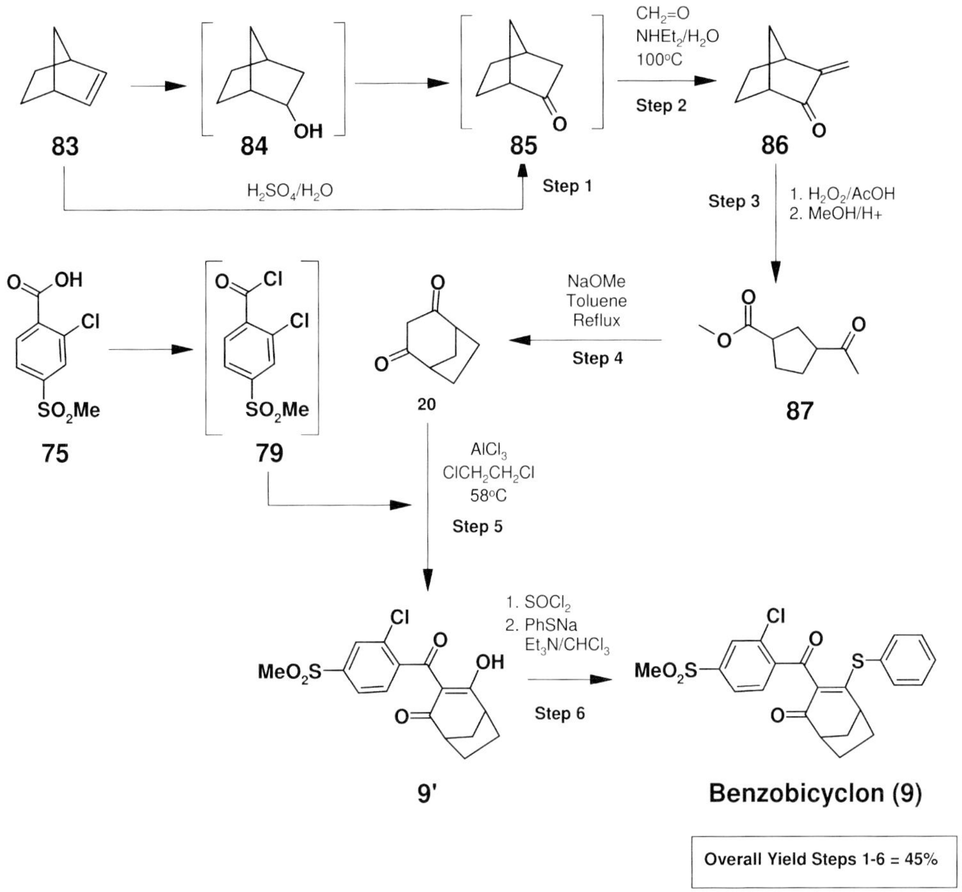

Scheme 4.3.9 Synthesis of benzobicyclon.

to C-acylation was prohibited by competitor patents [16]. Scheme 4.3.9 shows the industrial synthesis of benzobicyclon [78, 79].

Selected physical chemical, toxicological, and environmental properties of benzobicyclon are listed in Table 4.3.1, column 3.

4.3.8
Summary

Since their discovery in the early 1980s, the triketone herbicides have been extensively studied over the last two and half decades. In view of this, it may surprise the reader that only three commercial products have appeared to date. However, as has been described, other triketone products are due to appear on the market (e.g., tembotrione), and related compounds with this mode of action (see Chapter

4.4) arc likely to play an important role in weed control over the coming years, as there is to date no known case of resistance to HPPD inhibitors.

References

1 G. Mitchell, D.W. Bartlett, T.F. Fraser, T.R. Hawkes, D.C. Holt, J.K. Townson, R.A. Wichert, *Pest. Manage. Sci.*, **2001**, 57, 120–128.

2 R.O. Hellyer, *Aust. J. Chem.*, **1968**, 21, 2825–2828.

3 R.A. Gray, R.J. Rusay, C.K. Tseng, (Stauffer, now Syngenta AG) US 4 202 840, **1980**.

4 D.L. Lee, C.G. Knudsen, W.J. Michaelay, H. Chin, N.H. Nguyen, C.G. Carter, T.H. Cromartie, B.H. Lake, J.M. Shribbs, T. Fraser, *Pestic. Sci.*, **1988**, 54, 377–384.

5 J.W. Michaely, G.W. Kraatz, (Stauffer, now Syngenta AG), EP 90262, **1983**.

6 See Chapter 4.2.

7 K.E. Pallett, S.M. Cramp, J.P. Little, P. Veerasekaran, A.J. Crudace, A.E. Slater, *Pest Manage. Sci.* **2001**, 57, 133–142.

8 H. Matsumoto, M. Mizutani, T. Yamaguchi, J. Kadotani, *Weed Biol. Manage.*, **2002**, 2, 39–45.

9 K. Sekino, J. Wang, T. Nakano, S. Nakahara, T. Asami, H. Koyanagi, Y. Yamada, S. Yoshida, *Abstracts of Annual Meeting of Pesticide Science Society of Japan*, **2001**, p. 97.

10 H. Kakidani, K. Hirai, *J. Pest. Sci.* **2003**, 28, 409–415.

11 C. Yang, J.W. Pflugrath, D.L. Camper, M.L. Foster, D.J. Pernich, T.A. Walsh, *Biochem.*, **2004**, 43, 10414–10423.

12 J.M. Brownlee, K. Johnson-Winters, D.H.T. Harrison, G.R. Moran, *Biochemistry*, **2004**, 43, 6370–6377.

13 L. Willms, A. Van Almsick, H. Bieringer, T. Auler, F. Thurwachter, Felix. (Aventis, now Bayer CropScience). WO 01/007422, **2001**.

14 W. Zhang, G. Pugh, *Tetrahederon Lett.*, **2001**, 42, 5617–5620.

15 T. Isobe, T. Ishikawa, *J. Org. Chem.*, **1999**, 64, 6984–6988.

16 H.M. Chin (Zeneca, now Syngenta AG) US 4 781 751, **1988**; C.G.

Knudsen (Zeneca, now Syngenta AG) US 4 838 932, **1989**; N.H. Nguyen (Stauffer, now Syngenta AG), US 4,997,473, **1991**; E. Bay (Stauffer, now Syngenta AG), US 4,774,360, **1988**.

17 A. Akhrem, F.A. Lakhvich, S.I. Budai, T.S. Khlebnicova, I.I. Petrusevich, *Synthesis*, **1978**, 925–927.

18 P. Brown, M. Stephen, T.W. Bentley, R.O. Jones (Zeneca, now Syngenta AG), WO 99/28282, **1999**.

19 P.W. Wojtkowski, (E.I. Du Pont de Nemours and Co.), US 03/232984, **2003**.

20 K. Coppola, (Syngenta AG), WO 01/010806, **2001**.

21 See, for example, R. Beaudegnies, A.J.F. Edmunds, C. Luethy, R.G. Hall, S. Wendeborn, J. Schaetzer (Syngenta AG.), WO 04/058712, **2004**.

22 S.M. Brown, H. Rawlinson, J.W. Wiffen, (Zeneca, now Syngenta), WO 96/22957, **1996**.

23 (a) T. Inoue, M. Takata, T. Suzuki, I. Kenji, (Nippon Soda), WO 93/20035, **1993**. (b) W.J. Michaely, G.W. Kraatz, (Stauffer, now Syngenta). US 4780127, **1988**.

24 D.A. Jackson, A.J.F. Edmunds, M.C. Bowden, B. Brockbank (Syngenta AG), WO 05/105745, **2005**.

25 S. Kudis, U. Misslitz, E. Baumann, W. Von Deyn, K. Langemann, (BASF AG), WO 02/016305, **2002**.

26 D.B. Rubinov, I.L. Rubinova, A.A. Akhrem, Aphanasy A., *Chem.Rev.* **1999**, 99, 1047–1065.

27 (a) N.M. Berry, M.C.P. Darey, L.M. Harwood, *Synthesis* **1986**, 6, 476–80. (b) R.J. Anderson, J. Grina, F. Kuhnen, S.F. Lee, G.W. Luehr, H. Schneider, K. Seckinger, (Sandoz AG, now Syngenta AG), DE 3902818, **1989**.

28 D.L. Lee, M.P. Prisbylla, T.H. Cromartie, D.P. Dagarin, S.W.

Howard, W.M. Provan, M. Ellis, T. Fraser, L.C. Mutter, *Weed Sci.* **1997**, 45, 601–609.

29 D.L. Lee, C.G. Knudsen, W.J. Michaely, J.B. Tarr, H.-L. Chin, N.H. Nguyen, C.G. Carter, T.H. Cromartie, B.H. Lake, J.M. Shribbs, S. Howard, S. Hanser, D. Dagarin, *ACS Symposium Series*, **2001**, 774 (Agrochemical Discovery), 8–19.

30 J.B. Tarr, Book of Abstracts, 219th ACS National Meeting, San Francisco, March 26–30, **2000**.

31 D.L. Lee (ICI Americas now Syngenta) WO 93/03009, **1993**.

32 C.G. Carter, (Stauffer, now Syngenta AG), EP 186118, **1986**; W.J. Michaely, G.W. Kraatz, (Stauffer, now Syngenta AG), EP 137963, **1985**.

33 (Stauffer, now Syngenta AG), JP 62298584, **1987**.

34 J.E.D. Barton, D. Cartwright, C.G. Carter, J.M. Cox, D.L. Lee, G. Mitchell, F.H. Walker, F.X. Woolard, (ICI PLC UK and ICI Americas Inc., now Syngenta AG), EP 283261, **1988**.

35 D.L. Lee, F.H. Walker, F.X. Woolard, C.G. Carter, (Stauffer, now Syngenta AG), EP 316491, **1989**; C.G. Carter. (Stauffer, now Syngenta), US 4708732, **1987**.

36 S.F. Lee, (Sandoz AG., now Syngenta AG), EP 338992, **1989**.

37 H. Schneider, C. Luethy, A.J.F. Edmunds, (Syngenta AG), EP 1352901, **2003**.

38 W. Rueegg, (Novartis AG, now Syngenta AG), WO 00/000029, **2000**.

39 S.F. Lee, (Sandoz AG, now Syngenta AG), WO 92/07837, **1992**.

40 S.F. Lee, (Sandoz AG, now Syngenta AG), EP 394889, **1990**.

41 K. Komatsubara, T. Sato, K. Mikami, J. Yamada, M. Sato, (SDS Biotech), JP 06025144, **1994**.

42 H. Adachi, K. Tanaka, T. Kawana, H. Hosaka, (Nippon Soda), US 5294598, **1994**; H. Adachi, T. Aihara, K. Tanaka, T. Kawana, S. Yadama, H. Hosaka, (Nippon Soda). WO 93/01171, **1993**.

43 T. Sagae, M. Yamaguchi, H. Adachi, K. Tomida, A. Takahashi, T. Kawana,

(Nippon Soda). WO 96/14285, **1996**; A. Ueda, S. Suga, H. Adachi, K. Tomita, H. Yamagishi, H. Hosaka, Hideo, (Nippon Soda), JP 04029973, **1992**; A. Ueda, S. Suga, K. Tomita, H. Hosaka, (Nippon Soda), JP 03052862, **1991**.

44 M. Baba, N. Tanaka, T. Tsunoda, E. Oya, T. Igai, T. Nawamaki, S. Watanabe, (Nissan), JP 01052759 **1989**; N. Tanaka, E. Oya, M. Baba, (Nissan), EP 344775, **1989**.

45 T. Tsunoda, N. Tanaka, E. Oya, M. Baba, T. Igai, K. Suzuki, T. Nawamaki, S. Watanabe, (Nissan), JP 02000222, **1990**; T. Tsunoda, N. Tanaka, E. Ooya, M. Baba, K. Suzuki, T. Nawamaki, S. Watanabe, (Nissan), JP 01143851, **1989**.

46 J. Kast, W. Von Deyn, C. Nuebling, H. Walter, M. Gerber, K.O. Westphalen, (BASF AG), EP 666254, **1995**.

47 U. Misslitz, E. Baumann, W. Von Deyn, S. Kudis, K. Langemann, G. Mayer, U. Neidlein, M. Witschel, R. Gotz, M. Rack, P. Plath, M. Otten, K.-O. Westphalen, H. Walter, (BASF AG), WO 00/53590, **2000**; H. Walter, P. Plath, U. Kardorff, M. Witschel, R. Hill, W. Von Deyn, S. Engel, M. Otten, U. Misslitz, K.-O. Westphalen, (BASF AG), WO 98/40366, **1998**; P. Plath, H. Rang, K.-O. Westphalen, M. Gerber, H. Walter, (BASF AG), DE 4427996, **1996**; U. Plath, U. Kazrdorff, W. von Deyn, S. Engel, J. Kast, H. Rang, H. Koenig, M. Gerber, H. Walter, K.-O. Westphalen, (BASF AG), DE 4427995, **1996**.

48 G. Mayer, U. Misslitz, E. Baumann, W. Von Deyn, S. Kudis, M. Hofmann, T. Volk, M. Witschel, C. Zagar, A. Landes, K. Langemann, (BASF AG), WO 02/048121, **2002**.

49 M. Otten, Martina, W. Von Deyn, S. Engel, R. Hill, R.U. Kardorff, M. Vossen, P. Plath, H. Walter, K.-O. Westphalen, U. Misslitz, (BASF AG), WO 97/30986, **1997**.

50 W. Von Deyn, R. Hill, U. Kardorff, E. Baumann, S. Engel, G. Mayer, M. Witschel, M. Rack, N. Gotz, J. Gebhardt, U. Misslitz, H. Walter, K.-O. Westphalen, M. Otten, J.

Rheinheimer, (BASF AG), WO 98/31681, **1998**.

51 C. Boerboom, *Wisconsin Crop Manager*, **2006**, 13, 10–11.

52 L. Willms, A. Van Almsick, H. Bieringer, T. Auler, F. Thurwachter (Aventis, now Bayer CropScience) WO 01/007422, **2001**.

53 S. Kudis, E. Baumann, W. Von Deyn, K. Langemann, G. Mayer, U. Misslitz, U. Neidlein, M. Witschel, Matthias, K.-O. Westphalen, H. Walter, (BASF AG), WO 01/040199, **2001**.

54 H.-P. Krause, J. Kocur, J. de Una Martinez, U. Bickers, E. Hacker, G. Schnabel, (Aventis, now Bayer CropScience), WO 01/097614, **2001**; H. Bieringer, A. Van Almsick, E. Hacker, L. Willms, (Aventis, now Bayer CropScience), WO 01/028341, **2001**; A. Van Almsick, L. Willms, E. Hacker, H. Bieringer (Bayer CropScience), WO 02/085121 **2002**; F. Ziemer, A. Van Almsick, L. Willms, T. Auler, H. Bieringer, E. Hacker, C. Rosinger, (Bayer CropScience), WO 02/085120, **2002**.

55 (a) I. Nasuno, M. Shibata, M. Sakamoto, K. Koike, (Idemitsu), WO 94/08988, **1994**. (b) S. Tomita, M. Saito, H. Sekiguchi, S. Okawa, (Idemitsu), JP 2002114776, **2002**. (c) M. Saitou, H. Sekiguchi, S. Ogawa, (Idemitsu). WO 01/074802, **2000**. (d) M. Saitou, H. Sekiguchi, S. Ogawa, (Idemitsu), WO 00/069853, **2000**. (e) M. Saitou, H. Sekiguchi, S. Ogawa, (Idemitsu), WO 00/020408, **2000**.

56 I. Nasuno, M. Sakamoto, K. Nakamura (Idemitsu), WO 96/25412, **1996**; K. Nakamura, K. Koike, M. Sakamoto, I. Nasuno, (Idemitsu), US 2002016262, **2002**.

57 K. Koike, S. Abe, M. Kubota, Y. Takashima, (Idemitsu), JP 2004043397, **2004**; I. Nasuno, K. Koike, Kazuyoshi, (Idemitsu), JP 2004018416, **2004**.

58 C.-p. Tseng (E. I. Du Pont de Nemours & Co), WO 98/28291, **1998**.

59 T. Morita, T. Shimozono, K. Hirayama, H. Ishikawa, H. Yoshizawa, M. Yoshihara, (Hokko Chem Ind Co,) JP 07206808, **1995**; T.

Morita, T. Oono, Y. Kido, S. Maehara, H. Ishikawa, H. Yoshizawa, Hirokazu, H. Yamamura, Hiroshi. (Hokko Chem Ind Co.) JP 06271562, **1994**.

60 Y. Nakamura, M. Sano, (Ishihara Sangyo Kaisha), WO 05/118530, **2005**; Y. Nakamura, C.J. Palmer, H. Kikugawa, M. Sano (Ishihara Sangyo Kaisha), WO 01/014303, **2001**.

61 H.-P. Krause, J. Kocur, J. de Una Martinez, U. Bickers, E. Hacker, G. Schnabel, (Aventis, now Bayer CropScience), WO 01/097614, **2001**; H. Bieringer, A. Van Almsick, E. Hacker, L. Willms, (Aventis, now Bayer CropScience), WO 01/028341, **2001**; A. Van Almsick, L. Willms, E. Hacker, H. Bieringer (Bayer CropScience), WO 02/085121 **2002**; F. Ziemer, A. Van Almsick, L. Willms, T. Auler, H. Bieringer, E. Hacker, C. Rosinger, (Bayer CropScience), WO 02/085120, **2002**.

62 Bayer's new postemergence corn herbicide slated for 2009. G. Gullickson in *Agriculture Online*, http://www.agriculture.com/ag/story.jhtml?storyid=/templatedata/ag/story/data/1136309833704.xml&catref=ag1001.

63 K. Endo, S. Ito, N. Seishi, T. Nakajima, (Bayer CropScience), JP 2005306813, **2005**; K. Endo, S. Ito, Seishi, H. Mukaida, (Bayer CropScience), WO 04/105482, **2004**.

64 Z.L. Benko, J.A. Turner, (Dow Agrosciences LLC), WO 98/42648, **1998**.

65 H.G. Brunner (Ciba-Geigy AG, now Syngenta AG). EP 353187, **1990** and US 4,995,902, **1991**.

66 A.J.F. Edmunds, K. Seckinger, C. Luethy, W. Kunz, A. De Mesmaeker, J. Schaetzer, (Novartis, now Syngenta AG), WO 00/015615, **2000**; C. Luethy, A.J.F. Edmunds, R. Beaudegnies, S. Wendeborn, J. Schaetzer, (Syngenta AG), WO 05/058831 **2005**; C. Luethy, A.J.F. Edmunds, R. Beaudegnies, S. Wendeborn, J. Schaetzer, W. Lutz, (Syngenta AG), WO 05/058830, **2005**.

67 W.T. Rueegg (Syngenta AG), DE 102004012192, **2004**; W.T. Rueegg,

(Syngenta AG), WO 03/047343, **2003**;
W.T. Rueegg, (Syngenta AG.), WO
01/054501, **2001**; W.T. Rueegg,
(Syngenta AG) WO 03/047344, **2003**;
W.T. Rueegg (Syngenta AG) WO
03/047342, **2003**.

68 D.A. Jackson, M.C. Bowden, A.J.F.
Edmunds, B. Brockbank, (Syngenta
AG), WO 05/105718, **2005**.

69 M. Beraud, J. Claument, A. Montury,
ICIA0051, A new herbicide for the
control of annual weeds in maize, in
Proc Brighton Crop Prot Conf – Weeds,
BCPC, Farnham, Surrey, UK, pp 51–
56, **1993**.

70 Phillips McDougal, AgriService.
Products Section, 2004 Market,
November, **2005**.

71 J. Rouchaud, O. Neus, R. Bulcke, K.
Cools, H. Eelen, *Bull. Environ.
Contam. Toxicol.*, **1998**, 61, 669–676.

72 The technical properties of ICIA0051,
a new herbicide for maize and
sugarcane. T.J. Purnell, in *Proceedings
of the Annual Congress – South African*

Sugar Technologists' Association, **1991**,
65th 30–32. (CAN 116: 53566)

73 S. Guo, F. Yang, L. Zhang, *Nongyao*,
2001, 40, 20–21.

74 G.R. Armel, D.J. Mayonado, K.K.
Hatzios, H.P. Wilson, *Weed Technol.*,
2004, 18, 211–214.

75 B.H. Lake, T.J. Purnell, (Zeneca, now
Syngenta AG). WO 95/28839, **1995**.

76 Beilstein, EIII Band. 11, p. 685.

77 R.W. Brown, (ICI, now Syngenta AG),
WO 90/06301, **1990**.

78 K. Komatsubara, *Fain Kemikaru*,
2005, 34, 38–48; K. Sekino, K.
Komatsubara, H. Koyanagi, Y.
Yamada, *Shokubutsu no Seicho
Chosetsu*, **2004**, 39, 23–32.

79 K. Komatsuhara, K. Sano, T. Tabuchi,
J. Iwasawa, J.H. Kishi, (SDS Biotech
Corp.), JP 10265441, **1998**; H. Kishi,
T. Tabuchi, K. Komatsuhara, (SDS
Biotech Corp.), JP 2002003467, **2002**;
K. Komatsubara, M. Sano, T. Tabuchi,
J. Iwasawa, H. Kishi (SDS Biotech
Corp.), JP 10265432, **1998**.

4.4
Hydroxyphenylpyruvate Dioxygenase (HPPD) Inhibitors: Heterocycles

Andreas van Almsick

4.4.1
Introduction

As already mentioned in Chapter 4.2, all known HPPD inhibitors are chelating agents. To exhibit not only *in vitro* but also *in vivo* activity additional requirements such as uptake, transport and metabolic stability in plants (especially weeds) are necessary. The market compounds and the long list of published HPPD molecules with the general structure **1** (Fig. 4.4.1) fulfill, normally, all these needs [1].

There are many Q moieties but 1,3-cylohexanediones, pyrazolones and diketonitriles are the most important examples. It is essential to know that all compounds of the general structure **1** could exist in different tautomeric forms, as shown in Fig. 4.4.2 for Q = 1,3-cyclohexanedione.

Regarding the substitution pattern of **1**, the 2,4-disubstitution and the 2,3,4-trisubstitutions are particularly important (see Edmunds, Chapter 4.3, Section 4.3.5). *In vitro* activity is strongly connected with a substitution in 2-position ($R^1 \neq H$).

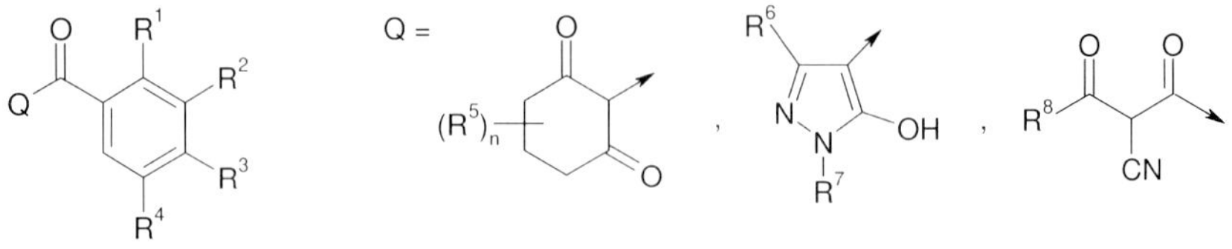

R$^{1-4}$ = H, alkyl, haloalkyl, halogen, etc.
R$^{5-6}$ = H, alkyl, etc.
R^7 = alkyl
R^8 = alkyl, cycloalkyl

Fig. 4.4.1. Markush structure of many HPPD inhibitors.

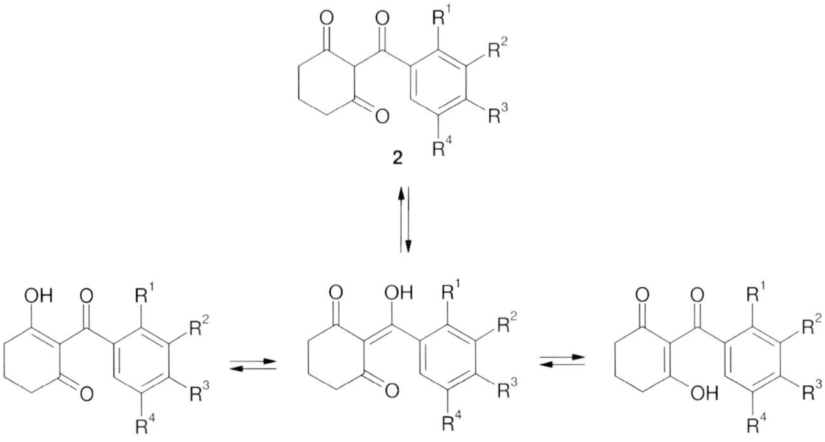

Fig. 4.4.2. Tautomeric forms of HPPD inhibitors of type 1,3-cyclohexanedione.

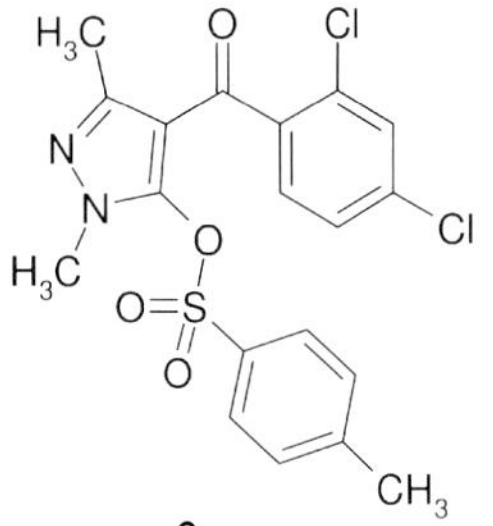

pyrazolynate

Fig. 4.4.3. Structure of pyrazolynate (**6**).

Whilst compounds of the general structure **1** are active *in vitro* and, therefore, drugs, for every Q different prodrugs are also known (see Edmunds, Chapter 4.3, Fig. 4.3.2).

For simplification, the HPPD inhibitors with Q different from 1,3-cyclohexanedione and its prodrugs are summarized here as heterocycles; and it was the heterocyclic HPPD-inhibitor pyrazolynate (**6**, Fig. 4.4.3) that was the first HPPD product launched into the market.

4.4.2
Market Products

4.4.2.1 Pyrazolynate (Pyrazolate)

When pyrazolynate (pyrazolate) (**6**, Fig. 4.4.3) was launched in 1980 by Sankyo Co., Ltd in Japan the world's first HPPD compound entered the herbicide market even though at that time the target site was unknown. Two years earlier, Sankyo had presented its activity on this area at the Fourth International Congress of Pesticide Chemistry in Zurich, Switzerland [2] but had already patented the main compounds in 1974 [3]. Interestingly, this all happened without the knowledge of the precise mode of action. Pyrazolynate and two analogues were previously classified as Protox inhibitors [4]. Pyrazolynate is not new and modern but it is included in this review as it is relatively unknown outside of Japan.

The herbicide with the trade name Sanbird® is able to control both annual and perennial weeds in paddy fields [5] with application rates of 3–4 kg ha^{-1}. As a very selective herbicide in rice it was a good innovation for the Japanese rice market. This product reached peak sales in Japan in 1986 with 650 000 ha (28.6% market share) [6]. This declined with the introduction of sulfonylurea herbicides such as bensulfuron-methyl in the year 1987. In 2005 Sanbird® was only used on 101 200 ha (5.9% market share) in Japan [7].

Pyrazolynate is a prodrug and itself not herbicidally active. It has low solubility in water [0.056 mg L^{-1} (25 °C)] and in solution it is hydrolyzed to give *p*-toluenesulfonic acid (**7**) and 4-(2,4-dichlorobenzoyl)-1,3-dimethyl-5-hydroxypyrazole (**8**), the herbicidal entity of pyrazolate [8–10] (Scheme 4.4.1).

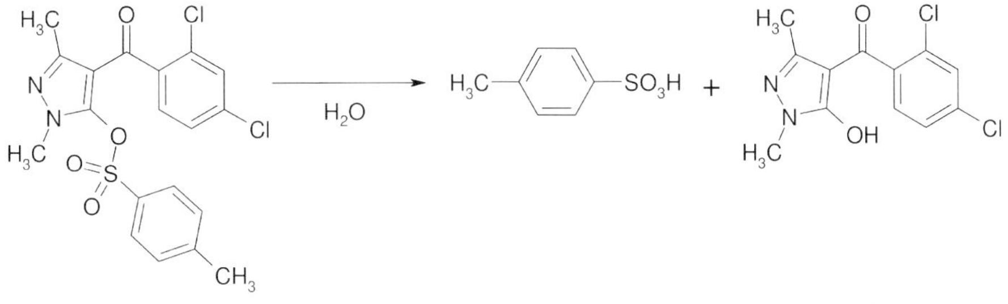

| **6** | **7** | **8** |

Scheme 4.4.1

The half-lives of pyrazolynate in water at 25 °C are: 52.7 h at pH 3; 17.5 h at pH 1; 25.0 h at pH 7; and 4.3 h at pH 9 [11]. In soil a DT_{50} of 8–10 days is observed [12].

Schemes 4.4.2 and 4.4.3 show the synthesis of pyrazolynate. 1,3-Dimethyl-5-pyrazolon (9) and 2,4-dichlorobenzoyl chloride (10) react in the presence of calcium hydroxide in isopropanol to give 4-(2,4-dichlorobenzoyl)-1,3-dimethyl-5-hydroxypyrazole (8) [3].

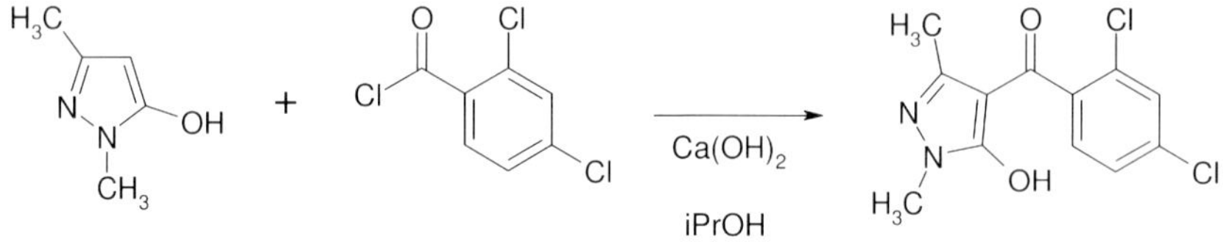

Scheme 4.4.2

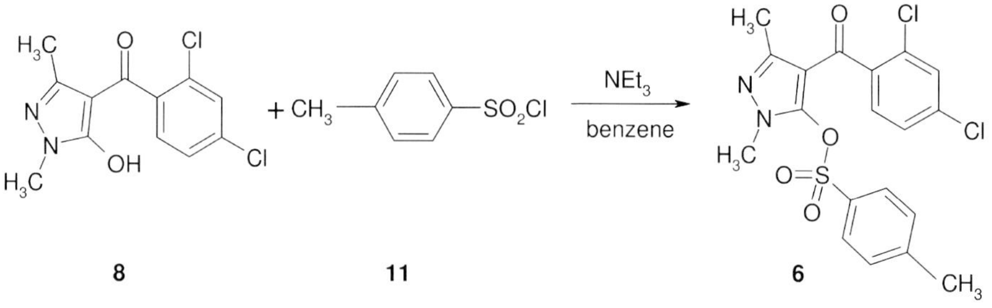

Scheme 4.4.3

4-Methylbenzenesulfonyl chloride (11) is then added to a solution of 8 and triethylamine in benzene [3].

Two points should be highlighted. Firstly, today instead of benzene other solvents such as toluene are used and, secondly, for the formation of substituted 4-benzoyl-1-alkyl-5-hydroxyxpyrazole like 8 other routes are also known. Scheme 4.4.4 shows the most popular one, with 8 as an example [13].

Both, 1,3-dimethyl-5-pyrazolon (9) and 2,4-dichlorobenzoic acid are commercial available, which allows a few steps synthesis of pyrazolynate. However, owing to the high application rate of 3–4 kg ha^{-1}, the treatment costs are very high. In theory the application rate could be lower by using the drug 4-(2,4-dichlorobenzoyl)-1,3-dimethyl-5-hydroxypyrazole (8) instead of the prodrug. Another important factor for the Japanese rice market is season-long weed control of a herbicide, which is not possible with the more polar and more water-soluble

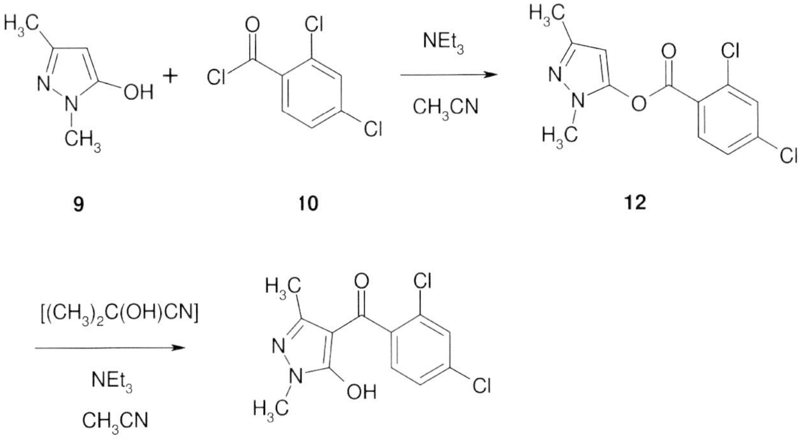

Scheme 4.4.4

drug **8** but is with the prodrug pyrazolynate. The effect is similar to that of a slow release formulation, the active ingredient is released over a long period of time and is, therefore, present at lethal dose rates for the weeds for a longer period of time.

4.4.2.2 Pyrazoxyfen

Pyrazoxyfen (**13**, Fig. 4.4.4) is a very close analogue of pyrazolynate and was lunched by Ishihara Sangyo Kaisha Ltd in 1985 for the Japanese rice market.

The herbicide was patented in 1977 [15] and reported in 1984 by F. Kimura [14]. The trade name is Paicer[R] and the herbicide has a broad weed control spec-

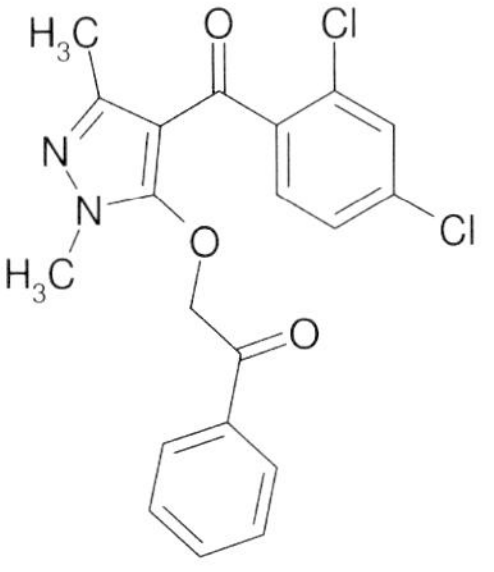

pyrazoxyfen

Fig. 4.4.4. Structure of pyrazoxyfen (**13**).

trum under flooded field condition, including for many annual and perennial weeds, with application rates of 3 kg ha^{-1}. It is very selective to transplanted rice and also to direct-seeded rice at temperatures < 35 °C. At higher temperatures temporary crop damage may occur [14].

Paicer$^{\circledR}$ reached peak sales in Japan in 1988 with 45 000 ha (2.2% market share) [6]. In 2005 the product was only used on 6911 ha (0.4% market share) in Japan [7]. As the second product to reach the Japanese market for the same segment as pyrazoynate and with the same mode of action, Paicer$^{\circledR}$ was and remains much less successful.

To synthesize pyrazoxyfen (**13**), 2-bromoacetophenone **14** is added to a solution of **8** and anhydrous potassium carbonate in methyl ethyl ketone (Scheme 4.4.5) [15].

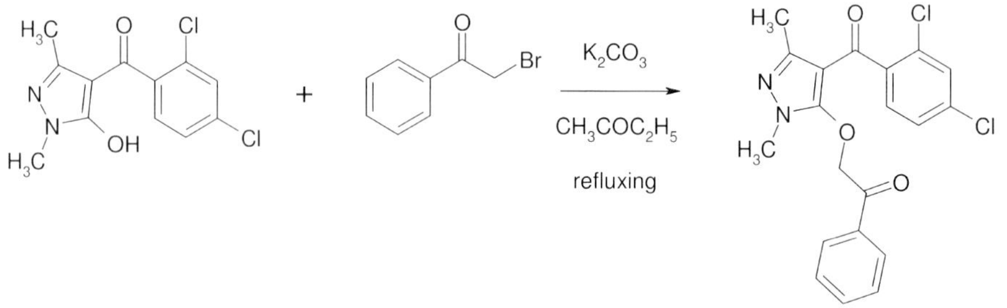

8 **14** **13**

Scheme 4.4.5

The difference between pyrazoxyfen and pyrazolynate is only the chosen prodrug system. In plants, both herbicides were metabolized to 4-(2,4-dichlorobenzoyl)-1,3-dimethyl-5-hydroxypyrazole (**8**). Pyrazolynate is only slightly soluble in water, but, once dissolved, is rapidly hydrolyzed to the herbicidally active metabolite [11]. In contrast, pyrazoxyfen shows considerable stability in aqueous solutions [16].

4.4.2.3 Benzofenap

As a third compound of this series benzofenap, (**15**, Fig. 4.4.5) was lunched by Mitsubishi Petrochemical Co. Ltd. (now Mitsubishi Chemical Corp.) and commercialized by Rhône-Poulenc Yuka Agro KK, a joint venture of Mitsubishi Chemical Corp. and Rhône-Poulenc Agro (now part of Bayer CropScience) in 1987 for the rice market. Interestingly, benzofenap is not only applied in Japan as Yukawide$^{\circledR}$ but also in Australia as Taipan$^{\circledR}$. The new herbicide was patented 1982 [17] and reported 1991 [18].

Yakawide$^{\circledR}$ reached peak sales 1998 in Japan with 180 000 ha (10% market share) [6]. In 2005 the product was only used on 62 000 ha (3.6% market share)

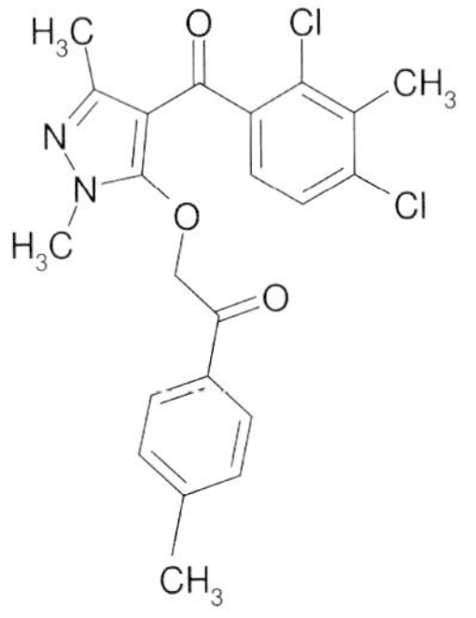

15

benzofenap

Fig. 4.4.5. Structure of benzofenap (**15**).

in Japan [7]. This third product for the Japanese rice market with HPPD mode of action Yakawide™ was much more successful than its closest analogue Paicer™.

The differences between benzofenap and pyrazoxyfen are the additional methyl groups on the biologically active metabolite 4-(2,4-dichloro-3-methylbenzoyl)-1,3-dimethyl-5-hydroxypyrazole (**16**) (Fig. 4.4.6) and the prodrug moiety 4′-methylacetophenone.

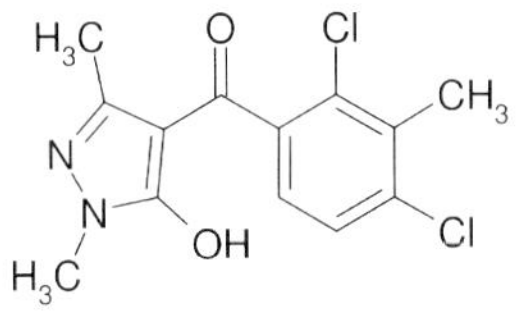

16 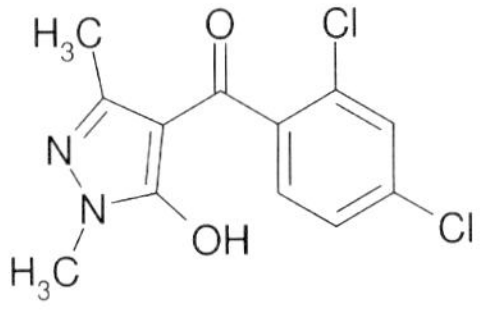   **8**

Fig. 4.4.6. Biologically active metabolites, **16** and **8**, of benzofenap and pyrazoxyfen, respectively.

These result in a different environmental behavior and different herbicidal activity [14, 18]. The half-lives in paddy field soil rose from 4 to 15 days for pyrazoxyfen and to 38 days for benzofenap. The application rate of 3 kg ha^{-1} is as high as for both the other rice herbicides but benzofenap allows a longer weed control of up to 50 days compared with 21–35 days with pyrazoxyfen. Importantly, benzofenap is a more crop selective herbicide. Another advantage of benzofenap over pyrazoxyfen is that it is not temperature-dependent. Even at higher temperatures no phytotoxicity is observed.

None of the three HPPD rice herbicides are able to control all annual and perennial weeds in rice, thus they need mixture partners, especially to fill gaps such as barnyard grasses or *Cyperus* spp. Common mixture partners are butachlor, pretilachlor, thiobencarb [19], piperophos [14], pyribaticarb and bromobutide [18].

4.4.2.4 Isoxaflutole

With the introduction of isoxaflutole (IFT) (**17**, Fig. 4.4.7) new crops, corn and sugarcane, came in the focus of HPPD inhibitors of type heterocycles. IFT is not the first HPPD compound for corn, this was sulcotrione in 1990, but it was the first for pre-emergence application. Reported 1995 by Luscombe et al. [20] the compound had been initially patented 1991 [21] by Rhône-Poulenc Agriculture Limited (now Bayer CropScience).

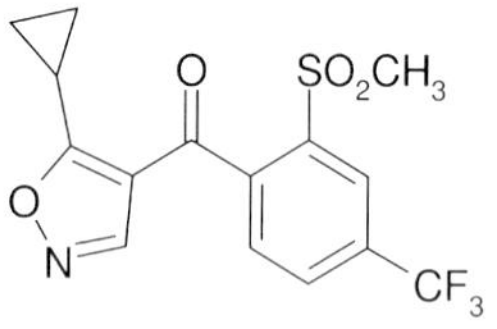

17

isoxaflutole

Fig. 4.4.7. Structure of isoxaflutole (**17**).

The herbicide with the trade names Merlin®, Balance®, Provence® and others was first launched 1996 in South America for broadleaf weed and grass control in corn and sugar cane. In corn, IFT is a selective pre-emergence herbicide. Applications are usually made in spring in post sowing/pre-emergence of the crop, but it is also possible to apply isoxaflutole in early pre-plant up to 3 weeks before the planting of the crop. The application rate of 75 g ha^{-1} is very low compared with other conventional pre-emergence herbicides for corn (e.g., S-metolachlor 0.8–1.6 kg ha^{-1}).

Common mixture partners in corn are flufenacet, aclonifen, terbuthylazine and, especially, atrazine to complete the weed spectrum.

In sugar cane, isoxaflutole controls annual grasses and some key annual broadleaf weeds. It may be applied pre- or post-emergence but normally pre-emergence is the preferred option. The application rate of 140 g ha^{-1} is still very low compared with other pre-emergence products [22, 23].

In sugarcane IFT may be tank-mixed with paraquat formulations, diuron, atrazine and Actril® DS.

In some countries isoxaflutole is also registered for weed control in other crops such as chick peas, poppy seed and some nurseries.

IFT is a much more complex compound than the three previously described rice compounds and needs, therefore, a longer synthesis route [21, 24]

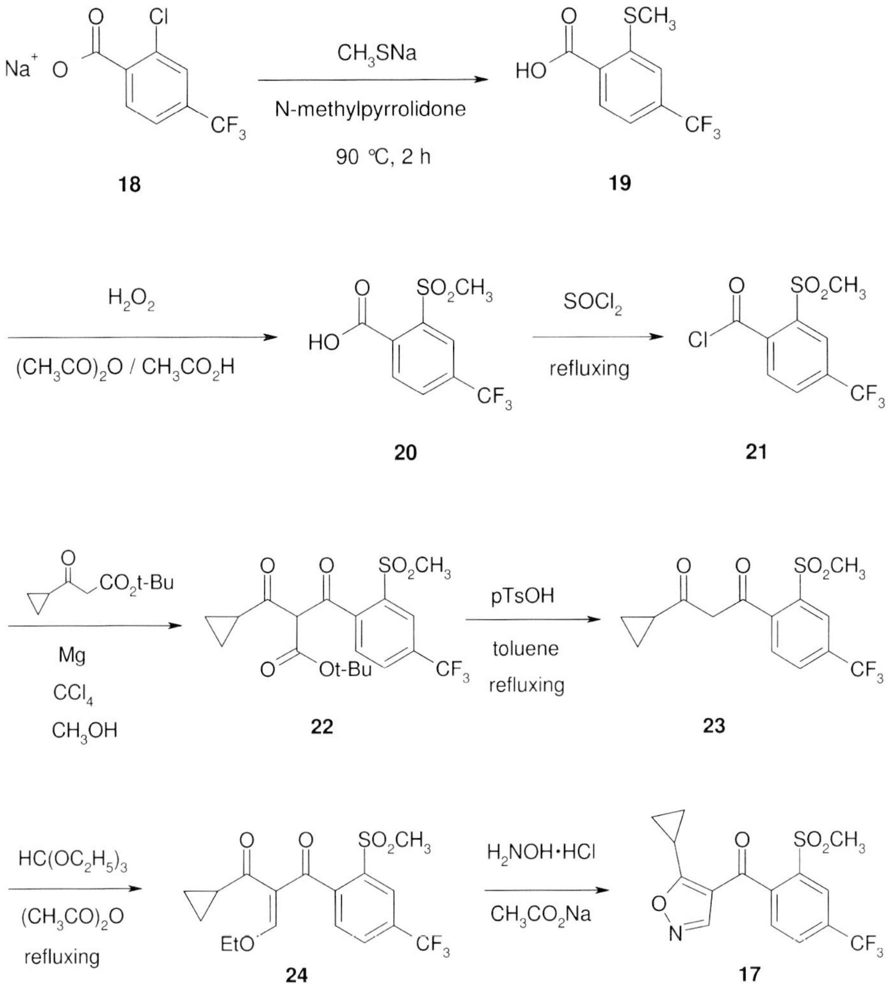

Scheme 4.4.6 Synthesis of isoxaflutole.

(Scheme 4.4.6). One possible educt is 2-chloro-4-trifluoromethylbenzoic acid sodium salt (18) to get 2-methylthio-4-trifluoromethylbenzoic acid (19). Further treatment with hydrogen peroxide and acetic anhydride in acetic acid yields 2-methylsulfonyl-4-trifluoromethylbenzoic acid (20). With thionyl chloride the corresponding benzoyl chloride 21 is available, which will be transformed into *t*-butyl 2-(2-methylsulfonyl-4-trifluoromethylbenzoyl)-3-cyclopropyl-3-oxopropionate (22) via the magnesium enolate of *t*-butyl 3-cyclopropyl-3-oxopropionate in methanol. To remove the *t*-butyl carboxylate group, 22 is refluxed in toluene in the presence of toluenesulfonic acid. The so-formed 1-(2-methylsulfonyl-4-trifluoromethylphenyl)-3-cyclopropylpropan-1,3-dione (23) is used to obtain 1-(2-methylsulfonyl-4-trifluoromethylphenyl)-3-cyclopropyl-2-ethoxymethylenepropan-

1,3-dione (**24**) in a mixture of triethylorthoformate and acetic anhydride. Finally, the addition of sodium acetate and hydroxylamine hydrochloride yields IFT (**17**).

As already mentioned, isoxazoles such as IFT are prodrugs and are not sufficiently persistent in plants to inhibit the HPPD enzyme. It is the first metabolite of isoxaflutole, the so-called DKN (diketonitrile) 3-cyclopropyl-2-[2-(methylsulfonyl)-4-(trifluoromethyl)benzoyl]-3-oxopropanenitrile (**25**) that is the herbicidally active entity. In soil, and also in plants, IFT undergoes rapid conversion into DKN [25]. In aqueous solutions there is an influence of temperature and pH on the chemical hydrolysis of IFT to DKN. The hydrolysis increases with increasing pH and temperature: for 295 K and pH 9.3 the rate of degradation was 100-fold faster than at pH 3.8. (Scheme 4.4.7) [26]. The DT_{50} in water is 11 days at pH 5, 20 h at pH 7 and 3 h at pH 9 [22].

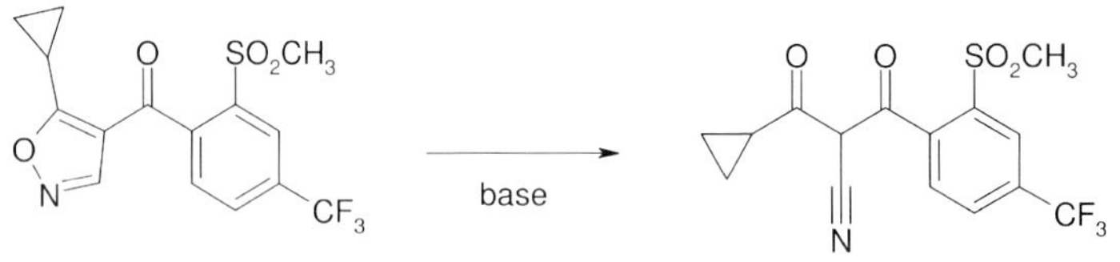

Scheme 4.4.7

The DT_{50} of IFT in soil is also very low and in the range of 12 h to 3 days under laboratory conditions. This is, however, once more dependant on several factors such as temperature, pH, moisture and soil type [27]. Moreover, the half-life of IFT in aqueous sterile solutions is higher than in soil at the same temperature and pH and confirms the catalytic effect of the soil reported by Taylor-Lovell et al. [28]. Used under normal agricultural conditions the rate of DKN formation will be affected by the quantity and frequency of rainfall. The log *P* of IFT is 2.19 and the water solubility is 6.2 mg L^{-1} compared with values for DKN of 0.4 and 326 mg L^{-1}, respectively. These properties restrict the mobility of IFT, which is retained at the soil surface, where it can be taken up by surface-germinating weed seeds. DKN, which has a laboratory DT_{50} of 20–30 days, is more mobile

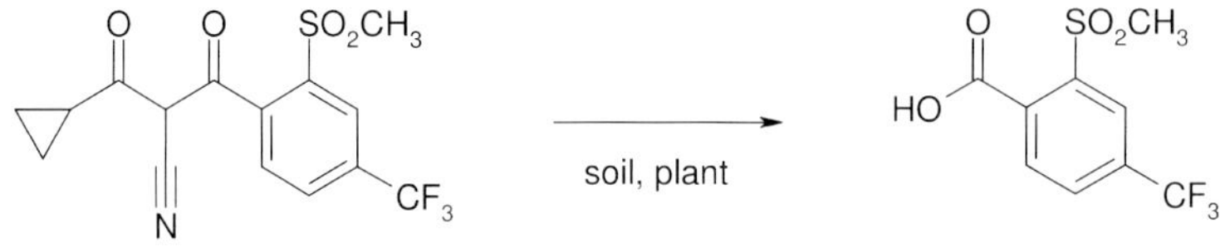

Scheme 4.4.8

and is taken up by the roots. In addition to influencing the soil behavior of IFT and DKN, the greater lipophilicity of IFT leads to greater uptake by seed, shoot and root tissues. In both plants and soil, DKN is converted into the herbicidally inactive benzoic acid **26** (Scheme 4.4.8). This degradation is more rapid in corn than in susceptible weed species and this contributes to the mechanism of selectivity, together with the deeper sowing depth of the crop [27].

4.4.2.5 Topramezone

The launch of topramezone (**27**, Fig. 4.4.8) for the post application corn market was for 2006 under the trade names Impact® in USA and Canada and Clio® in Germany and Austria. The compound is based on a BASF patent from 1995 [29].

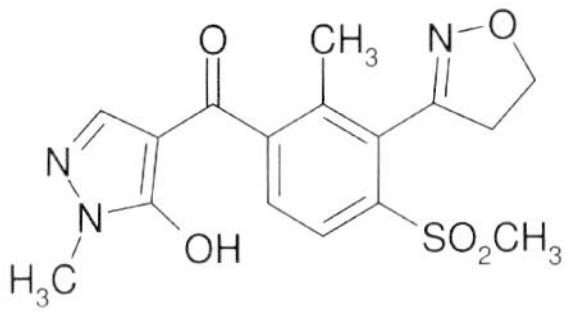

27

topramezone

Fig. 4.4.8. Structure of topramezone (**27**).

In 2005 BASF granted rights to develop, register and commercialize topramezone in North America to Amvac Chemical, whilst rights in Japan have been granted to Nippon Soda. The new corn compound will only be marketed in Latin America and Europe [30] by BASF.

Topramezone is aimed at the post-emergence control of major grass and broadleaf weeds in corn crops worldwide. This means that this new corn compound differentiates itself from sulcotrione and mesotrione in that it shows real cross spectrum activity like isoxaflutole and it is not limited to mainly broadleaf weed control.

Clio® is a 336 g L^{-1} SC-formulation with recommended application rates of 50–75 g ha^{-1} topramezone [32]. Like IFT for pre-application topramezone defines a new level of biological activity for HPPD compounds in post-application.

Topramezone is, like pyrazolynate, pyrazoxyfen and benzofenap, a pyrazolone but without a protective group and is therefore not a prodrug. Also noticeable is the 4,5-dihydroisoxazol group in 3-position of the benzoyl moiety.

Different synthesis routes have been published [29, 31]. Scheme 4.4.9 shows only one of them.

Starting with 3-nitro-*o*-xylene (**28**), 3-(2-methyl-6-nitrophenyl)-4,5-dihydro-isoxazole (**31**) is synthesized via the benzaldehyde oxime **29**. Subsequent reduction of the nitro group, replacement of the corresponding amino group by methyl

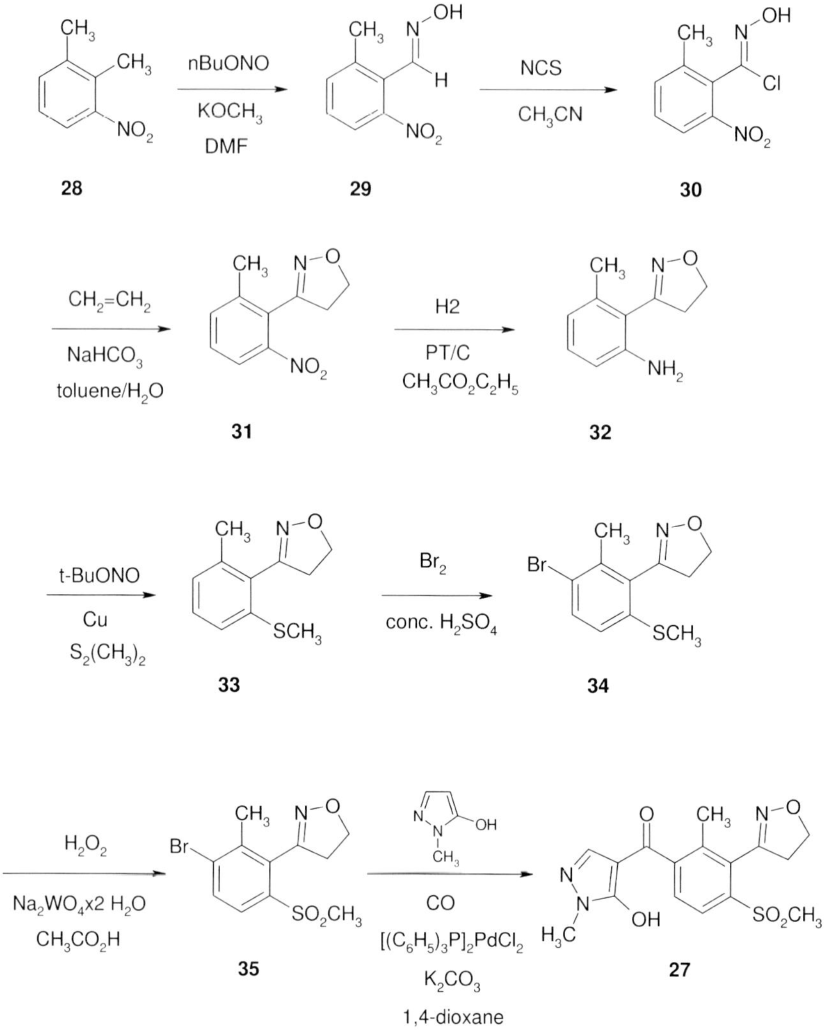

Scheme 4.4.9 Synthesis of topramezone.

sulfide, bromination to 3-[3-bromo-2-methyl-6-(methylthio)phenyl]-4,5-dihydro-isoxazole (34), and then oxidation affords the sulfone 35. Finally, topramezone (27) is available by conversion of 35 with 1-methyl-5-hydroxypyrazol in the presence of carbon monoxide and a suitable palladium catalyst, an alternative process to those described in Schemes 4.4.2 and 4.4.4.

4.4.2.6 Pyrasulfotole

During the Analyst & Investor Days in Lyon on September 5–6, 2005 Bayer CropScience announced the development of a new pyrazolone called pyrasulfo-

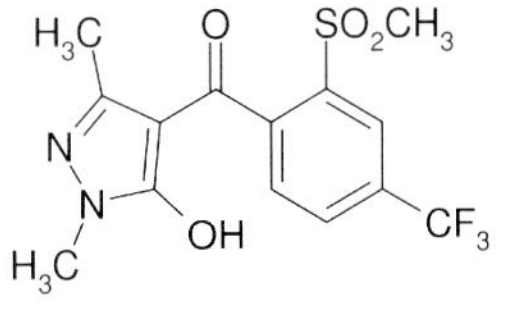

36

pyrasulfotole

Fig. 4.4.9. Structure of pyrasulfotole (**36**).

tole (**36**, Fig. 4.4.9) for the cereals market. The compound had been initially patented 2000 [33] by Aventis CropScience (now Bayer CropScience).

Pyrasulfotole would be the first HPPD compound for cereals and therefore a new mode of action for this crop. It is described as an innovative tool for resistance management with excellent broad-spectrum activity and excellent crop compatibility due to combination with proprietary safener technology [34]. Application rates and environmental behavior are so far not reported.

Interestingly, pyrasulfotole uses the same benzoic acid **20** like IFT and the also the well-known 1,3-dimethyl-5-pyrazolon (**9**) and, like topramezone, it is not a prodrug.

4.4.3
Conclusion

HPPD inhibitors of the heterocycle type are represented in rice, corn, sugar cane and, in future, also in the cereals market. Even if the three rice compounds have passed their commercial peak, HPPD inhibitors are very successful, especially in corn.

Interestingly, most of the compounds described here are prodrugs, but this does not mean that it is a prerequisite for this type of HPPD inhibitor, as can be seen with topramezone and pyrasulfotole. Moreover, the prodrug concepts used are chemically quite different. Whereas the three rice compounds leave their prodrug moiety as waste in the environment, IFT undergoes a conversion by opening its isoxazole ring without changing its molecular mass. All these HPPD inhibitors share the feature that they have a relatively higher log P and, therefore, lower water solubility and are metabolized to an active metabolite with a much lower log P and higher water solubility.

Other more important differences are the application rates. The three rice compounds are used on the kg scale whereas IFT and topramezone are used about 100 g ha^{-1} and lower.

Chemically, the compounds described are quite similar, with the exception of topramezone. The substitution patterns of the benzoyl moieties bare resemlance, even though there is of course a big difference between Cl, CH_3, CF_3 or SO_2CH_3.

Benzofenap and, in particular, topramezone also show that in 3-position a substitution is allowed and obviously important for good biological activity. Also important are the different Q moieties, but, as already mentioned, both pyrazolone and diketonitrile are chelating agents for Fe(II).

It is also fascinating to see the quite different biological activity already achieved with the shown variations. The future will show what will be at the end of the HPPD story regarding crop, application rate and profile.

References

1 *The Pesticide Manual*, 12th Edition, British Crop Protection Council, **2000**, pp. 602, 793, 797 and 848; Patent examples: W. J. Michaely, G. W. Kraatz, Stauffer Chemical Company, EP 0090262; L. Z. Benkö, J. A. Turner, M. R. Weimer, G. M. Garvin, J. L. Jackson, S. L. Shinkle, J. D. Webster, DOW Agrosciences LLC, WO 98/42678, **1998**, W. von Deyn, R. L. Hill, U. Kardorff, S. Engel, M. Otten, M. Vossen, P. Plath, H. Rang, A. Harreus, H. Koenig; BASF WO 96/26193, **1996**.

2 T. Konotsune, K. Kawakubo, T. Yanai, *Advances in Pesticide Sciences*, Zurich **1978**, Part 2, pp. 94–98.

3 T. Konotsune, K. Kawakubo, Sankyo Co. LTd., JP 50126830, **1975**, DE 2513750, **1975**, GB 1463473, **1975**.

4 D. Cole, K. Pallett, M. Rodgers, Discovering new Modes of Action for herbicides and the impact of genomics, *Pesticide Outlook* **2000**, 12, 223–229.

5 M. Ishida, T. Matsui, T. Yanai, K. Kawakubo, T. Honma, K. Tanizawa, M. Nakagawa, H. Okudaira, *Sankyo Kenkyusho Nempo* **1984**, 36, 44.

6 The Japan Association for Advancement of Phyto-Regulators collected data.

7 *Statistical Yearbook of Ministry of Agriculture, Forestry and Fisheries, Japan.*

8 K. Kawakubo, M. Shindo, T. Knonotsune, *Plant Physiol.* **1979**, 64, 774.

9 T. Matsui, T. Konotsune, K. Kawakubo, M. Ishida, *Pesticide Chemistry: Human Welfare and the Environment* ed. J. Miyamoto, P. C. Kearney, Vol. 1, Pergamon Press, Oxford, **1983**, pp. 327–332.

10 G. Sandmann, H. Reck, P. Boeger, *J. Agric. Food. Chem.* **1984**, 32, 868.

11 K. Yamaoka, M. Nakagawa, M. Ishida, *J. Pesticide Sci.* **1987**, 12, 209–212.

12 *The Pesticide Manual*, 12th Edition, British Crop Protection Council, **2000**, pp. 793–794.

13 For example, S. K. Gee, M. A. Hanagan, W. Hong, R. Kucharczyk, Du Pont, WO 97/08164, **1997**.

14 F. Kimura, *Jpn. Pesticide Information* **1984**, 45, 24.

15 R. Nishiyama, F. Kimura, T. Haga, N. Sakashita, T. Nishikawa, Ishihara Syngyo Kaisha Ltd., GB 2002375, **1978**.

16 M. Hiroshi, *ACS Symposium Series* **2005**, 892, 161–171.

17 Kokai Tokkyo Koho, Mitsubishi Petrochemical Co., Ltd., JP 57072903, **1980**.

18 K. Ikeda, A. Goh, *Japan Pesticide Information* **1991**, 59, 1991.

19 M. Miyahara, *Jpn. Pesticide Information* **1986**, 49, 15.

20 B. M. Luscombe, K. Pallett, P. Loubiere, J. C. Millet, J. Melgarejo, T. E. Vrabel, *Proc. Br. Crop Prot. Conf. – Weeds*, **1995**, 1, 35.

21 P. A. Cain, S. M. Cramp, G. M. Little, B. M. Luscombe, Rhône-Poulenc Agricultutes Ltd., EP 0527036, **1991**.

22 ''Isoxaflutole – Herbicide for broadleaf weed and grass control in maize and sugar cane, Technical Information'', Bayer CropScience AG, Alfred-Nobel-Str. 50, 40789 Monheim.

23 B. M. Luscombe, K. E. Pallett, *Pesticide Outlook*, **1996**, 29.

24 D. Bernard, A. Viauvy, Rhône-Poulenc Agricultutes Ltd, WO 99/02489, **1999**.

25 K. E. Pallett, J. P. Little, P. Veerase-karan, F. Viviani, *Pestic. Sci.*, **1997**, 50, 83.

26 E. Beltran, H. Fenet, J. F. Cooper, C. M. Coste, *J. Agric. Food Chem.* **2000**, 48, 4399.

27 K. E. Pallett, S. M. Cramp, J. P. Little, P. Veerasekaran, A. J. Crudace, A. E. Slater, *Pest. Manag. Sci.* **2001**, 57, 133.

28 S. Taylor-Lovell, G. K. Sims, L. M. Wax, J. J. Hassett, Hydrolysis and soil adsorption of the labile herbicide Isoxaflutole. *Environ. Sci. Technol.* **2000**, 34, 3186.

29 W. von Deyn, R. L. Hill, U. Kardorff, S. Engel, M. Otten, P. Plath, H. Rang, A. Harreus, H. König, H. Walter, K.-O. Westphalen, BASF AG, WO 96/26206, **1996**.

30 *Agrow*, **2005**, No 427, p 20.

31 J. Rheinheimer, W. von Deyn, J. Gebhardt, M. Rack, R. Lochtman, N. Götz, M. Keil, M. Witschel, H. Hagen, U. Misslitz, E. Baumann, BASF AG, WO 99/58509, **1999**.

32 A. Schönhammer, J. Freitag, H. Koch, BASF AG, 46. Österreichischen Pflanzenschutztagen, 30.11.–01.12.2005, Stadthalle Wels.

33 M. Schmitt, A. van Almsick, R. Preuss, L. Willms, T. Auler, H. Bieringer, F. Thuerwaechter, Aventis CropScience GmbH, WO 2001074785, **2001**.

34 B. Garthoff, Bayer CropScience AG, Innovation Driving Future Growth, Analyst & Investor Days, 05.09.–06.09.2005, Lyon.

5
Safener for Herbicides

Chris Rosinger and Helmut Köcher

5.1
Introduction

Herbicide safeners (also referred to as herbicide antidotes or protectants) fulfill an important role in crop protection. Safeners are chemicals that protect crop plants from unacceptable injury caused by herbicides. Either by placement on the crop seed or by way of a physiological selectivity mechanism, safeners in commercial use do not negatively impact the weed control of the herbicide. Although many herbicides have been developed for use without a safener, some of the strongest and most broad-spectrum herbicides tend towards border-line crop selectivity, which may completely preclude use in a particular crop or at least limit maximum use rates or the crop varieties that can be safely treated. It is for such situations that safeners have been developed. Several books and reviews of safeners have been written over the past 20 years [1–3]. It is not the intention of this chapter to cover in detail older safeners, but rather to focus on more recently developed commercial safeners as well as some of the older compounds still in wide commercial usage.

The story of herbicide safeners began in 1947 with an accidental observation by Otto Hoffmann, a researcher in the Gulf Oil Company. On entering his greenhouse on a hot summer afternoon he saw that tomato plants had suffered injury that he presumed was from 2,4-D vapor drift. However, plants treated with 2,4,6-trichlorophenoxyacetic acid showed no symptoms of this injury [4]. Hoffmann recognized the potential use of such an effect and started research into compounds that could protect crops from herbicide injury.

A fundamental problem for safener discovery and development is to find safeners that do not also antagonize weed control. The fruits of Gulf Oil Company research (reported by Hoffmann in 1969) was 1,8-naphthalic anhydride (NA), which works best as a seed treatment, whereby antagonism of weed control is not an issue. To the authors' knowledge just over a dozen further safeners have been commercialized in the years since NA was introduced, although several of the early safeners have since been superseded and/or withdrawn. This subse-

Modern Crop Protection Compounds. Edited by W. Krämer and U. Schirmer
Copyright © 2007 WILEY-VCH Verlag GmbH & Co. KGaA, Weinheim
ISBN: 978-3-527-31496-6

quent period of safener commercialization may be informally split into three phases; the first, mainly seed treatment safeners; the second, pre-emergence tank mix safeners, and the third post-emergence tank mix safeners. Table 5.1 shows the chemical structures and usage of these safeners. In all these cases the crops are monocotyledons (maize, sorghum, rice, and cereals such as wheat and barley). Figure 5.1 shows the effect of one of them, mefenpyr-diethyl. To date no comparable safeners have been commercialized for broad-leaved crops. However, the "extender" dietholate is used by FMC to help protect cotton against the herbicide clomazone (US patent application 20050009702). In addition, several compounds (daimuron, cumyluron and dimepiperate) generally considered herbicidal are included in some products principally because they reduce crop injury from another herbicidal component. These are of relevance, particularly in rice and the structures are also shown in Table 5.1. Because they are relatively old compounds they will not be covered here.

To ensure maximum crop safety, safeners that are applied in mixture with the herbicides need to act quicker than the herbicide injury develops. The mechanism of action of safeners has received much scientific attention and will be dealt with in some detail in this chapter (Section 5.3).

Safeners, like pesticides, must be registered before use. However, the regulatory situation for safeners is complex, in particular when considered on a global basis. For example, whereas several European countries require for safeners full data packages like those for pesticides, safeners do not fall under Annex I of the European Union pesticide directive 91/414. In the USA safeners are treated under inert legislation as opposed to pesticide legislation. However, full data sets (like those for active ingredients) are actually required for evaluation by the environmental protection agency (EPA) to establish a residue limit for the federal food, drug and cosmetic act. In Canada and Australia, safeners are now treated legally as pesticides. In other parts of the world safeners are legally treated as formulation additives.

Fig. 5.1. Post-emergence safening of wheat by mefenpyr-diethyl against mesosulfuron-methyl. (A) Untreated; (B) mesosulfuron-methyl at 60 g-a.i. ha^{-1}; and (C) mesosulfuron at 60 g-a.i. ha^{-1} plus mefenpyr-diethyl at 30 g-a.i. ha^{-1}.

Table 5.1 Structures of commercial safeners.

Common name (development codes)	Chemical structure	Application/Crop/Herbicides
1,8-Naphthalic anhydride (NA) Protect®		Seed treatment Maize Thiocarbamates
Cyometrinil (CGA43089) Concep I®		Seed treatment Sorghum Thiocarbamates/chloroacetamides
Oxabetrinil (CGA92194) Concep II® Superseded Concep I®		Seed treatment Sorghum Thiocarbamates/chloroacetamides
Fluxofenim (CGA133205) Concep III® Superseded Concep II®		Seed treatment Sorghum Thiocarbamates/chloroacetamides
Flurazole (MON4606) Screen®		Seed treatment Sorghum Chloroacetamides
Benoxacor (CGA154281)		Spray Pre, PPI Maize Chloroacetamides
Dichlormid (R25788)		Spray Pre, PPI Maize Thiocarbamates/Chloroacetamides
Furilazole (MON13900)		Spray pre-emergence Maize Chloroacetamides (Acetochlor)
AD-67 MON4660		Spray pre-emergence Maize Chloroacetamides (Acetochlor)

Table 5.1 (continued)

Common name (development codes)	Chemical structure	Application/Crop/Herbicides
Fenclorim (CGA123407)		Spray pre-emergence Rice Pretilachlor
Daimuron (K223, SK-23)		Water application post-emergence Rice Sulfonylureas
Cumyluron (JC-940)		Water application post-emergence Rice Sulfonylureas
Dimepiperate (MY-93)		Water application post-emergence Rice Sulfonylureas
Cloquintocet-mexyl (CGA185072)		Spray post-emergence Cereals Clodinafop-propargyl
Fenchlorazole-ethyl (AE F070542)		Spray post-emergence Cereals Fenoxaprop-ethyl
Mefenpyr-diethyl (AE F107892)		Spray post-emergence Cereals ACCase and sulfonylureas
Isoxadifen-ethyl (AE F122006)		Spray post-emergence Maize/rice ACCase and sulfonylureas

5.2
Overview of Selected Safeners

5.2.1
Dichloroacetamide Safeners

This class contains several important commercial safeners as well as a range that were reported (and patented) but not launched. They are all of relatively low molecular weight (MW $\leq$ 300) with the -N-C(O)-CHCl$_2$ substituent in common. Although, the first member of this class was commercialized in the early 1970s, three compounds are still of considerable commercial importance; benoxacor, dichlormid and furilazole.

5.2.1.1 Benoxacor

Commercially, benoxacor is one of the most important members of this class of safeners. It is included in products containing metolachlor as racemate or as the single isomer S-metolachlor (subsequently "(S-)metolachlor" indicates both are being referred to). These products are principally used in maize pre-plant, pre-plant incorporated and pre-emergence. Benoxacor was developed under the code CGA 154281 by Ciba-Geigy AG (now Syngenta) and was first reported in 1988 [5]. It was specifically claimed in the US patent US4601745 (filed 18th March 1985) but a priority date of 12th December 1983 relates back to general claims for the structure class (EP 149974). The synthesis of benoxacor, as disclosed in WO2001090088, involves a three-step process (Scheme 5.1).

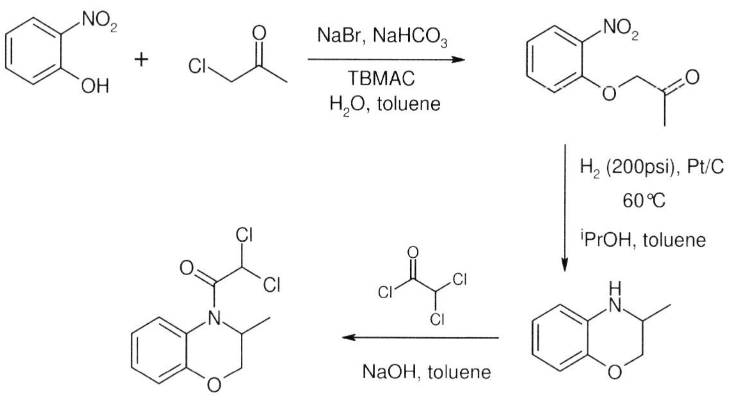

Scheme 5.1. Synthesis of benoxacor.

There are now numerous products that contain benoxacor and (S-)metolachlor, with and without further herbicide components (Table 5.2). As the patents for

Table 5.2 Product examples containing benoxacor.

Herbicide(s)	Trade name examples
S-metolachlor	Dual II magnum®, Cinch®
S-metolachlor + atrazine	Bicep II magnum®, Cinch® ATZ, Cinch® ATZ lite
Metolachlor + atrazine	Stalwart® Xtra
Metolachlor	Stalwart®, Parallel® Me-Too-Lachlor II®
S-metolachlor + mesotrione	Camix®
S-metolachlor + mesotrione + atrazine	Lexar®, Lumax®

Table 5.3 Physicochemical properties of benoxacor and metolachlor.

Property	Benoxacor	Metolachlor
Log P	2.6	2.9
K_{oc}	42–176	121–309

benoxacor and metolachlor have both expired, this product list also contains several from generic producers.

The use of benoxacor with (S)-metolachlor is particularly necessary under stress conditions for maize. Injury to corn from (S)-metolachlor is greater under cool or wet soil conditions [6–8] where both the availability of the herbicide may be increased and the ability of maize to metabolize metolachlor reduced [9]. Benoxacor and metolachlor have similar chemical properties, influencing there behavior in soil, and this tends to ensure that the safener and herbicide are taken up together, hence providing safening under various weather conditions.

Products such as Dual-II-magnum® and Cinch® are also labeled for use on sorghum seed treated with fluxofenim or flurazole. This is a notable example of safeners used in sequence so as to obtain optimal crop safety. Table 5.4 indicates that benoxacor has a favorable toxicological profile.

Table 5.4 Toxicological and soil degradation data for benoxacor.

Rat, oral	$LD_{50} > 5000$ mg kg^{-1}
Rat, inhalation	$LC_{50} > 2000$ mg m^{-3}
Rabbit, skin and eye irritation	Not irritant
Guinea pig, skin sensitizing	Slightly sensitizing
DT_{50} in soil	Rapid, ca. 5 days

5.2.1.2 Dichlormid

Of the safeners covered separately in this chapter, dichlormid is the oldest still in use. It was developed under the code number R25788 by Stauffer (now Syngenta) and first reported in 1972 [10]. It is used to safen maize against injury from acetochlor. Products include Surpass®, TopNotch®, Volley®, and Confidence®. Stalwart C® is a metolachlor product that contains dichlormid instead of benoxacor. Dichlormid is also present in several acetochlor products that also contain atrazine (e.g., Confidence Xtra®, Keystone®, Volley® ATZ).

The simple one-step synthesis of dichlormid (claimed in US 4278799) is shown in Scheme 5.2.

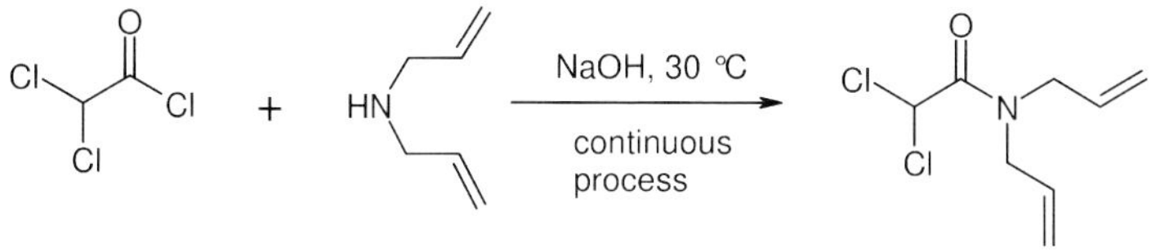

Scheme 5.2. One-step synthesis of dichlormid.

As described for benoxacor, dichlormid also has similar physicochemical properties to those of the herbicide components, allowing for similar plant uptake profiles for good safening potential. Further extensive coverage of dichlormid can be found in *Crop Safeners for Herbicides* [1].

5.2.1.3 Furilazole

Furilazole was developed by Monsanto Co. under the code number MON13900 and first reported in 1991 [11]. In that publication it was claimed to safen many herbicides from diverse classes, but detailed efficacy was only presented for the combination with the sulfonylurea herbicide halosulfuron-methyl (NC-319). Since its launch in 1995 furilazole has been marketed with halosulfuron-methyl in the products such as Battalion® and Permit® used pre- and post-emergence in corn and sorghum. It is also used in pre-emergence maize products containing acetochlor (e.g., Degree®, Degree Extra®, Harness®, Guardian®).

Note – Acetochlor can be safened by several dichloroacetamide safeners other than dichlormid and furilazole. For example, the product Acenit® contains the safener AD67 (MON4660) which has no assigned common name (see Table 5.1 for chemical structure).

The two-step synthesis of furilazole (claimed in patent EP 648768) is shown in Scheme 5.3.

The toxicological profile of furilazole is quite favorable (Table 5.5).

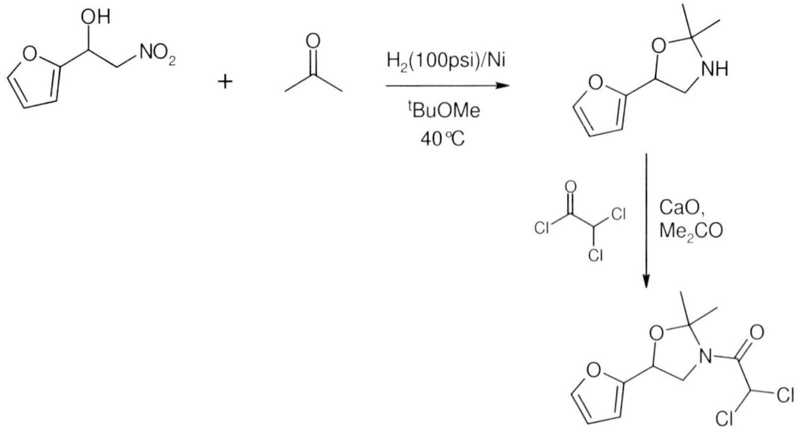

Scheme 5.3. Two-step synthesis of furilazole.

Table 5.5 Toxicological and soil degradation data for furilazole.

Rat oral	$LD_{50} > 869$ mg kg^{-1}
Rat inhalation	$LC_{50} > 2300$ mg m^{-3}
Rabbit skin and eye irritation	Not irritant to skin/slight eye irritant
Guinea pigskin sensitizing	Non-sensitizing
DT_{50} in soil	Rather rapid, ca. 10–20 days

5.2.2
Oxime Ethers

Three oxime ethers have been commercialized by Ciba Geigy (now Syngenta) as seed treatment safeners for sorghum; protection being provided against thiocarbamate and chloroacetamide herbicides (in particular metolachlor). The first (cyometrinil) was launched in 1978 as Concep I[R]. It was replaced in 1982 by oxabetrinil (Concep II[R]), which had less potential for negative crop effects from the seed treatment. Concep II[R] was in turn superseded by fluxofenim (Concep III[R]), which is still in commercial use. In this case, the reason for replacement is not fully clear, but was reportedly due to an undesirable interaction of Concep II[R] with downy mildew disease in sorghum [1]. Fluxofenim was developed under the code CGA133205 and first reported in 1986 [12]. The physical chemistry of fluxofenim (log $P = 2.9$) allows rapid uptake into seeds at use rates of 0.3–0.4 g-a.i. kg^{-1}.

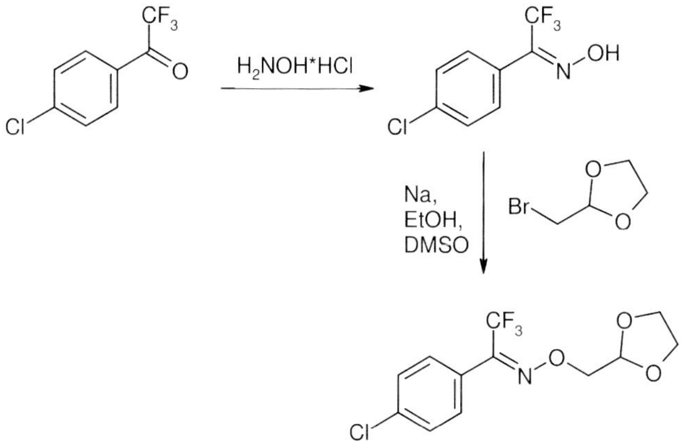

Scheme 5.4. Synthesis of fluxofenim.

Nonetheless the commercial market is limited by use only in the relatively minor crop sorghum. Scheme 5.4 shows the two-step synthesis route for fluxofenim.

5.2.3
Cloquintocet-mexyl

Cloquintocet-mexyl was developed under the code CGA 185072 by Ciba-Geigy (now Syngenta) and is used post-emergence in cereals. The basic patent (EP 94349) has a priority date of 7th May 1982. Various other country patents followed (e.g., US4902340 and US 5102445). It was first reported in 1989 [13] alongside the ACCase inhibitor clodinafop-propargyl, and till now the main use of

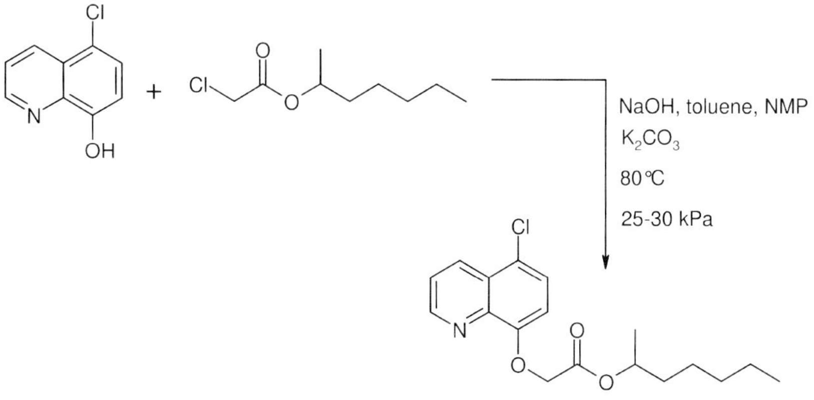

Scheme 5.5. Synthesis of cloquintocet-mexyl.

Table 5.6 Toxicological data for cloquintocet-mexyl.

Rat, oral	$LD_{50} > 2000$ mg kg^{-1}
Rat, dermal	$LD_{50} > 2000$ mg kg^{-1}
Rat, inhalation	$LD_{50} > 935$ mg m^{-3}
Rabbit, skin and eye irritation	Not irritant

cloquintocet-mexyl is still in mixtures with this ACCase-inhibiting herbicide. Products include Topik®, Horizon®, Discover® (US), Celio®, Hawk®, Magestan®. The first launch was in 1991 on Switzerland, South Africa and Chile, with the US registration of the safener/herbicide combination in 2000. The greatest safening is observed in wheat, with less safening in barley. Rye and triticale can also be safened. WO2002000625 claims a single-step synthesis route for cloquintocet-mexyl (Scheme 5.5).

Cloquintocet-mexyl has a favorable toxicological profile (Table 5.6).

In the soil cloquintocet-mexyl degrades rapidly to the free acid ($DT_{50} < 3$ days) with further degradation and mineralization within weeks or a few months. The parent safener and major metabolites are reported to bind strongly to soil and hence have low leaching potential.

5.2.4
Mefenypr-diethyl

Mefenpyr-diethyl is, like cloquintocet-mexyl, used post-emergence to safen cereals. It is used in combination with various aryloxyphenoxypropionates and sulfonylurea herbicides in wheat, rye, triticale and some varieties of barley. It was developed under the code AE F107892 by AgrEvo (now Bayer CropScience) and was first reported alongside iodosulfuron-methyl in 1999 [14] and has replaced its predecessor fenchlorazole-ethyl. Mefenpyr-diethyl had the advantage over fenchlorazole-ethyl of providing post-emergence selective grass weed control not only in wheat and rye but also in spring barley. A further, very important advantage of mefenpyr-diethyl was the property to act as a safener for a wider range of herbicides used post-mergence in cereal crops. The priority date for patent coverage of the pyrazoline safeners was November 1989 (WO9107874) and the first registration of mefenpyr-diethyl was in 1994. It is prepared using a two-step synthesis (Scheme 5.6).

As already pointed out, mefenpyr-diethyl is a versatile safener and it has been commercialized in combinations with several single or mixed herbicides, including fenoxaprop-P-ethyl (e.g., Puma S®), iodosulfuron-methyl-sodium (e.g., Hussar®) and mesosulfuron-methyl (Atlantis®). In general, the quantity of mefenpyr-diethyl required to provide adequate safening lies between 20 and 100 g-a.i. ha^{-1}, and there is no set ratio between the rates of the herbicides and

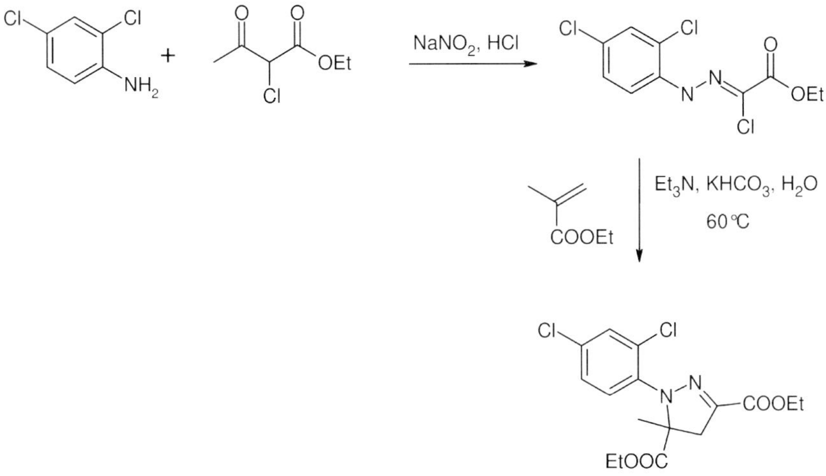

Scheme 5.6. Two-step synthesis of mefenypr-diethyl.

mefenpyr-diethyl. At this point it is worth mentioning some general considerations with regards to the dose rates required for safeners. Of course, from a commercial and safety standpoint the safener rate should be the lowest needed to obtain crop safety. Seed treatment rates can be selected independent of the subsequent herbicide dose. However, the maximum rate on the seed may sometimes be limited by negative phytotoxic effects. This is exemplified well by the germination inhibition in sorghum caused by cyometrinil, which eventually lead to its replacement by oxabetrinil. For products containing a mixture of safener and herbicide, significant development effort is needed to define the required herbicide/safener ratio. This ratio should be adequate to ensure crop safety and weed control at all recommended rates. A farmer that reduces the product rate to below the minimum that is recommended on the product label runs the risk of not only inadequate weed control (due to insufficient herbicide) but also possible crop injury due to insufficient safener. For mefenpyr-diethyl, a wide range of products exist globally, in which this critical herbicide/safener ratio is tuned to the specific herbicide(s) and agronomic conditions.

Mefenpyr-diethyl has a highly favorable toxicological and ecotoxicological profile (Table 5.7).

In the environment, mefenpyr-diethyl dissipates rapidly with a soil DT_{50} of <10 days. Complete mineralization occurs due to photolysis, hydrolysis and microbial degradation. There is no leaching risk, with the parent compound and soil metabolites not exceeding 0.1 ppb at 1 m soil depth in lysimeter trials.

Mefenpyr-diethyl is most probably a pro-safener, a term introduced by Rubin in 1985 [15]. With mefenpyr-diethyl a decarboxylation occurs rapidly in plants and soil and it is likely that the safening activity comes from mefenpyr-ethyl. How-

Table 5.7 Toxicological data for mefenpyr-diethyl.

Rat, oral	$LD_{50} > 5000$ mg kg^{-1}
Rat, dermal	$LD_{50} > 4000$ mg kg^{-1}
Rabbit, skin and eye irritation	Not irritant
Guinea pig, skin sensitizing	Not sensitizing
Mutagenicity *in vitro* and *in vivo*	Non-mutagenic

ever, the good post-emergence performance of mefenpyr-diethyl depends upon its physicochemical characteristics (log $P = 3.83$ @ pH 6.3, 21 °C), which lead to better leaf uptake than from mefenpyr-ethyl. The biochemical mode of action of mefenpyr-diethyl is covered in Section 5.3.

5.2.5
Isoxadifen-ethyl

The most recently commercialized safener is isoxadifen-ethyl. It is used post-emergence to safen maize and rice. It was developed under the code AE F122006 by AgrEvo (now Bayer CropScience) and was first reported in 2001 [16–18]. It was launched in US in 2002 in maize in combination with foramsulfuron (Option®). It is also used in combinations with foramsulfuron plus iodosulfuron-methyl-sodium (Equip®, Maister®). In rice it is used with fenoxaprop-P-ethyl (Ricestar®, Starice®) and ethoxysulfuron (Tiller Gold®). From this it can be seen that isoxadifen-ethyl takes safeners to a new level; being able to safen multiple herbicides (of various modes of action) in multiple crops. The priority date for patent coverage of the isoxazoline safeners was 16th September 1993 (DE4331448, US9507897, US5516750).

The synthesis of isoxadifen-ethyl claimed in WO 1995007897 is via a one-step route (Scheme 5.7).

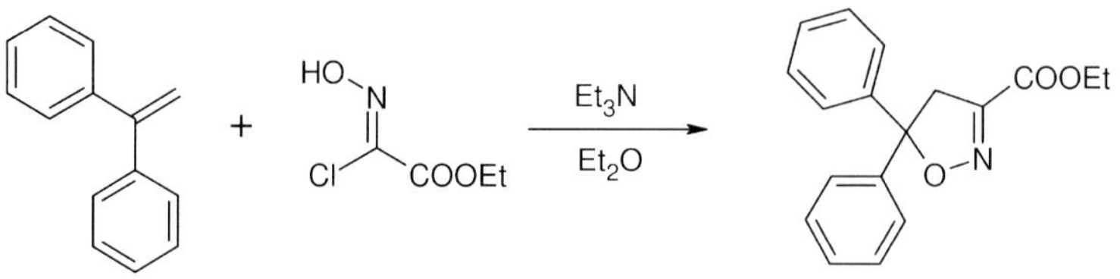

Scheme 5.7. One-step route to isoxadifen-ethyl.

Isoxadifen-ethyl has a favorable toxicological profile according to the U.S. Environmental Protection Agency notice of filing (Table 5.8) [19].

Table 5.8 Toxicological data for isoxadifen-ethyl.

Rat, oral	LD_{50} 1740 mg kg^{-1}
Rat, dermal	$LD_{50} > 2000$ mg kg^{-1}
Rat, inhalation	$LD_{50} > 5000$ mg m^{-3}
Rabbit, skin and eye irritation	Not irritant/slightly irritating

5.3
Mechanisms of Herbicide Safener Action

When applied alone, safeners generally have little visible effects on crop or weed species. This was found, for example, for the safeners fenchlorazole-ethyl and mefenpyr-diethyl [20, 21]. In contrast, fenchlorazole-ethyl exerted an immediate protective effect on wheat and prevented even a transient inhibition of leaf growth by fenoxaprop-ethyl [22]. The same was observed subsequently for combinations of mefenpyr-diethyl with fenoxaprop-P-ethyl.

Potentially, a safener could increase the tolerance of the crop by reduction of herbicide uptake and translocation, or by enhancement of metabolic herbicide in-activation in the crop tissue. Furthermore, a safener could counteract the effect of a herbicide at its biochemical target site, with a resultant reduction of crop sus-ceptibility. Evidence for and against these potential modes of action is presented in the following sub-sections. In addition, aspects of safener specificity (crop versus weed) are covered for situations where the safener is applied in tank mix with the herbicide.

5.3.1
Safener Interactions with the Herbicide Target Site

Potentially a safener could exert its effect in crop species by interference with her-bicide binding at the herbicidal target site. This possibility was tested, for exam-ple, in wheat chloroplast suspensions with combinations of the ACCase inhibitor fenoxaprop (herbicidally active free acid of the herbicide fenoxaprop-ethyl) and the safener fenchlorazole-ethyl. Even very high concentrations of fenchlorazole-ethyl (100 μM) did not alter the IC_{50} for fenoxaprop at the target enzyme (0.6 μM). The same result was obtained when instead of fenchlorazole-ethyl the corre-sponding free acid fenchlorazole was tested in this assay. This showed that no herbicide/safener interaction occurred at the herbicidal target enzyme [22]. Anal-ogous *in vitro* assays with the ALS inhibitor chlorsulfuron and the safener 1,8-naphthalic anhydride (NA) were carried out with target enzyme extracts from maize tissue. Also in this case, no herbicide/safener interaction was found at the target enzyme [23].

These findings were in contrast to a report of competitive binding of the tritiated dichloroacetamide safener R-29148 and the herbicides EPTC or alachlor at a proteinaceous component of maize seedling extracts. In addition, a good correlation was observed between competitive inhibition of [^{3}H]R-29148 binding by other dichloroacetamide compounds and their effectiveness as safeners. This was taken as support for the hypothesis that dichloroacetamide safeners act as receptor antagonists for the herbicides EPTC and alachlor [24].

One may also postulate that a safener could stimulate the activity of the herbicidal target enzyme and thus overcome phytotoxic effects of the herbicide in the crop species. In fact, Rubin and Casida [25] found a 25% increase in ALS activity in maize root or shoot tissue after application of the safener dichlormid. An increase of ALS levels in maize tissue was also reported after application of the safeners NA and oxabetrinil [26, 27]. This was in contradiction, however, to work of Barrett [28], who could not find any enhancement of ALS activity in maize and sorghum seedlings after treatment with the safeners NA, oxabetrinil, flurazole or dichlormid. Also, *in vivo* measurements of ALS activity in wheat after application of the sulfonylurea herbicide iodosulfuron-methyl sodium and the safener mefenpyr-diethyl did not indicate that the safener action could be attributed to a stimulation of target enzyme activity or to an interaction directly at the target enzyme [14].

Overall, with some exceptions, literature data suggest that herbicide/safener interactions at the herbicide target site or safener-induced effects on the activity of the herbicide target site are not the mechanism responsible for safener action. In addition, there are examples which, for circumstantial reasons, speak against a major involvement of the herbicide target site in the mechanism of safener action. One such example is the broad action of the safener NA against a spectrum of herbicides with different mechanisms of action. A more recent example is the safener mefenpyr-diethyl, which is not only a safener for ALS inhibitors of the sulfonylurea class but also an excellent safener for the ACCase inhibitor fenoxaprop-P-ethyl [21, 29]. Another point, worth mentioning in context with ALS and ACCase inhibitors, is the high similarity of herbicide binding characteristics to the herbicide target enzyme from cereal crops and from grass weed species, while the safeners for these herbicides act specifically only in cereal crop species, but not in the grass weeds.

5.3.2
Influence on Herbicide Uptake and Translocation

It is usually part of the investigations on the mechanism(s) of safener action to look for possible safener interactions with the herbicide partner at the process of herbicide uptake into the crop. Looking through the relevant literature gives a complex picture. This can also be seen in a review of Davies and Caseley [2], who present an exhaustive compilation of safener effects on herbicide uptake for relevant herbicide/safener combinations developed up to that time. Only in 20% of the cases was the uptake of the herbicide reduced in combination with the safener; 40% showed no influence of the safener on herbicide uptake, and in the remaining 40% of cases herbicide uptake was even stimulated by the safener. But, also in the cases of reduced herbicide uptake by the safener, the question remained whether this effect was the basis for the safener action. As an example, root uptake of the imidazolinone herbicide AC 263222 by chlorophyllous maize seedlings was reduced by 19% after seed dressing with NA, suggesting a contribution of this effect to the protective action of NA. Follow-up work, however, showed that NA exhibited also a protective effect when an interaction with herbicide uptake was excluded by application of the safener one day after the herbicide. This observation, but also contradictory results of other studies, which showed either no effect or a stimulatory effect of NA on herbicide uptake, made it questionable that an interference with herbicide uptake plays a significant role for the mechanism of action of this safener [30, 31]. It should be added that contradictory results (inhibition, stimulation or no effect on herbicide uptake) can also be found in the literature for other herbicide/safener combinations.

Uptake studies were also carried out with the recently developed combinations of the safener mefenpyr-diethyl with the sulfonylurea herbicides mesosulfuron-methyl and iodosulfuron-methyl-sodium, which are used for selective post-emergence weed control in wheat crop. In both combinations the safener had no influence on herbicide uptake [29].

In summary, it can be said that from present experience only in a few cases was herbicide uptake by the crop reduced in combination with a safener, and even then doubts remained as to whether the reduction of herbicide uptake was the mechanism of safener action. It is, therefore, concluded that interference with herbicide uptake by the crop has no importance as a mechanism of safener action, though it cannot be excluded that there may be cases where it plays an auxiliary role.

5.3.2.1 Translocation

Many of the modern herbicides, which are used in combination with a safener for selective post-emergence weed control in cereal crops, are ALS or ACCase inhibitors. The most sensitive morphological sites of action of these herbicides are the meristematic tissues, which in the early stage of development are located at the shoot base of the grass weed as well of the gramineous crops. After foliar spraying of these herbicides, long-distance transport to the basal meristems is a requirement for herbicidal action in grass weeds, as well as the phytotoxic effects in cereal crops. Theoretically, such phytotoxic effects could therefore be prevented, if a safener would act by specific inhibition of herbicide translocation in the phloem to the site of action. So far no case is known were a safener directly interferes with the long-distance translocation of these herbicides. However, there can be indirect effects on translocation due to a safener-induced enhancement of herbicide metabolism in the leaf mesophyll, which in turn may influence the amount of herbicide and metabolites transferred into the long-distance transport system. As an example, after foliar application of ^{14}C-labeled fenoxaprop-ethyl to wheat, translocation of ^{14}C-labeled material was not influenced by combination with the safener fenchlorazole-ethyl soon after application. However, after a period of three days the percentage of translocated ^{14}C-labeled material was lowered in the presence of the safener. This was interpreted as an effect of differential kinetics of herbicide metabolism and hence differential mobility characteristics of ^{14}C in the presence and absence of the safener [22].

Indirect effects on mobility were also reported for combinations of herbicides with safeners applied pre-emergence or by seed dressing. In corn seedlings treated with [^{14}C]metazachlor the amount of ^{14}C in the developing leaves was lowered when the seedlings had been incubated with the dichloroacetamide safener BAS 145138 (Dicyclonon). Analytical data suggested that safener-enhanced metabolism of metazachlor to a polar non-mobile metabolite in the adjacent seedling tissues reduced the amount of ^{14}C reaching the developing leaves [32]. In maize seedlings treated with the ^{14}C-labeled imidazolinone herbicide imazapic (AC 263222), the acropetal movement from root to shoot was markedly less in seedlings that had received a seed dressing with the safener 1,8-naphthalic anhydride (NA). This was attributed to the safener-enhanced formation of an immobile metabolite being retained in the seedling root [31].

5.3.3
Effects of Safeners on Herbicide Metabolism

With few exceptions, herbicides are subject to metabolic transformations both in weed and crop species, after they have penetrated the plant tissue and are under way to their target site. As a rule, the herbicide metabolites are more polar than the herbicidal parent compound, and they exhibit reduced phytotoxicity or are completely non-phytotoxic. While often the first step of herbicide metabolism entails a partial or total detoxification of the parent compound, there are other cases where the herbicidally active form is generated in the first metabolic reaction (e.g., the hydrolysis of the inactive fenoxaprop-P-ethyl to the herbicidally active free acid fenoxaprop-P) followed by detoxification of the molecule in the subsequent metabolic step.

The most important mechanisms for the detoxification of herbicides in weeds and crops are oxidative reactions (e.g., hydroxylations, oxidative dealkylations) catalyzed by the cytochrome-P450 mono-oxygenase system, and glutathione-S-transferase catalyzed conjugation reactions, which result in a nucleophilic displacement of aryloxy moieties, chlorine or other substituents by the tripeptide glutathione. It is known that the selective action of a herbicide in a certain crop is mostly based on a faster rate of herbicide detoxification in this crop than in the target weeds. Therefore, it is easy to speculate that the safener action could be due to an enhancement of herbicide detoxification in the crop. In the following this will be reviewed for the different safeners used in agricultural practice.

5.3.3.1 1,8-Naphthalic Anhydride (NA), Flurazole, Fluxofenim

These chemically diverse safeners all need to be applied to the crop (maize, sorghum) by seed dressing to obtain the selective safener effect. The oldest and best examined of these compounds is NA, which acts as a safener in combination with several classes of herbicides. Early studies of Sweetser [33] already suggested that the action of NA as safener for chlorsulfuron and other sulfonylureas in maize was due to an enhancement effect on the oxidative detoxification of these herbicides. Later it was demonstrated with the sulfonylurea compound triasulfuron that the effect of NA as a safener for this herbicide in maize seedlings was due to induction of a specific cytochrome P450-monooxygenase that catalyzes the hydroxylation of the parent compound to the detoxification product 5-hydroxytriasulfuron [34].

NA had also a stimulatory effect on the oxidative metabolism of the herbicide bentazone. Microsomal preparations of etiolated shoots from maize, which had received a seed treatment with NA, showed activity of a bentazone hydroxylase, which was not detectable in extracts from controls without safener pre-treatment [35]. Also, the improved tolerance of maize to the imidazolinone AC263222 after NA seed treatment could be related to enhanced AC 263222 hydroxylation by stimulation of a cytochrome P450 monooxygenase [31].

Gronwald et al. [36] reported that NA and flurazole substantially increased the glutathione-S-transferase activity in corn and sorghum (17- and 30-fold, respectively), when the herbicide metolachlor was used as substrate. This was well correlated to the protective effect of these safeners against metolachlor injury. In contrast, stimulation of GST activity was less than two-fold when, instead of metolachlor, the unspecific substrate CDNB (1-chloro-2,4-dinitrobenzene) was used as substrate. Flurazole had very similar effects in combination with the herbicide metazachlor. In particular, in sorghum it strongly stimulated the conjugation of this herbicide with glutathione [37].

Seed treatment of wheat with fluxofenim increased GST activity nine-fold, when assayed with the herbicide dimethenamid as a substrate. This increase correlated well with accelerated herbicide metabolism in wheat shoots, which was observed as a response to fluxofenim treatment [38].

5.3.3.2 Dichloroacetamides

The safeners of the dichloroacetamide family are usually applied in combination with the herbicide, either pre-plant incorporated or pre-emergence. Ekler and Stephenson [37] investigated the mode of interaction of the dichloroacetamide dichlormid, BAS 145138 and MG-191 with the herbicide metazachlor in maize and sorghum. They found an increase of the GST-catalyzed conjugation rate of metazachlor with glutathione (5- to 11-fold), and in addition an increase in GSH levels. The influence of BAS 145138 on the behavior of metazachlor in maize was also studied by Fuerst and Lamoureux [32], who concluded that the safener protected from metazachlor injury by acceleration of the enzymatic glutathione conjugation of the herbicide. Similarly the safener benoxacor was found to induce GST isoenzymes in maize. The increase in GST activity, assayed with metolachlor as substrate, was closely correlated with the protection of maize from metolachlor injury. Resolution of total GST activity by fast protein liquid chromatography (FPLC) resulted in four major activities, which to different degrees were all stimulated by the safener. One of them was only detectable in safener-treated plants [39]. Induction of GST activity, determined with alachlor as herbicide substrate, was also reported for the dichloroacetamide safener R-29148 [40].

Though this group of safeners appears to influence predominantly the GST system, Lamoureux and Rusness [41] reported that the safener BAS 145138 stimulated in maize not only the GSH conjugation but also the hydroxylation of the herbicide chlorimuron-ethyl.

5.3.3.3 Fenclorim

The safener fenclorim is used to prevent injury of the herbicide pretilachlor in paddy rice. Deng and Hatzios [42] analyzed GST extracts from several rice cultivars, with and without fenclorim pre-treatment. In all tested cultivars fenclorim increased GST activity with pretilachlor as substrate. FPLC elution patterns revealed multiple glutathione-S-transferases and mass spectrometry confirmed the formation of a pretilachlor conjugate with GSH. Apart from increasing the activity of the constitutive GST peaks, fenclorim also induced the formation of up to five new peaks, depending on the cultivar, which had activity towards pretilachlor.

5.3.3.4 Fenchlorazole-ethyl, Cloquintocet-mexyl

Fenchlorazole-ethyl/fenoxaprop-ethyl and cloquintocet-mexyl/clodinafop-propargyl were the first safener/herbicide combinations for selective post-emergence weed control in cereals. Studies of fenoxaprop-ethyl metabolism showed a more rapid decline in the level of fenoxaprop-ethyl and the free acid fenoxaprop in wheat, when applied in combination with the safener fenchlorazole-ethyl. Further studies in wheat suggested that the safener stimulated a GST-catalyzed detoxification reaction of the free acid fenoxaprop, which resulted in the formation of a glutathione conjugate with the 6-chloro-benzoxazolone moiety of the herbicide molecule. This effect was already apparent just a few hours after plant treatment and occurred only in the cereal crop, but not in the target grass weed species. The results suggested that the specificity of safener action is based on differential induction of the detoxification reaction in the crop versus the grass weed species [22, 43, 44]. Subsequently, multiple isoenzymes of GST were purified by Cummins et al. [45] from wheat shoots treated with fenchlorazole-ethyl, and it was found that only the safener-inducible isoenzymes catalyzed the detoxification of fenoxaprop-ethyl.

Also the protective effect of the safener cloquintocet-mexyl against phytotoxicity of clodinafop-propargyl in wheat was found to be based on an enhancement of herbicide detoxification in this crop. After ester hydrolysis the free acid clodinafop was metabolized in wheat by ring hydroxylation and ether cleavage with subsequent conjugate formation, while in the grass weed species metabolism was by malate ester formation. The safener specifically enhanced only herbicide metabolism in wheat, not in grass weed species [46, 47].

5.3.3.5 Mefenpyr-diethyl

In the above-mentioned combination, fenoxaprop-ethyl/fenchlorazole-ethyl the racemic form of the herbicide was subsequently replaced by the biologically active optical isomer fenoxaprop-P-ethyl and fenchlorazole-ethyl was replaced by the new safener mefenpyr-diethyl. Mefenpyr-diethyl alone did not have any phyto-toxic effects, even when applied in very high dosages. It was readily taken up by the foliage of the cereal crop and acted systemically. When fenoxaprop-P-ethyl, alone or in combination with mefenpyr-diethyl, was applied to the foliage of wheat, durum wheat or barley, it was – after foliar penetration – in both cases rapidly hydrolyzed to the free acid fenoxaprop-P. However, the rate of the subsequent conversion of the herbicidally active free acid into polar non-phytotoxic products was significantly faster in the presence than in absence of the safener [21]. The key step leading to the detoxification of fenoxaprop-P was again (as described above for the racemate fenoxaprop) the GST-catalyzed attack of glutathione at the fenoxaprop-P molecule, resulting in the formation of 4-hydroxyphenoxypropanoic acid and of a glutathione conjugate of 6-chlorobenzoxazolone (Fig. 5.2A). Both products were subject to further transformation reactions. Notably, mefenpyr-diethyl, as with the older safener fenchlorazole-ethyl, acted exclusively by en-hancement of the detoxification reaction, but did not alter the pathway of herbi-cide metabolism or the metabolite pattern in the crop species. Furthermore, mefenpyr-diethyl did not significantly influence the rate of fenoxaprop-P metabo-lism in wild oats (*Avena fatua*), as an example of a representative grass weed spe-cies, hence it acted specifically only in the cereal crops.

Determinations of GST activity against 1-chloro-2,4-dinitrobenzene (CDNB) and fenoxaprop in barley plants revealed that the exposure to mefenpyr-diethyl in-creased the conjugation rate with the unspecific substrate CDNB about two-fold, while a 12-fold increase was determined for the conjugation rate with fenoxaprop [48]. It was suggested that this was due to the specific induction by mefenpyr-diethyl of GST isoenzymes with fenoxaprop-conjugating ability. Analogous find-ings were previously described after application of the safener fenchlorazole-ethyl.

As already mentioned, it is a major advantage of mefenpyr-diethyl to act as a safener also in combination with sulfonylurea herbicides in cereal species. Com-binations have been developed with the sulfonylureas iodosulfuron-methyl-sodium and mesosulfuron-methyl.

Studies on the mode of safener action in wheat indicated that the safener en-hanced the metabolic degradation of both herbicides in the crop species, while it did not significantly alter their rate of degradation in the target weed species wild oats and blackgrass (*Alopecurus myosuroides*) [14, 21, 49].

In analogy to the findings for fenoxaprop-P-ethyl, mefenpyr-diethyl influenced in wheat only the rate of metabolism of the sulfonylurea compounds, but did not lead to any changes in the metabolite pattern. However, in contrast to fenoxaprop-P-ethyl, GSTs were not found to be involved in the metabolic detoxification of iodosulfuron-methyl sodium or mesosulfuron-methyl. The results of plant metab-olism studies suggested instead that specific cytochrome P450 monooxygenases are responsible for catalyzing early detoxification reactions. A metabolite of mesosulfuron-methyl, which appeared first after application of the herbicide to wheat plants, was identified as methyl 2-[3-(4-hydroxy-6-methoxypyrimidin-2-yl)ureidosulfonyl]-4-methanesulfonamidomethyl-benzoate and was likely formed

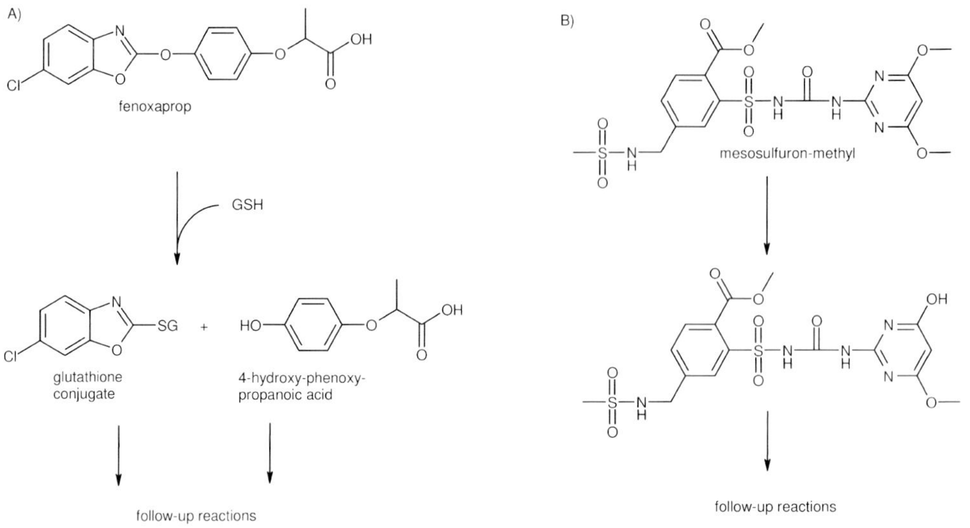

A)

fenoxaprop

GSH

glutathione
conjugate

4-hydroxy-phenoxy-
propanoic acid

follow-up reactions

B)

mesosulfuron-methyl

follow-up reactions

Fig. 5.2. Examples of herbicide detoxification reactions stimulated by the safener mefenpyr-diethyl. (A) Conjugation reaction of fenoxaprop with glutathione (GSH). (B) Oxidative demethylation of mesosulfuron-methyl.

by a P450 monooxygenase-catalyzed oxidative demethylation of the parent compound. The formation of this metabolite was markedly stimulated in wheat by the safener mefenpyr-diethyl [29] (Fig. 5.2B).

5.3.3.6 Isoxadifen-ethyl

This compound was recently developed as a safener for the sulfonylurea herbicide foramsulfuron in maize. Metabolism studies revealed that isoxadifen-ethyl also acts by enhancement of foramsulfuron metabolism in maize, while it does not influence the rate of metabolism of this herbicide in susceptible weed species [17, 18].

5.3.4
Conclusions

From the presented research data herbicide safeners obviously act in crops predominantly by enhancement of herbicide metabolism to non-phytotoxic degradation products. Notably, all safeners investigated so far only influenced the rate of herbicide metabolism, but did not alter the metabolic pathway. Hence safeners never altered the pattern of herbicide metabolites, but only led to quantitative shifts in the ratios between the phytotoxic parent compound and the metabolites of the herbicide, when compared with control plants without safener application. These quantitative differences between plants with and without safener treatment are mainly apparent the first days after plant treatment with the herbicide/safener combination. However, since the metabolic degradation of the herbicide takes place not only in safener-treated plants but also in plants without safener treatment, though there with lower rate, the quantitative differences in metabolite levels between the two treatments will become less and less with progressing disappearance of the phytotoxic parent compound.

Theoretically, a safener could act by direct activation of metabolic enzymes, but there are no data to support such a mechanism. Experimental findings rather suggest that safeners are a group of chemical compounds that act at a transcriptional level by regulating gene expression, as pointed out in reviews by Gatz [50] and by Davies and Caseley [2]. An early example of research along this line was the work of Wiegand et al. [51], who reported that in maize treatment with the safener flurazole caused a three- to four-fold increase of mRNA levels coding for a subunit of the GST I enzyme. In a further example, Hershey and Stoner [52] reported on two cDNAs (In2-1 and In2-2) from a 2-chlorobenzenesulfonamide (2-CBSU) induced maize cDNA library. In2-1 and In2-2 mRNA was found in the roots and shoot of maize seedlings after 2-CBSU application via hydroponic solution, maximum levels being reached after 6 h in the roots and after 12 h in the leaves. Both mRNAs were undetectable in seedlings without 2-CBSU induction.

While it appears now well established that safeners act at the gene expression level, it still needs to be elucidated what happens in detail at the molecular level after safener treatment of a plant. This may also help to better understand the molecular basis of the crucially required safener specificity, which results in full protection of the crop while retaining the efficacy of the herbicide on target weeds.

5.4
Concluding Remarks

Since their discovery and first introduction in the 1960–1970s herbicide safeners have provided a valuable tool for agriculture, enabling highly effective herbicides to be used in situations that would otherwise be impossible. Although this technology now competes with herbicide tolerant genetically modified (or naturally selected) crops, safeners still underpin an important part of the herbicide market in maize, cereals and rice. As described in Section 5.3, studies to identify the mechanism of safener action have also provided valuable information to help increase our understanding of herbicide metabolism in crops and weeds. Many of the commercial safeners are now off-patent, offering a chance for generic suppliers to enter the market together with off-patent herbicides. In contrast, recent mixture patents with new herbicides still allow exclusive usage by the patent holder. The number of patents for new safener classes has declined dramatically over the past 10 years, suggesting either a diminishing research success rate or, more likely, discontinued safener research in most research-based ag-chem companies. Nonetheless, because crop safeners allow the use of highly active and thus commercially competitive herbicides in situations not otherwise possible, it is expected that safeners will continue to play a valuable role in world agriculture for the foreseeable future.

References

1 K. Hatzios, R. Hoagland (Eds.), *Crop Safeners for Herbicides*, Academic Press, San Diego **1989**.

2 J. Davies, J. Caseley, *Pestic. Sci.* **1999**, 55, 1043–1058.

3 J. Davies, *Pesticide Outlook* **2001**, Feb., 10–15.

4 O. Hoffman, *Plant Physiol.* **1953**, 28, 622–628.

5 J. Peek, H. Collins, P. Porpiglia, J. Ellis, *Abstr. Annu. Weed Sci. Soc. Am.* **1988**, 28, 13.

6 L. Boldt, M. Barrett, *Weed Technol.* **1989**, 3, 303–306.

7 L. Rowe, J. Kells, D. Penner, *Weed Sci.* **1991**, 39, 78–82.

8 P. Viger, C. Eberlein, E. Fuerst, *Weed Sci.* **1991**, 39, 227–231.

9 P. Viger, C. Eberlein, E. Fuerst, J. Gronwald, *Weed Sci.* **1991**, 39, 324–328.

10 F. Chang, J. Bandeen, G. Stephenson, *Can. J. Plant Sci.* **1972**, 52, 707–714.

11 B. Bussler, R. White, E. Williams, *Proc. Br. Crop Prot. Conf. – Weeds,* **1991**, 1, 39–44.

12 T. Dill, *Abs. Annu. Southern Weed Sci. Soc. 39th Meeting,* **1986**.

13 J. Amrein, A Nyffeler, J. Rufener, *Proc. Br. Crop Prot. Conf. – Weeds,* **1989**, 1, 71–76.

14 E. Hacker, H. Bieringer, L. Willms, O. Ort, H. Koecher, H. Kehne, *Proc. Br. Crop Prot. Conf. – Weeds,* **1999**, 1, 15–22.

15 B. Rubin, O. Kirino, J. Casida, *J. Agric. Food Chem.* **1985**, 33, 489–494.

16 B. Collins, D. Drexler, M. Merkl, E. Hacker, H. Hagemeister, K. Pallett, C. Effertz, *Proc. Br. Crop Prot. Conf. – Weeds,* **2001**, 1, 35–42.

17 K. Pallett, P. Veerasekaran, M. Crudace, H. Koecher, B. Collins, *Proc. NCWSS Conf.* **2001**, Abs 77.

18 E. Hacker, H. Bieringer, L. Willms, G. Schnabel, H. Koecher, H. Hage-meister, W. Steinheuer, *Z. Pflanz.*

Pflanzenschutz, Sonderheft XVIII, **2002**, 747–756.

19 Notice of filing Pesticide Petition with EPA, Federal Register, **1999**, Vol 64 No. 110, 30997–31000.

20 H. Bieringer, K. Bauer, E. Hacker, G. Heubach, K. Leist, E. Ebert, *Proc. Br. Crop Prot. Conf. – Weeds,* **1989**, 1, 77–82.

21 E. Hacker, H. Bieringer, L. Willms, W. Rösch, H. Köcher, R. Wolf, *Z. Pflanz. Pflanzenschutz, Sonderheft XVII,* **2000**, 493–500.

22 H. Köcher, B. Büttner, E. Schmidt, K. Lötzsch, A. Schulz, A. *Proc. Br. Crop Prot. Conf. – Weeds,* **1989**, 495–500.

23 N. Polge, A. Dodge, J. Caseley, *Proc. Br. Crop Prot. Conf. – Weeds,* **1987**, 1113–1120.

24 J. Walton, J. Casida, *Plant Physiol.,* **1995**, 109, 213–219.

25 B. Rubin, J. Casida, *Weed Sci.,* **1985**, 33, 462–468.

26 H. Milhomme, J. Bastide, *J. Plant Physiol.,* **1990**, 93, 730–738.

27 H. Milhomme, C. Roux, J. Bastide, *Z. Naturforsch.,* **1991**, 46c, 945–949.

28 M. Barrett, *Weed Sci.* **1989**, 37, 34–41.

29 H. Köcher, *Pflanz.-Nachrichten – Bayer,* **2005**, 58, 179–194.

30 J. Davies, J. Caseley, O. Jones. *Proc. Brighton Crop Prot. Conf. – Weeds,* **1995**, 275–280.

31 J. Davies, J. Caseley, O. Jones, M. Barrett, N. Polge, *Pesticide Sci.,* **1998**, 52, 29–38.

32 E. Fuerst, G. Lamoureux, *Pesticide Biochem. Physiol.,* **1992**, 42, 78–87.

33 P. Sweetser, *Proc. Br. Crop Prot. Conf. – Weeds,* **1985**, 1147–1154.

34 M. Persans, M. Schuler, *Plant Physiol.,* **1995**, 109, 1483–1490.

35 J. McFadden, J. Gronwald, C. Eberlein, *Biochem. Biophys. Res. Commun.,* **1990**, 168, 206–213.

36 J. Gronwald, E. Fuerst, C. Eberlein, M. Egli, *Pesticide Biochem. Physiol.*, **1987**, 29, 66–76.

37 Z. Ekler, G. Stephenson, *Weed Res.*, **1989**, 29, 181–191.

38 D. Riechers, G. Irzyk, S. Jones, E. Fuerst, *Plant Physiol.*, **1997**, 114, 1461–1470.

39 F. Fuerst, G. Irzyk, K. Miller, *Plant Physiol.*, **1993**, 102, 795–802.

40 D. Holt, V. Lay, E. Clarke, A. Dinsmore, J. Jepson, S. Bright, A. Greenland, *Planta*, **1995**, 196, 295–302.

41 G. Lamoureux, D. Rusness, *Pesticide Biochem. Physiol.*, **1992**, 42, 128–139.

42 F. Deng, K. Hatzios, *Pesticide Biochem. Physiol.*, **2002**, 72, 24–39.

43 T. Yaakoby, J. Hall, G. Stephenson, *Pesticide Biochem. Physiol.*, **1991**, 41, 296–304.

44 H. Köcher, K. Trinks, E. Schmidt, *Abstracts SCI Meeting*, Interactions between pesticides in mixtures, London **1993**.

45 I. Cummins, D. Cole, R. Edwards, *Pesticide Biochem. Physio.*, **1997**, 59, 35–49.

46 K. Kreuz, J. Gaudin, J. Stingelin, E. Ebert, *Z. Naturforsch.*, **1991**, 46c, 901–905.

47 K. Kreuz, *Proc. Br. Crop Prot. Conf. – Weeds*, **1993**, 1249–1258.

48 R. Scalla, A. Roulet, *Physiol. Plantarum*, **2002**, 116, 336–344.

49 E. Hacker, H. Bieringer, L. Willms, K. Lorenz, H. Koecher, H. Huff, G. Borrod, R. Brusche, *Proc. Br. Crop Prot. Conf. – Weeds*, **2001**, 43–48.

50 C. Gatz, *Annu. Rev. Plant Physiol. Plant Mol. Biol.*, **1997**, 48, 89–108.

51 R. Wiegand, D. Shah, T. Mozer, E. Harding, J. Diaz-Collier, C. Saunders, E. Jaworski, D. Tiemeier, *Plant Mol. Biol.*, **1986**, 7, 235–243.

52 H. Hershey, T. Stoner, *Plant Mol. Biol.*, **1991**, 17, 679–690.

6
Genetically Modified Herbicide Resistant Crops

6.1
Overview

*Claire A. CaJacob, Paul C.C. Feng, Steven E. Reiser, and
Stephen R. Padgette*

6.1.1
Introduction

Herbicides are classified as either selective or broad spectrum. Selective herbicides can be used in-crop to control weeds without significant crop damage. Broad spectrum herbicides such as glyphosate and glufosinate are limited to pre-plant or post-directed applications. The technology to engineer herbicide resistance has enabled in-crop use of broad spectrum herbicides for improved weed control and yield.

Herbicide resistance can be generated through introduction of a gene or through a selection process. Crops generated via introduction of a gene are referred to here as genetically modified (GM) crops. Crops generated through a selection process were developed by identification of the desired herbicide resistant trait from a natural or mutagenized population of cells or plants. Where data is available, we will consider all herbicide resistant crops in this section regardless of the process by which they were generated.

In 2005, GM crops were cultivated in 21 countries with 71% of those acres being accounted for by herbicide resistant traits in soybean, corn, canola, and cotton. This percentage increases to 82% if one includes herbicide resistance trait acres that are stacked with other biotechnology traits. Globally, GM herbicide resistant soybean, cotton, canola, and corn were grown on 134.4 (60%), 12.1 (14%), 11.4 (18%), and 24.5 (7%) million acres, respectively. (Fig. 6.1.1) [1].

The growing global use of GM crops has had several positive agronomic, economic, and environmental impacts. In the United States in 2004 alone, the use of GM crops reduced pesticide use by 62 million pounds, with 55.5 million pounds of that accounted for by the use of herbicide resistant crops. GM crops also produced significant environmental benefits. In addition to reduced pesticide use, increased no-till practices have reduced water runoff, greenhouse gas emissions,

Modern Crop Protection Compounds. Edited by W. Krämer and U. Schirmer
Copyright © 2007 WILEY-VCH Verlag GmbH & Co. KGaA, Weinheim
ISBN: 978-3-527-31496-6

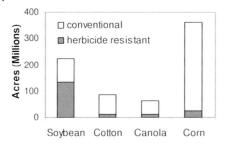

Fig. 6.1.1. Global cultivation of conventional or GM herbicide resistant crops in 2005.

and improved habitats for birds and animals [2]. These benefits are even greater if one includes all herbicide resistant crops and are expected to continue to expand the adoption of biotechnology crops.

6.1.2
Mechanisms for Engineering Herbicide Resistance

In general, herbicide resistance can be achieved through four primary strategies: detoxification of the herbicide to a non-phytotoxic metabolite; expression of an herbicide insensitive target; overexpression of the herbicide target; and cellular sequestration of the herbicide away from the target. Of these, only the first two strategies have been successfully used to develop commercial products to date. Readers are referred to other reviews on herbicide resistance [3, 4].

6.1.2.1 **Detoxification of Herbicide**
This strategy has led to commercial development of herbicide resistance for glufosinate, glyphosate and bromoxynil. Glufosinate and glyphosate resistance will be discussed in detail in later sections of this chapter (see also Chapter 6.2). Bromoxynil's herbicide activity is due to inhibition of electron transport in photosystem II. Crops engineered with bromoxynil nitrilase metabolize the herbicide to a non-phytotoxic compound [5].

Detoxification has also been utilized to engineer resistance against several other herbicides, including various auxins such as 2,4-D and dicamba, diphenyl ethers (DPEs) such as oxyfluorfen and acifluoren, pyridines such as thiazopyr, and chloroacetanilides such as alachlor. To date, none of these have been developed into commercial products [3].

6.1.2.2 **Expression of an Insensitive Herbicide Target**
This strategy was used to engineer resistance against glyphosate, and imidazolinones and sulfonylureas that inhibit acetolactate synthase (ALS), a key enzyme in the biosynthesis of branched chain amino acids. ALS resistant crops have primarily been generated through selection for an herbicide insensitive ALS allele from natural or mutagenized cell or plant populations [3].

Expression of an herbicide insensitive target has also been reported to provide resistance to diclofop and sethoxydim (ACCase inhibitors), various dinitroanilines, and inhibitors of phytoene desaturase, lycopene cyclase and hydroxyphenylpyruvate dioxygenase. None of these traits are currently incorporated into commercial products.

There are no commercial herbicide resistant crops that function by increased expression of the protein target, although some level of plant resistance has been reported for glyphosate, glufosinate, some DPEs and inhibitors of hydroxyphenylpyruvate dioxygenase. Similarly, cellular sequestration of the herbicide from the target has been reported with some DPEs, auxins and photosystem I inhibitors, but none have been developed commercially [3].

6.1.3
Commercialized Herbicide Resistant Crops

This section includes data for herbicide resistant crops generated by both selection and biotechnology processes. The first commercially available herbicide resistant crop in the United States was imidazolinone resistant corn introduced in 1992. This was followed by glyphosate resistant soybean and canola in 1996. Since then, the cultivation of herbicide resistant crops has grown globally with multiple herbicide resistant traits available in many large-acre crops (Table 6.1.1). The acreages of herbicide resistant traits in wheat and rice are insignificant and therefore not included in the table. The United States cultivates the greatest acreage of herbicide resistant crops and will be the primary focus of our discussions.

Table 6.1.1 Herbicide resistance traits by crop, the associated trade names, and manufacturers.

Crop	Herbicide resistance	Trade name	Company
Soybean	glyphosate	Roundup Ready	Monsanto
	sulfonylurea	STS	DuPont
Cotton	glyphosate	Roundup Ready	Monsanto
	glufosinate	Liberty Link	Bayer
	bromoxynil	BXN	Stoneville Cotton
Corn	glyphosate	Roundup Ready	Monsanto
	glufosinate	Liberty Link	Bayer
	imidazolinone	Clearfield	BASF
	sethoxydim	SR	BASF
Canola	glyphosate	Roundup Ready	Monsanto
	glufosinate	Liberty Link	Bayer
	imidazolinone	Clearfield	BASF

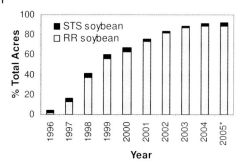

Fig. 6.1.2. Percentage of total acres of herbicide resistant soybeans by trait. *The value given for 2005 is a forecast.

6.1.3.1 Herbicide Resistant Soybeans

There are two commercially available herbicide resistant traits in soybean, glyphosate resistance (Roundup Ready®, RR) and sulfonylurea resistance (STS®). Herbicide resistance now accounts for over 90% of approximately 73 million total soybean acres grown in the United States (Fig. 6.1.2) [6].

RR soybean was developed using a glyphosate insensitive CP4-EPSPS (5-enolpyruvyl shikimate 3-phosphate synthase). Since its introduction in 1996, RR soybean has increased steadily from 2 to 89% of the total soybean acres grown in 2005 (Fig. 6.1.2). The widespread adoption of RR soybeans has resulted in significant grower and environmental benefits. In 2004 alone, RR soybeans reduced grower production costs by $1.37 billion and pesticide use by 22.4 million pounds. Furthermore, there has been about a 64% increase in the number of no-till soybean acres that decreased soil erosion, dust and pesticide run-off, and improved soil moisture, air and water quality [2].

STS soybean was introduced in 1993. Crop resistance was achieved by traditional breeding for an insensitive ALS allele. Since its introduction in 1993, STS soybean has accounted for 2 to 4% of the total soybean acres in production (Fig. 6.1.2).

6.1.3.2 Herbicide Resistant Cotton

There are three commercially available herbicide resistant traits in cotton: glyphosate resistance (Roundup Ready®, RR), glufosinate resistance (Liberty Link®, LL), and bromoxynil resistance (BXN®). In 2005, herbicide resistance traits were cultivated on over 80% of approximately 14 million total cotton acres in the United States (Fig. 6.1.3) [6]. This acreage represents a combination of herbicide resistance alone or stacked with other traits such as insect control traits.

RR cotton, engineered with CP4-EPSPS, was commercialized in 1997 and has since grown from 5 to 80% of the total cotton acres in 2005 (Fig. 6.1.3). Roundup Ready Flex cotton, a second generation product, was commercialized in 2006. The new technology enables glyphosate applications over-the-top from emergence through seven days prior to harvest and represents a significant improvement from the first generation product that limited glyphosate application through the four-leaf stage.

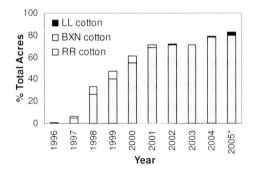

Fig. 6.1.3. Percentage of total acres of herbicide resistant cotton by trait. *The value given for 2005 is a forecast.

Glufosinate or phosphinothricin (Liberty™ and Basta®) inhibits glutamine synthetase that results in the toxic build-up of ammonia in plant cells. LL cotton, commercialized in 2004, was achieved via detoxification by phosphinothricin acetyl transferase (PAT). LL cotton grew from 1 to 3% of the total cotton acres in 2005 (Fig. 6.1.3).

Bromoxynil (Buctril®) inhibits electron transport in photosystem II by binding to the D1 protein. BXN cotton, achieved via detoxification and introduced in 1996, peaked at about 7% of the total cotton acres in 1998 but has steadily declined in use and was last sold in 2004 (Fig. 6.1.3).

As with soybean, the adoption of herbicide resistant cotton has resulted in significant grower and environmental benefits. Use of these traits in 2004 alone has resulted in a reduction in crop production costs of $264 million and pesticide use of 14 million pounds. A major effect of herbicide resistant cotton has been the increase in the adoption of no-till production. The percent increase in no-till acres has been higher in cotton than any other crop and resulted in about $20 per acre savings in fuel and labor costs [2].

6.1.3.3 Herbicide Resistant Corn

There are four commercially available herbicide resistant traits in corn: glyphosate resistance (Roundup Ready®, RR), glufosinate resistance (Liberty Link®, LL), imidazolinone resistance (Clearfield®, CF), and sethoxydim resistance (SR). In 2005 herbicide resistant traits were grown on almost 45% of approximately 81 million total corn acres in the United States (Fig. 6.1.4) [6]. This total represents a combination of herbicide resistance alone or stacked with other traits.

RR corn was commercialized in 1998. Resistance was achieved via expression of a glyphosate insensitive TIPS-EPSPS, which is a maize enzyme with two amino acid mutations that conferred glyphosate insensitivity (see Chapter 6.2, Section 6.2.2). A second-generation RR corn trait with CP4-EPSPS showed improved glyphosate resistance and was introduced in 2001. Since its introduction in 1998, RR corn has grown to 31% of the total corn acres in production for 2005 (Fig. 6.1.4).

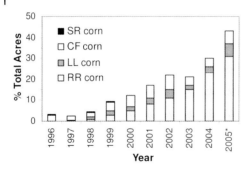

Fig. 6.1.4. Percentage of total acres of herbicide resistant corn by trait.
*The value given for 2005 is a forecast.

Imidazolinones (e.g., imazethapyr and imazapyr), like sulfonylureas, inhibit ALS. CF corn, commercialized in 1992, was developed by mutagenesis and selection of an imidazolinone insensitive ALS allele. Since its introduction in 1992, CF corn has been used in approximately 4 to 7% of the total corn acres (Fig. 6.1.4).

LL corn was developed via detoxification with the PAT gene and was introduced commercially in 1997. LL corn has grown to about 6% of the total corn acres in 2005 (Fig. 6.1.4).

Sethoxydim (e.g., Poast®) inhibits acetyl CoA carboxylase (ACCase) which is the first committed step in *de novo* fatty acid synthesis. SR corn was achieved by traditional breeding and selection for the herbicide insensitive ACCase allele and was introduced in 1996. SR corn accounted for less than 0.3% of corn acres in any one year and is no longer commercially available in field corn (Fig. 6.1.4).

Until 2004, the adoption of herbicide resistant traits in corn has been slower than in other crops mainly due to trade restrictions in export markets for GM products. Nevertheless, the adoption of herbicide resistant corn has resulted in significant grower and environmental benefits. Use of glyphosate and glufosinate resistance traits in 2004 alone resulted in a reduction in crop production costs of $139 million and pesticide use of 18.5 million pounds with numerous positive environmental attributes from increased no-till acres [2].

6.1.3.4 Herbicide Resistant Canola

Canada cultivates 90% of the canola acres in North America and is the focus of this survey. The three primary commercial herbicide resistant traits in canola are glyphosate resistance (Roundup Ready®, RR), glufosinate resistance (Liberty Link®, LL), and imidazolinone resistance (Clearfield®, CF). There are two other herbicide resistant traits that are used on a relatively small number of acres and these are bromoxynil resistance (BXN®) and triazine tolerance (TT). In 2004, herbicide resistant traits were grown on over 90% of the 12–13 million total canola acres in Canada (Fig. 6.1.5) [7, 8].

RR canola was introduced in Canada in 1996. Unlike other RR crops, RR canola was achieved by a combination of an insensitive enzyme (CP4-EPSPS)

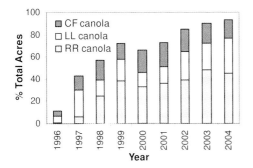

Fig. 6.1.5. Percentage of total acres of herbicide resistant canola by trait.

and a detoxification enzyme (GOX, glyphosate oxidase) that catalyzes the degradation of glyphosate to aminomethylphosphonic acid (AMPA) [9]. Since commercialization, RR canola has grown from about 1 to 45% of the total Canadian canola acres in 2004 (Fig. 6.1.5).

LL canola, introduced commercially in 1995, was achieved via detoxification by PAT. Use of LL canola has grown from about 6 to 30% of the total Canadian canola acres in 2004 (Fig. 6.1.5).

CF canola, introduced commercially in 1995, was achieved by mutagenesis and selection of an imidazolinone insensitive ALS allele. CF canola has grown from about 4 to 20% of the total Canadian canola acres (Fig. 6.1.5).

BXN canola was used in less than 100 000 acres at its peak and has since been withdrawn from the market. Triazine resistant canola, achieved through a selection process, has been used on a limited number of acres and never gained popularity.

The adoption of herbicide resistant canola has resulted in significant grower and environmental benefits. Herbicide resistant varieties allow farmers to plant earlier and control weeds better, resulting in greater yields. GM canola varieties have allowed farmers to save an estimated 8.2 million gallons of fuel and earned, on average, $10 more per acre. Their use has also increased the adoption of low and no-till framing, savings millions of acres from soil erosion [10].

There are several other commercially available herbicide resistant crops. The imidazolinone resistant trait is also available in rice, wheat, sunflower, and lentil. Glufosinate resistance is available in rice and glyphosate resistance was recently commercialized in alfalfa in 2005.

References

1 C. James, Global status of commercialized biotech/GM crops: 2005. ISAAA Briefs, pp. 34–2005, www.isaaa.org.

2 S. Sankula, E. Blumenthal, Biotechnology-derived crops planted in 2004 – Impacts on US Agriculture, **2005**, available at http://www.ncfap.org.

3 C. A. CaJacob, P. C. C. Feng, G. R. Heck, M. F. Alibhai, R. D. Sammons, S. R. Padgette, in *Handbook of Plant Biotechnology*, pp. 353–372, P. Christou and H. Klee (Eds), John Wiley & Sons, Chichester, **2004**.

4 J. Gressel, *Molecular Biology of Weed Control*, pp. 219–77, Taylor and Francis, London & New York, **2002**.

5 G. Freyssinet, B. Leroux, B. Pelissier, M. Lebrun, A. Sailland, *Serv. Biol. Mol. Cell Veg.*, **1989**, 75, 49–55.

6 Doane Marketing Research Inc. http://www.doanemr.com/Index.aspx7

7 Canola Council of Canada. http://www.canola-council.org

8 The Context Network, Biotech Traits Commercialized (Global 2005 edition). http://www.contextnet.com

9 G. F. Barry, G. M. Kishore, S. R. Padgette, M. Taylor, K. Kolacz, M. Weldon, D. Re, D. Eichholtz, K. Fincher, L. Hallas, in *Biosynthesis and Molecular Regulation of Amino Acids in Plants*, pp. 139–145, Singh B. K., Flores H. E., Shannon J. C. (Eds), American Society of Plant Physiologists, Madison, WI, **1992**.

10 BioteCanada, http://www.biotech.ca

6.2
Inhibitors of 5-enolpyruvyl Shikimate 3-phosphase Synthase (EPSPS)

Claire A. CaJacob, Paul C.C. Feng, Steven E. Reiser, and Stephen R. Padgette

6.2.1
Introduction

EPSPS is the sixth enzyme in the shikimate pathway that leads to the biosynthesis of aromatic amino acids, tryptophan, tyrosine, and phenylalanine. These aromatic amino acids along with intermediates of the pathway give rise to important secondary metabolites commonly referred to as phenylpropanoids that include phenolics, lignins, tannins, phytoalexins, etc. [1]. The shikimate pathway is localized in plastids and EPSPS is a key enzyme in regulating the flux through the pathway.

EPSPS catalyzes the transfer of phosphoenol pyruvate (PEP) to shikimate 3-phosphate (S3P) to produce 5-enolpyruvyl shikimate 3-phosphate (EPSP). This reaction starts with the binding of S3P to EPSPS to form a binary complex followed by binding of PEP to produce EPSP. Glyphosate (N-phosphonomethyl glycine) is an uncompetitive inhibitor of EPSPS and competes favorably with PEP for binding to the EPSPS/S3P binary complex to form a dead end complex. The binding specificity of glyphosate to EPSPS is extremely high [2]. Thousands of structural homologs, analogs, and derivatives were synthesized and screened for inhibition of EPSPS with most of them showing little to no activity. To date, glyphosate is the only commercialized molecule whose herbicidal activity is attributed to inhibition of EPSPS.

EPSPS is found in plants, fungi and bacteria [3]. As a result glyphosate shows broad spectrum activity against most plant species. Owing to the absence of EPSPS, glyphosate exhibits little to no toxicity in mammals, birds or fish [1]. Glyphosate was commercialized in 1975 and is the active ingredient of numerous commercial formulations, including Roundup[R]. The mechanism of action of glyphosate is attributed to inhibition EPSPS, resulting in the buildup of shikimate, and depletion of aromatic amino acids and phenylpropanoid metabolites.

6.2.2
Factors that Impact Glyphosate Efficacy

The success of glyphosate as an herbicide goes well beyond its ability to inhibit EPSPS. Glyphosate as a salt is highly water soluble; it is one of the most systemic herbicides and is readily translocated in the phloem thereby accessing difficult-to-control underground tissues such as the roots [4]. Glyphosate is applied post-emergence to foliar tissues and its efficacy is dependent on the efficiency of absorption. Glyphosate exhibits little pre-plant activity as the molecule is tightly bound to soil and not available to plants or leaching through the soil. Glyphosate in soil is quickly metabolized by microfilaria, primarily to AMPA, with an average half-life of less than a week to months depending on soil type [2].

6.2.2.1 Foliar Absorption
Studies have consistently shown that foliar absorption through the cuticle is the biggest barrier to glyphosate efficacy [5–7]. Over the years, glyphosate formulations have been developed to facilitate its absorption into the plant. Proprietary formulations have been developed using surfactants to maximize the amount and the speed of absorption; however, overly aggressive surfactants that cause excessive local tissue injury may antagonize glyphosate translocation [5, 8].

The standard method for studying plant uptake is to apply droplets of radiolabeled glyphosate to a single leaf in a plant. This method of application is drastically different from field application where a formulation after dilution to the desired volume is atomized and sprayed over-the-top of plants through a nozzle (Fig. 6.2.1). The leaf droplet method ignores potential variables in foliar spray interception, retention, and coverage. Furthermore, studies have shown that foliar absorption of glyphosate is affected by concentrations of glyphosate and surfactant, as well as spray droplet size [6, 8]. The leaf droplet method typically uses droplet size of 1 μL, which is equivalent to a diameter of 1200 μm. In comparison, a typical flat fan nozzle generates a broad range of droplet sizes from less than 100 to greater than 1000 μm with a considerably smaller volume median diameter of only 173 μm [9]. Studies have shown that glyphosate uptake is improved with large size droplets; however, this is offset by reduced foliar retention and coverage during spray application. Increasing the spray volume increases the foliar coverage but at the expense of reduced surfactant and glyphosate concentrations. It is apparent that plant absorption of glyphosate is affected by numerous interdependent variables that cannot be adequately modeled by the leaf droplet method [8].

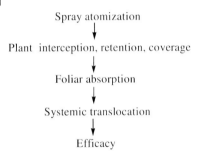

Fig. 6.2.1. Schematic representation of events that characterize an over-the-top spray application of glyphosate formulation *in planta*.

Our initial study in velvetleaf plants employed the standard leaf droplet method to compare glyphosate absorption among commercial formulations [5]. Realizing the limitation of the leaf droplet method, we cautiously began using a track sprayer for over-the-top spray application of formulations augmented with radio-labeled glyphosate [8, 10–13]. The track sprayer method required much greater care and preparation during experimentation but allowed us to generate data on absorption of glyphosate under realistic field use rates and parameters.

In our subsequent study, we used the track sprayer method to compare the absorption of ^{14}C-glyphosate among different commercial formulations in velvetleaf [10]. Plants were sprayed with a field use rate (0.2 kg-a.e. ha^{-1}) at a volume of 93 L ha^{-1} using a commercially available flat fan nozzle. We observed similar spray retention in plants among the three formulations, indicating that differential spray retention contributed little to no difference in efficacy. Following spray application, plants were harvested at various times to measure the levels of glyphosate remaining on the foliar surfaces, absorbed into the leaves, and translocated to the roots. With spray application, the applied dose varied, depending on plant size; therefore, the absorption efficiency was expressed as percentage of total recovered dose from each plant. Significant differences were observed among the formulations in the rate of glyphosate absorption (Fig. 6.2.2). The most efficient formulation (formulation A) rapidly absorbed 28% of the applied dose by 24 h after treatment (HAT) and plateaued thereafter. Formulation B showed much reduced absorption at only 16% by 24 HAT. Formulation C showed slow initial (13% at 24 HAT) but more prolonged absorption [10]. These results illustrated the subtle differences in glyphosate absorption among commercial formulations in velvetleaf under realistic field application parameters.

Not surprisingly, glyphosate absorption is also affected by species differences in cuticle structure and leaf morphology. Using the track sprayer method, foliar absorption of ^{14}C-glyphosate at field use rates with a flat fan nozzle ranged from 20 to 36% in velvetleaf, prickly sida, kochia and RR corn [10, 14]. These results indicated that studies on foliar absorption of glyphosate not only need to employ a relevant method (i.e., track sprayer) but also be conducted in the species of interest.

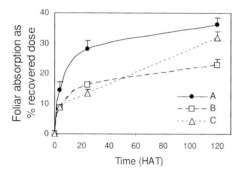

Fig. 6.2.2. Comparison of foliar absorption of ^{14}C-glyphosate with time (hours after treatment, HAT) in commercial formulations A, B, and C using over-the-top track spray application in young velvetleaf plants.

6.2.2.2 Systemic Translocation

Most plant species do not metabolize glyphosate [2], with the exception of soybean, which slowly metabolizes glyphosate to AMPA [15]. As a result, once glyphosate enters the phloem, it is translocated along the sucrose gradient from source to sink tissues [16]. The accumulation of glyphosate in sink tissues produces local injuries that diminish the sink strength and sucrose demand, resulting in reduced glyphosate translocation; thus glyphosate imposes a self-limitation on its own efficacy [17, 18]. Sink tissues can sense the effect of glyphosate within a few hours after application, although visible injury may take 7 to 10 days to appear [17]. For these reasons, efficacious glyphosate formulations generally exhibit rapid uptake and translocation to avoid self-limitation [10]. At the same time, fast absorption also results in favorable rainfastness for greater application flexibility.

With the track sprayer method, systemic translocation of absorbed glyphosate among commercial formulations was measured in roots that were shielded from the spray in velvetleaf plants. Formulation A, which showed the highest absorption (28%, Fig. 6.2.2), showed 6% translocation to roots at 24 HAT (Fig. 6.2.3) [10]. Root translocation was proportional to foliar absorption and followed the ranking of formulation A > C > B, which is also the ranking of overall plant efficacy. These results showed that, even with efficient absorption, only about one-third of the applied dose was absorbed, and only a fraction of that was translocated to the roots at 24 HAT. Since the amount translocated was proportional to that absorbed, increasing absorption would increase overall efficacy as long as translocation is not hindered in the process.

Our initial microscopy studies in velvetleaf plants showed that large 1-µL droplets used in the leaf droplet method caused localized spot necrosis on the leaf [5, 19, 20]. In contrast, smaller droplets as encountered in spray application caused little to no visible local injury [21]. A recent study in RR cotton also showed that glyphosate distribution to bolls differed between over-the-top spray versus manual

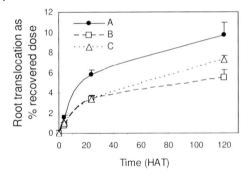

Fig. 6.2.3. Comparison of root translocation of ^{14}C-glyphosate with time (hours after treatment, HAT) in commercial formulations A, B, and C following over-the-top spray application in young velvetleaf plants.

leaf droplet application [12]. This difference was caused by the fact that different leaves, depending on age and position, source different bolls in plants; therefore, glyphosate translocation was dependent on which leaves intercepted the spray. These results further demonstrate that glyphosate application to a single leaf may produce misleading data on absorption and translocation of glyphosate with little relevance to spray application in the field. Nevertheless, the leaf droplet method continues to be used by most researchers when studying herbicide absorption and translocation in plants.

6.2.3
Development of Glyphosate Resistant Crops

A major milestone occurred in 1996 with the commercial introduction of glyphosate resistant soybean [22]. Since then, the glyphosate resistant trait (Roundup Ready®, RR) has been introduced into canola, cotton, corn, alfalfa, sugar beet, and others. All RR crops thus far contain a glyphosate insensitive EPSPS derived either from the plant or bacteria.

The X-ray crystal structures of *E. coli* EPSPS showed that the enzyme consists of two domains that undergo a conformational change upon ligand binding [23, 24]. The resulting closure of the two domains forms the catalytic pocket at the interface [25]. Glyphosate binding appeared to be modulated by several key amino acids near the vicinity of the catalytic pocket [23]. These key amino acids are highly conserved across species and in fact have been used to predict the sensitivity of the EPSPS to glyphosate [26].

Kinetic analysis showed that the endogenous maize EPSPS has a K_m-PEP of 27 μM with a K_i of 0.5 μM (Table 6.2.1). Mutations were introduced at the T102I and P106S positions to produce the variant TIPS [26]. The TIPS-EPSPS showed a K_m of 10.6 μM with a K_i of 58 μM. The double mutant enzyme preserved the EPSPS function while reducing sensitivity to glyphosate. Analysis of single mutations

Table 6.2.1 Comparison of steady state kinetics of EPSPS enzymes from plant and bacteria that are used to engineer glyphosate resistance in crops.

EPSPS	K_m-PEP (μM)	K_i (μM)	K_i/K_m
Maize wild type	27.0 ± 4.0	0.50 ± 0.06	0.02
Maize TIPS	10.6 ± 1.6	58.0 ± 14	5.5
Agrobacterium CP4	12	2720	227

(T102I or P106S) showed increased K_m or sensitivity to glyphosate, thus the desired kinetic properties were obtained only in the double mutant enzyme [26].

Bacterial sources of EPSPS were screened for insensitivity to glyphosate [27]. The EPSPS that is most insensitive to glyphosate was isolated from *Agrobacterium* species CP4. This enzyme showed desirable K_m-PEP (12 μM) but much greater K_i (2720 μM) with a K_i/K_m ratio that is 41× higher than that of TIPS-EPSPS [26]. CP4-EPSPS, which is kinetically superior to TIPS-EPSPS, is currently utilized in all RR crops. The TIPS-EPSPS was utilized in the first generation RR corn (GA21) [28], which has since been replaced by CP4-EPSPS in the second generation RR corn (NK603) [29].

6.2.3.1 Alternative Mechanisms for Engineering Glyphosate Resistance

Attempts were made to engineer glyphosate resistance by increased expression of a glyphosate sensitive EPSPS in *E. coli*, petunia and *Arabidopsis* [30–32]. Transgenic petunia with >20-fold increased expression of EPSPS showed limited resistance to glyphosate with growth inhibition at field use rates [31]. These results suggested that increased expression of a sensitive EPSPS is not likely to generate commercial level resistance to glyphosate in crops.

An alternative mechanism is to use enzymes that catalyze glyphosate detoxification. Glyphosate oxidoreductase (GOX) was cloned from *Achromobacter* sp. strain LBAA and was shown to catalyze the degradation of glyphosate to AMPA (Fig. 6.2.4) [33]. Glyphosate resistance was observed in plants expressing the GOX gene; however, at a level insufficient for commercialization. GOX is currently utilized in conjunction with CP4-EPSPS in RR canola. Recent reports have shown that AMPA may exhibit some plant toxicity of its own [15], which makes GOX less desirable for engineering glyphosate resistance.

Reports by Castle et al. [34] and Siehl et al. [35] described a glyphosate acetyl transferase (GAT, Fig. 6.2.4) useful for engineering plant resistance to glyphosate. GAT was originally cloned from *Bacillus licheniformis* but conferred no resistance to glyphosate when expressed in any host. The catalytic efficiency of GAT was improved through 11 iterations of DNA shuffling and the resulting gene, when expressed in tobacco and maize, conferred resistance to field use rates of glyphosate. A recent patent publication [36] describes yet another bacterial enzyme

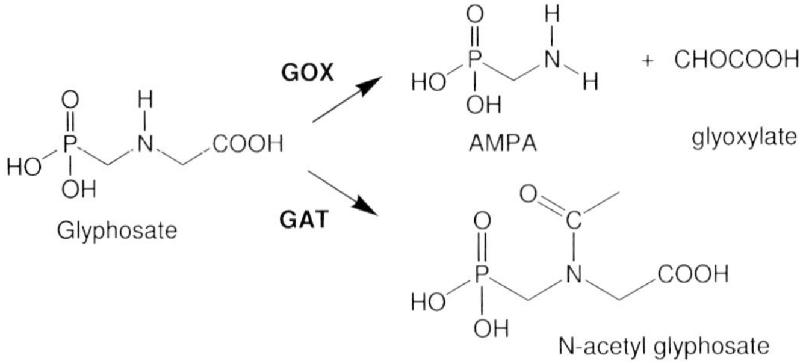

Fig. 6.2.4. Detoxification enzymes that may be useful for engineering glyphosate resistance in crops.

(GDC) suitable for engineering glyphosate resistance. This enzyme is described as having homology to decarboxylases and presumably catalyzes the decarboxylation of glyphosate. Both GAT and GDC are reportedly under development; however, neither gene has been utilized in a commercial crop to date.

6.2.3.2 Disease Control Benefits of Glyphosate Resistant Crops

Although EPSPS is also found in fungi and bacteria, glyphosate was previously shown to have very weak fungicidal activity (ED_{50} 100 to >1000 mg g^{-1}). Recent reports by Feng et al. [13], and Anderson and Kolmer [37] showed that glyphosate reduced the incidence of leaf and stripe rusts in RR wheat and of Asian rust in RR soybean. These fungi (*Puccinia triticina*, *P. striiformis*, and *Phakopsora pachyrhizi*) are obligate pathogens and have been responsible for major yield losses in wheat and soybean.

Studies in RR wheat showed that glyphosate, at a spray dose typically recommended for weed control (i.e., 0.84 kg-a.e. ha^{-1}), provided both preventive and curative activities for a period of 2–4 weeks against leaf and stripe rusts. Disease control was minimal in formulation blanks without glyphosate and was directly correlated to the level of systemic glyphosate in leaves. Field tests under natural heavy rust pressure further confirmed the activity of glyphosate [13]. Current results suggest that glyphosate may provide beneficial effects of rust control in RR wheat and RR soybean. Studies are underway to determine whether glyphosate has activity against other diseases in other RR crops.

From the stand point of engineering glyphosate resistance, either an insensitive EPSPS or a detoxification gene should be equally feasible; however, the disease control benefits of glyphosate is expected to be mostly associated with the use of an insensitive EPSPS due to preservation of glyphosate in crops. This has been observed in glufosinate resistant crops. Glufosinate has also shown fungicidal and disease control activities in glufosinate resistant plants [38–40]. Glufosinate showed only a 2–3 day disease control window [38], which is much shorter than that observed with glyphosate [13]. This can simply be explained by the fact that glufosinate resistant plants are engineered with PAT, which effectively detoxified the herbicidal and fungicidal activities of glufosinate. Presumably plants engineered with a glufosinate insensitive glutamine synthetase would demonstrate a longer disease control window.

6.2.4
Effects of CP4 Expression on Plant Resistance

The lack of plant metabolism and the use of a glyphosate-insensitive EPSPS translate to persistence of glyphosate, which continues to mobilize from source to sink tissues in RR crops [26]. The sink tissues in a plant vary, depending on the growth stage, as a result, the timing of glyphosate spray, which is determined by best weed control, will impact which sink tissues are at risk for glyphosate injury. Our experiences in developing RR crops have taught us that male reproductive tissues are strong sinks and vulnerable to glyphosate injury [26].

Monsanto's approach to second generation RR traits in crops such as corn, cotton, and soybean has been to improve upon first generation products by coordinating expression of the highly glyphosate insensitive CP4-EPSPS in tissues that are at-risk to glyphosate injury. Therefore, the development of second generation RR cotton, corn, and soybean has shifted from strong constitutive viral promoters to regulatory elements with enhanced expression, both spatial and temporal, in the at-risk tissues such as the developing pollen and tapetum. These regulatory expression elements have been engineered as part of a second CP4 EPSPS expression cassette in the case of second generation RR corn and RR Flex cotton. These second generation products have shown enhanced field performance compared to their forerunners. The different promoters used in first and second generation RR crops are highlighted in Table 6.2.2.

Table 6.2.2 Genetic elements used to engineer glyphosate resistance in first and second generation Roundup Ready crops.

Roundup ready crops	Event(s)	Expression cassette 1		Expression cassette 2	
		Promoter/ intron	Coding region	Promoter/ intron	Coding region
RR Corn-1	GA21	Os.Act1/ Os.Act1	TIPS-EPSPS	None	None
RR Corn-2	NK603	Os.Act1/ Os.Act1	CP4-EPSPS	e35S/hsp70	CP4-EPSPS
RR Cotton-1	1445	FMV	CP4-EPSPS	None	None
RR Flex Cotton-2	MON88913	FMV/TSF1	CP4-EPSPS	35S/ACT8	CP4-EPSPS
RR Soybean	40-3-2	e35S	CP4-EPSPS	None	None
RR Canola	RT73	FMV	CP4-EPSPS	FMV	GOX
RR Alfalfa	J101 & J163	eFMV	CP4-EPSPS	None	None

6.2.4.1 **Roundup Ready Cotton**

The second generation RR Flex cotton was commercialized in 2006 to provide growers with greater flexibility in the amount and timing of glyphosate application [41]. Both first and second generation products employ CP4-EPSPS; however, RR Flex cotton will employ two CP4-EPSPS expression cassettes [42–45]. In particular, the first cassette uses a chimeric promoter composed of the *Arabidopsis thaliana* TSF1 gene promoter that encodes elongation factor EF-1alpha [46–48] and enhancer sequences from the Figwort Mosaic virus 35S promoter [49], together with a *cis*-acting TSF1 intron. The second cassette utilizes another chimeric promoter composed of the ACT8 gene promoter of *Arabidopsis thaliana* [50] combined with the enhancer region of the cauliflower mosaic virus (CaMV) 35S promoter [51] together with intron sequences from the ACT8 gene. These chimeric promoters provide strong vegetative expression from the viral enhancer elements and at the same time boost expression in key male reproductive organs via the promoter elements from TSF1 and ACT8 (Fig. 6.2.5B and C). The result is that RR Flex cotton plants are able to withstand glyphosate applications with excellent boll retention throughout the growing season (Fig. 6.2.5A).

Fig. 6.2.5. Field performance and tissue expression profiles of CP4-EPSPS in first- and second-generation Roundup Ready (RR) cotton. (A) Comparison of boll retention between RR Flex cotton-2 (right) and RR cotton-1 (left) treated with Roundup (2.5 kg-a.e. ha^{-1}) at 4, 6, 10, and 14-node stages. (B) Immunolocalization of CP4-EPSPS protein in the anther wall (AW), but not in the mature pollen (MP), in event 1445 of RR cotton-1. (C) Strong CP4-EPSPS expression in both AW and MP in event RR60, a predecessor of RR Flex cotton-2.

6.2.4.2 **Roundup Ready Corn**

Expression of CP4-EPSPS by the e35S promoter produced corn plants that exhibited vegetative tolerance but poor reproductive tolerance (i.e., male sterility) when challenged with commercially applicable rates of glyphosate [26, 29]. The use of rice actin 1 promoter (Os Act1) and intron elements [52] boosted expression in key male reproductive tissues and produced male fertility. The first generation RR corn (GA21) employs the Os Act1 promoter and introns with the TIPS-EPSPS [28]. The second generation RR corn (NK603) [29] employs CP4-EPSPS in two expression cassettes driven by Os Act1 and e35S promoters for high expression in both male reproductive and vegetative tissues, thus giving rise to robust and consistent field performance.

6.2.4.3 **Roundup Ready Soybean**

The current RR soybean event utilizes CP4-EPSPS under the transcriptional regulation of the 35S promoter [22]. RR soybeans demonstrate excellent reproductive fertility from application of labeled rates of glyphosate; however, reproductive fertility could not be demonstrated from 35S expression of CP4-EPSPS in *Arabidopsis*, tomato or tobacco. The second generation RR soybean is being developed using a modified version of the gene and a new chimeric promoter in a single expression cassette to enhance expression in male reproductive and vegetative tissues.

6.2.5
Stacking Traits in Roundup Ready Crops

New RR traits will likely be commercialized as part of a stack with other biotechnology traits to deliver multiple attributes and benefits simultaneously. Traits can be stacked by breeding via cross pollination of lines containing different traits. Alternatively, traits can be stacked in the transformation vector generating multiple traits in one transformation event. Breed stacking has the advantage of utilizing existing events without the need to generate new events; however, as the number of genetic loci increases so does the complexity of event management and selection. Vector stacking contains multiple traits in one genetic loci, which simplifies breeding but requires *de novo* transformation. Trait stacking is a complicated decision that needs to take into account the market demand and grower needs.

Bollgard® (BG) cotton, which protects cotton plants from lepidopteran pests, was introduced in 1996. RR cotton was commercialized in 1997 as a single trait but also with limited availability of the RR/BG stack. The adoption of RR cotton peaked at about 2001 while RR/BG stack has continued to grow and accounted for more acres in 2005 than any of the single trait events (Fig. 6.2.6) [53, 54]. A second-generation lepidopteran product in cotton, Bollgard® II, was introduced exclusively with a RR stack in 2004. These stacked products were generated though breeding. Undoubtedly new trait stacks can be expected from the introduction of RR Flex cotton in 2006.

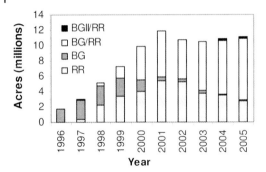

Fig. 6.2.6. Distribution of RR cotton acres as single or stacked traits with Bollgard (BG) since its commercial introduction in 1997.

In RR corn, three different trait stacks are commercially available. RR corn was stacked with YieldGard® Corn Borer (YGCB) for control of lepidopteran pests. In 2005, RR/YGCB was grown on more than 11 million acres, which was similar to that of RR alone [55]. RR corn has also been stacked with YieldGard® Root Worm (YGRW) for rootworm control. The first triple stack, which combines RR, YGCB and YGRW, was introduced in 2005 for glyphosate resistance with lepidopteran and rootworm control. These stacked products were produced through breeding. There are no RR stacks available to date in other crops such as soybean and canola, but that is expected to change in the future.

References

1 B. B. Buchanan, W. Gruissem, R. L. Jones, *Biochemistry Molecular Biology of Plants*, pp. 1286–1311, American Society of Plant Physiol, Rockville, **2000**.

2 J. E. Franz, M. K. Mao, J. A. Silorski, in *Glyphosate: A Unique Global Herbicide*, pp. 65–101, ACS Monograph 189, Whashington D.C., **1997**.

3 G. M. Kishore, D. M. Shah, *Annu. Rev. Biochem.*, **1998**, 57, 627–663.

4 R. H. Bromilow, K. Chamberlain, S. G. Patil, *Pestic. Sci.* **1990**, 30, 1.

5 P. C. C. Feng, J. S. Ryerse, R. D. Sammons, *Weed Technol.* **1998**, 12, 300–307.

6 S. H. Liu, R. A. Campbell, J. A. Studens, R. G. Wagner, *Weed Sci.* **1996**, 44, 482–488.

7 R. J. L. Ramsey, G. R. Stephenson, J. C. Hall, *Pest. Biochem. Physiol.*, **2005**, 82, 162–175.

8 P. C. C. Feng, T. Chiu, R. D. Sammons, J. S. Ryerse, *Weed Sci.*, **2003**, 51, 443–448.

9 R. E. Etheridge, A. R. Womac, T. C. Mueller, *Weed Technol.* **1999**, 13, 765–770.

10 P. C. C. Feng, J. J. Sandbrink, R. D. Sammons, *Weed Technol*, **2000**, 14, 127–132.

11 P. C. C. Feng, M. Tran, T. Chiu, R. D. Sammons, G. R. Heck, C. A. CaJacob., *Weed Sci.*, **2004**, 52, 498–505.

12 P. C. C. Feng, T. Chiu, *Pest. Biochem. Physiol.*, **2005**, 82, 36–45.

13 P. C. C. Feng, G. J. Baley, W. P. Clinton, G. J. Bunkers, M. F. Alibhai,

T. C. Paulitz, K. K. Kidwell, *Proc. Natl. Acad. Sci. U.S.A.*, **2005**, 48, 17290–17295.

14 P. C. C. Feng, J. J. Sandbrink, J. E. Cowell. Abstract #40, National meeting of WSSA, Toronto, Canada, **2000**, 17.

15 K. N. Reddy, A. M. Rimando, S. O. Duke, *J. Agric. Food Chem.*, **2004**, 52, 5139–5143.

16 P. C. C. Feng, T. Chiu, R. D. Sammons, *Pest. Biochem. Physiol.*, **2003**, 77, 83–91.

17 D. R. Geiger, H. D. Bestman, *Weed Sci.*, **1990**, 38, 324–329.

18 D. R. Geiger, W. J. Shieh, M. A. Fuchs, *Pest. Biochem. Physiol.*, **1999**, 64, 124–133.

19 P. C. C. Feng, J. S. Ryerse, C. R. Jones, R. D. Sammons, *Pest. Sci.*, **1999**, 55, 385–386.

20 J. S. Ryerse, P. C. C. Feng, R. D. Sammons, *Microsc. Today*, **2001**, 1, 22–24.

21 J. S. Ryerse, R. A. Downer, R. D. Sammons, P. C. C. Feng, *Weed Sci.*, **2004**, 52, 302–309.

22 S. R. Padgette, K. H. Kolacz, X. Delannay, D. B. Re, B. J. Lavallee, C. N. Tinius, W. K. Rhodes, Y. I. Otero, G. F. Barry, D. A. Eichholtz, G. M. Peschke, D. L. Nida, N. B. Taylor, G. M. Kishore, *Crop Sci.*, **1995**, 35, 1451–1461.

23 E. Schönbrunn, S. Eschenburg, W. A. Shuttleworth, J. V. Schloss, N. Amrhein, J. N. S. Evans, W. Kabsch, *Proc. Natl. Acad. Sci. U.S.A.*, **2001**, 98, 1376–1380.

24 W. C. Stallings, S. S. Meguid-Abdel, L. W. Lim, H. S. Shieh, H. E. Dayringer, N. K. Leimgruber, R. A. Stegeman, K. S. Anderson, J. A. Sikorski, S. R. Padgette, G. M. Kishore, *Proc. Natl. Acad. Sci. U.S.A.*, **1991**, 88, 5046–5050.

25 M. Alibhai, W. C. Stallings, *Proc. Natl. Acad. Sci. U.S.A.*, **2001**, 98, 2944–2946.

26 C. A. CaJacob, P. C. C. Feng, G. R. Heck, M. F. Alibhai, R. D. Sammons, S. R. Padgette, in *Handbook of Plant Biotechnology*, pp. 353–372, P. Christou, H. Klee (Eds), John Wiley & Sons, Chichester, **2004**.

27 S. R. Padgette, D. B. Re, G. F. Barry, D. A. Eichholtz, X. Delannay, R. L. Fuchs, G. M. Kishore, R. T. Fraley, in *Herbicide-Resistant Crops: Agricultural, Environmental, Economic, Regulatory, and Technical Aspects*, pp. 53–84, Duke S. (ed), CRC Lewis Publishers, Boca Raton, FL, **1996**.

28 R. S. Sidhu, B. G. Hammond, R. L. Fuchs, J. N. Mutz, L. R. Holden, B. George, T. Olson, *J. Agric. Food Chem.*, **2000**, 48, 2305–2312.

29 G. R. Heck, C. L. Armstrong, J. D. Astwood, C. F. Behr, J. T. Bookout, S. M. Brown, T. A. Cavato, D. L. DeBoer, M. Y. Deng, C. George, J. R. Hillyard, C. M. Hironaka, A. R. Howe, E. H. Jakse, B. E. Ledesma, T. C. Lee, R. P. Lirette, M. L. Mangano, J. N. Mutz, Y. Qi, R. E. Rodriguez, S. R. Sidhu, A. Silvanovich, M. A. Stoecker, R. A. Yingling, J. You, *Crop Sci.*, **2005**, 44, 329–339.

30 H. J. Klee, Y. M. Muskopf, C. S. Gasser, *Mol. Gen. Genet.*, **1987**, 210, 347–442.

31 D. M. Shah, R. B. Horsch, H. J. Klee, G. M. Kishore, J. A. Winter, N. E. Tumer, C. M. Hironaka, P. R. Sanders, C. S. Gasser, S. Aykent, N. R. Siegel, S. G. Rogers, R. T. Fraley, *Science*, **1986**, 233, 478–481.

32 S. G. Rogers, L. A. Brand, S. B. Holder, E. S. Sharps, M. J. Brackin, *Appl. Environ. Microbiol.*, **1983**, 46, 37–43.

33 G. F. Barry, G. M. Kishore, S. R. Padgette, M. Taylor, K. Kolacz, M. Weldon, D. Re, D. Eichholtz, K. Fincher, L. Hallas, in *Biosynthesis and Molecular Regulation of Amino Acids in Plants*, pp. 139–145, Singh B. K., Flores H. E., Shannon J. C. (Eds), American Society of Plant Physiologists, Madison, WI, **1992**.

34 L. A. Castle, D. L. Siehl, R. Gorton, P. A. Patten, Y. H. Chen, S. Bertain, H. Cho, N. Duck, J. Wong, D. Liu, *Science*, **2004**, 304, 1151–1154.

35 D. L. Siehl, L. A. Castle, R. Gorton, Y. H. Chen, S. Bertain, H. Cho, R. Keenan, D. Liu, M. W. Lassner, *Pest. Manag. Sci.*, **2005**, 61, 235–240.

36 P. E. Hammer, T. K. Hinson, N. B. Duck, M. G. Koziel, Methods to confer herbicide resistance, WO2005003362, **2005**.

37 J. A. Anderson, J. A. Kolmer, *Plant Dis.*, **2005**, 89, 1136–1142.

38 Y. Wang, M. Browning, B. A. Ruemmele, J. M. Chandlee, A. P. Kausch, *Weed Sci.*, **2003**, 51, 130–137.

39 C. A. Liu, H. Zhong, J. Vargas, D. Penner, M. B. Sticklen, *Weed Sci.* **1998**, 46, 139–146.

40 H. Uchimiya, M. Iwata, C. Nojiri, *Bio/Technol.*, **1993**, 11, 835–836.

41 M. Lloyd, S. Culpepper, E. Cerny, B. Coots, C. Corkern, T. Cothren, K. Croon, K. Ferreria, J. Hart, B. Hayes, S. Huber, A. Martens, B. McCloskey, M. Oppenhuizen, M. Patterson, Z. Shappley, J. Subramani, D. Reynolds, T. Witten, A. York, Yield and fruiting behavior of Roundup Ready Flex cotton in ten environments in 2001. *Proceedings of Beltwide Cotton Conference*, Nashville, Tennessee, **2003**.

42 Y. S. Chen, C. Hubmeier, M. Tran, A. Martens, R. E. Cerny, R. D. Sammons, C. CaJacob, *Plant Biotech. J.*, **2006**, 4, 477–487.

43 K. L. Fincher, S. Flasinski, J. Q. Wilkinson, Plant expression constructs, US Patent 6,660,911, **2003**.

44 K. L. Fincher, S. Flasinski, J. Q. Wilkinson, Chimeric cauliflower mosaic virus 35S-arabidopsis actin 8

promoters and methods of using them, US Patent 6,919,495, **2005**.

45 K. L. Fincher, S. Flasinski, J. Q. Wilkinson, Chimeric figwort mosaic virus-elongation factor 1 α promoters and methods of using them, US Patent 6,949,696, **2005**.

46 C. Curie, T. Liboz, E. Gander, C. Medale, C. Bardet, M. Axelos, B. Lescure, *Nucleic Acids Res.*, **1991**, 19, 1305–1310.

47 C. Curie, M. Axelos, C. Bardet, R. Atanassova, N. Chaubet, B. Lescure, *Mol. Gen. Genet.*, **1993**, 238, 428–436.

48 M. Axelos, C. Bardet, T. Liboz, A. Le Van Thai, C. Curie, B. Lescure, *Mol. Gen. Genet.* **1989**, 219, 106–112.

49 R. Richins, H. Scholthof, R. Shepard, *Nucleic Acids Res.* **1987**, 15, 8451–8466.

50 Y. Q. An, J. M. McDowell, S. Huang, E. C. McKinney, S. Chambliss, R. B. Meagher, *Plant J.*, **1996** 10, 107–121.

51 R. Kay, A. Chan, M. Daly, J. McPherson, *Science*, **1987**, 236, 1299–1302.

52 D. McElroy, W. Zhang, J. Cao, R. Wu, *Plant Cell*, **1990**, 2, 163–71.

53 USDA, NASS; Crop Production Acreage supplement. http://usda. mannlib.cornell.edu/reports/nassr/field/pcp-bba/

54 USDA, AMS; Cotton Varieties Planted, United States, years 1995–2005. http://www.ams.usda.gov/cotton/cnpubs.htm

55 Personal communication.

6.3
Glutamine Synthetase Inhibitors

Günter Donn

6.3.1
Introduction

Despite the fact that the atmosphere consists of 78% of nitrogen, plants evolved in contrast to animals under conditions where accessible nitrogen sources were growth limiting due to the chemical inertness of the molecule. Whereas in animals effective pathways evolved to detoxify and to excrete surplus ammonia as urea or ureides, plants are dependent on perfect mechanisms of ammonia recycling. The key enzyme in plants to assimilate, reassimilate and to detoxify ammonia is glutamine synthetase which converts ammonia and glutamate into glutamine under consumption of ATP. Especially in photosynthetically active cells, considerable amounts of ammonia are released in the photorespiratory C_2 cycle which have to be recycled with high efficiency to prevent the build up of high ammonia levels that eventually are toxic or may cause the loss of the volatile NH_3.

Phytopathogenic *Pseudomonas* strains were the first organisms that exploited this Achilles heel of plants: *P. syringae* pv tabaci produce the glutamine synthetase inhibitor tabtoximine-β-lactam, which enables the pathogen to colonize the host tissue killed by the toxin.

In the late 1960s/early 1970s *Streptomyces* strains were discovered that produce a tripeptide consisting of two molecules alanine and an unusual amino acid containing a phosphino group. The latter compound was named L-phosphinothricin and the tripeptide is known as bialaphos (syn. bilanaphos). L-Phosphinothricin was recognized as a glutamate analogue and potent inhibitor of bacterial glutamine synthetases. In the mid-1970s it was recognized that the natural tripeptide as well as the amino acid L-phosphinothricin and the synthetic racemate named glufosinate reveal high herbicidal potential as post-emergent nonselective herbicides. For two decades glufosinate as well as the natural tripeptide have been commercialized and recognized as valuable tools in post-emergent weed control strategies.

Twenty years ago it became evident that the phosphinothricin producing *Streptomyces* strains have in their genomes highly specific acetyltransferase genes that after transfer into transgenic crop plants protect these in a perfect manner from herbicidal activity of phosphinothricin and glufosinate. This opened up fascinating opportunities to use these glutamine synthetase inhibitors as selective herbicides in transgenic crops.

6.3.2
Role of Glutamine Synthetase in Plant Nitrogen Metabolism

Amongst the plant enzymes that use ammonia as substrate, glutamine synthetase (GS; E.C. 6.3.1.2) has the highest affinity (K_m 3–5 µM) for this nitrogen source. Ammonia is released in plant tissues by nitrite reduction and amino acid catabolism but the highest amount, up to 90%, originates in photosynthetic tissues from the photorespiratory C2 cycle [1].

In photosynthetic tissues under atmospheric conditions the oxygenase activity of Rubisco leads to the formation of 2-phosphoglycolate in the chloroplasts, which is cleaved into inorganic phosphate and glycolate. In peroxisomes this intermediate is oxidized by glycolate oxidase to glyoxylate and H_2O_2. Glyoxylate is rapidly metabolized by the enzymes glutamate-glyoxylate-aminotransferase and serine-glyoxylate-aminotransferase. In both cases glycine is the end product. In

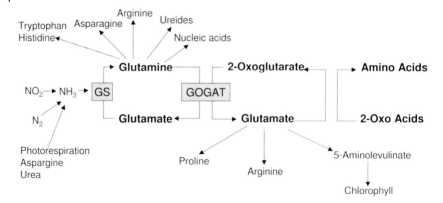

Fig. 6.3.1. Central role of the GS/GOGAT cycle in plant N-metabolism.

mitochondria two molecules of glycine are converted into one molecule of serine and $CO_2 + NH_3$ are released, which are reassimilated in the chloroplasts [2].

Glutamine synthetase (GS) uses glutamate and ammonia as substrates (Fig. 6.3.1). The resulting glutamine is the substrate for glutamate synthase (glutamine-2 oxoglutarate-aminotransferase, GOGAT), which transfers the amido group from glutamine to 2-oxoglutarate, synthesizing two molecules of glutamate [3]. The GS-GOGAT cycle enables plants to assimilate and to recycle ammonia with high efficiency. The end products of both enzymes are substrates for the respective partner enzyme as well as amino donors for the synthesis of amino acids, purines and pyrimidines [4].

Due to its central role in nitrogen metabolism, plants typically confer several GS genes. They code for GS isoforms that are differentially expressed. Plant GS enzymes consist, like all known eukaryotic GS enzymes, of eight subunits [5]. The molecular weight of the subunits varies in the range 38–45 kDa, depending on the species and the subcellular localization of the respective isoform. At least one cytosolic isoform (GS1) and a chloroplast specific (GS2) can be distinguished in most higher plants, whereas their relative abundance varies considerably between species [6]. The expression of the gene is enhanced by high light intensity [7] and high sucrose levels. In some species a root specific isoform (GSR) can be distinguished and in legumes at least one nodule specific isoform has been discovered [8].

Each subunit of the enzyme has an active center with high binding affinity for the substrates. Glutamate is activated by the enzyme via formation of glutamyl phosphate and consequently this intermediate is amidated with ammonia. For the activation, ATP and Mg^{2+} are required (Scheme 6.3.1).

GS_2 deficient barley mutants isolated under conditions that suppress photorespiration grow without phenotypic aberrations under nonphotorespiratory conditions (2% O_2, 0.7% CO_2), but mutants with less than 40% of the wild-type GS_2 activity show severe phytotoxic symptoms, mainly chlorophyll destruction, when

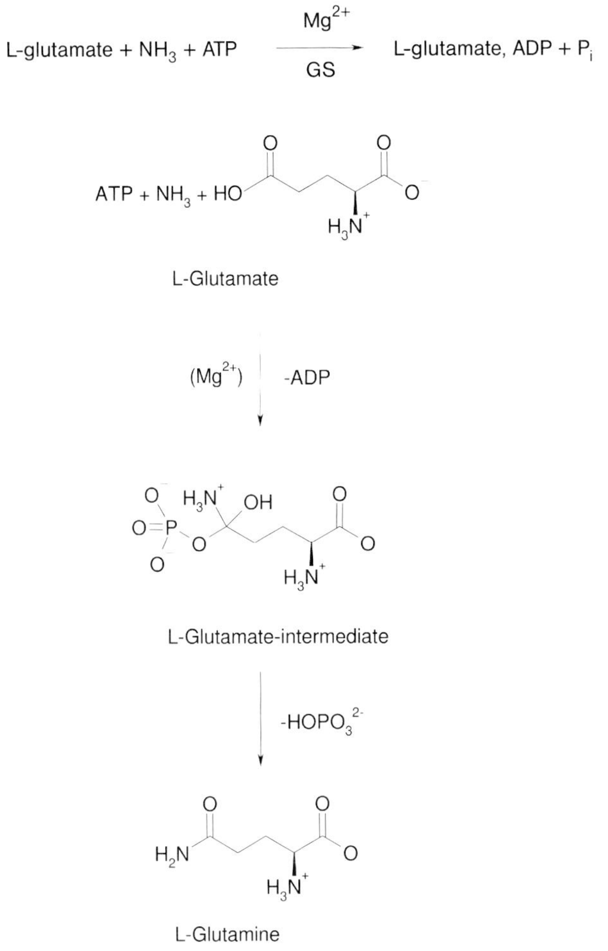

$$\text{L-glutamate} + NH_3 + ATP \xrightarrow[\text{GS}]{Mg^{2+}} \text{L-glutamate, ADP} + P_i$$

ATP + NH$_3$ + HO

L-Glutamate

(Mg^{2+})   -ADP

L-Glutamate-intermediate

-HOPO$_3$$^{2-}$

L-Glutamine

Scheme 6.3.1. Reaction catalyzed by glutamine synthetase (GS).

grown under normal atmospheric conditions (20% O_2, 0.03% CO_2) in full light [9]. The mutants show a significant increase in the level of free ammonia in their leaves, depending on the light intensity. Interestingly, under photorespiratory conditions this increase of the ammonia level is correlated with the development of phytotoxic symptoms, whereas these symptoms were not observed under conditions where photorespiration was suppressed, even though in both cases the ammonia level was elevated [10].

These mutants demonstrate that glutamine synthetase is a potential target for herbicidal compounds and they indicate that photosynthetic tissue is most vulnerable for herbicidal damage caused by GS inhibitors.

Some 50 years ago it was discovered that certain phytopathogenic *Pseudomonas* strains, namely *P. syringae* pv. tabaci, release a toxic metabolite at the site of leaf

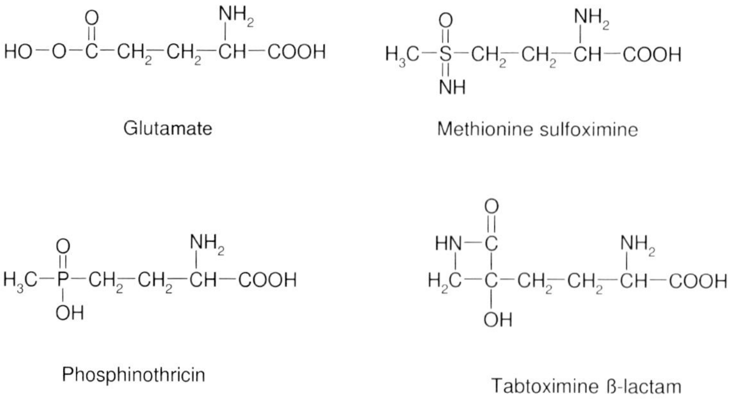

Glutamate

Methionine sulfoximine

Phosphinothricin

Tabtoximine ß-lactam

Fig. 6.3.2. Glutamate and some analogues described as GS inhibitors.

infection. The metabolite causes the formation of a chlorotic halo at the infection site and the damaged tissue of the host is then colonized by the bacteria. The molecular structure of the toxic metabolite was identified [11]. The toxic compound is known as tabtoxinin-β-lactam (Fig. 6.3.2), a strong inhibitor of plant GS [12]. If the toxin is applied on tobacco leaves it causes the same symptoms as virulent toxin producing bacteria. Similar symptoms develop after local administration of methionine sulfoximine (Fig. 6.3.2), which was known at that time as a strong GS inhibitor [13].

6.3.3
Phosphinothricin, a Potent GS Inhibitor

In 1972 the team of Professor Zaehner at Tübingen described a *Streptomyces* strain producing a novel compound with antibiotic properties. The antibiotic tripeptide produced by *Streptomyces viridochromogenes* consists of two alanine residues and a novel amino acid that was named phosphinothricin [14] (Fig. 6.3.2). Owing to its structural analogy to glutamate Bayer et al. [14] tested and proved the hypothesis that phosphinothricin acts as an inhibitor of bacterial GS enzyme, whereas the tripeptide phosphinothricyl-alanyl.alanine did not inhibit the isolated GS enzyme. Nevertheless the tripeptide was 1.000–10.000-fold more active in its growth inhibitory effect on different bacteria. The discrepancy is explained by the observation that free phosphinothricin cannot be taken up efficiently by bacteria, whereas the tripeptide is taken up into the bacteria by peptide carriers and, subsequently, the tripeptide is cleaved.

Independent of the research activities in Germany, a Japanese research team discovered the same tripeptide produced by a *Streptomyces* isolate from a Japanese soil sample. This isolate was named *S. hygroscopicus* and the tripeptide was named bialaphos (syn. bilanaphos) [15].

In 1984 a third phosphinothricin-producing microorganism was discovered and described as *Kitasatospora phosalacinea* (syn. *Streptomyces phosalacineus*) which produces phosphinothricyl alanyl-leucine (phosalacine) [16, 17].

6.3.4
Discovery of the Herbicidal Activity of Phosphinothricin

In the mid-1970s, DL-phosphinothricin was synthesized in Hoechst central research laboratories and tested for its biological activity in the biological research unit of the Agricultural Division. Whereas the compound did not show a significant herbicidal activity in the screening for preemergent herbicides, it showed a strong and broad activity against almost all weeds after foliar application in the PO screening, but the compound did not show selectivity in field crops. Field experiments confirmed the excellent broad spectrum weed control potential of DL-phosphinothricin and the development of the compound as a non-selective post-emergent herbicide was initiated [18].

In 1984 the ammonium salt of the compound was introduced to the market under the common name glufosinate-ammonium as a post-emergent herbicide for directed spray application in vineyards. In the following years the label was extended for using glufosinate-ammonium in orchards and plantation crops and subsequently further uses were developed [19].

In Japan, Meiji Seika discovered the herbicidal activity of bialaphos. Owing to its good performance as a natural product for weed control after foliar application, the tripeptide was developed as a foliar non-selective herbicide for the Japanese market, where it was introduced in 1984 under the trade name Herbiace [20].

To date, all attempts to synthesize more potent structural analogues of glutamate with herbicidal activity failed despite the research efforts dedicated to the herbicide target glutamine synthetase [19].

6.3.5
Mode of Glutamine Synthetase Inhibition

In 1968 Ronzio and Meister had already developed a model for GS inhibition by methionine sulfoximine (MSO) [21]. They postulated that MSO inhibits GS in two steps. The first step is reversible when the inhibitor competes with glutamate at the binding site. In the second step the substrate analogue is phosphorylated (Scheme 6.3.2) and then irreversibly bound to the enzyme. Manderscheid and Wild [22] confirmed the two-step hypothesis using L-phosphinothricin in their inhibition studies with GS from wheat. They found that the phosphorylated phosphinothricin was irreversibly bound to the enzyme. Furthermore, they concluded that each subunit of the enzyme is able to bind one molecule of phosphinothricin.

Only the L-enantiomer of the racemic glufosinate (DL-homoalanin-4-yl(methyl)phosphonic acid) acts as an GS inhibitor. The tripeptide bialaphos

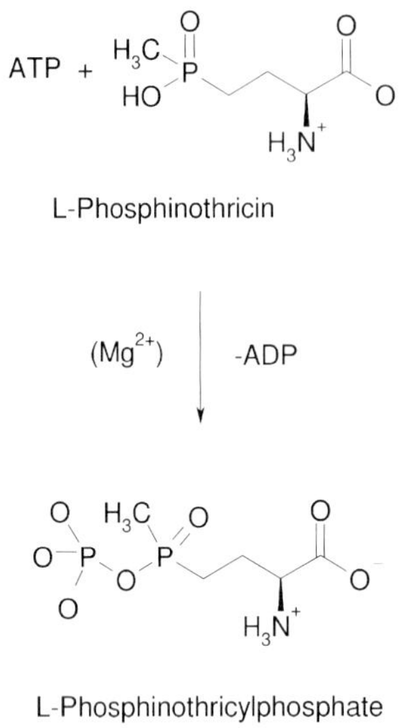

L-Phosphinothricin

(Mg^{2+}) -ADP

L-Phosphinothricylphosphate

Scheme 6.3.2. Formation of the phosphorylated intermediates of
L-phosphinothricin.

does not inhibit the GS enzyme directly. After foliar uptake the tripeptide is
cleaved and then the released L-phosphinothricin inhibits the GS enzyme. Therefore, at the GS target site both herbicides act identically.

6.3.6
Physiology of the Herbicidal Activity of Phosphinothricin

6.3.6.1   Herbicidal Symptoms of Phosphinothricin

One–three days after the herbicide is applied the first symptoms become visible,
depending on weed species and climatic conditions. Chlorotic spots and necrotic
zones increase rapidly. These symptoms develop either simultaneously as in most
dicot weeds or subsequently as in grasses. In grasses intensive chlorosis usually
precedes wilting and desiccation. Usually, the treated plants are killed within 7–
10 days. Low temperatures delay the herbicidal activity significantly. Sublethal
doses or unfavorable climatic conditions may lead to regrowth, especially on older
plants.

6.3.6.2   Physiological Effects of GS Inhibition in Plants

When plants are kept in the light, already 1 h after foliar application of glufosinate an increase in free ammonia is measurable. Within 24 h the ammonia level
is 100-fold increased, whereas in plants kept in constant darkness, the level of
free ammonia is only slightly increased after 24 h [23]. This observation is in
agreement with the fact that the vast majority of the released ammonia is generated in the photosynthetic C_2 carbon cycle and to a smaller degree by nitrite re-

duction in the light or amino acid metabolism in darkness. It is known from growth chamber experiments as well as from field observations that glufosinate-ammonium causes fast and strong symptoms at high light intensities, whereas under low light and in darkness the symptom development is delayed [24]. The amino acid pools of glufosinate treated plants undergo dramatic changes in parallel to the ammonia accumulation. Glutamine, glutamate, asparagine, aspartate, glycine, serine and alanine are depleted shortly after glufosinate treatment, while arginine and aromatic amino acids increase in parallel [25]. It was concluded that the relative increase of the latter amino acids is a consequence of the depleted *de novo* synthesis of the amino acids, showing a rapid turnover as well as a result of protein breakdown, especially of proteins showing a high turnover rate like Rubisco. A decrease in protein content in treated plants was indeed observed [26]. Photosynthetic carbon fixation is inhibited by glufosinate within hours as well, whereas the photosynthetic electron transport in chloroplasts prepared from glufosinate treated plants did not decrease within 48 h after herbicide application. Ammonia at high concentrations is regarded as toxic for plants [27], leading to a perturbation of membrane transport processes, most probably due to a collapse of the pH gradient normally maintained by membranes [28]. Originally it was thought that ammonia accumulation as the consequence of glutamine and glutamate depletion caused by GS inhibition is the main reason for phytotoxicity of GS inhibitors in plants, but not all experimental data can be explained by this hypothesis. The results of chlorophyll fluorescence measurements on glufosinate treated plants do not support the hypothesis of ammonia induced uncoupling of photophosphorylation [29]. In addition, *Sinapis* plants kept under nonphotorespiratory conditions (0.1% CO_2, 2% O_2) did not show an inhibition of photosynthesis even though ammonia accumulated to levels that were strongly inhibitory under normal atmospheric conditions causing photorespiration [30]. Furthermore, in detached *Sinapis* leaves kept under photorespiratory conditions, feeding of glutamine or glutamate drastically reduced the inhibition of photosynthesis even though the ammonia accumulation was more pronounced than in leaves that did not get additional glutamine or glutamate [31]. These observations indicate that interrupting the GS-GOGAT cycle causes glyoxylate accumulation due to the depletion of glutamate [32] which acts as substrate for glutamate-glyoxylate-aminotransferase which converts glyoxylate into glycine. Either glyoxylate itself or the arrested glycine–serine conversion, arresting the re-import of C_3 skeletons back into the Calvin cycle, may be the cause for the rapid breakdown of photosynthetic CO_2 fixation [33]. Wild and Wendler [34] showed evidence that glyoxylate inhibits, at very high concentrations, Rubisco directly and inhibits at lower concentration Rubisco-activase, which would explain the rapid breakdown of photosynthetic carbon fixation as well. The latter hypothesis together with the observation the rapid depletion of the pools of crucial amino acids necessary for purine and pyrimidine synthesis as well as for protein synthesis explain the severe and eventually irreversible metabolic disturbance, leading to inhibition of photosynthetic carbon fixation, *de novo* protein synthesis, and finally to the death of the plant tissue.

6.3.6.3 Modulation of Herbicidal Activity of Glufosinate by Environmental Conditions

As a highly polar and water-soluble compound that is insoluble in epicuticular wax and lipid bilayers it is explicable that environmental factors as air humidity and temperature strongly influence the uptake and herbicidal efficacy of phosphinothricin. Uptake is significantly higher at high air humidity [35] even though this effect is less pronounced in the formulated commercial product. The compound is more active above 25 °C than below 10 °C. Hot, humid weather conditions at high light intensities give excellent weed control results even for weed species that are hard to control under less favorable conditions. When glufosinate is applied on plants at temperatures below 10 °C, the translocation as well as the development of herbicidal symptoms is delayed [36], which eventually may lead to reduced herbicidal activity under adverse environmental conditions.

6.3.6.4 Uptake and Translocation of Glufosinate-ammonium

More than 50% of the active ingredient that can penetrate into the leaf is taken up within the first 4–6 h after foliar application and more than 90% is taken up within 24 h [37, 38]. Ullrich et al. [39] showed evidence for active uptake of the compound which is mediated by amino acid carriers. As already mentioned, air humidity modulates the uptake. Under conditions that favor rapid symptom development (high temperatures and high light intensity), the translocation of the compound is limited, whereas in plants kept in the dark after application the active ingredient shows a considerable phloem mobility. Under field conditions glufosinate-ammonium is regarded as a contact herbicide with partial phloem mobility [40].

6.3.7
Use of Phosphinothricin-containing Herbicides in Agriculture and Horticulture

Its mode of action and its slow metabolization in plants explains why phosphinothricin has a very broad herbicidal activity and lacks any selectivity. Herbicides containing this active ingredient originally were developed and brought to market as non-selective post-emergent herbicides for vegetation management in orchards, vineyards, plantation crops, reforested areas and tree nurseries. The selectivity can be generated by directed spraying of the herbicide on the weed canopy and careful avoidance of drift on leaves of the respective crop. Also, in field crops or in horticultural indications in vegetables and ornamentals the crop can be protected from herbicidal damage either by shielded spraying or by application before planting of the crop. In both cases, exposure of the crop to the active ingredient is prevented.

Further registered applications of glufosinate-ammonium cover its use as harvest aid [19], especially for pre-harvest leaf and vine desiccation in potatoes [41].

6.3.8
Attempts to Generate Crop Selectivity for Glufosinate

Due to its activity against a broad weed spectrum, its unique mode of action, its complete biodegradability and low toxicity against non-target organisms [42], attempts were initiated to explore approaches that may allow the use of glufosinate as a selective herbicide in major field crops. In parallel to the genetic approaches outlined in the following paragraph, special spraying devices were developed that allow directed spraying between the crop rows whilst protecting the crop from the sprayed herbicide. Even though the selective use of glufosinate in conventional maize varieties with the help of a directed spraying device was registered in 1993, the necessity to invest in the specific application equipment limited the use of this system considerably.

6.3.8.1 Genetic Approaches to generate Glufosinate-Selectivity in Crops: Target-based Approaches

In the mid-1980s attempts were initiated to select glufosinate tolerant mutants by exposure of regenerable tissue cultures from crops to inhibitory concentrations of the herbicide. When plants were regenerated from *in vitro* selected tobacco and maize cell cultures conferring a 4–8-fold increased glufosinate tolerance, the regenerants showed only marginally improved tolerance to glufosinate after foliar application, which was not worthwhile to be used in breeding programs (Donn, unpublished). No significant changes in GS activity were observed in regenerants of both crops. In contrast to these negative results a phosphinothricin tolerant alfalfa cell line was obtained by stepwise increase of the inhibitor concentration in the culture medium [43]. The resulting mutant cell population tolerated a 20-fold increased herbicide dose, but failed to regenerate to plants. The resistant alfalfa cell line overexpressed GS up to 10-fold compared with the wild type cells due to an amplification of a GS_1 gene.

The constitutive overexpression of the alfalfa GS1 gene in transgenic tobacco led to a significant accumulation of alfalfa GS protein in the tobacco plants, leading to a up to 10-fold increased GS activity in these plants but, nevertheless, these plants were only partially tolerant against foliar application of glufosinate [44]. These plants did show symptoms of leaf chlorosis after treatment with agronomical relevant glufosinate doses of 1–2 kg-a.i. ha^{-1}.

In parallel, attempts were made to mutate the alfalfa GS c-DNA. Even though it was possible to complement a GS deficient *E. coli* mutant by the alfalfa GS_1 cDNA [45], all attempts to generate in the *E. coli* system glufosinate tolerant GS_1 mutants failed. The few mutants that showed a reduced binding affinity for glufosinate lost their binding affinity for glutamate as well.

In summary, to date, all attempts to generate glufosinate tolerance either by target overexpression or by target mutation were unsuccessful. This fact, together with the observation that in weed populations which were exposed to glufosinate for two decades no target based mutants were found, is a strong hint that also in future it will be unlikely that glufosinate resistant weeds based on target site mutations will evolve.

6.3.8.2 Crop Selectivity by Expression of Phosphinothricin Acetyl Transferase

Twenty years ago Japanese and German research groups characterized, independently of each other, two genes belonging to the biosynthesis gene clusters of phosphinothricin producing *Streptomyces* strains. Both genes conferred bialaphos resistance to *E. coli* and consequently both genes were successfully used as selectable marker genes in gene transfer experiments in crops. The bialaphos resistance (bar) gene from *S. hygroscopicus* has been described by Thompson et al. [46] and has been widely used in plant transformation experiments. A similar widespread use as selectable marker experienced the homologous gene from *S. viridochromogenes*, which was described by Wohlleben et al. [47] as phosphinothricin-acetyl-transferase (pat) gene in 1988. The two genes and their gene products share a high degree of homology. On the DNA level they show 85% homology and they code for proteins that share 87% homology. The biochemical properties of the two proteins in respect of pH and temperature optimum, and their substrate specificity, are very similar [48]. Both enzymes N-acetylate with a high specificity phosphinothricin (Scheme 6.3.3) and desmethyl-phosphinothricin, a precursor molecule in the biosynthetic pathway of this natural substance, whereas they do not acetylate proteinagenous amino acids.

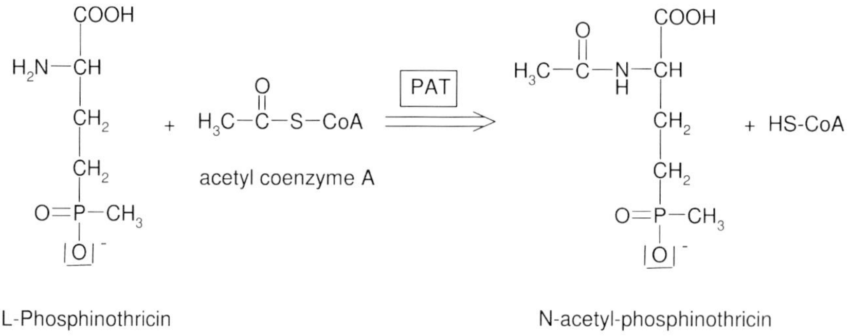

$$
\text{L-Phosphinothricin} + \text{acetyl coenzyme A} \xrightarrow{\text{PAT}} \text{N-acetyl-phosphinothricin} + \text{HS-CoA}
$$

Scheme 6.3.3. Inactivation of L-phosphinothricin by N-acetylation.

Owing to the high substrate affinity of both enzymes, trace amounts of the proteins are sufficient to protect the transgenic plants from herbicidal damage. Even if less than 0.1% of the total protein consists of Bar or Pat protein, the respective plants efficiently acetylate phosphinothricin quantitatively when it enters the plant cells. These plants do not show any signs of GS inhibition even after application of high doses of glufosinate, which exceed the field application rate 5–10-

fold. Natural evolution provided the phosphinothricin producing *Streptomyces* strains with a perfect mechanism to keep the level of free phosphinothricin within their cells extremely low. The responsible enzymes protect crop plants as well against this herbicidal substance in a perfect manner when the responsible *Streptomyces* gene is transferred and expressed in crops under the control of appropriate promoters.

Because *Streptomyces* show a different codon usage profile than higher plants, a synthetic pat gene was synthesized, coding for the same protein but using the preferred plant codons [49]. The synthetic gene is characterized by a GC content of 50% whereas the natural pat gene has a GC content of 70%. The expected advantage of the synthetic gene was to minimize the risk of gene silencing due to the lower GC content. After 20 years of coexistence of both gene versions in transgenic crops it is evident that the natural gene did not reveal a higher probability of pat gene silencing than with the synthetic version. Both genes allowed the breeding of glufosinate tolerant crop varieties that are expressing the transgene since more than 20 generations.

6.3.8.3 Bar and Pat Gene in Plant Breeding

Both genes facilitated the development of efficient gene transfer protocols for various crops. They were successfully used to establish gene transfer protocols for maize [50–52] and rice [53, 54], regardless of the transformation method used by the respective researchers. These genes are still attractive in crop transformation experiments because they are good selectable marker genes *in vitro* to select the transgenic offsprings of the few transgenic cells scattered in the cultured plant tissue in media that contain inhibitory concentrations of phosphinothricin either in the form of the tripeptide or glufosinate. The regenerated transgenic plants are easily distinguishable from nontransgenic siblings by leaf application of the GS inhibitor, which leaves the transgenic regenerants unaffected whilst the nontransgenic siblings develop severe herbicide symptoms. Both genes enabled researchers to develop clean gene constructs conferring solely agronomical useful genes to crops and avoiding the use of antibiotic resistance genes in plant transformation experiments.

Since 1995, transgenic glufosinate tolerant (Liberty Link) canola varieties have been grown commercially in Canada and in 1997 Liberty Link maize was introduced to North American agriculture. Whereas in 2005 Liberty Link canola was grown on more than 25% of the Canadian canola acreage, approximately 6% of the maize acreage in the US was planted with Liberty Link corn at that time. In maize the ratio reflects the predominance of Bt corn amongst the transgenic varieties. In canola the Liberty Link varieties are high yielding hybrid varieties sold under the brand name In Vigor canola.

Furthermore the success story of transgenic canola in Canada which is currently grown on more than 80% of the Canadian canola acreage is explained by the fact that weedy cruciferous relatives of canola are controlled by the 2 complementary herbicides more precisely than with traditional selective canola herbicides. The improved control of the weedy relatives helps to produce a high quality oil which is almost free of glucosinolates and erucic acid [55].

6.3.9
Use of N-Acetyl-Phosphinothricin as Proherbicide

Whereas N-acetyl-phosphinothricin is not deacetylated in plants, bacterial enzymes have been described that can remove the acetate residue from the molecule [56, 57]. Transgenic plants conferring a bacterial deacetylase gene under the control of a constitutive promoter show herbicidal symptoms when sprayed with N-acetyl-phosphinothricin. Therefore, transgenic plants expressing an appropriate deacetylase gene can be selectively eliminated in plant canopies [58]. If the deacetylase gene is linked to tissue specific promoters, specific cells can be ablated in transgenic plants, conferring the gene construct. For example, a deacetylase gene from *Stenotrophomonas* sp. linked to a tapetum specific promoter was successfully used to generate facultatively male sterile tobacco flowers after treatment of the plants with N-acetylphosphinothricin in the flower bud stage (Fig. 6.3.3).

6.3.10
Conclusions

The natural glutamine synthetase inhibitor phosphinothricin as well as its synthetic racemate glufosinate are broad spectrum post-emergent herbicides that will play a role in future agriculture due to the unique mode of action. Because these GS inhibitors fully control weeds that have evolved resistances against other types of herbicides, the use of phosphinothricin-containing herbicides in tolerant crops will remain an important option for future sustainable agriculture.

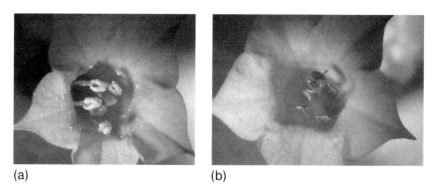

(a)   (b)

Fig. 6.3.3. Tobacco flowers treated with N-acetylphosphinothricin.
(a) Nontransgenic control; (b) transgenic tobacco conferring a bacterial
deacetylase gene under the control of a tapetum specific promoter.

References

1 A. J. Keys, J. E. Bird, M. J. Cornelius, P. J. Lea, R. M. Wallsgrove, B. J. Miflin, *Nature*, **1978**, 275, 741–743.

2 J. N. Siedow, D. A. Day, B. B. Buchanan, W. Gruisson, R. L. Jones, *Biochem. Mol. Biol. Plants*, **2000**, 676–729.

3 B. J. Miflin, P. J. Lea, Ammonia assimilation, Miflin B. J. (ed), *The Biochemistry of Plants*, **1980**, Ch. 5: amino acids and derivatives. Academic Press, New York, 169–202.

4 P. J. Lea, *Plant Biochemistry* (eds. P. M. Dey, I. Harborne), Academic Press, San Diego **1997**, 273–313.

5 S. F. McNally, B. Hirel, Glutamine synthetase in higher plants, *Physiol. Vég.*, **1983**, 21, 761–774.

6 S. M. Ridley, S. F. McNally, *Plant Sci.*, **1985**, 39, 31–36.

7 T. K. Peterman, H. M. Goodman, *Mol. Gen. Genet.*, **1990**, 230, 145–154.

8 B. G. Forde, H. M. Day, J. F. Turton, W. J. Shen, J. V. Cullimore, I. E. Oliver, *Plant Cell*, **1989**, 1, 391–401.

9 R. M. Wallsgrove, J. C. Turner, N. P. Hall, A. C. Kendall, S. W. Bright, *Plant Physiol.*, **1987**, 83, 155–158.

10 P. J. Lea, S. M. Ridley, A. D. Dodge (ed). *Herbicides and Herbicide Metabolism*, Cambridge University Press, Cambridge, **1989**, 137–170.

11 T. F. Uchytil, R. D. Durbin, *Experentia*, **1980**, 36, 301–302.

12 M. D. Thomas, P. J. Langston-Unkefer, T. F. Uchytil, R. D. Durbin, *Plant Physiol.*, **1983**, 71, 912–915.

13 M. Leason, D. Cunliffe, D. Parkin, P. J. Lea, B. Miflin, *J. Phytochem*, **1982**, 21, 855–857.

14 E. Bayer, K. Gugel, K. Haegele, H. Hagenmayer, S. Jessipow, W. A. Koenig, H. Zaehner, *Helv. Chim. Acta*, **1972**, 55, 224–239.

15 Y. Ogawa, T. Tsuruoka, S. Inouye, T. Niida, *Sci. Rep. Meiji Seika Kaisha*, **1973**, 13, 42–53.

16 S. Omura, K. Hinotozawa, N. Imanura, M. Murata, *J. Antibiot.*, **1984**, 37, 939–940.

17 S. Omura, M. Murata, H. Hanaki, K. Hinotozawa, R. Oiwa, H. Tanaka, *J. Antibiot.*, **1984**, 37, 829–835.

18 F. Schwerdtle, H. Bieringer, M. Finke, *Z. Pflanz., Pflanzenschutz Sonderheft*, 9, 431–440.

19 G. Hörlein, *Rev. Environ. Contam. Toxicol.*, **1994**, 138, 73–145.

20 S. Mase, *Jpn. Pestic. Inform.*, **1984**, 45, 27–30.

21 R. A. Ronzio, A. Meister, *Proc. Natl. Acad. Sci. U.S.A.*, **1968**, 59, 164–170.

22 R. Manderscheid, A. Wild, *J. Plant Physiol.*, **1986**, 123, 135–142.

23 H. Köcher, *Proc. Soc. Chem. Ind. Pesticide Group Meet. Monogr.*, **1989**, 42, 173–182.

24 H. Köcher, *Aspects Appl. Biol.*, **1983**, 4, 227–233.

25 C. Wendler, M. Barniske, A. Wild, *Photosyn. Res.*, **1990**, 24, 55–61.

26 M. Lacuesta, B. González-Moro, C. González-Murua, T. Aparicio-Tejo, A. Monoz-Rueda, *J. Plant Physiol.*, **1989**, 1234, 304–307.

27 A. Jungk, in: B. Hock, E. F. Elstner (eds), *Planzentoxikologie*, BI Wissenschaftsverlag Mannheim, **1984**, 224–229.

28 J. K. M. Roberts, M. K. L. Pang, *Plant Physiol.*, **1992**, 100, 1571–1574.

29 M. Lacuesta, A. Munoz-Rueda, C. Gonzalez-Murua, M. N. Sivak, *J. Exp. Bot.*, **1992**, 43, 159–165.

30 A. Wild, H. Sauer, W. Ruehle, *Z. Naturforsch.*, **1987**, 42c, 263–269.

31 H. Sauer, A. Wild, W. Ruehle, *Z. Naturforsch.*, **1987**, 42c, 270–278.

32 A. Wild, C. Wendler, *Pestic. Sci.*, **1990**, 30, 422–424.

33 C. Wendler, A. Putzer, A. Wild, *J. Plant Physiol.*, **1992**, 139, 666–671.

34 A. Wild, C. Wendler, *Z. Naturforsch.*, **1993**, 48c, 369–373.

35 C. Coetzer, K. Al-Khatib, T. M. Longhin, *Weed Sci.*, **2001**, 49, 8–13.

36 A. R. Kumaratilake, C. Preston, *Weed Sci.*, **2005**, 53, 10–16.

37 H. Köcher, K. Lötzsch, *Proc. Asian-Pacific Weed Sci. Soc. Conf.*, **1985**, 10, 193–198.

38 G. J. Steckel, S. E. Hart, L. M. Wax, *Weed Sci.*, **1997**, 45, 378–381.

39 W. R. Ullrich, C. I. Ullrich-Eberius, H. Köcher, *Pestic. Biochem. Physiol.*, **1990**, 37, 1–11.

40 J. N. Beriault, G. P. Horsman, M. D. Devine, *Plant Physiol.*, **1999**, 121, 619–628.

41 www.bayercropscienceus.com/crops/view/potatoes/product/rely.

42 E. Dorn, G. Goerlitz, R. Heusel, K. Stumpf, *Z. Pflanz. Pflanzenschutz Sonderh*, **1992**, 13, 459–468.

43 G. Donn, E. Tischer, J. Smith, H. Goodman, *J. Mol. Appl. Genet.*, **1984**, 2, 621–635.

44 P. Eckes, P. Schmitt, W. Daub, F. Wengenmayer, *Mol. Gen. Genet.*, **1989**, 217, 263–268.

45 S. Dassarma, E. Tisher, H. M. Goodman, *Science*, **1986**, 232, 1242–1244.

46 C. J. Thompson, N. R. Movva, R. Tizard, R. Crameri, J. E. Davies, M. Lauwereys, J. Bottermann, *EMBO J.*, **1987**, 6, 2519–2523.

47 W. Wohlleben, W. Arnold, J. Broer, D. Hillmann, E. Strauch, A. Pühler, *Gene*, **1988**, 70, 25–37.

48 A. Wehrmann, A. VanVliet, C. Opsomer, J. Botterman, A. Schulz, *Nat. Biotechnol.*, **1996**, 14, 1274–1278.

49 P. Eckes, B. Uijtewaal, G. Donn, *J. Cell. Biochem.*, **1989**, 13D, 334.

50 W. J. Gordon-Kamm, T. M. Spencer, M. L. Mnangano, T. R. Adams, R. J. Daines, W. G. Start, J. V. O'Brian, S. A. Chambers, W. R. Adams Jr., N. G. Willets, T. B. Rice, C. J. Mackey, R. W. Krueger, A. P. Kausch, P. G. Lemaux, *Plant Cell*, **1990**, 2, 603–618.

51 M. Golovkin, M. Abraham, S. Morocz, S. Bottka, A. Feher, D. Dudits, *Plant Sci.*, **1993**, 90, 41–52.

52 B. R. Frame, P. R. Drayton, S. V. Bagnall, C. J. Lewnau, W. P. Bullock, H. M. Wilson, J. M. Dunwell, J. A. Thompson, K. Wang, *Plant J.*, **1994**, 6, 941–948.

53 P. Christou, *Biotechnology*, **1991**, 9, 957–962.

54 S. K. Datta, K. Datta, N. Soltanifar, G. Donn, I. Potrykus, *Plant Mol. Biol.*, **1992**, 20, 619–629.

55 M. D. Devine, J. L. Buth, *Proc. BCPC Conf. – Weeds*, **2001**, 367–372.

56 K. Bartsch, G. Kriete, I. Broer, A. Pühler, **1996**, WO 9827201.

57 G. Kriete, K. Niehaus, A. M. Perlik, A. Pühler, *Plant J.*, **1996**, 9, 809–818.

58 X. Chen, W. Yang, E. Sivamani, A. H. Bruneau, B. Wang, R. Qu, *Mol. Breed.*, **2005**, 15, 339–347.

7
Microtubulin Assembly Inhibitors (Pyridines)

Darin W. Lickfeldt, Denise P. Cudworth, Daniel D. Loughner, and Lowell D. Markley

7.1
Introduction

Herbicides with the microtubulin assembly inhibitor [1, 2] (MAI) mode of action are generally applied pre-emergence for control of annual grasses and small-seeded broadleaf weeds, causing swelling in meristematic regions such as root tips. Susceptible plants may show thickened or swollen hypocotyls or internodes [3]. The MAIs are grouped into five chemical families: the dinitroanilines, the phosphoroamidates, the pyridines, the benzamides, and the benzoic acids (Herbicide Resistance Action Committee class K1). The most popular family of the MAIs is the dinitroanilines, which includes herbicides such as trifluralin, benefin, oryzalin, pendimethalin, and prodiamine. In the 1980s the demand for pre-emergence herbicides that were more efficacious, colorless and dependable at lower application rates led to the investigation of potential compounds in the pyridine family. In the pyridine family, there are only two herbicides being marketed today – dithiopyr and thiazopyr – so they are the focus of this chapter. Both herbicides were initially patented by Monsanto [4, 5] before subsequently being sold to Rohm & Haas Company (1994). Ultimately the products became the property of Dow AgroSciences through the acquisition of the Rohm & Haas Agricultural Chemical business by The Dow Chemical Company (2001). These products can be used by professional turf, ornamental, perennial tree & vine and *Oryza* growers to control a broad range of troublesome broadleaf and grass weeds.

7.2
Biology of Microtubulin Assembly Inhibitors (Pyridines)

Dithiopyr is a pre-emergence and early postemergence herbicide primarily used in turf, ornamentals and *Oryza* in the United States, Canada, Japan, China, Aus-

Modern Crop Protection Compounds. Edited by W. Krämer and U. Schirmer
Copyright © 2007 WILEY-VCH Verlag GmbH & Co. KGaA, Weinheim
ISBN: 978-3-527-31496-6

tralia, Egypt, South Korea, Taiwan and Puerto Rico. It is applied pre-emergence or postemergence to turf at 150–560 g-a.i. ha^{-1} per application. Early postemergence applications can be utilized to control *Digitaria* spp. seedlings in their early stage and prior to emergence of a second tiller [6]. Adjuvants have low influence on postemergent control because translocation from treated leaves is minimal [7].

Another turf use pattern is selective pre-emergence control of *Poa annua* L. in overseeded warm-season turf. A common practice in warm climates is to overseed warm-season turfgrass species with cool-season turfgrass species such as *Lolium perenne* L. to maintain a green color through the winter months when warm-season grasses typically go dormant. Dithiopyr has been proven effective for selective control of *Poa annua* for 4–6 months after treatment while not injuring *Lolium perenne* that was seeded 8 weeks prior to treatment.

Applications of Dithiopyr to paddy grown *Oryza* are targeted to control *Echinochloa* spp. Dithiopyr can be formulated into several different formulations, including an emulsion in water (EW) containing 240 g a.i. L^{-1}, an emulsifiable concentrate (EC) up to 120 g a.i. L^{-1}, and a wettable powder (WP) with 40% a.i. In addition, granular formulations are available.

Dithiopyr controls key annual monocot species and many dicot species, including *Digitaria* spp., *Poa annua* L., *Eleusine indica* (L.) Gaertn., *Oxalis* spp., *Euphorbia* spp., *Medicago lupulina* L., and *Stellaria media* (L.) Vill. In warmer climates or while seeking control of more challenging weed species such as *Eleusine indica* (L.) Gaertn., sequential applications may be necessary [8, 9].

Most species of cool-season and warm-season turfgrasses are tolerant when the root system is well established. However, some species (such as *Agrostis tenuis*) and some varieties (such as *Cynodon. dactylon* × *C. transvaalensis* "Tifgreen") are not tolerant. Dithiopyr should not be applied to new perennial turf until the root system is well established [10]. It should also not be applied to sod within three months of harvest. Dithiopyr's effect on rooting of established turfgrass species was shown to be minimal and not significantly different than most other pre-emergent herbicides with an MAI mode of action [11, 12].

Thiazopyr is a pre-emergence herbicide that is currently used in non-crop areas, tree, vine and *Oryza* crops and has demonstrated selectivity in *Medicago* spp., *Gossypium* spp., *Arachis* spp., *Glycine* spp., and *Saccharum* spp. [13–15]. It is effective on most annual grasses and certain broadleaf weeds. Thiazopyr is presently registered in 13 countries in North America, Latin America, Europe, Australia and Asia.

A key strength of thiazopyr is its long residual control of annual grass weeds when used at its typical rate range of 0.56–1.12 kg-a.i. ha^{-1}. Also of note is the high level of *Cyperus* spp. suppression when applied pre-emergence. The use of thiazopyr in the United States is as a residual herbicide in permanent crops and in non-crop areas. Citrus, tree-nuts, vines, pomefruit and stonefruit are of primary importance, primarily for control of *Panicum maximum*.

7.3
Environmental Fate of Microtubulin Assembly Inhibitors (Pyridines) [16]

Dithiopyr is strongly adsorbed to soil (K_{oc} average: 1638 mL g^{-1}), but can be desorbed in soils low in organic matter. Soil half-life in field studies ranged from 3 to 49 days (17 day average) with degradation resulting mostly from microbial activity [17]. The major metabolites detected were the diacid and two forms of a monoacid. These metabolites dissipated within one year. Dissipation from field soils can also occur through volatilization. Dithiopyr is stable to photolysis on treated soil. Leaching or runoff, even from highly permeable golf course putting greens, has been shown to be minimal [18–20].

The photolytic half-life in water was 17.6 days, indicating a moderate rate of degradation and a potential for degradation in surface water. The two monoacids and the diacid were the primary metabolites observed. The potential movement in water would be limited due to the low water solubility of dithiopyr and its strong adsorption to soil particles and plants.

Thiazopyr is considered to be relatively immobile in soils due to a low water solubility and high affinity for soil organic matter. Soil microorganisms and hydrolysis are the primary routes of degradation in soil. The average DT_{50} was 64 d (8–150 d) following soil dissipation studies on various soils. The monoacid metabolite applied at normal use rates also had limited mobility. In aqueous solutions the DT_{50} was 15 d, indicating surface water contamination should not be an issue.

In plants, oxygenases metabolize the dihydrothiazole ring to the sulfoxide, sulfone, hydroxyl derivative and thiazole. Thiazopyr is also deesterified to its carboxylic acid.

Table 7.1 Toxicology of dithiopyr and thiazopyr [16].[a]

Organism	Administered	Measure	Dithiopyr value	Thiazopyr value
Rats/mice	Oral	LD_{50} (mg kg^{-1})	>5000	>5000
Rats	2 yr	NOEL (mg kg^{-1})	≤10	0.36
Dogs	1 yr	NOEL (mg kg^{-1})	≤0.5	0.5
Bobwhite quail	Acute oral	LD_{50} (mg kg^{-1})	2250	1913
Rainbow trout	96 h	LC_{50} (mg L^{-1})	0.46	3.2
Honeybee	Topical	LD_{50} (µg per bee)	81	>100
Earthworm	14 d	LC_{50} (mg kg^{-1})	>1000	>1000

[a] Dithiopyr and thiazopyr are non-mutagenic and non-genotoxic. EPA toxicity class is III.

7.4
Toxicology of Microtubulin Assembly Inhibitors (Pyridines) [16]

In rats, dithiopyr is rapidly absorbed, extensively metabolized and rapidly excreted (Table 7.1). Eye irritation in rabbits was slight while skin irritation was non-irritating. Following three weeks of repeated skin exposure to dithiopyr technical, mild transient skin irritation and increased liver weights were the only effects observed in rats.

Animals quickly metabolize and eliminate thiazopyr (Table 7.1). Rat-liver microorganisms use sulfur and carbon oxidation along with deesterification for degradation. Studies in bluegill sunfish demonstrated 98% elimination within 14 days.

7.5
Mode of Action of Microtubulin Assembly Inhibitors (Pyridines)

Dithiopyr is not systemic and is absorbed by roots and to some degree by the foliage of susceptible plants. The most important site of uptake appears to be the meristematic regions since dithiopyr translocation is limited and the primary site of action is meristematic tissues. Efficacy symptoms are most evident by a swelling of the meristematic regions such as root tips in susceptible plants where mitosis is inhibited. This mode of action is disrupting spindle microtubule formation in late prometaphase. Dithiopyr does not bind to tubulin but to another protein that may be a microtubule associated protein (MAP) [1, 2]. These MAPs function in microtubule stability and the action of this molecule results in shortened microtubules that cannot form spindle fibers normally responsible for separating chromosomes to the poles of the cell during mitosis. Cortical microtubules, which normally prevent isodiametric cell expansion, are also essentially absent, resulting in club-shaped roots tips of susceptible plants. Thiazopyr also inhibits microtubule assembly in roots of emerging seedlings but is not effective as an early postemergence treatment, like dithiopyr.

There have been no cases of weed resistance to dithiopyr or thiazopyr reported. In one study, *Digitaria ischaemum* that was resistant to fenoxaprop-p was controlled by dithiopyr [21]. However, cross-resistance to other biotypes resistant to the MAI mode of action could probably occur.

7.6
Synthesis: Dithiopyr and Thiazopyr [22, 23]

Dithiopyr (**1**) and thiazopyr (**2**, Fig. 7.1) are pyridine-based herbicides. These compounds, which are accessed via the pyridine **3**, are synthesized by a similar route (Scheme 7.1).

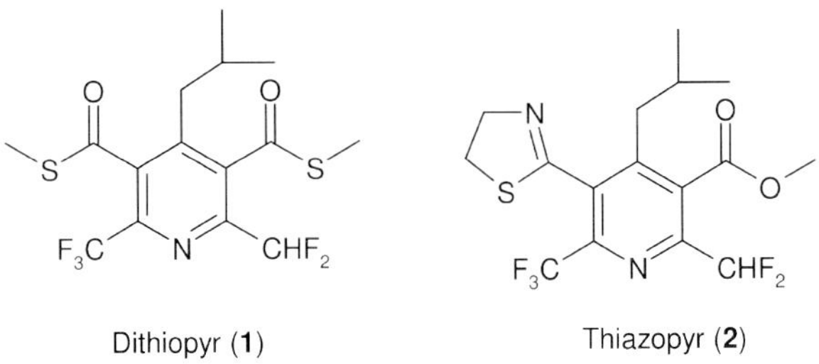

Fig. 7.1. Structures of dithiopyr (1) and thiazopyr (2).

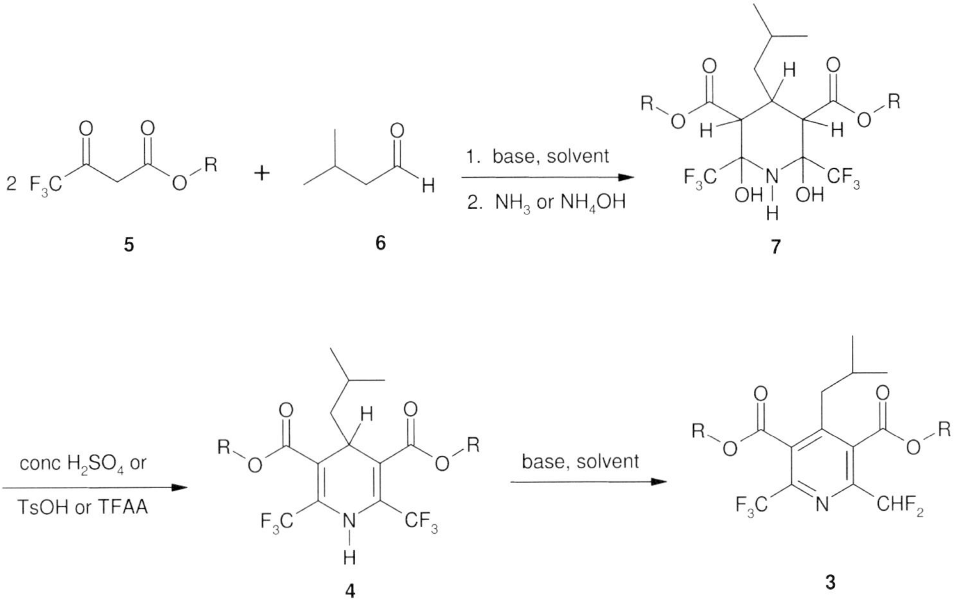

Scheme 7.1. Synthetic route to the common pyridine intermediate 3 [4, 5].

The syntheses of dithiopyr (1), thiazopyr (2) and related compounds [4, 5] begin with a Hantzsch-type base-catalyzed intermolecular cyclization [22], which provides the dihydropyridines 4 (R = Me or Et). Two equivalents of methyl or ethyl trifluoroacetoacetate (5, R = Me or Et) are allowed to react with one equivalent of isovaleraldehyde (6) in the presence of a base, like piperidine, in a suitable solvent at temperatures varying from room temperature to reflux. The intermediate dihydroxytetrahydropyran (structure not shown) is converted into the dihydroxypiperidine 7 by reaction with a nitrogen source, such as ammonium hydroxide or ammonia gas. Reaction of 7 with a dehydrating agent, such as concentrated sulfuric acid, toluenesulfonic acid, or trifluoroacetic anhydride, gives a mixture of the 1,4-dihydropyridine **4** and its 3,4-isomer. In the case of both **1** and **2**, the major isomer is the 1,4-isomer and it is isolated cleanly. Regiochemical preference for the 3,4-isomer is determined by the choice of dehydrating agent as well as the group in the 4-position of the dihydroxypiperidine **7**. The dihydropyridine **4** is then allowed to react with a base, such as DBU, tributylamine, triethylamine,

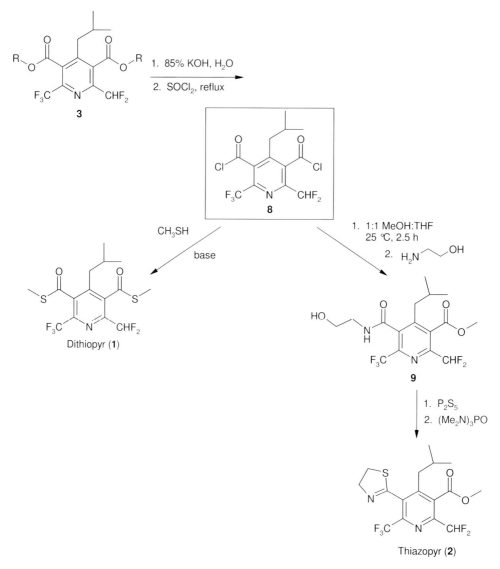

Scheme 7.2. Syntheses of dithiopyr (**1**) and thiazopyr (**2**) from the pivotal intermediate bis-acid chloride **8** [4, 5].

or 2,6-lutidine, either neat or in a suitable solvent to provide the dehydrofluori-nated [23] pyridine **3**, a common intermediate in the syntheses.

Saponification of the esters is accomplished with 85% potassium hydroxide in aqueous media, providing the 3,5-diacid. The diacid is converted into the pivotal intermediate, the bis-acid chloride **8**, by reaction with neat thionyl chloride at re-flux. The acid chloride is treated with methanethiol in the presence of a base to give dithiopyr (**1**) (Scheme 7.2).

Thiazopyr (**2**) is synthesized in a similar fashion. Treatment of the bis-acid chloride **8** in methanol:THF (1:1) at room temperature for 2.5 h affords the 5-chlorocarbonyl-3-methyl ester selectively [24], which is allowed to react with 2-hydroxyethyl amine to form the corresponding 2-hydroxyethyl amide **9**. The hy-droxyethyl amide **9** is subsequently treated with phosphorus pentasulfide and hexamethyl phosphoramide, resulting in sulfurization and cyclization to form the 4,5-dihydrothiazole in thiazopyr (**2**).

References

1 K.C. Vaughn, L.P. Lehnen, Jr., *Weed Sci.*, **1991**, 39, 450–457.

2 L.P. Lehnen, Jr., K.C. Vaughn, *Pestic. Biochem. Physiol.*, **1991**, 40, 58–67.

3 Weed Science Society of America, *Herbicide Handbook*, 8th edn. W.K. Vencill, Ed. Lawrence, KS, **2002**.

4 L.F. Lee, U.S. Patent 4,692,184, **1987**.

5 Y.-L.L. Sing, L.F. Lee, U.S. Patent 4,988,384, **1991**.

6 B.J. Johnson. *Weed Technol.*, **1997**, 11, 144–148.

7 S.J. Keeley, B.E. Branham, D. Penner, *Weed Sci.*, **1997**, 45, 205–211.

8 G. Wiecko, *Weed Technol.*, **2000**, 14, 686–691.

9 B.J. Johnson, *Weed Technol.*, **1997**, 11, 693–697.

10 Z.J. Reicher, D.V. Weisenberger, C.S. Throssell, *Weed Technol.*, **1999**, 13, 253–256.

11 P.H. Dernoeden, N.E. Christians, J.M. Krouse, R.G. Roe, *Aronomy J.*, **1993**, 85, 560–563.

12 P.J. Landschoot, T.L. Watschke, B.F. Hoyland, *Weed Technol.*, **1993**, 7, 123–126.

13 L.J. Kuhns, T.L. Harpster, *Northeastern Weed Sci. Soc. Proc.*, **1998**, 52, 127–129.

14 S.E. Crane, J.A. Holmdal, R.E. Murray, *Southern Weed Sci. Soc. Proc.*, **1998**, 51, 234.

15 L.J. Kuhns, T.L. Harpster, *Northeastern Weed Sci. Soc. Proc.*, **1997**, 51, 115–117.

16 British Crop Protection Council, *The Pesticide Manual*, 12th edn., C.D.S. Tomlin, Ed. **2000**.

17 S. Hong, A.E. Smith, *J. Agric. Food Chem.*, **1996**, 44, 3393–3398.

18 S. Hong, A.E. Smith, *J. Environ. Quality*, **1997**, 26, 379–386.

19 S. Gupta, V.T. Gajbhiye. *J. Environ. Sci. Health, Part B*, **2002**, 37, 573–586.

20 S. Hong, A.E. Smith, *J. Environ. Sci. Health, Part B*, **2001**, 36, 529–543.

21 J.F. Derr, *Weed Technol.*, **2002**, 16, 396–400.

22 A. Hantzsch, *Justus Liebigs Ann. Chem.*, **1882**, 215, 1–82.

23 L.F. Lee, G.L. Stikes, J.M. Molyneaux, Y.L. Sing, J.P. Chupp, S.S. Woodard, *J. Org. Chem.* **1990**, 55, 2872–2877.

24 L.F. Lee, G.L. Stikes, L.Y.L. Sing, M.L. Miller, M.G. Dolson, J.E. Normansell, S.M. Auinbauh, *Pestic. Sci.*, **1991**, 31, 555–568.

8
Inhibition of Cell Division (Oxyacetamides, Tetrazolinones)

Toshio Goto, Akihiko Yanagi, and Yukiyoshi Watanabe

8.1
Introduction

Oxyacetamides and tetrazolinones are new classes of herbicides characterized by excellent efficacy against many major annual grass weeds and certain dicotyledonous weeds, with pre- and post-emergent activity and long lasting weed control.

Oxyacetamides and tetrazolinones inhibit early plant development by disturbing cellular and biochemical level functions. The induced morphological and physiological symptoms are very similar to those of the well-known chloroacetamide herbicides. According to the symptomatic similarity, the Herbicide Resistance Action Committee (HRAC) classifies the herbicides into an action group K_3. The K_3 herbicides are described as inhibitors of cell division or inhibitors of very long-chain fatty acid (VLCFA, >18 carbon chain in length) synthesis. The mode of action of the K_3 group remains unclear.

The selected herbicides flufenacet and mefenacet from the class of oxyacetamides and fentrazamide from tetrazolinones are introduced here.

8.2
Mode of Action

Oxyacetamides and tetrazolinones taken up via the soil provide a strong effect on meristem bearing cell division in the root and shoot tips. Complete arrest of cell division results in cessation of growth and distortion of elongated tissue, leading to plant death.

The mode of action of K_3 herbicides has been reported from biochemical and physiological studies with chloroacetamide herbicides [1]. The findings propose the involvement of the inhibition of VLCFA biosynthesis through a reaction involving covalent binding between herbicide and target enzyme. However, the target site of group K_3 is not sufficiently clarified by binding studies.

Modern Crop Protection Compounds. Edited by W. Krämer and U. Schirmer
Copyright © 2007 WILEY-VCH Verlag GmbH & Co. KGaA, Weinheim
ISBN: 978-3-527-31496-6

In plants, VLCFAs are synthesized by the membrane-bound, multienzyme acyl-CoA elongase system on the endoplasmic reticulum [2]. The synthesis involves sequential addition of a C2-unit from malonyl-CoA to a fatty acid acceptor by a four-step reaction analogous to *de novo* fatty acid synthesis in the plastid. The first step is the condensation of an acyl-CoA primer (fatty acids > 16 carbon long) with malonyl-CoA to form β-ketoacyl-CoA followed by reduction to β-hydroxyacyl-CoA, dehydration to 2-enoyl-CoA, and a second reduction forming longer chain acyl-CoA . The substrates of acyl elongation are esterified to CoA rather than to acyl carrier protein (ACP) by fatty acid synthase [3]. VLCFAs are essential biological components or precursors of cuticular waxes [4], seed storage triacylglycerols [5], and glycosphingolipids in the plasma membrane [6].

Much investigation with chloroacetamides has focused on fatty acid metabolism, especially fatty acid elongation to elucidate the mode of action.

Phytotoxic chloroacetamides provided a linear relationship between severe inhibition of growth and inhibition of the incorporation of [^{14}C]oleic acid into VLCFAs in *Scenedesmus acutus* [7]. In higher plants, the incorporation of [^{14}C]stearic acid or malonyl-CoA into VLCFAs was inhibited by chloroacetamides while the formation of fatty acids up to C_{18} was not influenced [8]. Acyl elongation with 20:0-CoA and 18:0-CoA primer substrates was inhibited by the active (S)-enantiomer of metolachlor but not by the (R)-isomer [1, 9]. Inhibition of VLCFA formation was also observed in metazachlor-resistance mutant (Mz-1) cells of *S. acutus* [9]. Thus, the phytotoxic action of chloroacetamide herbicides is most likely by the inhibition of VLCFA synthesis.

Inhibition of 20:0-CoA elongation increased with time- and temperature-dependency on preincubation. The findings indicate that formation of the enzyme–inhibitor complex is as an irreversible chemical reaction [10]. The enzyme–inhibitor bond is formed by nucleophilic attack of an enzyme. Chloroacetamides bind covalently to cysteines *in vitro* [11]. Condensing enzymes contain one essential, highly reactive cysteine, which covalently binds the acyl primer substrate before the condensing reaction; mutagenesis studies show the enzymatic similarity of the fatty acid elongase [12]. Based on the peptide mapping analysis of the covalent binding between chloroacetamide and chalcone synthase or stilbene synthase, the active site cysteine residue in condensing enzymes was recently concluded to be the primary common target of the herbicides [13].

The above investigations imply (1) a high affinity of the condensing elongation enzyme to its inhibitors in each step, (2) an increase of inhibition of elongation step with the decrease of acyl-CoA substrate concentration, and (3) a tight binding of inhibitors with the target enzyme [10].

The inhibition reaction is due to the nucleophilic attack of the elongase-condensing enzyme. Inhibitors should have an electrophilic C-atom. Chloro- or oxyacetamides have an active methylene formed by the leaving Cl or heterocycle-oxy. Tetrazolinones bind with a target enzyme through nucleophilic addition eliminating the tetrazolinone moiety. Nucleophilic interaction of the elongase-condensing enzyme with inhibitors is assumed to be an inhibitor–enzyme binding mechanism [1].

Genomics studies with gene encoding VLCFA-elongases from *Arabidopsis* and heterologous expression in *Saccharomyces* support the biochemical and physiological arguments for the molecular target of K_3 herbicides [14, 15].

8.3
Chemistry and Biology of Oxyacetamides and Tetrazolinones

8.3.1
Chemistry of the Compounds

8.3.1.1   Oxyacetamides/Flufenacet, Mefenacet

The first compound of the heteroaryloxyacetamide class (simplified as oxyacetamides) launched in 1986 was mefenacet (FOE 1976; Fig. 8.1), as a paddy rice herbicide. Whereas FOE 1976 was synthesized at Bayer (now Bayer CropScience), its good performance was investigated by biologists of Nihon Tokushu Nouyaku Seizou K.K. (now Bayer CropScience K.K.) through primary, secondary and field trial tests.

The physicochemical (water solubility of 4 mg L^{-1} at 20 °C) and biological properties of FOE 1976 were confirmed as highly suitable for paddy rice [16].

Continuous study of oxyacetamide chemistry shifted research from the paddy herbicide to an upland herbicide with increasing water solubility that is suitable for such upland use. To this end, benzanellated analogues such as the benzothiazole moiety of mefenacet were changed to simple five-membered heterocycles that contain at least one nitrogen atom to increase water solubility, and sulfur or oxygen atom to decrease lipophilicity, for instance thiazoles, thiadiazoles, oxazoles and oxadiazoles (Fig. 8.2). Consequently, many patent applications of the new class of heteroxyacetamide herbicides were disclosed [17, 18].

Through structure–activity correlation studies with the new oxyacetamide substances, only 1,3,4-thiadiazole derivatives with specified substituents provided high herbicidal activity. Requisite properties of the compound for selection were (1) very good efficacy against grassy weeds, (2) very good compatibility for maize and soybeans, and (3) suitable water solubility (56 mg L^{-1} at 25 °C). Based on these results, flufenacet (FOE 5043; Fig. 8.1) was selected and developed as a second-generation heteroxyacetamide class for use as an upland herbicide.

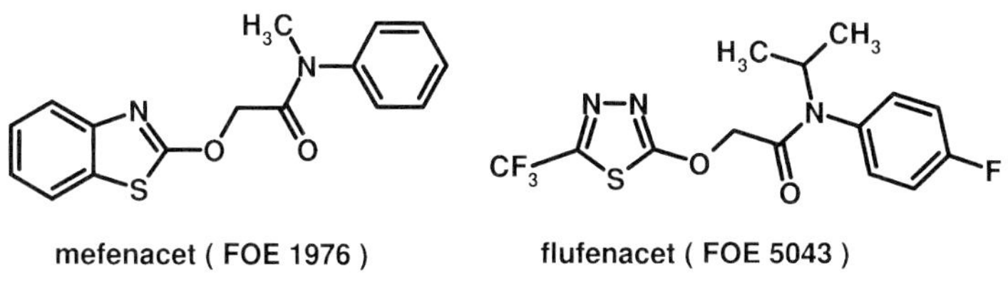

mefenacet (FOE 1976)   flufenacet (FOE 5043)

Fig. 8.1. Products from oxyacetamides.

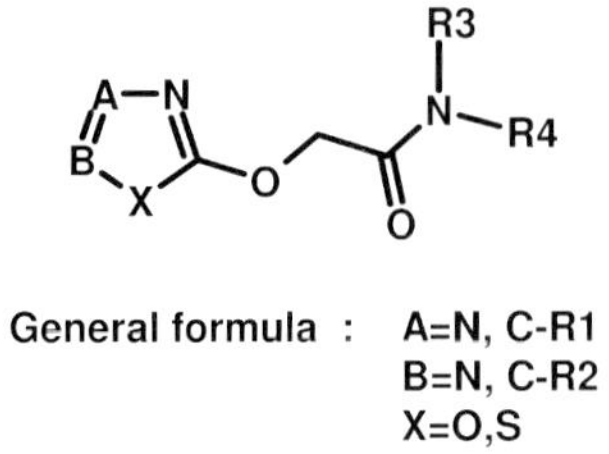

General formula : A=N, C-R1
B=N, C-R2
X=O,S

Fig. 8.2. General formula based on the concept of using oxyacetamides as upland herbicides.

Figures 8.3 and 8.4 show the synthetic pathways for mefenacet [19] and flu-fenacet [20], respectively. The new key intermediates to produce flufenacet are acetoxyacetamide derivative 3, derived from acetoxyacetylchloride (2) and N-isopropyl-4-fluoroaniline (1), and 2-methylsulfonyl-5-trifluoromethyl-1,3,4-thiadiazole (5), derived from trifluoroacetic acid.

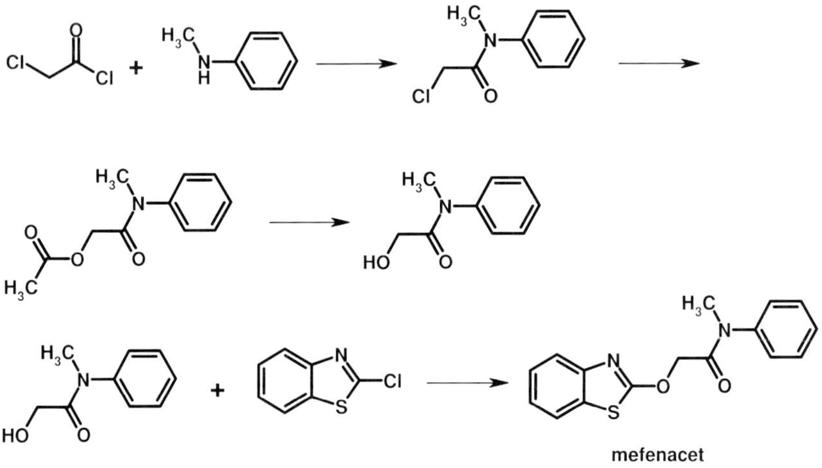

mefenacet

Fig. 8.3. Synthetic pathways to mefenacet.

In summary, to date, two heteroxyacetamide class compounds have been launched on the market, i.e., mefenacet and flufenacet.

8.3.1.2 Tetrazolinones/Fentrazamide

Tetrazolinones were relatively unknown in herbicide chemistry until 1985, when Uniroyal Chemical applied for a patent describing the herbicidal action of carba-moyl tetrazolinones [21]. Several companies have explored this chemistry, and in 1999 Bayer CropScience launched the first practical tetrazolinone herbicide, "fen-trazamide", for grass control in rice (Fig. 8.5).

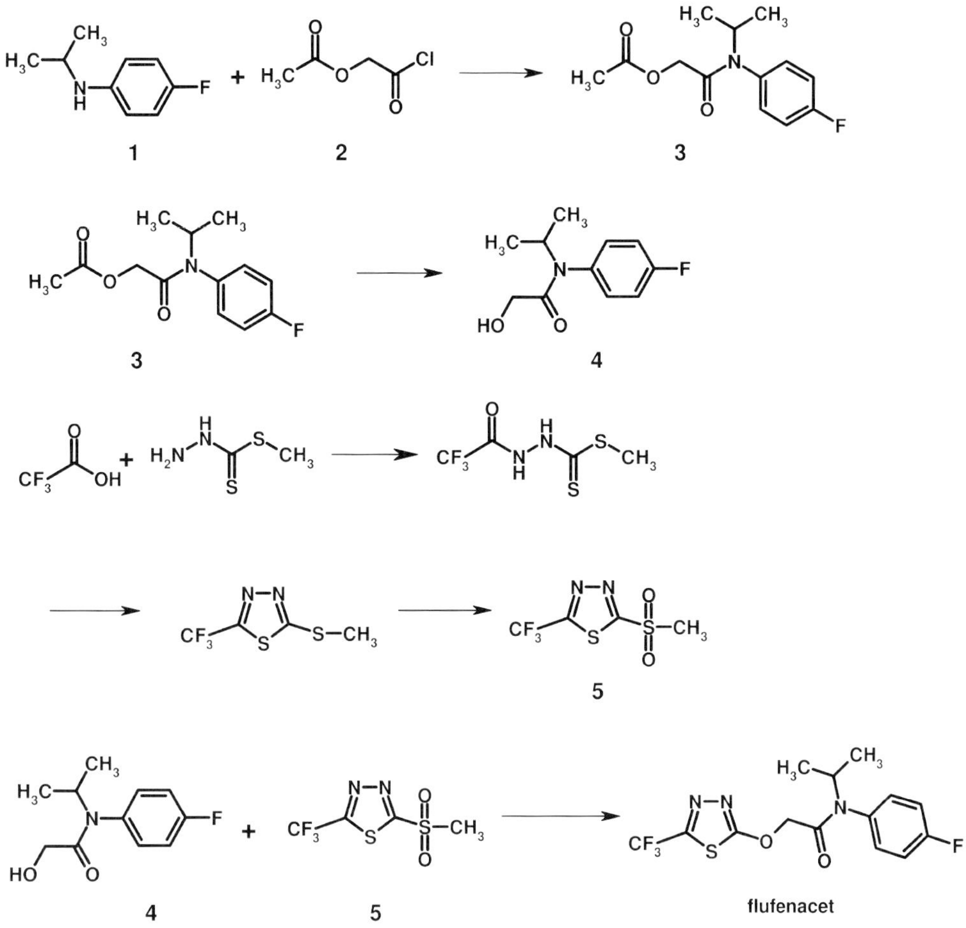

Fig. 8.4. Synthetic pathways to flufenacet.

In 1991, Nihon Bayer Agrochem K. K. (now Bayer CropScience K. K.) started a program for the synthesis and optimization of carbamoyl tetrazolinones. Early in the program, research focused on possible usage in rice because of the high activity of the chemical group to barnyard grass. In contrast, 4-phenyl analogs among

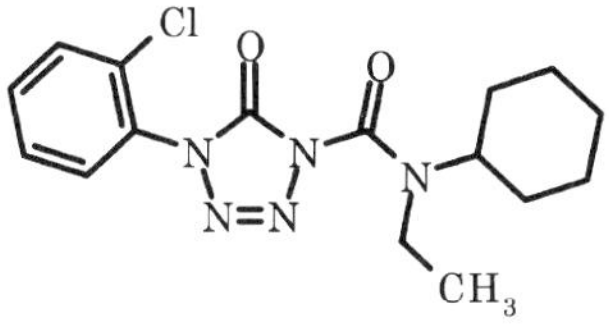

Fig. 8.5. Fentrazamide.

various 4-substituted-1-carbamoyl-tetrazolinones showed some selectivity to trans-planted rice. These findings led to 4-phenyl-1-carbamoyl tetrazolinones as a lead structure for the development of new rice herbicides.

A thorough investigation of phenyl substitution patterns revealed that (1) ortho-substitutions with one or two small group(s) such as methyl, ethyl, F, Cl and Br made a significant contribution to activity, whereas substitution at the 3 or 4 position had a weak effect; but (2) some substituents, such as electron-releasing groups (methyl, methoxy, ethyl) and certain electron-withdrawing groups (F, CF_3), resulted in phytotoxicity to rice. In other words, methyl and/or Cl group introduced at the ortho positions of the phenyl ring were most suitable for providing good herbicidal performance. Furthermore, evaluation of the various carbamoyl groups attached to ortho-substituted phenyl tetrazolinones indicated that (1) lower mobility in soil resulted in better crop compatibility, which is due to decreasing mobility with increasing total number of carbon atoms in the N-alkyl group; and (2) a significant decrease in activity was observed when either a linear alkyl group of C_4 or longer was introduced, or the total number of carbon atoms in an N-alkyl group exceeded eight, while the existence of a C_5 or C_6 cycloalkyl group had a positive effect on herbicidal action.

Based on the results, fentrazamide, with high activity to barnyard grass, an excellent safety to rice seedlings and a lower mobility in soil, was selected [22–24].

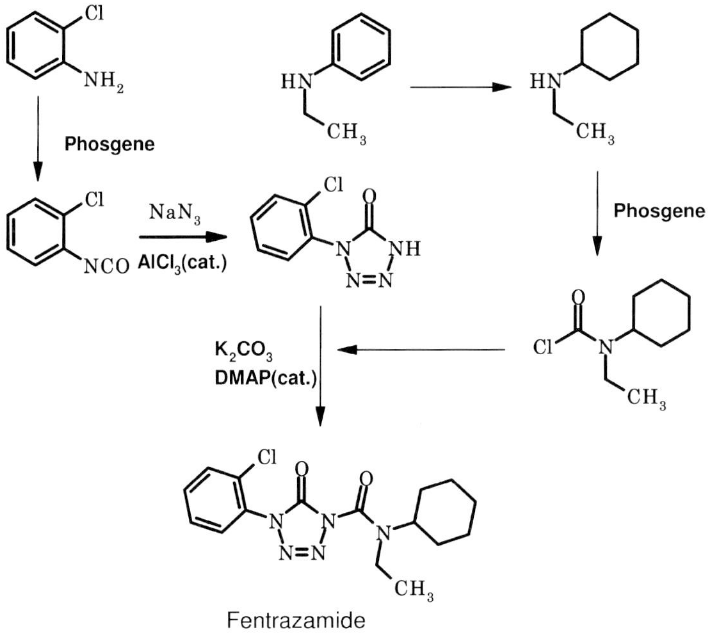

Fentrazamide

Fig. 8.6. Synthetic pathways to fentrazamide.

As shown in Fig. 8.6, the manufacturing processes of fentrazamide involve the conversion of two inexpensive run-of-the mill anilines, 2-chloroaniline and N-ethyl aniline, into 2-chlorophenyl isocyanate and N-cyclohexyl-N-ethylcarbamoyl chloride, respectively. 1-(2-Chlorophenyl)-5(4H)-tetrazolinone can be provided quantitatively by reacting equimolar amounts of 2-chlorophenyl isocyanate and sodium azide in the presence of catalytic amounts of aluminum trichloride in dimethylformamide [25]. The tetrazolinone reacted with N-cyclohexyl-N-ethylcarbamoyl chloride, in the presence of a catalytic amount of 4-dimethylaminopyridine (DMAP), to afford fentrazamide with no formation of its O-carbamoylated isomer [26].

Subsequent investigations by our research group have revealed that, in general, non-aromatic substituted tetrazolinones with an N-phenyl isopropyl carbamoyl group and phenyl- or heteroaryl-substituted tetrazolinones with a dialkyl carbamoyl group are active against barnyard grass [27, 28]. However, none of these tetrazolinones has reached the market as a K_3 herbicide.

8.3.2
Biology of the Compounds

8.3.2.1 Flufenacet

Flufenacet is a selective pre- and early post-emergence herbicide. It is taken up mainly through the root system and xylem-transported to the meristematic tissue of the roots and young shoot to cause growth inhibition. In the greenhouse, flufenacet at 250 g-a.i. ha^{-1} controls >95% of grasses, including *Echinochloa crusgalli*, *Digitaria sanguinalis*, *Setaria viridis*, *Panicum miliaceum* and *Alopecurus myosuroides*, and also >80% of dicots, such as *Amaranthus retroflexus*, *Chenopodium album* (CHEAL), *Galium aparine* and *Galinsoga parviflora* [29].

The crop tolerance of flufenacet is attributed to rapid detoxification by glutathione S-transferases [30].

8.3.2.2 Mefenacet and Fentrazamide

Mefenacet and fentrazamide are used at pre- and post-emergence of weeds, mainly in transplanted rice. They provide stable efficacy against *Echinochloa* sp. (ECHSS) and other dominant weeds in paddy with long-lasting control. The proper application timings are from before emergence of weeds up to the 3 leaf stage (LS) of ECHSS. Mefenacet at 1000–1200 g-a.i. ha^{-1} and fentrazamide at 200–300 g-a.i. ha^{-1} effectively control ECHSS and annual sedges with good compatibility to transplanted rice [31, 32]. The plant compatibility is derived from a low mobility of the herbicides in soil. Almost all active ingredients of mefenacet and fentrazamide applied are detected within 0.5 cm of the soil surface.

Possible application of fentrazamide at 0-DAT (0 Days After Transplanting, that is to say simultaneous application with transplanting of young rice seedlings before emergence of weeds) is basically due to the strong adsorption of the active ingredient to surface layer of the soil. The 0-DAT application technique achieves efficient labor saving in rice cultivation.

8.4
Biology of the Marketed Products and use Pattern

8.4.1
Marketed Products

8.4.1.1 Flufenacet Products

More than 80% of the value of flufenacet-containing products is currently generated in maize (US and Europe) and autumn uses in winter cereals (Europe mainly).

Flufenacet single product (Define™) can control most annual grasses and selected annual broadleaf weeds in maize and soybeans by the treatment of the herbicide alone or as its recommended tank mixes. Possible applications are preplant surface, preplant incorporated or pre-emergence. In cereals, flufenacet is used in ready mixtures with either diflufenican or pendimethalin. A premix with metribuzin (Axiom ᴿ) is more effective than other grass herbicides for early-season suppression of *Ambrosia elatior* and *Polygonum* sp. in maize. A premix with isoxaflutole (Epic ᴿ) controls major grasses and broadleaf weeds, including *Digitaria* sp., *Setaria* sp., *Panicum dichotomiflorum*, CHEAL, *Amaranthus* sp. and *Eriochloa villosa* control in maize. Epic acts season-long by recharge-action to provide one-pass weed control.

8.4.1.2 Mefenacet Products

Despite the launch of many new one-shot rice herbicides, mefenacet products maintained ca. 16% of Japanese total one-shot application area in 2005 PY (Japanese pesticide sales year from Oct. 2004 to Sep. 2005). The inherent performance of mefenacet – high activity to ECHSS, broad application period and long lasting efficacy – offered a platform for creating a so-called one-shot herbicide. To ensure wide weed control spectrum in paddy rice, mefenacet has been mixed with proper antidicotyledon partners, especially with sulfonylurea (SU) rice herbicides. Typical mefenacet combination products are as a plus bensulfuron-methyl (Zark ᴿ), a plus pyrazosulfuron-ethyl (Act ᴿ), and a plus imazosulfuron (Batl ᴿ). They stably control ECHSS, *Cyperus difformis* (CYPDI), *Scirpus juncoides* (SCPJU), *Monochoria vaginalis* (MOOVP), annual broad-leaved weeds (BBBBB), *Eleocharis acicularis* (ELOAL), *Sagittaria pygmaea* (SAGPY) and *Cyperus serotinus* (CYPSE) with a good safety to transplanted rice at application from 3 DAT up to the 3 LS of ECHSS. In contrast to known sequential application with certain other herbicides, such one-shot products reduced weeding time in rice fields.

The combinations formulated to GR type are conventionally used.

8.4.1.3 Fentrazamide Products

The application area of fentrazamide products has constantly increased since its launch and was up to ca. 14% of total one-shot application area in 2005 PY.

Fentrazamide products such as a plus bensulfuron-methyl (Innova ᴿ), a plus pyrazosulfuron-ethyl (Doublestar ᴿ) and a plus imazosulfuron (Leading ᴿ) provide the same performance as mefenacet–SU combinations. In addition, these products are applicable at 0-DAT due to their outstanding safety to young rice seedlings [33]. New combinations with HPPD inhibitors like benzobicyclon are under development as a countermeasure for SU-resistance weeds. GR formulations of the products are conventionally used. Special easy-to-use formulation types such as SC and floating granules (GF) packed in water-soluble poly(vinyl alcohol) (PVA) poach and throw-in type application technique as well as 0-DAT application satisfy the farmer's demand for labor saving [34].

In seeded paddy rice, fentrazamide mixture with propanil (Lecspro®) is used as an early post-emergence herbicide for controlling ECHSS, CYPDI, *Cyperus iria*, *Fimbristylis miliacea*, *Leptochloa chinensis* and *Sphenoclea zeylanica* [35].

8.5
The Future of Flufenacet, Mefenacet and Fentrazamide

Increasing generic pressure has influenced the use of flufenacet products, especially in soybeans. Although turning the tide is obviously difficult, flufenacet may be applied either alone or in tank mixtures in cereals, potatoes, sunflowers and vegetables. The occurrence of resistance weeds on using flufenacet is extremely rare, so that the combination of VLCFA-synthesis and HPPD inhibitors may allow the control of glyphosate-, triazine-, and ALS-resistant species of weeds.

The global rice herbicide market has steadily declined, which is due to changes in Japanese farming conditions, such as reducing rice acreage, reducing demand for rice, diversifying consumption patterns, aging and reducing farming populations. Such farming conditions seem to be common to other developing countries. Genetically modified herbicide-tolerant rice varieties are not likely to have significant effects on herbicide sales until GM-rice gains global social acceptance. An increasing demand for rice, reducing farming labor and changing land usage in populous and developing countries are noticeable trends. Therefore, low application volume, less-toxic, one-shot use and value-added rice herbicides seem to be essential in the longer term.

References

1 P. Böger, B. Matthes, J. Schmalfuß, *Pest Manag. Sci.* **2000**, 56, 497–508.

2 C. Cassagne, R. Lessire, J. J. Bessoule, P. Moreau, A. Creach, F. Schneider, B. Sturbois, *Prog. Lipid Res.* **1994**, 33, 55–69.

3 J. L. Harwood (Ed), *Plant Lipid Biosynthesis*, Cambridge University Press, UK, **1998**, 185–220.

4 E. Ebert, K. Ramsteiner, *Weed Res.* **1984**, 24, 383–389.

5 E. Fehling, D. J. Murphy, K. D. Mukherjee, *Plant Physiol.* **1990**, 94, 492–498.

6 E. B. Cahoon, D. V. Lynch, *Plant Physiol.* **1991**, 95, 58–68.

7 M. Couderchet, J. Schmalfuß, P. Böger, *Pestic. Sci.* **1998**, 52, 381–387.

8 B. Matthes, J. Schmalfuß, P. Böger, *Z. Naturforsch.* **1998**, 53c, 1004–1011.

9 J. Schmalfuß, B. Matthes, P. Mayer, P. Böger, *Z. Naturforsch.* **1998**, 53c, 995–1003.

10 J. Schmalfuß, B. Matthes, K. Knuth, P. Böger, *Pestic. Biochem. Physiol.* **2000**, 67, 25–35.

11 J. R. C. Leavitt, D. Penner, *J. Agric. Food Chem.* **1979**, 27, 533–536.

12 M. Ghanevati, J. G. Jaworski, *Biochim. Biophys. Acta*, **2001**, 1530, 77–85.

13 C. Eckermann, B. Matthes, M. Nimtz, V. Reiser, B. Lederer, P. Böger, J. Schröder, *Phytochemistry*, **2003**, 64, 1045–1054.

14 S. Trenkamp, W. Martin, K. Tietjen, *Proc. Natl. Acad. Sci. U.S.A.*, **2004**, 101, 11903–11908.

15 C. Lechelt-Kunze, R. C. Meissner, M. Drewes, K. Tietjen, *Pest Manag. Sci.* **2003**, 59, 847–856.

16 H. Förster, R. R. Schmidt, H. J. Santel, R. Andree, *Pflanz. Nachrichten Bayer*, **1997**, 50, 105–116.

17 H. Förster, W. Hofer, V. Mues, L. Eue, R. R. Schmidt, Ger. Pat. DE 2914003, **1980** (Bayer A.G.).

18 H. Förster, R. Andree, H. J. Santel, R. R. Schmidt, H. Strang, Ger. Pat. DE 3724359, **1989** (Bayer A.G.).

19 H. Förster, W. Hofer, V. Mues, L. Eue, R. R. Schmidt, Ger. Pat. DE 2822155, **1979** (Bayer A.G.).

20 H. Förster, R. Andree, H. J. Santel, R. R. Schmidt, H. Strang, Ger. Pat. DE 3821600, **1989** (Bayer A.G.).

21 R. A. Covey, P. J. Forbes, A. R. Bell, Eur. Pat. EP146279, **1985**.

22 A. Yanagi, Y. Watanabe, S. Narabu, S. Ito, T. Goto, *J. Pestic. Sci.* **2002**, 27, 199–209.

23 T. Goto, S. Ito, A. Yanagi, Y. Watanabe, K. Yasui, *Weed Biol. Manag.* **2002**, 2, 18–24.

24 A. Yanagi, *Pflanz. Nachrichten Bayer*, **2001**, 54, 2–12.

25 A. Yanagi, Y. Watanabe, S. Narabu, Eur. Pat. EP638561, **1995**.

26 A. Yanagi, Y. Watanabe, S. Narabu, Eur. Pat. EP646577, **1995**.

27 T. Goto, S. Ito, Y. Watanabe, PCT Int. WO2000-040568, **2000**.

28 T. Goto, K. Moriya, F. Maurer, S. Ito, K. Wada, K. Ukawa, R. Watanabe, A. Ito, N. Minegishi, Eur. Pat. EP695748, **1996**.

29 R. Deege, H. Förster, R. R. Schmidt, W. Thielert, M. A. Tice, G. J. Aagesen, J. R. Bloomberg, H. J. Santel, *Proc. Brighton Crop Prot. Conf. – Weeds*, **1995**, 43–48.

30 B. Bieseler, C. Fedtke, T. Neuefeind, W. Etzel, L. Prade, P. Reinemer, *Pflanz. Nachrichten Bayer*, **1997**, 50, 117–140.

31 M. Aya, K. Yasui, K. Kurihara, A. Kamochi, L. Eue, *Proceedings of the 10tth Asian-Pacific Weed Science Conference*, Chiangmai, Thailand, **1985**, 567–574.

32 R. R. Schmidt, L. Eue, H. Förster, V. Mues, *Med. Fac. Landbouww. Rijksuniv. Gent*, **1984**, 1075–1084.

33 K. Yasui, T. Goto, H. Miyauchi, A. Yanagi, D. Feucht, H. Fürsch, *Proc. Brighton Crop Prot. Conf. – Weeds*, **1997**, 67–72.

34 Y. Nishi, H. Miyauchi, *Pflanz. Nachrichten Bayer*, **2001**, 54, 43–50.

35 H. Fürsch, *Pflanz. Nachrichten Bayer*, **2001**, 54, 127–142.

9
Acetyl-CoA Carboxylase Inhibitors

Jean Wenger and Thierry Niderman

9.1
Introduction

Acetyl-CoA carboxylase (ACC) catalyzes the first step in fatty acid biosynthesis. Owing to its role it has been exploited as an important herbicide target. Two chemical classes, the aryl-oxy-phenoxy-propionate (AOPP or fop) and the cyclohexanedione (CHD or dim) herbicides are widely used to control a broad selection of grass weeds in dicot crops and some of them even in cereals or in rice [1, 2]. Their frequent use has resulted in the development of resistance in several grass species [3].

Dicot tolerance is based on the inherent insensitivity of broadleaves to these herbicides, whereas in monocot crops the selectivity is usually due to higher rates of herbicide detoxification [3, 4].

AOPP and CHD herbicides are well described in the literature and are known to inhibit the carboxylate transferase (CT) function of homomeric ACC found in the plastids of grasses [1, 5].

Over a decade ago, 2-aryl-1,3-diones emerged in the literature as a new, weakly active class of ACC inhibitors [6]. Recently, pinoxaden was reported as a novel cereal graminicide that belongs to this class [7].

This chapter presents an insight into recent developments in biochemistry and resistance mechanisms of ACC and gives an overview of the aryl-diones (ADs) as a novel class of ACC inhibitors. In addition, the industrial synthesis, biology and metabolism of pinoxaden is described.

Modern Crop Protection Compounds. Edited by W. Krämer and U. Schirmer
Copyright © 2007 WILEY-VCH Verlag GmbH & Co. KGaA, Weinheim
ISBN: 978-3-527-31496-6

9.2
Biochemistry

9.2.1
Overview

Acetyl-CoA carboxylase (ACC; EC 6.4.1.2) is a biotin-dependent carboxylase that produces malonyl-CoA from bicarbonate as a source of carboxyl group and ATP as a source of energy. The reaction catalyzes the conversion of acetyl-CoA into malonyl-CoA through the incorporation of a carboxyl group into the acetyl radical of the acetyl-CoA. This transcarboxylation reaction is performed following the three-step process followed by all biotin-dependent transcarboxylases (Scheme 9.1)

$$HCO_3^- + enzyme\text{-}biotin + ATP\text{-}Mg \rightarrow enzyme\text{-}biotin\text{-}CO_2^- + ADP\text{-}Mg + Pi \quad [1]$$

$$Enzyme\text{-}biotin\text{-}CO_2^- + acceptor \rightarrow acceptor\text{-}CO_2^- + enzyme\text{-}biotin \quad [2]$$

$$HCO_3^- + acceptor + ATP\text{-}Mg \rightarrow acceptor\text{-}CO_2^- + ADP\text{-}Mg + Pi \quad [3]$$

Scheme 9.1

The overall ACC transformation is the result of the cooperation of different catalytic activities: [1] carbamoyl-phosphate synthase, [2] biotin-carboxylase and [3] acetyl-CoA transcarboxylase.

In prokaryotes and in plastids of some plants, the ACC is a multisubunit enzyme, whereas in eukaryotes the cytosolic isozyme and, in some instances also the plastid isozyme, are multidomain proteins. The latter contain three major functional domains, which account for the biotin carboxylase (BT), biotin carboxyl-carrier (BCC) and carboxyltransferase (CT) activities and, which are organized in one large polypeptide.

The chloroplastic ACC is responsible for the synthesis of malonyl-CoA then metabolized to a fatty acid chain up to C_{18}. This is in part exported to the cytoplasm, thus contributing to the control of flux through the plant's *de novo* fatty acid biosynthetic pathway [8].

The cytoplasmic malonyl-CoA pool is dispatched into the following:

- Long and very long chain fatty acids, which are elongation products of the C_{18} lipids,
- A large group, consisting of flavonoids, pigments and stilbene derivatives through to the synthesis of naringenin.
- N-malonyl-D-amino acids.

Figure 9.1 summarizes the role of acetyl-CoA carboxylase in plants.

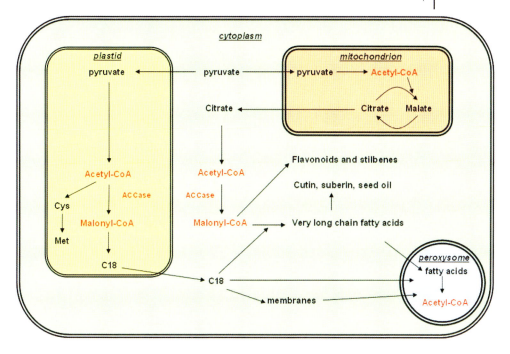

Fig. 9.1. Schematic drawing of the central metabolic role of acetyl-CoA in plants. Acetyl-CoA is the starting material for the biosynthesis of fatty acids, some amino acids, flavonoids, sterols, and isoprenoids. Acetyl-CoA does not cross membranes and is produced in the compartment where it is needed.

Grasses contain two multidomain ACC, one chloroplastic and one cytosolic, whereas dicots contain two well-differentiated forms, a cytosolic multidomain ACC and chloroplastic multisubunits ACC (Table 9.1) [9, 10].

The multisubunits enzyme is encoded by the nuclear DNA, with the exception of the β-subunit of carboxyltransferase that is encoded by a chloroplastic gene [11]. In grasses, the chloroplastic multidomain ACC is encoded by a nuclear gene, which is distinct from that coding for the cytosolic multidomain ACC.

Cytoplasmic and plastidic ACCs from wheat are 2260 amino acids and 2311 long, respectively, and their sequences are 67% identical [12]. A chloroplast targeting signal is present at the N-terminus of the multidomain plastid ACC from wheat [13], maize [14], and *Brassica napus* [15].

Numerous works have been published that attempt to analyze the quaternary structure of the multifunctional ACCs. It seems that the active enzyme has to be at least dimeric, either homodimeric or heterodimeric. Polymeric filaments (10–15 units) are also detected, which contain a heterodimeric subunit periodically interspersed throughout the otherwise homodimeric filamentous enzyme.

Table 9.1 Summary of different ACCs in plants.

Characteristics	Chloroplast	Cytoplasm
Grasses		
Type	Eukaryote type I, isoform 1	Eukaryote type I, isoform 2
Molecular mass	≈ 240 kDa	≈ 220 kDa
Native structure	Homodimer	Homodimer
Reaction	Three catalytic domains per protein	Three catalytic domains per protein
Major role	Fatty acid biosynthesis	Secondary metabolite biosynthesis
Dicots		
Type	Prokaryote type II	Eukaryote type I
Molecular mass	≈ 32–80 kDa	≈ 230 kDa
Native structure	Multicomponent enzyme	Homodimer
Reaction	One catalytic domain per enzyme	Three catalytic domains per protein
Major role	Fatty acid biosynthesis	Secondary metabolite biosynthesis

Furthermore, ACC isozymes are bound to the outer face of the mitochondria and in conjunction with other proteins, in particular carnitine phosphate transferase, a complex that regulates the flux of malonyl-CoA out of the mitochondria [16]. Recently, a specific ACC from yeast was shown to be targeted to mitochondria [17]. Moreover, Focke et al. [18] have presented biochemical evidence for a mitochondrial localized acetyl-coenzyme A carboxylase in barley.

It is not clear how many subunits and different proteins are contained in such complexes; the quaternary structure is likely to depend on the particular function of the ACC isozyme. In each organism, and even in each differentiated tissue, ACC quaternary structures will be dependent on genetic (isozymes and allelic variants), functional and activation factors. In others words, the biological role of the different isoforms of ACC determines the quaternary structure of the enzyme and associated proteins.

ACC regulation in cells is poorly understood, but redox control and phosphorylation are probably key factors. In pea, activation was found to be mediated through reduction of a disulfide bond between the α-CT and the β-CT subunits [19]. Plastid ACC is likely to be subject to redox regulation similar to that of several key enzymes of photosynthesis.

Phosphorylation of serine residue(s) of the β-subunit of the carboxyltransferase unit occurs in pea chloroplasts incubated in the light [20]. Alkaline phosphatase treatment reduces ACC activity in parallel to removal of phosphate groups from ACC. This activation by phosphorylation is opposite to the inhibition of animal ACC by phosphorylation but is consistent with the increase in ATP concentration and rates of fatty acid synthesis in chloroplasts in the light and the activation of other plastid enzymes by phosphorylation. These results suggest that the CT subunit reaction is rate determining for overall ACC activity, at least for the multisubunits enzyme of dicots.

In wheat cytosolic ACC none of the four conserved motifs containing serine residues corresponding to phosphorylation sites in rat, chicken, and human ACC [21] is present at a similar position.

9.2.2
Mode of Action of ACC Inhibitors

The first consistent study of an effect of ACC inhibitors on plant lipid biosynthesis was reported by Hoppe [22], and showed strong inhibition of incorporation of ^{14}C-labeled acetate into plant lipids. It was only in late 1987 that the two independent laboratories of Burton and Focke demonstrated that the site of action of these inhibitors was located in the acetyl-coenzyme A carboxylase. Burton et al. [23] found that ACC isolated from chloroplasts of corn seedling was inhibited by the herbicides sethoxydim and haloxyfop, with IC_{50} concentrations of 2.9 and 0.5 μM, respectively, whereas the ACC from pea chloroplasts was not inhibited by these inhibitors. Focke and Lichtenthaler [24] reported that the cyclohexane-1,3-dione derivatives cycloxydim, sethoxydim and clethodim inhibited fatty acid biosynthesis in a chloroplast enzyme preparation from barley when acetate and acetyl-CoA were the substrates, but not when malonate and malonyl-CoA were added. These results suggested that ACC was the site of action for these herbicides. Moreover, they showed that ACC from dicot species reported almost no inhibition, suggesting that the mechanism of selectivity between dicot and grass species was at the ACC site of action.

Two key papers [25, 26] established that the two types of ACC enzymes in plant correlated with the differential inhibition of the new herbicides represented by two classes: the AOPPs and the CHDs, which are strong inhibitors of the multidomain plastid ACC found in grasses. Prokaryotic-type multisubunit plastid ACC is resistant to these herbicides, as are eukaryote ACCs from animals and yeast.

Widely used commercial herbicides, represented by AOPPs and CHDs, are potent inhibitors of ACCs of sensitive plants and kill them by shutting down fatty acid biosynthesis, thus leading to metabolite leakage from the membranes and cell death [27]. AOPPs and CHDs inhibit the carboxyltransferase activity (Scheme 9.1, reaction [2]), thus blocking the transfer of the carboxyl group to acetyl-CoA [28]. They show nearly competitive inhibition with respect to the substrate acetyl coenzyme A [29].

This observation confirms that an inhibitor of the CT domain is sufficient to block the function of the ACC (Fig. 9.2), and it establishes this domain as a valid target for the development of inhibitors against these enzymes.

Interaction of AOPP and CHD inhibitors is an important tool to understand plant ACC biochemistry, and the use of chimeric genes was a significant step forward in the elucidation of differential activities for different chemical classes [12, 31]. Gornicki et al. showed that some determinants of sensitivity were located on a 400-amino acid fragment of wheat plastidic ACC in the CT domain [12]. The chimeric genes consisted of the yeast GAL10 promoter, the yeast ACC1 leader and the wheat acetyl-CoA carboxylase cDNA.

100 amino acids

Fig. 9.2. Schematic representation of a plastidic homomeric acetyl coenzyme A carboxylase (ACC) showing the three functional domains (BC, biotin carboxylase; BCC, biotin carboxyl-carrier; and CT, carboxyl transferase) and the transit peptide (TP) that is absent in cytosolic ACC. The five amino acid residues critical for sensitivity to ACC-inhibiting herbicides have been referenced after the sequence from black-grass plastidic ACC (EMBL accession AJ310767). (From Ref. [30].)

The yeast ACC1 3'-tail was used to complement ACC1 null mutation. These genes encode a full-length plastid enzyme, with or without the putative chloroplast transit peptide, as well as five chimeric cytosolic/plastid proteins (Fig. 9.3).

Combining this yeast gene replacement strains system with kinetics values from purified plastidic proteins together provide a convenient tool to study herbicide interaction with the enzyme and a powerful screening system for new inhibitors.

Pinoxaden [32], the leading compound of the Ads, acts primarily on the plastidic homomeric ACC, but also exhibits new features.

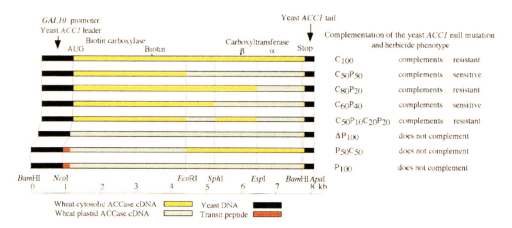

Fig. 9.3. Chimeric genes constructed for expression of wheat cytosolic, plastidic, and cytosolic/plastidic ACC in yeast. Construct names reflect the composition of the encoded proteins (C, cytosolic ACC; P, plastid ACC). Locations of key restriction sites used in the constructions are shown. (From Ref. [12].)

9.2.3
Resistance

The frequent use of AOPP and CHD graminicides has resulted in the development of resistance to these herbicides in some grass species throughout the
world [33]. Up to now, 35 resistant species [34] have been reported. The species
in which resistance has developed include the important grass weeds *Alopecurus
myosuroides*, *Avena fatua*, *Setaria viridis*, *S. faberi*, *Lolium rigidum* and *Eleusine
indica*.

Mechanisms of resistance to ACC-inhibiting herbicides can be divided into two
categories: ACC-related and metabolism-based. Metabolism-based resistance is
well described and reviewed in the literature [35, 36]. In most cases, resistance is
due to alteration of the target enzyme, making it less sensitive to inhibition, as
reviewed by Devine [37] and by Délye [38]; the latter gives an overview on homomeric plastidic ACC isoforms with altered sensitivities to AOPPs and CHDs or
both [38]. Furthermore, the identification of mutations involved in altered sensitivity was achieved recently (Table 9.2) [38].

An updated version of this table is maintained at the International Survey of
Herbicide Resistant Weeds Web site (http://www.weedscience.org).

A single point mutation leading to substitution of an isoleucine (Ile) by a leucine (Leu) residue at position 1781 within the CT binding domain of plastidic
ACC in *Alopecurus myosuroides* (blackgrass) has been found to confer resistance
to most CHDs and AOPPs [46]. A homologous mutation is responsible for target
site resistance in three other grass weeds, *Lolium sp.* (Rye-grass) [43, 47], *Avena
fatua* L. (wild-oat) [41], and *Setaria viridis* (green foxtail) [42].

The mutations leading to an isoleucine (Ile)-asparagine (Asn) exchange at position 2041, of tryptophan (Trp) in position 2027 to cysteine (Cys), as well as glycine
(Gly) to alanine (Ala) in 2096, affects mainly the AOPPs in blackgrass and in rye-
grass [44, 45]. In blackgrass again, aspartic acid (Asp) to glycine (Gly) mutation at
position 2078 leads to resistance on APPs and CHDs [44].

Three-dimensional models of homodimeric ACC were reconstructed for a
detailed evaluation of the effects of amino acid substitutions at positions 1781,
2027, 2041, 2078, and 2096 in black-grass ACC upon herbicide binding [48], using models built into maps obtained by electron crystallography of the yeast free
ACC CT domain as templates [49].

All five amino acids given in Table 9.2 are located within the active site cavity of
the ACC CT domain [48]. Only the substitution at position 2041 interferes directly with herbicide binding. It has been proposed that the other four mutations
cause resistance by hampering inhibitor access to its binding site or by altering
the spatial shape of the herbicide binding site [46]. Zhang et al. have determined
the crystal structures of the CT domain of yeast ACC in complex with haloxyfop
and diclofop [50]. The inhibitors are bound in the active site, at the interface of
the dimer of the CT domain. Unexpectedly, inhibitor binding requires large conformational changes for several residues in the interface, which create a highly
conserved hydrophobic pocket that extends deeply into the core of the dimer.

Table 9.2 Amino acid substitutions within plastidic, homomeric ACC and associated cross-resistance patterns observed at the whole plant level.[a]

Amino acid residue[b]		Weed species[e]	Resistance[a]									Ref.
			APPs[c]					CHDs[d]				
Wild-type	Resistant		Cd	Dc	Fx	Fz	Hx	Ct	Cx	Sx	Tk	
Ile$_{1781}$	Leu	*Alomy*	S	R	R	R	S	S	R	R	R	39, 40
	Leu	*Avefa*	ND	R	ND	ND	ND	ND	ND	R	ND	41
	Leu	*Setavir*	ND	R	R	ND	ND	ND	ND	R	R	42
	Leu	*Lol* sp.	S	R	R	ND	ND	ND	R	ND	ND	43
Trp$_{2027}$	Cys	*Alomy*	R	ND	R	ND	R	S	S	ND	ND	44
Ile$_{2041}$	Asn	*Alomy*	R	ND	R	ND	R	S	S	ND	ND	45
	Asn	*Lol* sp.	R	R	ND	ND	R	ND	S	ND	ND	45
	Val	*Lol* sp.	S	ND	ND	ND	R	ND	S	ND	ND	45
Asp$_{2078}$	Gly	*Alomy*	R	ND	R	ND	R	R	R	ND	ND	44
Gly$_{2096}$	Ala	*Alomy*	R	ND	R	ND	R	S	S	ND	ND	44

[a] S and R respectively indicate that plants containing at least one copy of the ACC mutant allele are sensitive or resistant to the corresponding herbicide either in the field or in bioassays (see text for comment). ND, not determined at the whole plant level.
[b] Amino acid number is standardized to *A. myosuroides* plastidic, homomeric ACC (EMBL accession AJ310767).
[c] Cd, clodinafop; Dc, diclofop; Fx, fenoxaprop; Fz, fluazifop; Hx, haloxyfop.

The mutation of two residues that are located in this binding site and affect herbicide sensitivity disrupts the structure of the domain.

9.2.4
Detection of Resistance

To date, detection and management of resistance has predominantly been carried out with bioassays. These are essentially based on comparative growth of seedlings or plants of suspected resistant and sensitive weed biotypes subjected to different herbicide treatments [51–53]. Such bioassays are simple, but do not differentiate between target site and metabolic resistance mechanisms.

ACC-based resistance is expressed in pollen, whereas metabolism based is not [53, 54].

The main I1781L mutation leading to resistance can only occur by substitution of an adenine (A) by thymine (T) or cytosine (C) at the first position in the cognate codon. As a result, it was possible to develop a polymerase chain reaction

(PCR)-based allele-specific amplification assay to detect the I1781L mutation in the plastidic ACCase of *L. rigidum* and *A. myosuroides* plants, providing a quick and efficient method for monitoring a key resistance mechanism to ACC inhibitors in these species [43, 55].

Kaundun and Windass [56] described an alternative derived Cleaved Amplified Sequence (dCAPS) method [57] that can be used on several grass weeds and that offers the additional advantage of easy discrimination between homozygous and heterozygous L1781 mutation bearing plants.

9.3
Aryl-diones as Novel ACC Inhibitors

9.3.1
Discovery

The first 2-aryl-1,3-diones (ADs) were reported in 1977 by Wheeler (Union carbide) [58]. He claimed biocidal aryl-cyclohexenyl esters **1**, **2** (Fig. 9.4) with pre- and post-emergence herbicidal effects and miticidal activity against *Tetranychus urticae*.

Ten years later R. Fischer et al. (Bayer) discovered 2-aryl-indolizine-2,4-diones with herbicidal and miticidal activity [59] and reported compound **3** (Fig. 9.5) to inhibit plastidic ACC of grasses [6].

Almost simultaneously Cederbaum (Ciba-Geigy) [60] as well as Fischer and coworkers [61] claimed the herbicidal activity of 2-mesityl-tetrahydro-pyrazolo-1,3-diones **4**.

The Bayer research group described further heterocyclic diones (Fig. 9.5), which all belong to this chemical class: 3-aryl-pyrrolidine-2,4-diones **5** [62, 63], 3-aryl-furan-2,4-diones **6** [64], 2 aryl-cyclopentan-1,3-diones **7** [65] 4-phenyl-[1,2]oxazin-3,5-diones **8** [66].

Although interesting, all these AD derivatives were substantially weaker herbicides than commercial AOPPs and CHDs. Many of these compounds with good miticidal activities are reported to be phytotoxic [67]. A major breakthrough with regard to the herbicidal activity was achieved as aryl moieties bearing ethyl, ethynyl or methoxy groups in the 2,6-positions were synthesized [68]. Such a sub-

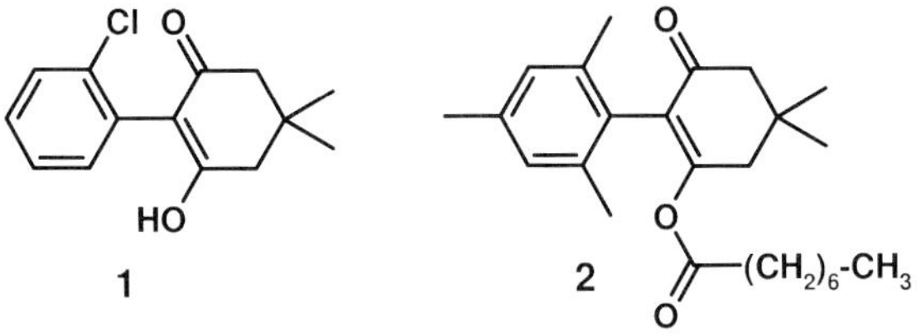

Fig. 9.4. Biocidal aryl-cyclohexenyl esters **1** and **2**.

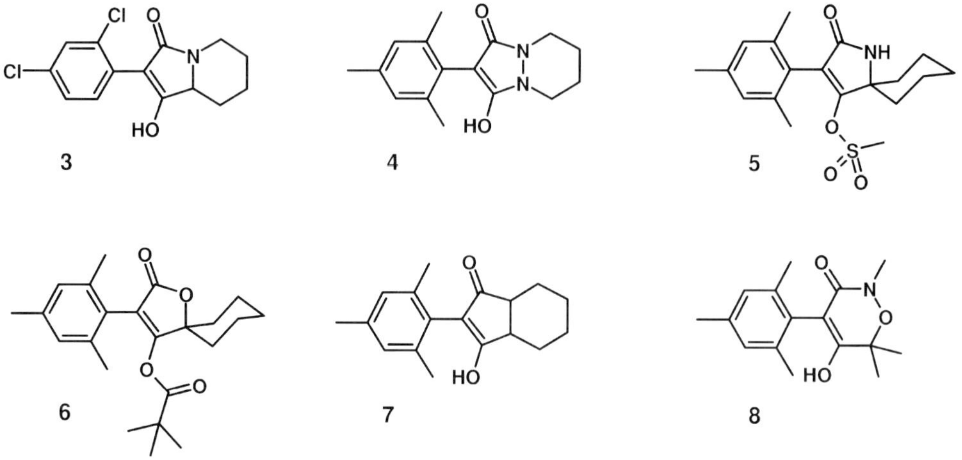

Fig. 9.5. Structures of compounds **3**–**8**.

stitution pattern boosts the herbicidal activity in combination with each type of 1,3-dione, whereas the miticidal activity is strongly reduced [69].

Further variations of the hydrazine moiety [70] ultimately led to pinoxaden (Scheme 9.5 below shows the synthesis of pinoxaden).

9.3.2
Syntheses

4-Aryl-pyrazolidin-3,5-diones **10** and their esters **11** were prepared as outlined in Scheme 9.2.

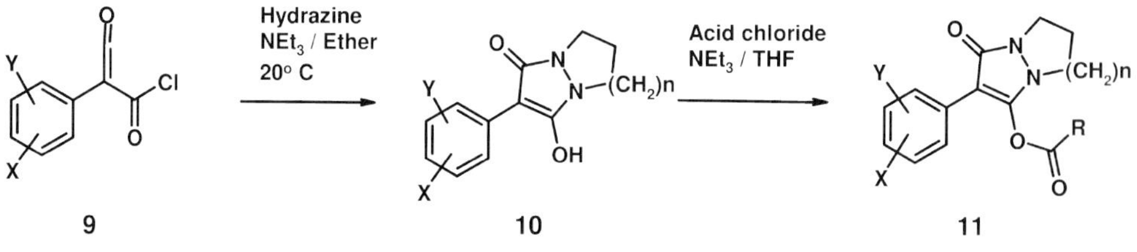

Scheme 9.2

Phenyl-substituted chlorocarbonylketenes **9**, first described by Nakanishi [71], represent a highly reactive equivalent of the phenyl-malonates. They react under mild conditions with hydrazines [65]. The ketoenol **10** is esterified with a standard method.

Scheme 9.3 depicts the synthetic route to tetramates **15** [60, 67].

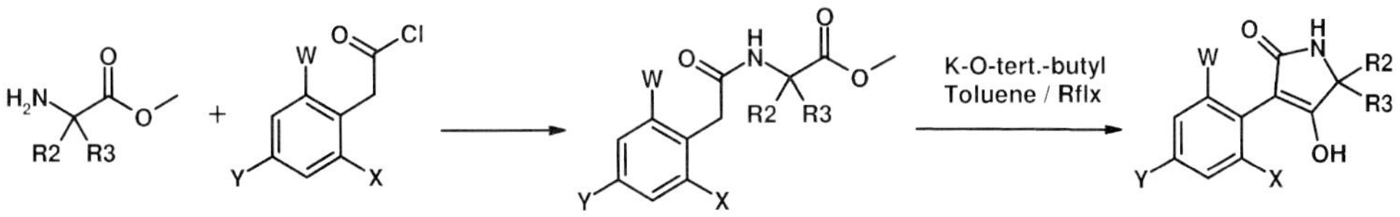

Scheme 9.3

Acylation of the amino acids **12**, (synthesized from a ketone via a Strecker amino acid synthesis) with aryl-acetyl chloride **13** leads to the intermediate **14**, which is cyclized to the tetramic acid **15** with potassium *tert*-butylate in refluxing toluene [62].

Indolizine-diones, tetronic acids, [1,2]-oxazin-3,5-diones and cyclopentanediones were obtained with similar cyclization steps.

Alternatively, most AD derivatives can be prepared with a Suzuki coupling reaction between iodonium-ylides **17** and phenylboronates **18** (Scheme 9.4).

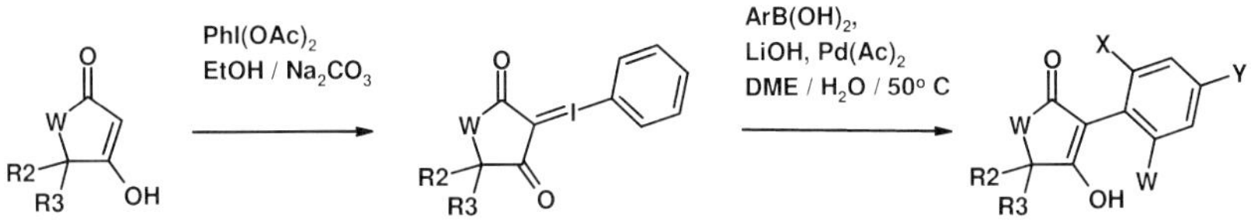

W = NH / O / S / N(Me)-O / methylene / ethylene

Scheme 9.4

The iodoniumylides **17** were obtained from diverse diones **16** and (diacetoxy)iodobenzene.

The yields of the cross-coupling reaction with sterically hindered arylboronates were modest [72]. However, many ADs were best prepared by this method since it allows convergent syntheses.

9.3.3
Structure–Activity Relationships

Very little data related to herbicidal activities of the ADs have been released [73, 74]. The overview given in this section reflects mainly the results based on studies with 4-aryl-pyrazolidin-3,5-diones [75], which were optimized towards activity and selectivity in small grain cereals.

The 2-aryl-1,3-diones can be separated into three parts for the analysis of the structure-activity relationships, namely the aryl-, the dione with its procidal forms

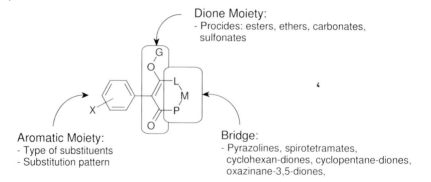

Dione Moiety:
- Procides: esters, ethers, carbonates, sulfonates

Aromatic Moiety:
- Type of substituents
- Substitution pattern

Bridge:
- Pyrazolines, spirotetramates, cyclohexan-diones, cyclopentane-diones, oxazinane-3,5-diones,

Fig. 9.6. Aryl-, dione (with its procidal forms), and bridging moieties of 2-aryl-1,3-diones.

and the bridging moiety (Fig. 9.6). Each part can be examined separately as they appear to play distinct and different roles in the overall expression of the herbicidal activity.

Only the free diones are active *in vitro* on ACC [76] and are responsible for the target site activity. With a pK_a of about 3.9, free diones are highly soluble in water.

Various prodrugs have been synthesized, with the aim of increasing the penetration into the leaves. Carbamates and most ethers are weakly active. However, carbonates are less active than the free dione. Since most esters hydrolyze easily, the free dione is usually equally active. A slight increase of activity and less variability was observed with aliphatic or aromatic sulfonates and the pivaloyl-esters and their homologs.

The substitution pattern of the aryl moiety strongly influences the overall herbicidal activity (Fig. 9.7). 2,4- and 2,6-Dihalo-aryl derivatives are all very weak compounds. 2,6-dimethyl-phenyl and 2,6-diethylphenyl substitution lead to some activity, whereas the 2,4,6-trimethyl pattern give a fair control of grasses at 100 g-a.i. ha^{-1} if combined with an optimized dione.

A 2-ethyl group leads to an increased activity and a 2,6-diethyl-4-methyl substitution pattern boosts the graminicidal activity. The level of activity can even be improved with a 2-ethynyl or a 2-methoxy-substituent in a 2,4,6-pattern. However, these two functionalities induce a higher level of phytotoxicity in cereals. Interestingly, 2,6-dimethoxy or 2,6-dibromoaryl ADs were found to be almost inactive.

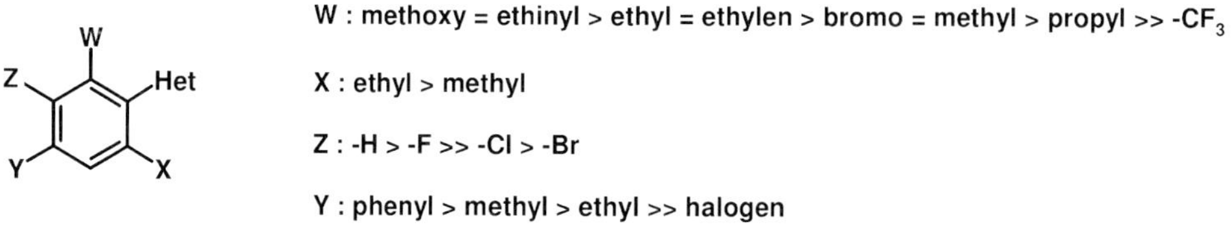

W : methoxy = ethinyl > ethyl = ethylen > bromo = methyl > propyl >> -CF$_3$

X : ethyl > methyl

Z : -H > -F >> -Cl > -Br

Y : phenyl > methyl > ethyl >> halogen

Fig. 9.7. Summary of the influence of the substitution pattern of the aryl moiety on overall herbicidal activity.

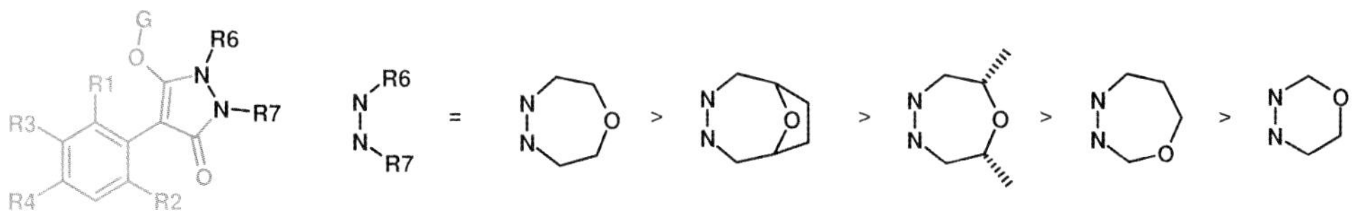

Fig. 9.8. Relative activity of cyclic hydrazines.

Position Y must be functionalized. Compared with methyl, a phenyl group or a thiophene leads to a broader spectrum, but also to phytotoxicity on cereals. In parallel, some activity on broadleaf weeds and a strong inhibition on cytosolic ACC are observed.

Introduction of a halogen in the position Z leads to a decrease of the herbicidal activity (Fig. 9.7).

High levels of herbicidal activity on grasses were found with many types of diones such as tetramates, cyclopentane-diones, cyclohexanediones or oxazin-3,5-diones linked to 2,6-diethyl-4-methyl-benzene.

With pyrazolines bridges, cyclic hydrazines are clearly the most active derivatives. There is not a great difference between five-, six- and seven-membered rings. However, [1,3,4]oxadiazinane and diverse oxadiazepanes derivatives were

Table 9.3 Compounds were applied with the adjuvant A12127 used at 0.5% at a rate of 60 g ha^{-1} on barley and wheat, 30 g ha^{-1} on Alomy (Alopecurus myosuroide), Avefa (Avena fatua), Lolpe (Lolium perenne), Setfa (Setaria faberi) at 2 leaf stage.

Structure	Barley	Wheat	Alomy	Avefa	Lolpe	Setfa
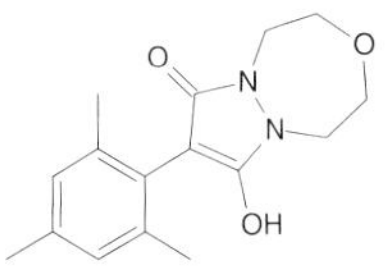	0	0	10	60	80	50
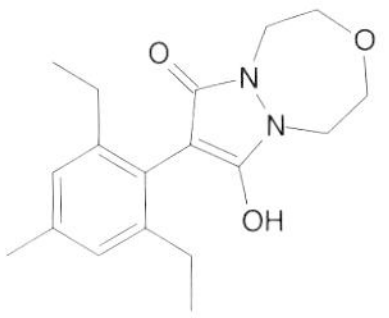	0	10	100	100	100	90
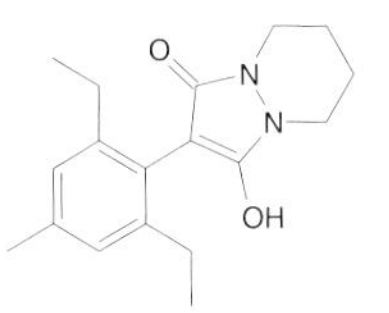	80	80	100	100	80	90

found to increase the activity and to have better selectivity in all cereal crops, the [1,4,5]oxadiazepane being the most active (Fig. 9.8).

Table 9.3 summarizes the optimization in the aryl-pyrazolines series. It demonstrates the effect of the aryl substitution with regard to the level of activity and the influence of the ring oxygen atom in the bridging moiety on the selectivity.

9.3.4
AD versus AOPP and CHD on ACC

AOPP and CHD herbicides are far more potent inhibitors of chloroplastic ACC than they are of the cytosolic ACC. Quizalofop is reported to inhibit maize chloroplastic ACC (IC50 0.03 μM) some 500-fold more strongly than the cytosolic form of maize ACC (IC50 $\sim$ 60 μM); a similar differential was reported in the same paper for fluazifop [77]. Similarly, Joachimiak et al. indicate that CHDs (sethoxydim) and AOPPs (haloxyfop) inhibit the chloroplastic form of wheat ACC at least 50-fold more potently than the cytosolic [31].

Our own unpublished data paint a similar picture in respect of clodinafop and propaquizafop, with both of these AOPPs inhibiting the cytosolic ACCase only relatively very weakly (with IC_{50}s > 300 μM) [78]. Interestingly, in contrast, some of the aryl diones appear to be quite potent inhibitors of the cytosolic enzyme. For example, the ADs SYN 271312 and SYN 436752 (Fig. 9.9) exhibit IC_{50}s of only 2 and 0.3 μM versus the cytosolic maize ACC [78]. The potency of these two compounds was further confirmed in gene replacement studies, according to Joachimiak et al. [31]. Unlike sethoxydim, clodinafop or propaquizafop, both ADs potently inhibited the growth of ACC1 null mutant yeast strains expressing the wheat cytosolic ACC (Table 9.4) [78].

At the primary sequence level the cytosolic ACC of dicot plants is similar to that in grasses and, accordingly, appears to exhibit an overall similar degree of sensitivity to inhibitors (for example quizalofop inhibits pea cytosolic ACC with a K_i of $\sim$7 μM versus $\sim$50 μM for maize cytosolic ACC [77]). Accordingly, the two ADs, SYN 271312 and SYN 436752 were also relatively potent inhibitors of pea cytosolic ACC, exhibiting IC_{50}s of 15 and 1 μm, respectively. Furthermore, these two compounds also exhibited significant glass house activity versus broadleaves,

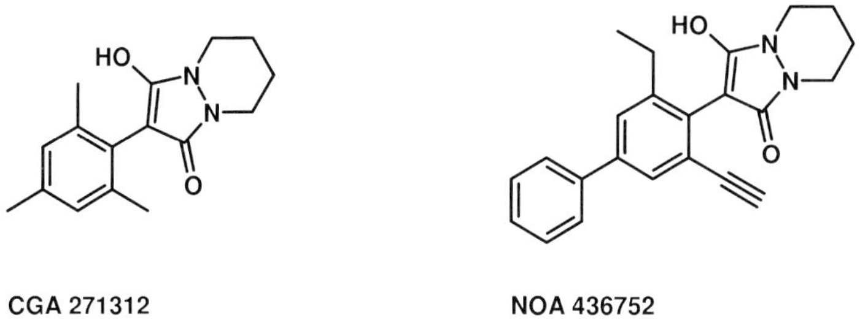

CGA 271312   NOA 436752

Fig. 9.9. Inhibitors of the cytosolic enzyme.

Table 9.4 Growth inhibition (in %) by herbicides (10 μM) of yeast gene-replacement haploid strains expressing chimeric wheat ACCs. Protein names reflect composition of C (cytosolic ACC) and P (plastid ACC). Work in collaboration with P. Gornicki.

Chimeral proteins	AOPP			CHD	AD	
	Clodinafop	Propaquizafop		Sethoxydim	SYN 271312	SYN 436752
C100	0	0		0	63	99
C50/P50	98	93		99	94	100
C80/C20	0	0		0	98	100
C60/P40	98	97		100	96	99
C50/P10/C20/P20	0	0		2	99	100

indicating that AD chemistry may also offer potential for control of dicot as well as of grass weeds.

9.3.5
AD on Herbicide Resistant ACC

As discussed in Section 9.2.3, resistance is a relevant feature for a novel chemical class.

In a recent paper Shukla attempted to identify molecules that target herbicide-sensitive and -resistant forms of ACC [80]. Among several experimental and commercial compounds, all the tested substances inhibited ACC from sensitive biotypes of *Setaria viridis* (green foxtail) and *Eleusine indica* (goosegrass). The I_{50}s of

Table 9.5 I_{50}s and R/S I_{50} ratios for ACC from herbicide-sensitive and -resistant biotypes of *Setaria viridis* (S1, sensitive; R1 and R2, resistant) and *Eleusine indica* (S3, sensitive; R3, resistant) and maize.

Herbicide	*Setaria viridis*					*Eleusine indica*			Maize		
	I_{50} (μM)			I_{50} (μM)		I_{50} (μM)			I_{50} (μM)		
	S1	R1	R1/S1	R2	R2/S1	S3	S4	R3/S3	S4	R4	R4/S4
Cpd. 8	51	67	1.3	54	1	76	227	3	18	40	2.2
Sethoxydim	54	>100	>1.9	89	1.7	39	407	11	0.9	>100	>111
Fluazifop						56	>500	>90	1.5	55	37
Diclofop	0.7	28	47								

most compounds assayed against resistant ACCs were higher than those against the corresponding sensitive ACC, indicating reduced binding to the resistant enzyme. However, compound **8** (Fig. 9.5), which belongs to the Ads, was almost equally effective against resistant and susceptible enzymes (Table 9.5) of green foxtail (*Setaria veridis*) and goosegrass (*Eleusine indica*). Additionally, this AD had similar I_{50}s on ACC from wild-type (S4) and sethoxydim-resistant biotypes of maize.

9.4
Pinoxaden

9.4.1
Characteristics

Pinoxaden is a new graminicide for cereal crops developed by Syngenta [7]. It belongs to the AD chemical class. Table 9.6 summarizes its physicochemical and toxicological data.

9.4.2
Technical Synthesis

A convergent synthetic route (Scheme 9.5) leads to the intermediate NOA 407854, which is esterified in the last step with pivaloyl chloride.

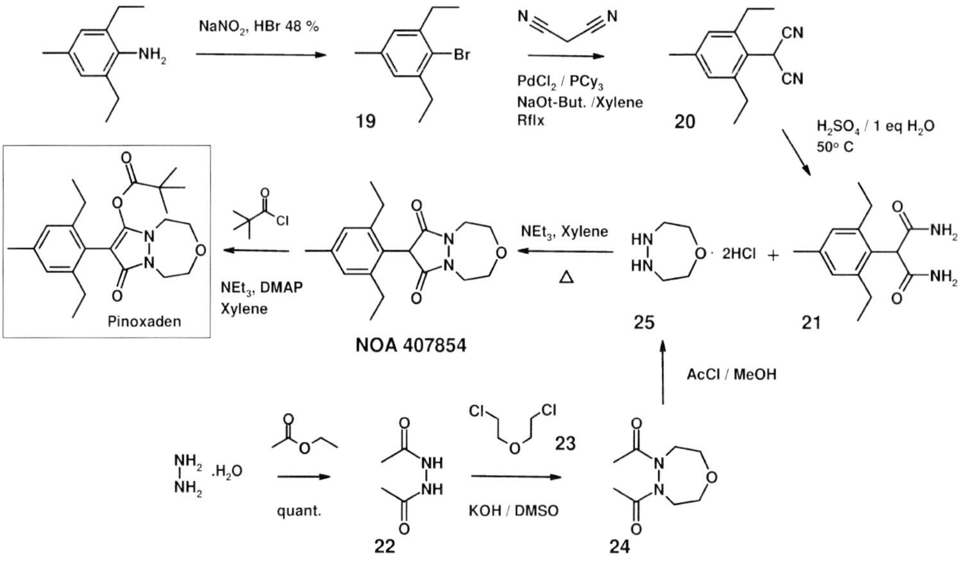

Scheme 9.5

Table 9.6 Properties of pinoxaden.

Physicochemical properties

Common name	Pinoxaden (provisionally approved by ISO)
Company code	NOA 407855
Melting point	121 °C
Partition coefficient octanol–water	Log P_{OW} = 3.2
Solubility	200 mg L^{-1} in water
Vapour pressure	4.6×10^{-7} Pa

Toxicological profile

Rat: Acute oral LD_{50} (mg kg^{-1} bw)	>5000
Rat: Acute dermal LD_{50} (mg kg^{-1} bw)	>2000
Rat: Inhalation LC_{50} (mg L^{-1})	5.2
Skin and eye irritation	Irritant

Ecotoxicological & environmental profile

Birds: acute LD_{50} (mg kg^{-1} bw)	>2250: negligible risk to birds
Earthworms: LC_{50} (mg kg^{-1} dry soil)	>1000: non-toxic
Bees: LD_{50} (µg per bee; contact)	>100: safe to bees
Aquatic organisms	No risk to algae, fish and daphnia
Non-target flora and fauna	No risk to dicot plants and no adverse effects against beneficial arthropods

The precursor aryl-malonamide **21** is prepared in a three-step procedure from 2,6-diethyl-toluidine. A technically feasible cross-coupling reaction has been developed for the synthesis of aryl malononitrile **20** starting from benzene derivative **19** and malononitrile. The optimized procedure with $PdCl_2$/tricyclo-hexylphosphine and sodium *tert*-butoxide as base in refluxing xylene [81] was improved even further using palladium dichloride/triphenylphosphine as catalyst and sodium hydroxide as base in 1-methyl-2-pyrrolidone at 125–130 °C [82]. The aryl-malononitrile **20** is hydrolyzed to the aryl-malonamide **21** in conc. sulphuric acid.

The [1,4,5]oxadiazepane dihydrochloride **25** is obtained in three steps. Reflux-ing hydrazine hydrate in ethyl acetate [83] generates N,N′-diacylhydrazine **22**. A cyclocondensation with ether **23** in DMSO [84] followed by the acid-catalyzed hy-drolysis of **24** provides the oxadiazepane **25**.

The pyrazoline-dione NOA 407854 is prepared by refluxing aryl-malonamide **21** with oxadiazepane **25** and triethylamine in xylene [85].

9.4.3
Biology

Pinoxaden is applied post-emergence at use rates of 30–60 g-a.i. ha^{-1} [7]. Interplay of the active compound with a safener proves essential to maximize the tolerance [7]. Methyl oleate as adjuvant enhances the level of activity without impairing the crop safety [86].

Pinoxaden is applied flexibly from the two-leaf up to the flag leaf stage of grasses [7]. Its weed spectrum covers a wide range of key annual grass species like *Alopecurus myosuroides* (blackgrass), *Apera spica venti* (silky bent grass), *Avena* spp. (wild oats), *Lolium* spp. (ryegrass), *Phalaris* spp. (canary grass), *Setaria* spp. (foxtails) and other monocot weed species commonly found in cereals [7].

In an uptake experiment, over 90% of the radiolabeled pinoxaden was incorporated into the crops within 5 h when treatment solutions were applied in droplets to the adaxial leaf surface of two-leaved plants of barley, winter wheat or durum wheat. After 24 h, about 20% is translocated out of the treated leaf by basipetal movement below the treated area [87]. Cloquintocet does not affect the absorption or the movement of the herbicide within the crop.

While active against certain ACC-resistant biotypes, both target site and metabolic resistant, pinoxaden is not active on all of them [7].

9.4.4
Metabolism and Selectivity

The total radioactive residues in winter wheat treated in autumn applications under out-door conditions declined rapidly in forage from 6.7 mg kg^{-1} on day 1 to 0.3 mg kg^{-1} 14 days after treatment (DAT). Ultimately, the total residues in grain, husks and straw at maturity were low. Scheme 9.6 gives the major detected metabolites and a proposed metabolic pathway [88].

Pinoxaden is hydrolyzed within a very short time to the parent acid, which is rapidly hydroxylated to the major metabolite found in plants, SYN 505164. The benzylic alcohol is oxidized to a large extent to the acid SYN 502836 or glycosylated and further conjugated. NOA 447204, which is the primary and main metabolite in soils, is also found in plants at lower levels. It is hydroxylated to SYN 505887. All the metabolites, except the parent acid NOA 407854 were inactive *in vitro* tests on plastidic wheat ACC and did not have any phytotoxic effect on emerged grasses and cereals in greenhouse trials, even at higher rates.

The effect of the safener cloquintocet-mexyl on the biokinetics and metabolism of pinoxaden in barley, winter and durum wheats, *Avena fatua* and *Lolium rigidum* was studied with the radiolabeled herbicide [89]. Safening is achieved by enhancing the metabolism of pinoxaden within the crop (Table 9.7).

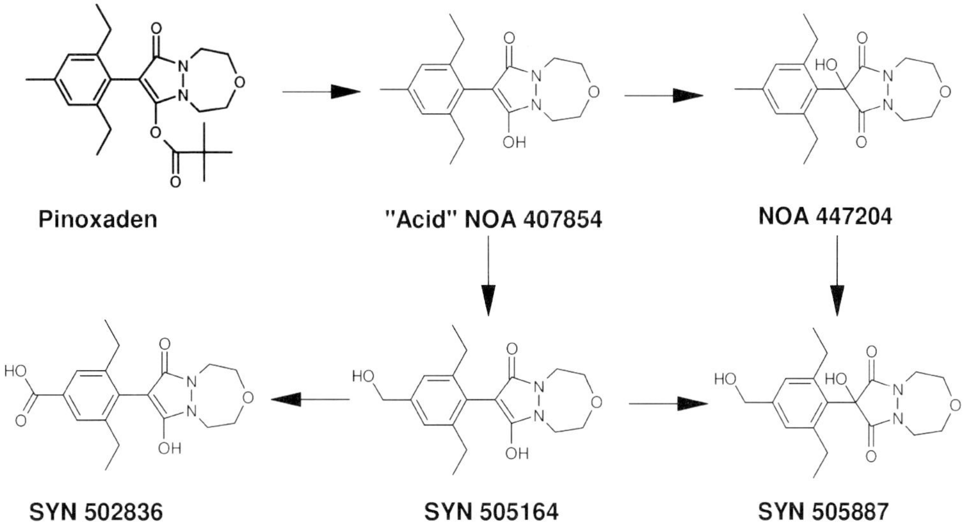

Scheme 9.6

The safener mainly triggers the hydroxylation of the methyl group of NOA 407854 to SYN 505164 in all cereal crops, but does not seem to affect the hydroxylation of the dione to NOA 447204. Cloquintocet has no relevant effect on the metabolism of pinoxaden in the grass weed *Lolium rigidum* or in *Avena fatua*.

Table 9.7 Pinoxaden was applied at a rate equivalent to 90 g ha^{-1} with the adjuvant A12127 used at 0.5%. Leaves treated with 20 × 0.2 mL droplets containing ^{14}C-labeled pinoxaden at 4000 dps. Treatments were made up with or without cloquintocet-mexyl (S) added at 25% the rate of the herbicide (H). Barley (*cv Manitou*), Winter wheat (*cv Soisson*), Durum wheat (*cv Colossea*).

	Barley		Winter wheat		Durum wheat		Lolium rigidum		Avena fatua	
	H & S	H	H & S	H	H & S	H	H & S	H	H & S	H
Pinoxaden	1.5	2.1	0	0	0.8	0.3	0	0	0	0
SYN407854	32.3	48.3	36	81.6	48.8	79.1	84.2	86.6	84.7	90.4
Total parent	33.8	50.4	36	81.6	49.7	79.4	84.2	86.6	84.7	90.4
SYN505164	38.3	25.5	59.8	13.8	35.6	14.7	3.8	3.5	11.1	7.1
SYN447204	8.3	8.7	4.2	4.1	5.8	3.1	4.2	3.5	3.2	2.5
Others	19.6	15.4	0	0.5	9	2.8	7.8	6.4	1	0
All metabolites	66.2	49.6	59.8	18.4	50.4	20.6	15.8	13.4	15.3	9.6

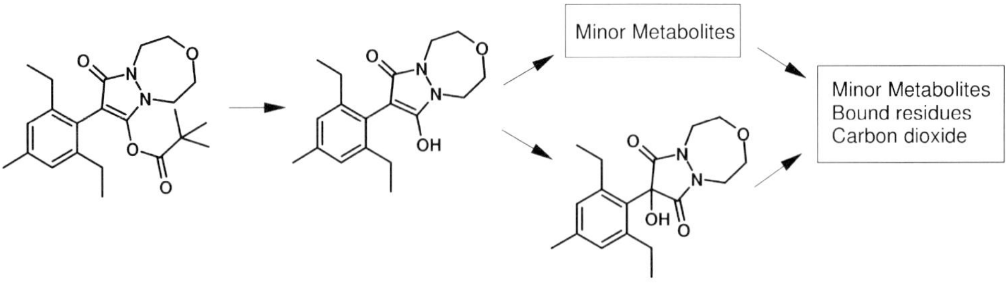

Minor Metabolites

Minor Metabolites
Bound residues
Carbon dioxide

Pinoxaden		NOA 407854	pH 5	pH 7.4	NOA 447204	
log P$_{ow}$	3,2	log P$_{ow}$	0,62	- 1.1	log P$_{ow}$	1,8
Water sol.	0.200 g / L	Water sol.	5.2 g / L	380 g / L	Water sol.	0.370 g / L
		pKa	3.82			

Fig. 9.10. Soil metabolism of pinoxaden.

Figure 9.10 summarizes the soil metabolism of pinoxaden [88].

Pinoxaden hydrolyses very rapidly in soil to NOA 407854 with half-lives below one day under aerobic, aerobic–anaerobic, and sterile–aerobic conditions.

NOA 407854 is highly soluble in water at neutral pH, but is rapidly hydroxylated to NOA 447204, which is almost insoluble. The half-life of NOA 407854 varies from 1.8 to 6.1 days, depending on the type of soil, whereas NOA 447204 degrades with a half-life of 6.2 to 37 days.

Diverse minor metabolites accounted for less than 5% of the applied radioactivity. The bound residues reach a maximum of 49% after 14 to 30 days.

The only identified volatile metabolite was carbon dioxide, demonstrating mineralization. Up to 47.6% of the applied radioactivity was mineralized after 100 days in laboratory soil metabolism studies. Finally, minimal dissipation has been observed in soils.

9.5
Summary and Outlook

Three classes of commercial herbicides, the AOPP, the CHD and the newly discovered AD derivatives, inhibit the CT function of the eukaryotic ACC found in the plastids of grasses [3, 4]. Interestingly, ACC is a target site of the novel insecticides spirodiclofen and spiromesifen [90, 91], which also belong to the chemical class of AD.

Progress in function elucidations of ACC inhibitors has largely contributed to the understanding and the differentiation of resistance mechanisms [30].

Considerable effort has been undertaken in recent years to elucidate the mode of action of herbicides on ACC at the molecular level. Point mutations in the chloroplastic ACC CT domain of different monocots, principally *Lolium rigidum*, have been correlated with resistant phenotypes [44, 92, 93]. Fine mapping of these mutations in the sequence has led to the identification of those amino acids important for herbicide action in the CT domain. At the same time, the CT domain of the protein, spanning the domain where the point mutations have been found, has been crystallized [49].

Taken together, both approaches have allowed the prediction of the herbicide binding to the CT domain. In particular, it was possible to determine amino acid changes responsible for herbicide resistance to AOPP and/or CHD analogues and localize the amino acid directly involved in the binding of herbicides, but only for this domain [50].

AOPPs were shown to bind inside the active site cavity of ACC CT dimers. Binding of one AOPP molecule inside one of the two active sites of a CT dimer caused conformational changes in the structure of the whole dimer. The general binding mode of AOPP and CHD inhibitors inside the CT active site cavity is very likely to be similar for both plastidic and cytosolic isoforms given the conservation among homomeric ACC proteins. However, the emergence of a new chemical class such as the AD shows differences, which have led to the inhibition of the dicot cytosolic enzyme and the opening of new paths for research.

References

1 K. Hirai, in *Herbicide Classes in Development*, pp 234–238, P. Böger, K. Wakabayashi, K. Hirai (Eds), Springer-Verlag, Berlin-Heidelberg, **2002**.

2 B. Hock, C. Fedtke, R. R. Schmidt, *Herbizide*, G. Thieme Verlag, Stuttgart, **1995**.

3 M. D. Devine, in *Herbicide Classes in Development*, pp 103–113, P. Böger, K. Wakabayashi, K. Hirai (Eds), Springer-Verlag, Berlin-Heidelberg, **2002**.

4 C. Alban, P. Balder, R. Douce, *Biochem. J.* **1994**, 300, 557–565.

5 B. J. Nicolau, J. B. Ohlrogge, E. S. Wurtele, *Arch. Biochem. Biophys.* **2003**, 414, 211–222.

6 P. Babenzinski, R. Fischer, *Pestic. Sci.* **1991**, 33, 455–466.

7 U. Hofer, M. Muehlebach, S. Hole, A. Zoschke, *J. Plant Diseases Prot.* Special Issue, **2006**, XX, 989–995.

8 J. B. Ohlrogge, J. G. Jaworski, *Annu. Rev. Plant Physiol. Mol. Biol.* **1997**, 48, 109–136.

9 B. J. Incledon, J. C. Hall, *Pest. Biochem. Physiol.* **1997**, 57, 255–271.

10 Y. Sasaki, Y. Nagano, *Biosci. Biotechnol. Biochem.* **2004**, 68, 1175–1184.

11 T. Konishi, K. Shinohara, K. Yamada, Y. Sasaki, *Plant Cell Physiol.* **1996**, 37, 117–122.

12 T. Nikolskaya, O. Zagnitko, G. Tevzadze, R. Haselkorn, P. Gornicki, *Proc. Natl. Acad. Sci. U.S.A.* **1999**, 96, 14647–14651.

13 P. Gornicki, J. Faris, I. King, J. Podkowinski, B. Gill, R. Haselkorn, *Proc. Natl. Acad. Sci. U.S.A.* **1997**, 94, 14179–14185.

14 M. Egli, S. Lutz, D. Somers, B. Gengenbach, *Plant Physiol.* **1995**, 108, 1299–1300.

15 W. Schulte, R. Topfer, R. Stracke, J. Schell, N. Martini, *Proc. Natl. Acad. Sci. U.S.A.* **1997**, 94, 3465–3470.

16 M. J. MacDonald, L. A. Fahien, L. J. Brown, N. M. Hasan, J. D. Buss, M. A. Kendrick, *Am. J. Physiol. Endocrinol. Metab.* **2005**, 288, E1–E15.

17 U. Hoja, S. Marthol, J. Hofmann, S. Stegner, R. Schulz, S. Meier, E. Greiner, E. Schweizer, *J. Biol. Chem.* **2004**, 279, 21779–21786.

18 M. Focke, E. Gieringer, S. Schwan, L. Jansch, S. Binder, H. P. Braun, *Plant Physiol.* **2003**, 133, 875–884.

19 A. Kosaki, K. Mayumi, Y. Sasaki, *J. Biol. Chem.* **2001**, 276, 39919–39925.

20 L. J. Savage, J. B. Ohlrogge, *Plant J.* **1999**, 18, 521–552.

21 J. Ha, S. Daniel, I. S. Kong, C. K. Park, H. J. Tae, K. H. Kim, *Eur. J. Biochem.* **1994**, 219, 297–306.

22 H. H. Hoppe, *Z. Pflanzenphysiol.* **1981**, 102, 189–197.

23 J. D. Burton, J. W. Gronwald, D. A. Somers, J. A. Conelly, B. G. Gengenbach, D. L. Wyse, *Biochem. Biophys, Res. Commun.* **1987**, 148, 1039–1044.

24 M. Focke, H. R. Lichtenthaler, *Z. Naturforsch.* **1987**, 42c, 1361–1363.

25 T. Konishi, Y. Sasaki, *Proc. Natl. Acad. Sci. U.S.A.* **1994**, 91, 3598–3601.

26 C. Alban, P. Baldet, R. Douce, *Biochem. J.* **1994**, 300, 557–565.

27 M. D. Devine, R. H. Shimabukuro, in *Herbicide Resistance in Plants*, pp 141–169, S. B. Powles, J. A. M. Holtum (Eds), FL. CRC Press, Boca Raton, **1994**.

28 A. R. Rendina, A. C. Craig-Kennard, J. D. Beaudoin, M. K. Breen, *J. Agric. Food Chem.* **1990**, 38, 1282–1287.

29 J. D. Burton, J. W. Gronwald, R. A. Keith, D. A. Somers, B. G. Gengenbach, D. L. Wyse, *Pestic. Biochem. Physiol.* **1991**, 39, 100–109.

30 C. Délye, *Weed Sci.* **2005**, 53, 728–746.

31 M. Joachimiak, G. Tevadze, J. Podkowinski, R. Haselkorn, P. Gornicki, *Proc. Natl. Acad. Sci. U.S.A.* **1997**, 94, 9990–9995.

32 D. Porter, M. Kopec, U. Hofer, *WSSA Abstracts*, **2005**, 95.

33 M. D. Devine, C. V. Eberlein, in *Herbicide Activity: Toxicology, Biochemistry and Molecular Biology*, pp 159–185, R. M. Roe, J. D. Burton, R. J. Kuhr (Eds), IOS Press, Amsterdam, **1997**.

34 I. M. Heap, in *International Survey of Resistant Weeds*, www.weedresearch. com, assessed March, **2006**.

35 L. L. Van Eerd, R. E. Hoagland, R. M. Zablotowicz, J. C. Hall, *Weed Sci.* **2003**, 51, 472–495.

36 K. Kreuz, R. Tommasini, E. Martinoia, *Plant Physiol.* **1996**, 111, 349–353.

37 M. D. Devine, Acetyl-CoA Carboxylase, in *Inhibitors in Herbicide Classes in Development*, pp104–113, P. Boeger, K. Wakabayashi, K. Hirai (Eds), Springer-Verlag, Berlin-Heidelberg, **2002**.

38 Ch. Délye, *Weed Sci.* **2005**, 53, 728–746.

39 C. Délye, A. Matéjicek, J. Gasquez, *Pestic. Manag. Sci.* **2002b**, 58, 474–478.

40 S. R. Moss, K. M. Cocker, A. C. Brown, L. Hall, L. M. Field, *Pestic. Manag. Sci.* **2003**, 59, 190–201.

41 M. J. Christoffers, M. L. Berg, C. G. Messersmith, *Genome*, **2002**, 45, 1049–1056.

42 C. Délye, T. Wang, H. Darmency, *Planta* **2002**, 214, 421–427.

43 C. Délye, A. Matéjicek, J. Gasquez, *Pest Manag. Sci.* **2002**, 58, 474–478.

44 C. Délye, X.-Q. Zhang, S. Michel, A. Matéjicek, S. B. Powles, *Plant Physiol.* **2005**, 137, 794–806.

45 C. Délye, X.-Q. Zhang, C. Chalopin, S. Michel, S. B. Powles, *Plant Physiol.* **2003**, 132, 1716–1723.

46 C. Délye, C. E. Calmes, A. Matéjicek, *Theor. Appl. Genetics* **2002**, 104, 1114–1120.

47 O. Zagnitko, J. Jelenska, G. Tevzadze, R. Haselkorn, P. Gornicki, *Proc. Natl. Acad. Sci. U.S.A.* **2001**, 98, 6617–6622.

48 C. Délye, X. Q. Zhang, S. Michel, A. Matéjicek, S. B. Powles, *Plant Physiol.* **2005**, 137, 794–806.

49 H. Zhang, Z. Yang, Y. Shen, L. Tong, *Science*, **2003**, 299, 2064–2067.

50 H. Zhang, B. Tweel, L. Tong, *Proc. Natl. Acad. Sci. U.S.A.*, **2004**, 101, 5910–5915.

51 P. Boutsalis, WO 98/55860, **1998**.

52 P. Boutsalis, *Weed Technol.* **2001**, 15(2), 257–263.

53 A. Letsouzé, J. Gasquez, *Weed Res.* **2000**, 40, 151–162.

54 J. Richter, S. B. Powles, *Plant Physiol.* **1993**, 102, 1037–1041.

55 N. Balgheim, J. Wagner, K. Hurle, P. Ruiz-Santaella, R. De Prado, *Congress Proceedings – BPCP Int. Congress: Crop Science & Technology*, Glasgow, UK, **2005**, 1, 157–162.

56 S. Kaundun, J. D. Windass, *Weed Res.* **2006**, 46, 34–39.

57 J. Faris, A. Sirikhachornkit, R. Haselkorn, B. Gill, P. Gornicki, *Mol. Biol. Evol.* **2001**, 18, 1720–1733.

58 Th. N. Wheeler, ZA 78174, **1979**.

59 R. Fischer, A. Krebs, M. Albrecht, H. J. Santel, R. R. Schmidt, K. Luerssen, H. Hagemann, B. Becker, K. Schaller, W. Stendel, EP 0 355 599, **1990**.

60 F. Cederbaum, H. G. Brunner, M. Boeger, WO 92/16510, **1992**.

61 B. W. Krueger, R. Fischer, H. J. Bertram, T. Bretschneider, S. Boehm, A. Krebs, T. Schenke, H. J. Santel, K. Luerssen, DE 41 09 208, **1992**.

62 R. Fischer et al. EP 0 456 063, **1991**.

63 R. Fischer et al. EP 0 596 268, **1993**.

64 R. Fischer et al. EP 0 528 156, **1993**.

65 R. Fischer et. al. WO 96/03366, **1996**.

66 R. Fischer et al. WO 03/48138, **2003**.

67 T. Bretschneider, R. Fischer, J. Benet-Buchholz, *Pflanz.-Nachrichten Bayer*, **58/2005**, 307–318.

68 Th. Maetzke, A. Stoller, S. Wendeborn, H. Szczepanski, WO 0117972, **2001**.

69 M. Muehlebach, personal communication.

70 M. Muehlebach, J. Glock, Th. Maetzke, A. Stoller, WO 99/47525, **1999**.

71 S. Nakanishi, S. Butler, *Org. Prep. Proced.* **1975**, 7, 155–158.

72 M. Muehlebach, J. Wenger, personal communication.

73 R. Fischer et al. WO 05/92897, **2005**.

74 R. Fischer et al., WO 06/355, **2006**.

75 M. Muehlebach, et al., to be published.

76 Th. Nidermann, personal communication.

77 D. Herbert, K. A. Walker, L. J. Price, D. J. Cole, K. E. Pallet, S. M. Riley, J. L. Harwood, *Pesict. Sci.* **1997**, 50, 67–71.

78 Th. Nidermann, in preparation.

79 S. Huang, A. Sirikhachornkit, X. Su, J. D. Faris, B. S. Gill, R. Haselkorn, P. Gornicki, *Plant Mol. Biol.* **2002**, 48, 805–820.

80 A. Shukla, C. Nycholat, M. V. Subbramanian, R. J. Anderson, M. Devine, *J. Agric. Food Chem.* **2004**, 54, 5144–5150.

81 A. Schnyder, WO 00/78712, **2000**.

82 M. Zeller, WO 04/050607, **2004**.

83 T. Maetzke, A. Stoller, S. Wendeborn, H. Szcepansky, WO 01/017973, **2001**.

84 B. Jau, M. Parak, WO 2003/051853, **2003**.

85 T. Maetzke, R. Mutti, H. Szczepanski, WO 2000/078881, **2000**.

86 J. Glock, A. Friedmann, D. Cornes, WO 01/017352, **2001**.

87 G. Hall, personal communication.

88 Extracted from submitted data sets to Registration Authorities.

89 G. J. Hall, P. Carter, A. Burridge, to be published.

90 R. Fischer, E. M. Franken, R. Nauen, U. Teuschel, WO 02/48321, **2002**.

91 R. Nauen, H.-J. Schnorbach, A. Elbert, *Pflanz.-Nachrichten Bayer* **58/2005**, 3, 417–440.

92 M. J. Christoffers, M. L. Berg, C. G. Messersmith, *Genome*, **2002**, 45, 1049–1056.

93 X. Q. Zhang, S. B. Powles, *Planta*, **2006**, 223, 550–555.

10
Photosynthesis Inhibitors: Regulatory Aspects, Reregistration in Europe, Market Trends and New Products

Karl-Wilhelm Münks and Klaus-Helmut Müller

10.1
Introduction

Herbicides acting as inhibitors of photosynthesis by blocking of electron transport in photosystem II belong to the eldest classes of plant protection agents. These compounds are still of market relevance, especially in developing countries, but they are out of the focus of modern herbicide research due to their high application rates in response to the high enzyme concentration for photosynthesis in plants and their cross-resistance behavior.

Photosynthesis inhibitors are divided into the compound classes of triazines, triazinones, the newest one triazolinones (see amicarbazone, Section 10.6.1), uracils and phenylcarbamates belonging to the C 1 group of HRAC classification scheme, the arylureas and amides belonging to the C 2 group, and the nitriles, benzothiadiazinones and phenylpyridazines in the C 3 group of the HRAC classification [1]. Photosynthesis, which takes place in the chloroplasts, was already recognized as the principle of "assimilation of carbon dioxide" by plants in the mid-19th century but the individual reaction steps were evaluated and well understood with the research of Hill in 1937 and starting in the 1950s [2, 3], mainly via the investigations with these inhibitors, especially the ureas [4–7], the triazines [8–10] and the triazinones [11–13] between 1956 and 1975. Already in 1961, M. Calvin, University of California, had won the Nobel Prize in chemistry "for his research on the carbon dioxide assimilation in plants" and investigations of the light-dark reactions in photosynthesis [14] and the synthesis of carbohydrates from CO_2. But also the modern protein structure chemistry and the investigations via X-ray with the description of binding niches and inhibitors binding in it were started with the ubichinon binding pocket in the photosynthesis and led to the Nobel prize for chemistry in 1988 being awarded to H. Deisenhofer, R. Huber and H. Michel [15].

Since the Hill reaction (Scheme 10.1) permits the quantitative determination of the inhibitory properties of photosynthesis blockers on chloroplast systems by measurement of O_2 evolution (oxygen electrode, Warburg manometer) and thus

Modern Crop Protection Compounds. Edited by W. Krämer and U. Schirmer
Copyright © 2007 WILEY-VCH Verlag GmbH & Co. KGaA, Weinheim
ISBN: 978-3-527-31496-6

$$H_2O + A \longrightarrow AH_2 + 1/2\,O_2$$

Scheme 10.1. Hill reaction scheme.

the 50% inhibitory concentration of a photosynthesis inhibitor in the Hill reaction, these values (in their negative logarithm as pI_{50} values) can be used in quantitative structure–activity studies (QSAR) regarding the *in vitro* activity ("QSAR; Hansch approach" [16, 17, 20–22]) and can be compared with their pI_{50} values in greenhouse trials to evaluate biochemical activities versus biological activities [describing and including transport, membrane and metabolism effects (ADME)]. Consequently, QSAR, as a method to improve the biological activities in synthesis programs for inventing new crop protection compounds, was investigated broadly and approved first in photosynthesis research programs [18, 19].

The 1,3,5-triazines were invented first in the Geigy laboratories [23, 24], with simazine (1955) as the first representative of this group followed by atrazine (Geigy, 1958), propazine (Geigy, 1960), trietazine (Geigy, 1960), terbutylazine (Geigy, 1966) and cyanazine (Shell, 1971) from the 2,6-diamino-4-chloro-1,3,5-triazines, prometryne (Geigy, 1962), ametryn (Geigy, 1964), desmetryne (Geigy, 1964) and terbutryne (Geigy, 1966) from the 2,6-diamino-4-methylmercapto-1,3,5-triazines, and terbumeton (Geigy, 1966) from the 2,6-diamino-4-methoxy-1,3,5-triazines.

The 1,2,4-triazinones [25], metamitron (Bayer, 1975) and metribuzin (Bayer, Du Pont, 1971) and the 1,3,5-triazine-2,4(1H,3H)-dione [26] hexazinone (Du Pont, 1975), resulted first from the resynthesis of university publications or analogue synthesis using uracils as starting ideas.

The uracils [27] bromacil (Du Pont, 1952), lenacil (Du Pont, 1974) and terbacil (Du Pont, 1966) were invented in the Du Pont laboratories, whereas the pyridazinones [28, 29], pyrazon (BASF, 1962) came out from research investigations in BASF and Sandoz laboratories. The phenylcarbamates [32] desmedipham and phenmedipham invented by Schering AG are also included in the C1 group (HRAC classification) (Fig. 10.1).

The herbicidal effect of aryl- and hetarylurea, systematically studied starting from first observations in 1946, was improved between 1951 [30] and 1973 [31]. From this chemistry today the compounds chloroxuron (Ciba, 1960), dimefuron (Hoechst, 1969), diuron (Du Pont, 1954), ethidimuron (Bayer, 1973), fenuron (Du Pont, 1957), fluometuron (Ciba, 1960), isoproturon (Hoechst, 1974), linuron (Hoechst, 1960), methabenzthiazuron (Bayer, 1968), metobromuron (Ciba, 1963), metoxuron (Sandoz, 1968), monolinuron (Hoechst, 1958), neburon (Du Pont, 1957), siduron (Du Pont, 1964) and tebuthiuron (Elanco, 1973) are still used.

The amides propanile [33] and pentanochlor [34], also belonging to the C2 group (HRAC classification) fulfill the general formula for photosynthesis inhibitors bearing an CONH group (Fig. 10.2).

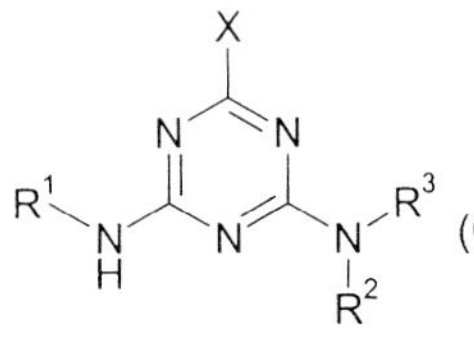

Fig. 10.1. PS II inhibitors, C1 group.

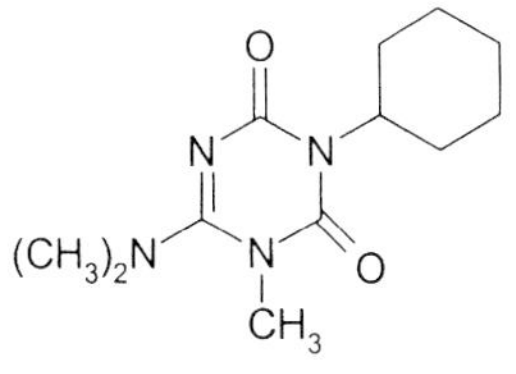

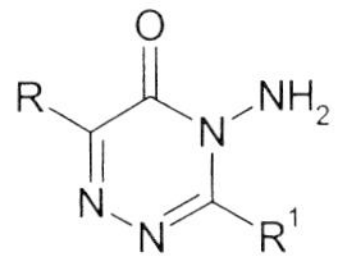

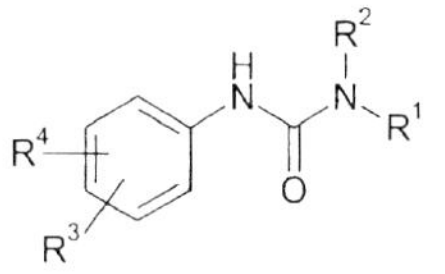

a) $R^1 = CH_3$, $R^2 = H$, CH_3

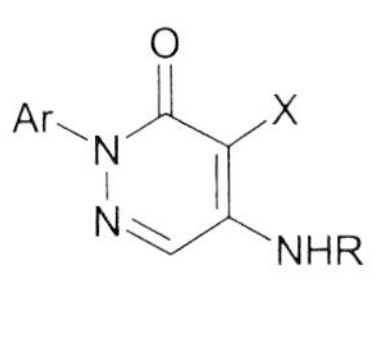

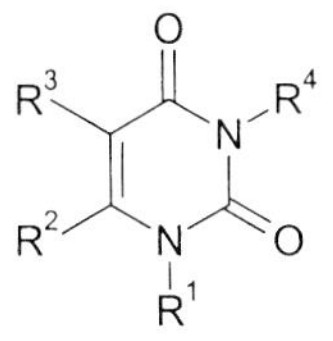

b) $R^1 = CH_3$, $R^2 = OCH_3$

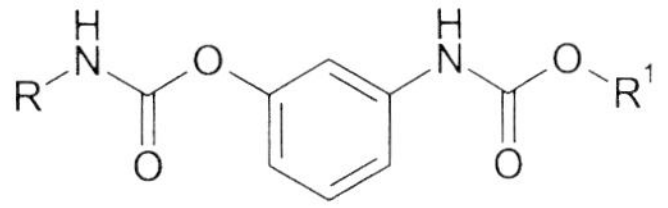

arylureas

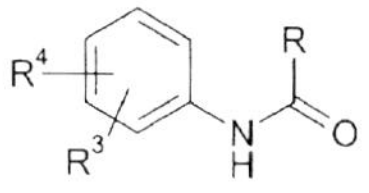

R = alkyl

R^3, R^4 = halogen, alkyl

amides

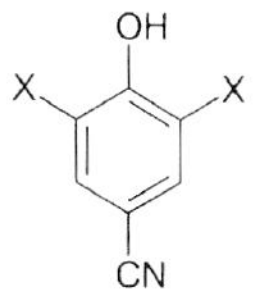

X = I, Br

nitriles

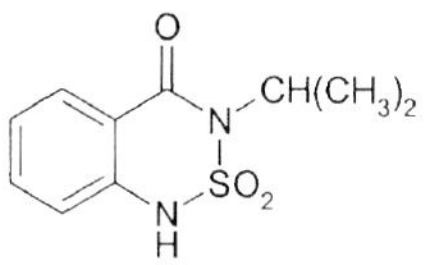

bentazone

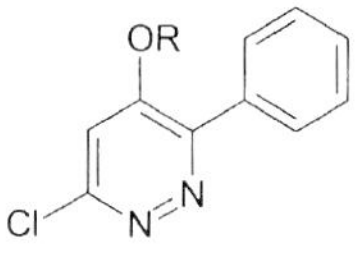

a) $R = COSC_8H_{17}$

b) $R = H$

phenylpyridazines

Fig. 10.2. PS II inhibitors, C2 and C3 group.

The photosynthesis inhibitors of the C3 group (HRAC classification), the nitriles [35] bromofenoxim, bromoxynil, ioxynil, as well as the benzothiadiazinone bentazone [36] and the phenylpyridazines [37] pyridate and pyridafol have completely different structures, without any CONH group (Fig. 10.2).

Whereas photosynthesis inhibitors represented nearly 50% of market share of all herbicides in 1980 [38] the situation has significantly changed, not only by the introduction of the newer ALS-inhibitors like the sulfonylureas, HPPD-inhibitors and the genetically modified crops resistant against EPSP-synthase and glutamine synthetase inhibitors but also through significant changes in reregistration requirements, especially in Europe.

10.2
The Reregistration Process in the European Union

The registration of agrochemicals falls under national laws of all the countries throughout the world were plant protection compounds are used. These national laws regulate the data requirements for active compounds as well as for formulations, mixtures etc., the risk assessment process and requirements for labeling the marketed plant protection product. Early on in the history of agrochemicals the companies inventing, developing and marketing plant protection compounds and products as well as the public were looking for harmonisation of data requirements and risk assessment for registration. Examples of supranational harmonisation activities are given in Tables 10.1–10.4.

Additionally, global harmonisation endeavors are undertaken by the FAO and WHO. The FAO supports harmonisation efforts, e.g., through the information system "Prior Informed Consent" (PIC). In this information system an exchange on certain hazardous pesticides and industrial chemicals in international trade takes place between member authorities. The members have agreed on an international code of conduct on the distribution and use of pesticides and on guidelines related to the development and evaluation of data considered in the registration process. Further, WHO (World Health Organisation) organizes joint meetings of their members together with the WHO on pesticide residues (JMPR) to define and organize the MRL Database on Pesticides, in which the maximum pesticide residue levels are documented. The WHO has developed the pesticides evaluation scheme "WHOPES" in which it establishes and publishes specifications for technical material and related formulations of public health pesticides. WHO reviews safety reports, issues, guidelines for laboratory and field evaluation of insecticides and repellents and gives recommendations on equipment and application manuals. It publishes health criteria (EHC) monographs on chemicals/pesticides, e.g., the WHO Classification of Pesticides by Hazard and the WHO/FAO Pesticide Datasheets (IPCS Inchem) [39].

The OECD published a vision document [on the occasion of the 14th meeting of the Working Group on Pesticides (WGP) held in Paris on 5th and 6th November 2002] with statements on achievements to-date in the international harmoni-

Table 10.1 Supranational Harmonisation Activities in EC, US and NAFTA.

Political union/ country	Responsible authority	Legislation	Object of registration	Time to registration
E.C./ countries of the E.C.	EU Commission, through the European Food Safety Authority (EFSA) National authorities of the different countries	Directives like Directive 91/414/ EEC, national laws like COPR, COP(A)R, PPPR, Deutsches Pflanzenschutzgesetz etc.	Active ingredients (a.i.) regulated by EEC Directives (adopted to national laws). Products regulated by national laws	Up to 4 years until Annex 1 inclusion for a new active substance
USA/States	EPA	Federal Insecticide, Fungicide, and Rodenticide Act (FIFRA); Food Quality Protection Act (FQPA; 1996)	A.i. and products. States may register a new end use product or an additional use of a federally registered pesticide product under specific conditions	Up to 3–4 years
NAFTA	Technical Working Group of Pesticides (TWG). US-EPA, Canadian Pest Management Regulatory Agency (PMRA), a consortium of Mexican agencies (CICOPLAFEST)	Common data submissions for manufacturers – electronic harmonisation. Joint reviews. Eliminating trade problems related to differences in MRL (maximum residue limits)	A.i. and products	Subject to national timelines

sation of the regulatory approaches for agricultural pesticides (chemical and biological) and in the use of work sharing arrangements in examining and reporting on data submissions (dossiers) provided by industry as well as the use of country evaluations (monographs) to support applications for their registration or the reregistration or to support the establishment of MRLs or import tolerances for particular active substances. It also published a statement of their vision for the next ten years, including details of the specific objectives, milestones to be reached along the way, and the indicators and measures of success to be used to record and document progress achieved [40].

Table 10.2 Supranational harmonisation activities in Central- and South-America.

Political union/country	Responsible authority	Legislation	Object of registration	Time to registration
Central America				
Belize, Costa Rica, El Salvador, Honduras, Mexico, Nicaragua, Panama	Technical Regional Pesticide Working Group OIRSA	Harmonized registration data and labeling requirements FAO specifications and Codex Alimentarius Data exchange on efficacy within region	Products	Up to 2 years
South America				
(a) Andean Community Bolivia, Columbia, Ecuador, Peru, Venezuela (b) Mercosur Argentina, Brazil, Paraguay and Uruguay, Chile and Bolivia	National authorities	Common Pesticide Registration Manual (July 2002) "Norma Andina para el Registro y Control de Plaguicidas Quimicos de Uso" Agricola-Decision 436. Comision Andina. Gaceta Oficial del Acuerdo de Cartagena Ano XIV-No. 347 Lima, 17 June 1998 (based on FAO principles) "International Code of conduct for the distribution and use of Pesticides"	Products	Up to 2 years

By working together, OECD governments and industry are "sharing the burden" of testing and assessing high production volume chemicals, pesticides and, most recently new chemicals". OECD programs on harmonisation are leading to exchange of documents used in reregistration and registration in OECD countries, beginning already in 1992, by comparing pesticide data reviews, by working out OECD databases on pesticide and biocide review schedules, by issuing guidance on the preparation of dossiers and monographs, by undertaking joint reviews on new compounds like, for example, Project "Cornelia" on Bayer's corn

Table 10.3 Supranational Harmonisation Activities in Asia.

Political union/ country	Responsible authority	Legislation	Object of registration	Time to registration
Asia				
(a) Japan	MAFF	National Specific data requirements and test protocols	A.i. and products	Up to 4 years
(b) P. R. China/ Vietnam	National	Harmonisation of MRLs	A.i. and products	Up to 2 years
(c) South Korea, other	National	National Efforts in harmonisation through the Regional Network on Pesticides in Asia and the Pacific (RENPAP)	A.i. and products	Up to 2 years
India	National CIBRC	National Data generated for Indian Registration v/s Data needs of most developing countries match very well	A.i. and products	Up to two years, incl. late fixation of MRL by Ministry of Health
Australia	National	National Comparable to EU requirements	A.i. and products	Up to 2 years
New Zealand	National	National Comparable to EU	A.i. and products	Up to 2 years

herbicide foramsulfuron (Joint review between US-EPA, Canadian PMRA and German BvL, 2000–2002), by surveying best practices in the regulation of pesticides in twelve OECD countries and by recommending the electronic protocols used for data submission. Progress in harmonisation of data requirements and test guidelines are also achieved through surveying test guideline program (TGP) priorities for pesticides, minimum data requirements for establishing MRLs and import tolerances, guidance notes for analysis and evaluation of chronic toxicity and carcinogenicity studies, etc. They stated the vision that by the end of 2014 the regulatory system for agricultural pesticides will have been harmonised to the extent that country data reviews (monographs) for pesticides

Table 10.4 Supranational harmonisation activities in Africa.

Political union/ country	Responsible authority	Legislation	Object of registration	Time to registration
Africa CSP (comite Sahelien des Pesticides) (Chad, Mali, Burkina Faso, Niger, Mauretania, Senegal, Cape Verde, Gambia and Guinea-Bissau)	CILSS		A.i. and products	Up to 2 years
SADC (Southern African Development Community) Angola, Botswana, Congo (DR), Lesotho, Malawi, Mauritius, Mozambique, Namibia, Seychelles, South Africa, Swaziland, Tanzania, Zambia and Zimbabwe		South Africa; Registration Act 36/ 1947 and Agricultural Remedies Registration Procedure Policy Document	A.i. and products	Up to 2 years

prepared in the OECD format on a national or regional basis (e.g., EU or NAFTA) can be used to support independent risk assessments and regulatory decisions made in other regions or countries.

In such a harmonisation process the EC enacted in 1991 the *"Council Directive of 15 July 1991 concerning the placing of plant protection products on the market"* *(91/414/EEC)*.

In this Directive the EC regulates the registration and reregistration of active ingredients and products for all countries in the EU. This Directive came into force on 26 July 1993 and must be implemented by national laws in all countries in the EU, e.g., in the UK by the Plant Protection Products Regulations 2003.

The main elements of the Directive are as follows:

- To harmonise the overall arrangements for authorization of plant protection products within the European Union.

This is achieved by harmonising the process for considering the safety of active substances at a European Community level by establishing agreed criteria for considering the safety of those products. Product authorization remains the responsibility of individual Member States.

The Directive provides for the establishment of a *positive list of active substances* (Annex I) that have been shown to be without unacceptable risk to humans or the environment.

New and existing active substances can be initially included to Annex I of the Directive for a period of 10 years pending their successfully passing the European Commission's (EC) review program.

Member States can only authorize the marketing and use of plant protection products after an active substance is listed in Annex I, except where transitional arrangements apply.

Before an active substance can be considered for inclusion in Annex I of Directive 91/414/EEC, companies must submit a complete data package (dossier) on both the active substance and at least one plant protection product containing that active substance. The data required is:

- Identification of an active substance and plant protection product.
- Description of their physical and chemical properties.
- Their effects on target pests.
- A comprehensive file of study reports to allow for a risk assessment to be made of any possible effects on workers, consumers, the environment and non-target plants and animals.

Detailed lists of the data required to be evaluated to satisfy inclusion in Annex I of the Directive, or the authorization of a plant protection product are set out in the Directive (Annexes II and III). Annex II data relate to the active substance and Annex III to the plant protection product. These data are submitted to one or more Member States for evaluation. A report of the evaluation is submitted to the European Food Safety Authority (EFSA). Following peer review of the report the EFSA makes a recommendation to the European Commission on whether Annex I inclusion is acceptable. This recommendation is then discussed by all Member States in the framework of the Standing Committee on the Food Chain and Animal Health (SCFA), previously the Standing Committee on Plant Health (SCPH). Where necessary, the Scientific Panel is consulted before the SCFA can deliver an opinion on whether an active substance should be included in Annex I of 91/414/EEC.

All member states are obliged to the "*Uniform Principles*".

The "Uniform Principles" (Annex VI of Directive 91/414/EEC) establishing common criteria for evaluating products at a national level were published on 27 September 1997 (OJ L265, p. 87). Application of the Uniform Principles ensures that authorizations issued in all Member States are assessed to the same standards.

The Directive states that all active ingredients should be reviewed periodically within 10 years.

This applies to all old agrochemical compounds (substances) used in a country of the EU prior to 1991 or before a country became a member of the EU (Reregistration). Thus all old photosynthesis inhibitors, for example, needed to be reviewed and the manufacturers had to apply for registration (listing on Annex I) by submitting dossiers prepared under the Directive 91/414/EEC [41]. A similar reregistration process was set from the US-EPA for all compounds on the market before 1984 in the US.

10.3
Main Changes in Guidelines regarding EU Registration

The following main changes have also to be applied on the preparing of registration data for such compounds, which have been registered in different countries based on dossiers regulated under the national laws before the reregistration procedure was enforced.

10.3.1
Good Laboratory Practice

Good Laboratory Practice (GLP) is concerned with the organizational process conditions under which studies are planned, performed, monitored, recorded; GLP ensures that the way the work is done is adequately standardized and of a sufficiently high quality to produce reliable results that can, with confidence, be compared with others carrying out the same work and applying the same general principles. Internationally accepted GLP guidelines, drawn up by the Organisation for Co-operation and Development, provided a reference point for later EU legislation [42]). The respective Directive applied to both active ingredients and formulated products, came into effect on 30 June 1988. The subsequent "Authorizations" Directive, 91/414/EEC, and others extended the scope of GLP, by requiring GLP compliance for all safety and efficacy studies whether conducted in the field or laboratory, and whether using formulated product or active ingredient.

10.3.2
Physical and Chemical Properties of Active Substance

Declaration of toxicological, ecotoxicological, or environmental significant *impurities* is needed. Especially hazardous chemicals like, for example, nitrosamines have to be declared and are regulated by a maximum admissible concentration. In this example, the total nitrosamine content of a pesticide formulation must not exceed 1 mg kg^{-1} of the active substance present.

Use and declaration of *analytical methods* have to be in correspondence with the "Technical Material and Preparations: Guidance" for generating and reporting methods of analysis in support of pre- and post-registration data requirements for Annex II (part A, Section 4) and Annex III (part A, Section 5). FAO guidelines (guidance on FAO specifications) and copies of specifications are available from www.fao.org/ag/agp/agpp/pesticid/ and GCPF (formally GIFAP) guidelines. Information on CIPAC can be obtained from www.cipac.org.

These guidelines apply to all studies started after *1st October 1999*.

Where an FAO specification is available for an active substance in a preparation, the tolerance limits must meet those in the FAO specification. However, where there is no appropriate FAO specification, tolerances must meet limits as accepted by the FAO Group of Experts [43, 44].

Where an active substance is present as an ester or a salt, the active substance content must be expressed as the amount of the ester or salt present (as the technical material) with a statement declaring the amount of the active principle.

The methods used for the determination of physical properties should be in accordance with the requirements of EC Directive 94/37/1.

10.3.3
Storage Stability

The data submitted must support the proposed shelf-life of the preparation. It is normally expected that a preparation should have a shelf-life of at least two years. Only where a preparation has a shelf-life of less than 2 years should the label include a "Use by . . ." date or other precautionary phrase.

Where a loss of $\geq 5\%$ of active substance occurs then the fate of the active substance must be addressed and the breakdown products identified [45].

10.3.4
Physical and Chemical Characteristics of Preparation

The physical and chemical characteristics of preparations via parameters (e.g., explosive properties, oxidizing properties, flashpoint and other indications of flammability, acidity/alkalinity and pH, surface tension, density, wettability, suspensibility, dilution stability, dry and wet sieve test, particle size distribution and other properties of the formulation) and the corresponding methods have to be determined and reported in detail [46].

Specific new tests on viscosity and surface tension are guided by the Commission Directive 98/98/EC of 15 December 1998.

10.3.5
Operator Exposure Data Requirements

New regulations to protect the applicant of the plant protection products were brought into force, regulating data requirements, experimental details for the measurement and model calculations [47, 48].

10.3.6
Residue Data Requirements

The guidance documents embrace also the following aspects:

- Metabolism and Distribution in Plants (Appendix A).
- General recommendations for the design, preparation and realization of residue trials (Appendix B).
- Testing of plant protection products in rotational crops (Appendix C).
- Comparability, extrapolation, group tolerances and Data requirements (Appendix D).
- Calculation of maximum residue levels and safety intervals (Appendix E).

These documents are available on the European Commission website, at http://europa.eu.int/comm/food/plant/protection/resources/publications_en.htm

10.3.7
Estimation of Dietary Intakes of Pesticides Residues

Estimates of pesticide intake need to be made to compare potential consumer dietary exposure with acceptable dietary intakes derived from toxicological studies. At its most basic level, if estimates of long- and short-term intake are less than the acceptable daily intake (ADI) and the acute reference dose (acute RfD), respectively, then the risks to the consumer may be regarded acceptable.

The Guidelines and Criteria for the Preparation and Presentation of Complete Dossiers and Summary Dossiers for the inclusion of Active Substances in Annex I of Directive 91/414/EEC (Article 5.3 and 8.2) (Document 1663/VI/94) require that an estimate is made regarding the theoretical intakes of pesticide residues by consumers. Consumer risk assessment is a vital part of the approval process and it is in the applicant's interest to estimate potential intakes since intake estimates can assist in assessing whether further information is required.

Two intake calculation models are now available, one for short-term (acute) intake calculations the other for long-term (chronic) intake calculations. These two models present updates of the previous versions, with new adult, vegetarian, elderly and more detailed child consumption data incorporated. As Excel spreadsheets, they are designed to be more user friendly than the previous versions and are available with accompanying guidance notes.

10.3.8
Fate and Behavior of Agricultural Pesticides in the Environment

The data provided by applicants must permit an assessment to be made of the fate and behavior of the pesticide in the environment. This information is subsequently used to assess the risk to non-target species (soil or aquatic organisms, plants, etc.) and following crops that will be exposed to the pesticide formulation, its active substance(s), and the metabolites, transformation and degradation products of the active substance(s). The information provided should therefore be sufficient to:

- Predict the distribution, fate and behavior of the pesticide in the environment, as well as the time courses involved, i.e., estimate the concentrations in soil, water and air and assess how these concentrations compare with any recognized limits or standards.
- In conjunction with other data, identify measures necessary to minimize contamination of the environment and impact on non-target species.
- In conjunction with other considerations, permit a decision to be made as to whether the pesticide can be approved, and the uses for which it can be approved.
- In conjunction with other data, classify the product as to risk.
- Specify relevant risk and safety phrases for the protection of the environment, which are to be included on labels.

As indicated above, the nature and amount of data required for pesticide approval depend on the properties and use of each active substance. A stepwise, tiered or triggered approach allows an efficient selection of tests essential to each individual contamination risk analysis. The environmental exposure to a pesticide depends primarily on the following factors.

Concentration of Chemical in the Relevant Environmental Compartment
The highest concentrations usually occur during and just after application.
 Following application, the concentration of residues declines due to:

- degradation
- movement into other compartments
- dilution.

Degradation can include such processes as hydrolysis, photolysis, microbial metabolism, etc. Movement reduces the concentration in the treated compartment but transports residues to untreated compartments, e.g., from plant surface to soil or soil to water.

Bioavailability of the Chemical
For substances entering surface waters, the availability of a chemical to organisms is primarily related to its concentration in the aqueous phase. When strongly adsorbed to sediment or soil, availability will often be significantly re-

duced. Under some circumstances it would also be necessary to consider exposure of organisms via the food chain or via the atmosphere.

Nature of the System or Organism
Exposure assessment for an organism requires information on such aspects as:

- Does it live in the treated area, or in an area to which the pesticide could be transported?
- Does it actively consume treated crops or become exposed via the dermal or inhalation routes etc?

To assess the risk of contamination of the environment or exposure of non-target organisms, the potential of the pesticide for movement through the environment must be addressed. For pesticides used in, on or over soils, a study of the pesticide's breakdown in soil is required. Similarly, for pesticides intended for use in or near water, or whose entry into water cannot be ruled out, information must be supplied on mobility in soil and on the fate in the aquatic environment, including natural water/sediment systems. Such requirements and any trigger values for performing different types of study are detailed in the data requirements guidance.

10.3.9
Specific Guidance regarding Water Limits according Annexes of the Authorizations Directive

Account must be taken of fate and behavior in relation to groundwater as well as surface water.

The behavior of any environmentally significant metabolites, transformation and degradation products of the pesticide, with significant potential to contaminate water or soil and cause harm to non-target organisms, must also be investigated. The EC Authorizations Directive stipulates that this requirement applies to those products formed from the pesticide active substance and occurring at levels above 10% of the added pesticide. It may be necessary to investigate such products formed at levels < 10% where they are known to have significant effects on target and or non-target organisms. The Annexes of the Directive and EU guidance documents should be referred to for more detailed guidance on this point. Unless otherwise stated, the term pesticide in this document will be deemed to include both the active substance and any significant metabolites, transformation and degradation products. However, in the context of the 0.1 μg L^{-1} drinking water limit being applied to groundwater, it is possible to make the case (with appropriate supporting data) that metabolites, transformation and degradation products are "not relevant" and this limit does not then apply [49, 50].

10.3.10
Ecotoxicology Requirements

The areas that need to be addressed are:

- Risk to birds and other terrestrial vertebrates;
- risk to aquatic life;
- risk to honeybees;
- risk to non-target arthropods;
- risk to earthworms;
- risk to soil microbial processes;
- risk to other soil macro-invertebrates (see above);
- risk to other non-target organisms (flora and fauna);
- risk to biological methods of sewage treatment (see above).

The guidance deals with each of these issues in turn. It addresses the basic data requirements and highlights appropriate risk assessment schemes and other sources of information that can be used in producing a good risk assessment. There may also be other references and information that can be used in support of the risk assessment. Notably, the use of such material should be scientifically justified.

EPPO Risk Assessment Schemes
The European and Mediterranean Plant Protection Organisation (EPPO) have produced several schemes that can be used to assess the risk to non-target organisms. These schemes aim to provide a basis for undertaking an appropriate risk assessment.

- Depending upon the proposed use pattern data are required on the acute, dietary and reproductive toxicity of an active substance and/or product to birds. Further details of when such studies are required are outlined in Annex II Section 8.1 and Annex III 10.1 and 10.3 of Directive 96/12/EC.
- Data are always required on the acute toxicity of an active substance to two fish species, Daphnia magna and algae.

If the active substance is a herbicide data are also required on an additional species of an alga as well as an aquatic plant. Full details of the appropriate studies are provided in Section 8.2 Directive 96/12/EC and the Aquatic Guidance Document.

Data are also required on the toxicity of the plant protection product . Further details on when these data are required can be found in the Aquatic Guidance Document, as well as Section 10.2.1 of Directive 96/12/EC.

- Depending on the persistence of the active substance in the water phase of a sediment water study, toxicity data are required to address the possible *chronic risk* of an active substance. Guidance on when these data are needed and on appropriate studies is provided in Section 8.2 and 10.2.4 of Directive 96/12/EC as well as in the Aquatic Guidance Document.
- Depending upon the partitioning and persistence of an active substance in the sediment phase of natural water sediment study, data may be required on its toxicity to sediment dwelling invertebrates. Details of when this study is required and choice of test method are given in the Aquatic Guidance Document. Information is also given in Section 8.2.7 of Directive 96/12/EC.
- Data are required on the bioconcentration potential of an active substance when the log P_{OW} is >3. Further details are given in Section 8.2.3 Directive 96/12/EC, together with the Aquatic Guidance Document.

Buffer zones and LERAPs

In certain instances it may be necessary for the product to have a buffer zone restriction added to the label to protect aquatic life.

Honeybee Risk Assessment

- Acute oral and contact toxicity tests are required in conjunction with a hazard quotient. Where the hazard quotient is greater than 50, further testing may be required. Details of the types of tests and calculation of the hazard quotient are given in Section 8.3 and 10.4 of Directive 96/12/EC. Guidance is also given in the Terrestrial Guidance Document.
- An appropriate risk assessment is required where the hazard quotient is >50, further testing may be required (see above).
- In certain cases, a bee brood feeding test may also be required. Reference should be made to Section 8.3.2 of Directive 96/12/EC and the Terrestrial Guidance Document.

Risk to Non-target Arthropods

The risk to non-target arthropods must be addressed, except where use is in situations where there is no exposure. Details of when the tests are not required are given in Section 8.3.2 of Directive 96/12/EC.

Initially, laboratory tests are undertaken, with further higher tier testing, e.g., extended laboratory tests, required if effects of >30% are seen. Tests are usually undertaken with a representative formulation of the active substance. Details of the tests required are given in Section 8.3.2 and 10.5 of Directive 96/12/EC.

Risk for Soil Non-target Microorganisms
Key guidance on the risk assessment for soil non-target microorganisms is given in the Guidance Document on Terrestrial Ecotoxicology. European Commission Working Document 2021/VI/98 describes the same risk assessment procedure as it is applied to *soil non-target macro-organisms (earthworms, beetles etc.)*

New MRL (Maximum Residue Levels) regulation for the European Union are being established [51]. Key features of the document are:

- Several foods will be subject to MRLs for the first time.
- It provides for MRL controls to be extended to animal feeds in the future.
- A default MRL of 0.01 mg kg^{-1} (set as a limit of determination) will apply to those commodities where no specific MRL is set, unless a different default level is agreed, or until such time as an MRL is set on the basis of the evaluation of data.

Annex I is necessary for the full implementation of controls under EC Regulation 396/2005, but the new commodity list will not be employed until the EC Regulation comes into force.

10.4
Situation of PS II Inhibitors in the EC Markets

The latest submission of data for an active substance being on the market two years after the Directive 91/414/EEC was published or an active substance that was on the market before 1 May 2004 in the Czech Republic, Estonia, Cyprus, Latvia, Lithuania, Hungary, Malta, Poland, Slovenia and Slovakia and which is not included in stages one to three of the program of work and which is not covered by Regulation (EC) No 1112/2002 was implemented inter alia by COMMISSION REGULATION (EC) No 2229/2004 of 3 December 2004, laying down further detailed rules for the implementation of the fourth stage of the work referred to in Article 8(2) of Council Directive 91/414/EEC at the *latest* by November 2005 [52].

The Directive 91/414/EEC stipulates according to article 5 for inclusion of an active substance in Annex I, the following shall be taken into particular account:

1. Where relevant, an acceptable daily intake (ADI) for man;
2. An acceptable operator exposure level if necessary;
3. Where relevant, an estimate of its fate and distribution in the environment as well as its impact on non-target species.

Table 10.5 EU listed PS II Inhibitors and Specific Provisions (Status November 2005).

Number	Common name, identification numbers	IUPAC name	Purity (1)	Entry into force	Expiration of inclusion	Specific provisions
11	Bentazone CAS No 25057-89-0 CIPAC No 366	3-Isopropyl-(1*H*)-2,1,3-benzothiadiazin-4(*3H*)-one-2,2-dioxide	960 g kg^{-1}	1.8.2001	31.7.2011	Only uses as herbicide may be authorized In their decision-making according to the uniform principles, Member States must pay particular attention to the protection of groundwater Date of Standing Committee on Plant Health at which the review report was finalized: 13.7.2000
16	Pyridate CAS No 55512-33-9 CIPAC No 447	6-Chloro-3-phenylpyridazin-4-yl S-octyl thiocarbonate	900 g kg^{-1}	1.1.2002	31.12.2011	Only uses as herbicide may be authorized For the implementation of the uniform principles of Annex VI, the conclusions of the review report on pyridate, and in particular Appendices I and II thereof, as finalized in the Standing Committee on Plant Health on 12 December 2000 shall be taken into account. In this overall assessment Member States: – must pay particular attention to the protection of groundwater – must pay particular attention to the potential impact on aquatic organisms and must ensure that the conditions of authorization include, where appropriate, risk mitigation measures

| 28 | Isoproturon CAS No 34123-59-6 CIPAC No 336 | 3-(4-Isopropylphenyl)-1,1-dimethylurea | 970 g kg^{-1} | 1 January 2003 | 31 December 2012 | Only uses as herbicide may be authorized For the implementation of the uniform principles of Annex VI, the conclusions of the review report on isoproturon, and in particular Appendices I and II thereto, as finalized in the Standing Committee on Plant Health on 7 December 2001 shall be taken into account. In this overall assessment Member States: – must pay particular attention to the protection of the groundwater, when the active substance is applied in regions with vulnerable soil and/or climatic conditions or at use rates higher than those described in the review report and must apply risk mitigation measures, where appropriate – must pay particular attention to the protection of aquatic organisms and must ensure that the conditions of authorization include, where appropriate, risk mitigation measures |
| 51 | Linuron CAS No 330-55-2 CIPAC No 76 | 3-(3,4-Dichlorophenyl)-1-methoxy-1-methylurea | 900 g kg^{-1} | 1 January 2004 | 31 December 2013 | Only use as herbicide may be authorized For the implementation of the uniform principles of Annex VI, the conclusions of the review report on linuron, and in particular Appendices I and II thereof, as finalized in the Standing Committee on the Food Chain and Animal Health on 3 December 2002 shall be taken into account. In this overall assessment Member States: – must pay particular attention to the protection of wild mammals, non-target arthropods and aquatic organisms. Conditions of authorization should include risk mitigation measures, where appropriate – must pay particular attention to the protection of operators. |

Under the "*Uniform Principle*" application, especially the maximum admissible concentration on drinking water of 0.1 µg L^{-1}, set by EC Directive on Drinking Water (98/83/EC) [53], was a hurdle for a listing into Annex I for many compounds belonging to the class of PS II inhibitors. Nevertheless, although this standard value does not reflect any risk under toxicological assessment it is binding for all EU member states as, for example, *The Pesticide Safety Directorate* stated [54]:

> Especially for the distribution in the environment the EC Directive
> on Drinking Water (98/83/EC) has set a maximum admissible
> concentration of 0.1 µg L^{-1} for any individual pesticide in drinking
> water. The figure is independent of any toxicological or environ-
> mental assessment and does not necessarily represent risk.
> Nevertheless it is UK Government policy to control the use of
> pesticides in such a way as to reduce the occurrence and levels
> of pesticide contamination found in drinking water (Annex VI
> of 91/414/EEC which is 97/57/EEC requires that this 0.1 µg L^{-1}
> standard for any individual pesticide applied to groundwater).

By searching under European Union, Reregistration of Plant Protection Agents, residues in Groundwater via Google.de, ≈ 21500 citations are found, indicating the political importance of this question caused by, for example, the "Grundwasserrichtlinie" in Germany. It is of eminent importance in public awareness [55].

Another politically important subject is carcinogenicity, which led to non-listing on Annex I under 91/414/EEC Directive of atrazine [56]:

> Unlike the EU the US-EPA has reregistered Syngenta's atrazine
> for use in maize, sugarcane, sorghum, cereals and other crops.
> Atrazine failed to get reregistration in Europe in October 2002
> because of suggestions that it could be linked to increased cancer
> risks. The US-EPA concluded that there have been no studies
> confirming increased risk.

Other major reasons for non-listing of PS II inhibitors in Annex 1 under 91/414/EEC Directive could be: changes in buffer zones listings, withdrawals for commercial reasons and failures to meet data submission deadlines.

Examples for PSII inhibitors that have already been included into Annex I are pyridate, isoproturon and bentazone. Table 10.5 exemplifies the provisions that have been imposed for Member States' Registration Authorities to address, e.g., by implementation of national use restrictions to ensure these substances are being used safely in the EU Member States.

Additionally from the C 1 group of PS II inhibitors the phenylcarbamates desmedipham and phenmedipham are listed in Annex I and from the C 3 group bromoxynil and ioxynil. The triazinones metamitron and metribuzin are applied for listing, the uracil lenacil, the pyridazon pyrazon/chloridazon, the ureas diuron, fluometuron, methabenzthiazuron, and the amide propanil. Off label for minor use (essential use) are applied for cyanazine, dimefurone and fenuron, metobromuron and metoxuron as well as for pentanochlor.

Table 10.6 describes the status of the reregistration process of PS II inhibitors in the EU (Status November 2005).

Evidently, from these data, the most important groups of chemistry in PS II inhibitors in the 1980s, i.e., triazines and, to a large extent, ureas will not be used anymore in the European Union, with some very small exceptions (Table 10.7).

10.5
Marketshare of PS II Compound Groups Today

Whereas photosynthesis inhibitors belonged in 1980 to the most important herbicide classes [38] the market situation changed, especially in Europe, at beginning of the 1980s through, especially, the introduction of new cereal, corn and oil seed rape herbicides from other herbicide classes but also through the reregistration process in Europe up to now.

The value of PSII inhibitors sold in the EU declined from ca. 745 Mio Euro in 1995 to only ca. 441 Mio Euro in 2004, i.e., by minus ca. 40%. At the same time, the value of the total herbicide market in the EU increased from ca. 1.600 Mio Euro to ca. 2.000 Mio Euro, i.e., by plus 25%. Thus, the value share of PSII inhibitors in the total EU herbicide market decreased from ca 45% in 1995 to only 22% in 2004.

Out of a total number of 50 PSII inhibitors, sales could only be recorded in the EU for some 40 compounds. Although the total number of PSII herbicides where sales could be recorded in the EU has only slightly declined from some 40 compounds in 1995 to about 36 compounds in 2004, only ten compounds have yet been included into Annex I as per end of 2005, while the future of a total number of slightly more than 30 PSII compounds still on sale in Europe is unclear or their life has come to an end, for one of the following reasons:

1. The substance has not even being notified, i.e., no dossier has been submitted (4 compounds where sales could be recorded).
2. The substance has not yet passed the EU Review Program successfully (ten compounds where sales could be recorded).
3. The substance has passed the EU Review Program with a negative outcome (so far 19 compounds where sales could be recorded), either because the notifier has not further supported the compound or the EU Commission has taken a negative decision on Annex I inclusion (see Tables 10.6 and 10.7).

Despite the fact that some 19 PSII compounds have already ended up in a non-inclusion, many of those are still on the market and sales can be recorded in the year 2004, mainly because either time-limited essential uses have been granted, or existing stocks are being sold out.

Table 10.6 Status of Reregistration process of PS II inhibitors (Nov 2005, Source: EU Commission).

Chemical Family	A.i.	No. in list of authorization	No. of authorizations in EU	Countries of authorization	List-no. (stage)	RMS	Registration status	Status list of peer review (State of main works)	Reasons for non-inclusion/withdrawl	Essential uses (according to regulation 2076/2002/EC authorization of PPPs until 30 June 2007	Dossier submitted by	List of uses supported by available data
Triazine	Cyanazine	not in EU-list	/	/	/	/	Not on list but essential uses (1336/2003/EC)			UK: Pea, bean, brassica, narcissi, oilseed rape, allium, forestry; SE: Oil seed rape, pickling cucumber; IR: Onion	Syngenta	
Triazinone	Metami-tron	370	23	FI, SE, DK, IE, UK, NL, BE, LU, DE, AU, FR, ES, IT, PT, EL, PL, CZ, HU, SK, SI, EE, LV, LT	3B	United Kingdom	Pending, data list	No information			BCS	

	Metribuzin	125	24	2	FI, SE, DK, IE, UK, NL, BE, LU, DE, AU, FR, ES, IT, PT, EL, PL, MT, CY, CZ, HU, SK, SI, EE, LV	Denmark	Pending, dossier DAR	Pending	BCS
Uracil	Lenacil	366	15	3B	IE, UK, BE, AU, FR, ES, PT, IT, EL, PL, CY, CZ, HU, SK, EE	Belgium	Pending, dossier	No information	
Phenyl-carbamate	Desmedi-pham	15	23	1	FI, SE, DK, IE, UK, NL, BE, LU, DE, AU, FR, ES, IT, PT, EL, PL, CZ, HU, SK, SI, EE, LV, LT	Finland	Inclusion in Annex I (expiration: 28. Feb. 2015)	Inclusion 2004/58/EC 6.10.2004, p. 26	Agrevo, main data submitter — Sugar & fodder beet

Table 10.6 (continued)

Chemical Family	A.i.	No. in list of authorization	No. of authorizations in EU	Countries of authorization	List-no. (stage)	RMS	Registration status	Status list of peer review (State of main works)	Reasons for non-inclusion/withdrawl	Essential uses (according to regulation 2076/2002/EC authorization of PPPs until 30 June 2007)	Dossier submitted by	List of uses supported by available data
	Phenmedipham	34	24	FI, SE, DK, IE, UK, NL, BE, LU, DE, AU, FR, ES, IT, PT, EL, PL, CZ, HU, SK, SI, EE, LV, LT	1	Finland	Inclusion in Annex I (expiration: 28. Feb. 2015)	Inclusion 2004/58/EC 6.10.2004, p. 26			Task Force on Phenmedipham (BCS, United Phosphorus Ltd.)	Sugar & fodder beet, red beet (beetroot)

Amide	Propanil	382	5	FR, ES, PT, IT, EL	3B	Italy	Pending, data list	No information	No information		
	Pentano-chlor	582	1	UK	3		Out 07/03; 2076/2002/EC; essential uses 1336/2003/EC		Not supported anymore by the notifier: With the available data no safe use could be assessed by the Commission.		UK: Umbellifers, herbs, ornamentals
Nitrile	Bromoxynil	11	18	DK, IE, UK, NL, BE, LU, DE, AU, FR, ES, PT, IT, EL, PL, CZ, HU, SK, SI	1	France	Inclusion in Annex I; 04/58/EC	Inclusion, 2004/58/EC, 6.10.2004, p. 26		BCS, Makhteshim Agan	Winter/spring cereals (barley, wheat, oats, rye, triticale), maize (member state: EU)

Table 10.6 (*continued*)

Chemical Family	A.i.	No. in list of authori-zation	No. of authori-zations in EU	Countries of authori-zation	List-no. (stage)	RMS	Regis-tration status	Status list of peer review (State of main works)	Reasons for non-inclusion/withdrawl	Essential uses (according to regulation 2076/2002/EC authorization of PPPs until 30 June 2007	Dossier submitted by	List of uses supported by available data
	Ioxynil	22	19	FI, SE, DK, IE, UK, NL, BE, LU, DE, AU, FR, ES, PT, IT, EL, PL, CY, HU, SI	1	France	Inclusion in Annex I; 04/58/EC	Inclusion, 2004/58/EC, 6.10.2004, p. 26			BCS, Makhteshim Agan, ACI International, CFPI Nufarm	Winter/ spring cereals
Urea	Metobro-muron	558	8	BE, DE, AU, FR, SP, PT, IT, EL	3		Out 07/03; 2076/2002/EC; essential use 1336/2003/EC	No infor-mation	Not supported anymore by the notifier: With the available data no safe use could be assessed by the Com-mission.	BE: Lambs lettuce. bean, potato: ES: Potato: DE: Lambs lettuce. bean, tobacco: FR: Lambs lettuce. artichoke		

Metoxuron	559	8	3	IE, UK, NL, BE, LU, FR, ES, IT	Out 07/03: 2076/2002/ EC. essential use 1336/ 2003 EC	No infor- mation	Not supported anymore by the notifier: With the available data no safe use could be assessed by the Commi- ssion.	BE: Carrot, potato; FR: Carrot; IR: Carrot; LU: Carrot, potato; NL: Carrot, potato, iris, gladiolus; UK: Carrot, parsnip

Table 10.7 Withdrawn PS II Inhibitors from Broad Reregistration in EU (Status Nov 2005, Source: EU Commission).

Chemical family	A.i.	Countries of authorization	RMS	Registration Status	Status list of peer review (State of main work)	Essential uses (according to regulation 2076/2002/EC authorization of PPPs until 30 June 2007
Triazines 1	Ametryn	FR, ES, IT		Out 07/03; 2076/2002/EC		
	Atrazine	IE, UK, BE, LU, FR, ES, PT, EL, PL, HU, SK	United Kingdom	Out 10/04; 04/247/EC	Withdrawn, 2004/248, EC, 16.03.2004, page 53	
	Desmetryne	UK, AU, ES, IT		Out 07/03; 2076/2002/EC		
	Prometryne	IE, UK, AU, FR, ES, PT, IT, EL, PL, MT, CY, HU, SK, EE, LV, LT		Out 7/03 essential use, 835/04; 2076/2002/EC bzw. 1336/2003/EC		
Triazines 2	Propazine			Out 07/03; 2076/2002/EC		
	Simazine	DK, IE, UK, BE, LU, AU, FR, ES, PT, IT, EL, PL, CY, SK	United Kingdom	Out 10/04; 04/247/EC	Withdrawn, 2004/247, EC, 16.03.2004, page 50	
	Simetryne				No information	
	Terbumeton	ES, IT		Out 07/03; 2076/2002/EC	No information	

Group	Compound	Countries		Status		Crop use
	Terbuthylazine		United Kingdom	Pending, data list	No information	
	Terbutryne	FI, SE, IE, UK, AU, FR, ES, PT, IT, EL, PL, HU, SK		Out 7/03; 2076/2002/EC	No information	
	Trietazine	IE, UK		Out 7/03; 2076/2002/EC	No information	
	Hexazinone	IE, AU, FR, ES, IT, CZ, HU, SK		Out 7/03; 2076/2002/EC	No information	
Uracil	Bromacil	IE, UK, AU, FR, ES, PT, IT, EL		Out 7/03 essential use; 2076/2002/EC	No information	
	Lenacil	IE, UK, BE, AU, FR, ES, PT, IT, EL, PL, CY, CZ, HU, SK, EE	Belgium	Pending, dossier	No information	
	Terbacil	UK, FR, ES, EL, PL, HU		Out 7/03; 2076/2002/EC	No information	
Urea 1	Chloroxuron	FR		Out 07/03; 2076/2002/EC	No information	
	Dimefuron	IE, DE, AU, FR		Out 07/03; 2076/2002/EC; essential use 1336/2003/EC	No information	**DE**: Oil seed rape
	Diuron	DK, IE, UK, BE, LU, DE, AU, FR, ES, PT, IT, EL, PL, CY, HU	Denmark	Pending, Dossier DAR	Pending, EFSA conclusion 14.01.2005	
	Ethidimuron (= sulfodiazol)	FR, ES, IT		Out 07/03; 2076/2002/EC	No information	
	Fenuron	UK, HU		Out 7/03 essential use; 2076/2002/EC bzw. 1336/2003/EC	No information	**UK**: Pea, bean, spinach
	Fluometuron	ES, EL	Greece	Pending, dossier	No information	

Table 10.7 (*continued*)

Chemical family	A.i.	Countries of authorization	RMS	Registration Status	Status list of peer review (State of main work)	Essential uses (according to regulation 2076/ 2002/EC authorization of PPPs until 30 June 2007
Urea 2	Metobromuron	BE, DE, AU, FR, SP, PT, IT, EL		Out 07/03; 2076/ 2002/EC; essential use 1336/2003/EC	No information	
	Metoxuron	IE, UK, NL, BE, LU, FR, ES, IT		Out 07/03; 2076/ 2002/EC; essential use 1336/2003/EC	No information	
	Monolinuron		United Kingdom	Out 9/01; 00/234/EC	**Withdrawn,** 2000/ 234/EC, 22.03.2000. p. 18	
	Neburon	IT		Out 07/03; 2076/2002/EC	No information	
	Siduron	FR		Out 07/03; 2076/2002/EC	No information	
	Tebuthiron			Out 07/03; 2076/2002/EC	No information	
Nitrile	Bromofenoxim	ES, IT		Out 07/03; 2076/2002/EC	No information	

The class of urea herbicides is widely being eliminated in Europe, but, more distinctively, the important class of triazine herbicides is disappearing from the EU herbicide market. None of their some ten representatives in Europe where sales can be recorded have made it into Annex I, except terbutylazine, which is still pending for the time being. The traditionally most important triazine representatives, atrazine and simazine, have not passed the EU Review Program and will have to be replaced by new chemistry in the EU, while both substances, particularly atrazine, still represent a significant importance in the US market, e.g., atrazine sales in the US accounted for some 165 Mio Euro in 2004, since this compound is widely and efficiently used in the US corn market, also in combination with Roundup-Ready.

Overall, the EU Review Program and the associated costs of maintaining substances in the market is leading to a significant streamlining in the number of PSII compounds by 50–75%, depending how many of the ten still pending compounds will be included into Annex I. The value share of the remaining PSII inhibitors in the total EU herbicide market will further decline in the years to come, and the downward trend recorded from ca. 45% market share in 1995 to only ca. 22% in 2004 will be further characteristic for this class of chemistry, since the phase-out is still ongoing.

10.6
A New Herbicide for Corn and Sugarcane: Amicarbazone – BAY MKH 3586

10.6.1
Introduction

Amicarbazone is a new herbicide for broad spectrum weed control in corn and sugarcane. It belongs to the chemical class of carbamoyl triazolinones and acts as an inhibitor of photosystem II. It was discovered 1988 by the former Plant Protection Division of Bayer AG (now Bayer CropScience) and developed under the internal code no. BAY MKH 3586 (Fig. 10.3).

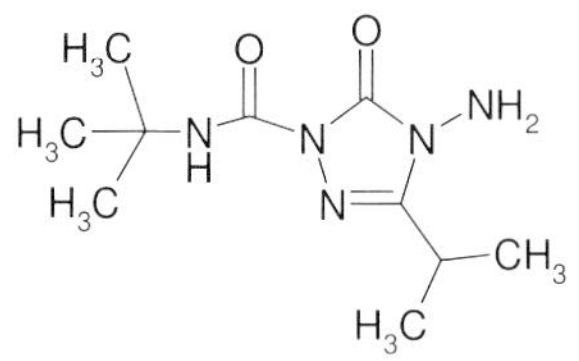

Fig. 10.3. Amicarbazone, BAY MKH 3586, Dinamic®.

Table 10.8 Physicochemical properties of amicarbazone.

Melting point	137.5 °C
Vapor pressure (Pa)	1.3×10^{-6} (20 °C) 3.0×10^{-6} (25 °C)
Dissociation constant (20 °C):	Amicarbazone has no acidic or basic properties in aqueous solutions. It is not possible to specify dissociation constants for water.
Solubility in water (g L^{-1}) (20 °C):	4.6 in unbuffered and buffered solutions; solubility not influenced by pH in the range pH 4–9
Volatility (Henry's law constant at 20 °C)	6.8×10^{-8} Pa m^3 mol^{-1}
Solubilities in organic solvents (g L^{-1}) (20 °C)	*n*-Heptane: 0.07 Xylene: 9.2 Poly(ethylene glycol) (Lutrol): 79 Dimethyl sulfoxide: >250 Dichloromethane: >250
Partition coefficient, Log P_{OW} in octanol–water (20 °C):	1.23 (pH 7) 1.14 (unbuffered)

10.6.2
Physicochemical Properties of Amicarbazone

Table 10.8 gives some physicochemical data.

10.6.3
Discovery of the Active Ingredient

Research is in most cases a continuous process and takes place in small steps. To better understand the discovery of amicarbazone we should go back to the year 1964 when Dornow published the first examples of the hitherto unknown class of 4-amino-1,2,4-triazin-5-ones [25] (Fig. 10.4).

Research chemists at the former Farbenfabriken Bayer AG identified these compounds in 1965 as herbicides [57] and specified the mode of action as inhibition of photosystem II [11, 58].

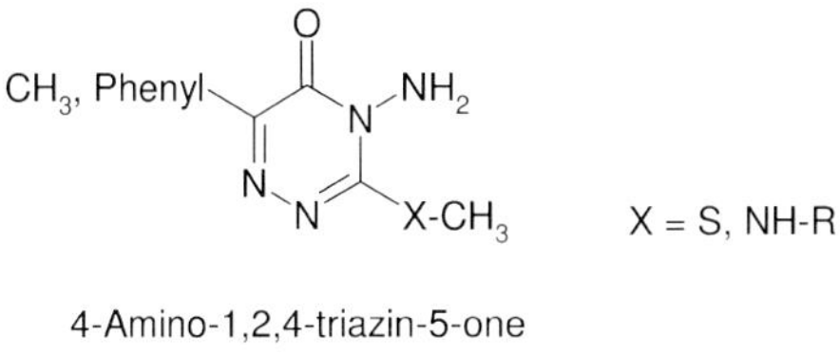

4-Amino-1,2,4-triazin-5-one

Fig. 10.4. 1,2,4-Triazinones, general formula.

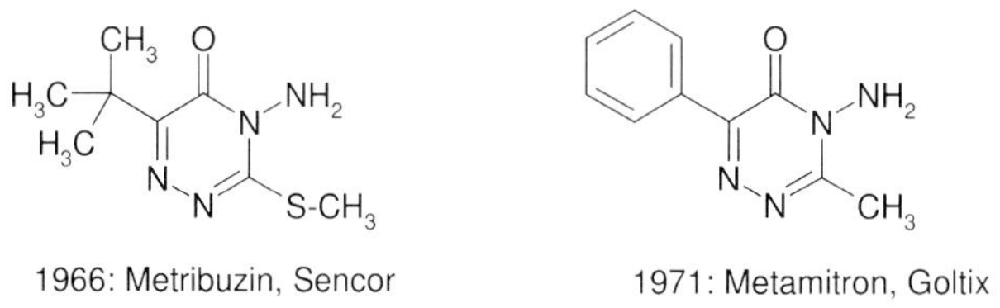

Fig. 10.5. 1,2,4-Triazinones, marketed compounds.

The optimization process led in 1966 to the discovery of metribuzin [57] and five years later to metamitron [59], two commercially very successful herbicides for soybeans and sugarbeet, respectively (Fig. 10.5).

Figure 10.6 gives a schematic representation of the essential atoms of a herbicide binding to the 32 kDa peptide of photosystem II, indicating the sp^2 hybrid with X (usually O, S or C) attached to a lipophilic group and the essential positive charge.

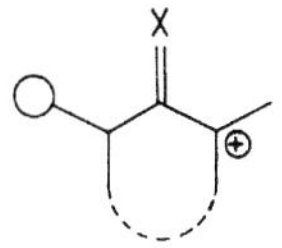

Fig. 10.6. Requirements for PS II inhibitors.

Following the above concepts of the structural requirements of PS II inhibitors [60] five-membered analogues of metamitron were synthesized and checked for their biological activity [61] (Scheme 10.2).

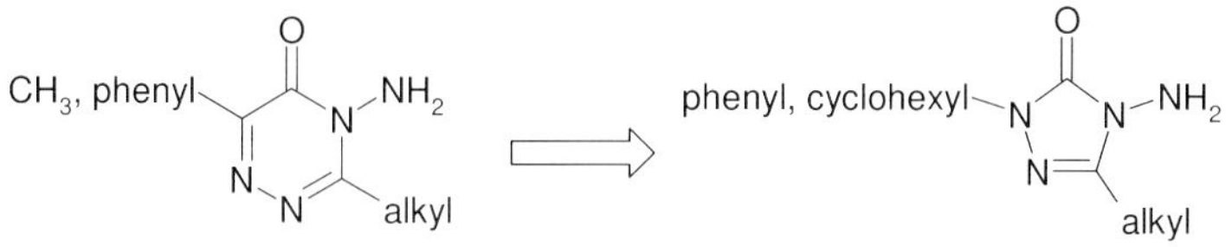

Scheme 10.2. Five-membered analogues of metamitron.

Although active *in vitro*, these compounds were rather inactive *in vivo* as herbicides.

With the goal of a corn herbicide, research efforts continued in the field of triazinones, and N-alkyl derivatives, and various sulfur [62], oxygen [62], nitrogen [62–65] and carbon substituents [66] were filed for patent. In 1981 a compound with the internal code number BAY KRA 4145 was synthesized and taken into development after intensive field tests [67] (Scheme 10.3).

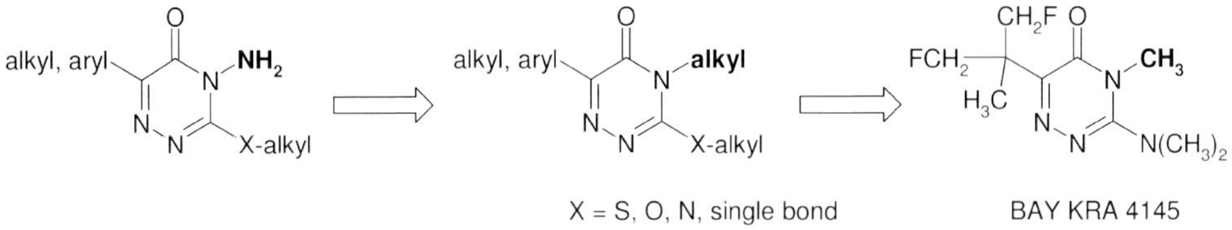

Scheme 10.3. Structure elucidation to BAY KRA 4145.

To reduce the high costs of a linear synthesis a convergent approach was evaluated using trimethylaminoguanidine as intermediate. Following the concept to use a new intermediate in different ways a lot of chemistry was performed, including reaction with phosgene (Scheme 10.4).

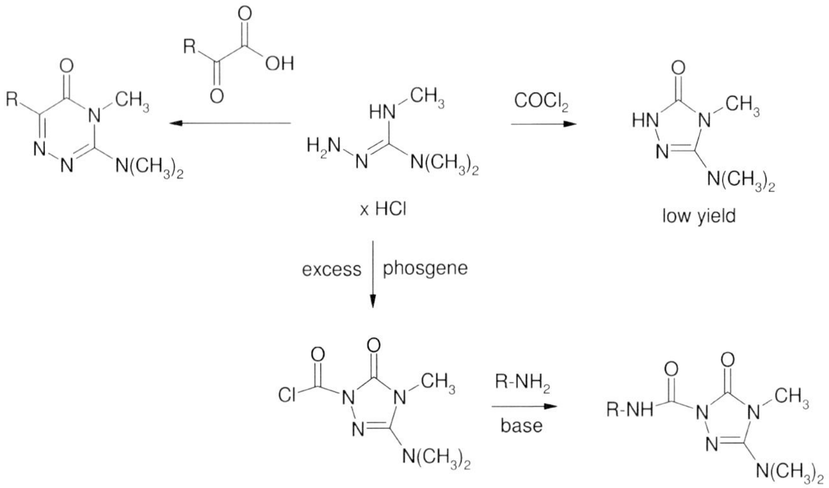

Scheme 10.4. Trimethylaminoguanidine as intermediate.

Excess phosgene generated in good yield the chlorocarbonyl triazolinone. Reaction with various amines produced carbamoyl triazolinones, the first derivative being isolated in May 1986 [68]. They are active *in vitro* as PS II inhibitors, but show, in contrast to the directly linked triazolinones, higher herbicidal *in vivo* activity. This was the starting point of a major synthesis program, generating more than 2500 compounds of the general type shown in Fig. 10.7.

All synthetic variations showing good herbicidal activity have been published in various patents (Table 10.9).

Regarding activity, selectivity and cost of goods, BAY MKH 3586 [77, 78] represents the optimum and was developed for selective weed control in corn and sugarcane.

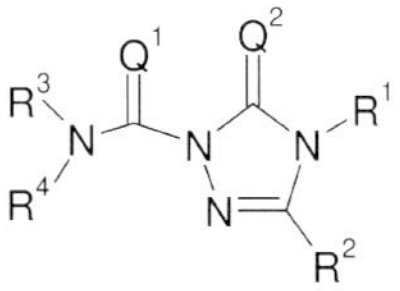

$Q^1, Q^2 = O, S$

$R^1 = H, C, N, O$
$R^2 = H, C, N, O, S, Hal$

Fig. 10.7. General formula of carbamoyl triazolinones.

Table 10.9 Synthetic variations of structure shown in Fig. 10.7 and associated patents.

R^1	R^2	Ref.
Alkyl	N(alkyl)$_2$	68–72
Alkyl	S-alkyl	68–72
Alkyl	NH-alkyl	72, 73
Alkyl	Alkyl	72, 74
Alkyl	Halogen	72, 75
Alkyl	O-alkyl	72, 76
NH$_2$	Alkyl	72, 77–80
NH-alkyl	Alkyl	72, 81, 82
NH$_2$	S-alkyl	72, 83, 84
NH$_2$, NH-alkyl, N(alkyl)$_2$	O-alkyl	72, 85
NH$_2$, NH-alkyl, N(alkyl)$_2$	NH$_2$, NH-alkyl, N(alkyl)$_2$	72, 86, 87

10.6.4
Synthesis

Final Product   There are various methods of synthesizing carbamoyl triazolinones [68–87]. In the case of N-amino triazolinones a protecting group like a Schiff base can be helpful [77–80]. Otherwise, under suitable conditions it is possible to add isocyanates directly in a kinetically controlled reaction to the amidic nitrogen of N-NH$_2$-triazolinones [77, 78, 88] (Scheme 10.5).

The first synthesis of the intermediate 4-amino-3-isopropyl-1,2,4-triazol-5-one was described by F. Malbec et al. [89]. The synthesis of this known intermediate can also be achieved by several other methods:

- Hydrazinolysis of acylated carbazates [90].
- Cyclization of carbohydrazide with carboxylic acids or ortho esters [90].
- Hydrazinolysis of ester carbalkoxy-hydrazones [89, 91, 92].

Owing to several disadvantages, such as low yield, long reaction time, formation of side products or number of synthesis steps, new synthetic methods were elaborated:

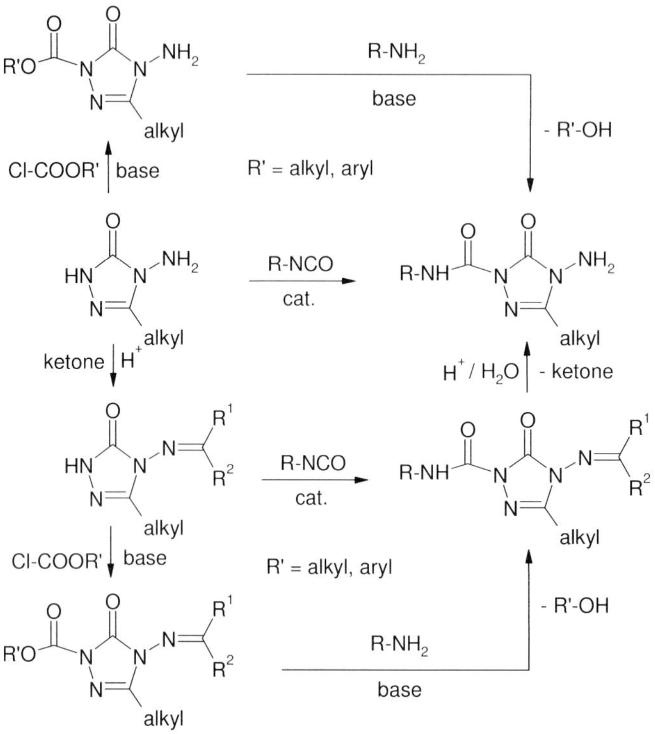

Scheme 10.5. Synthesis of N-amino-carbamoyltriazolinones.

- *In situ* preparation of carbohydrazide and cyclization with isobutyronitrile in the presence of a suitable tin compound as reaction auxiliary [93].
- Hydrazinolysis of 5-isopropyl-1,3,4-oxadiazol-2(3*H*)-one [94–98] (Scheme 10.6).

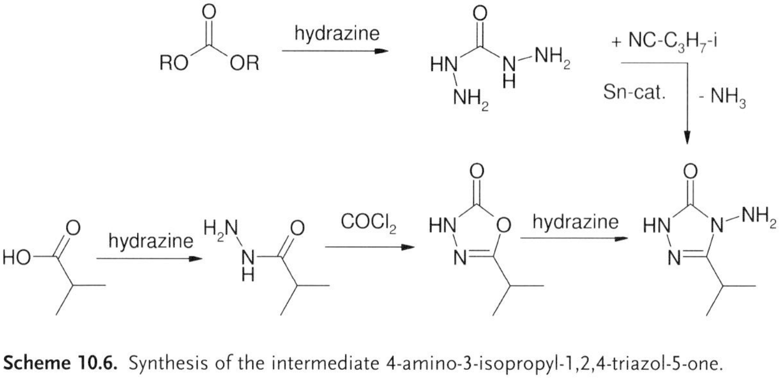

Scheme 10.6. Synthesis of the intermediate 4-amino-3-isopropyl-1,2,4-triazol-5-one.

10.6.5
Biological Behavior

Amicarbazone is tolerated by corn and sugarcane crops and shows excellent activity against many major annual dicotyledonous weeds that infest these crops.

In corn it may be applied up to a maximum rate of 500 g a.i. ha^{-1} to the soil at preplant or pre-emergence timings. In combination with other corn herbicides [99, 100] such as isoxaflutole the application rate can be reduced. Additionally, amicarbazone also shows contact activity on emerged weeds. The compound provides burndown as well as residual weed control, which is particularly useful in reduced and zero tillage corn production systems. Important weeds controlled by amicarbazone are, amongst others, velvetleaf (*Abutilon theophrasti*), common lambsquarters (*Chenopodium album*), pigweed (*Amaranthus* spp.), common cocklebur (*Xanthium strumarium*) and morning-glory species (*Ipomoea* spp.). In October 2005 amicarbazone was granted conditional registration by EPA in the United States [101]. From its biological spectrum and its mode of action it will compete mainly against atrazine (in all markets), but also replace the broadleaf part of the weed control spectrum of alachlor, acetochlor and metolachlor where grasses are not the dominant weeds [101].

In the Brazilian market amicarbazone was introduced in 2004 under the trade name Dinamic^{R} by Arysta LifeScience [102] for weed control in sugarcane.

It can be applied either pre-emergence or post-emergence at application rates up to 1500 g a.i. ha^{-1} solo or with 700 g a.i. ha^{-1} in combination [103, 104] with tebuthiuron (750 g a.i. ha^{-1}) or ametryn (1500 g a.i. ha^{-1}). In tank mixtures with metribuzin (960 g a.i. ha^{-1}) the rate can be reduced to 560 g a.i. ha^{-1} (post-emergence) or 800 g a.i. ha^{-1} (pre-emergence). Besides dicot weeds like painted spurge (*Euphorbia heterophylla*) and morning-glories, many annual grasses like marmeladegrass (*Brachiaria plantaginea*), southern sandbur (*Cenchrus echinatus*), bengal commelina (*Commelina benghalensis*) and guineagrass (*Panicum maximum*) are controlled [104]. More detailed information about the biological profile was revealed at the British Crop Protection Conference – Weeds 1999 [104].

10.6.6
Metabolites

In a corn metabolism study [105] amicarbazone and two degradation products were identified as the major components in corn matrices (Fig. 10.8).

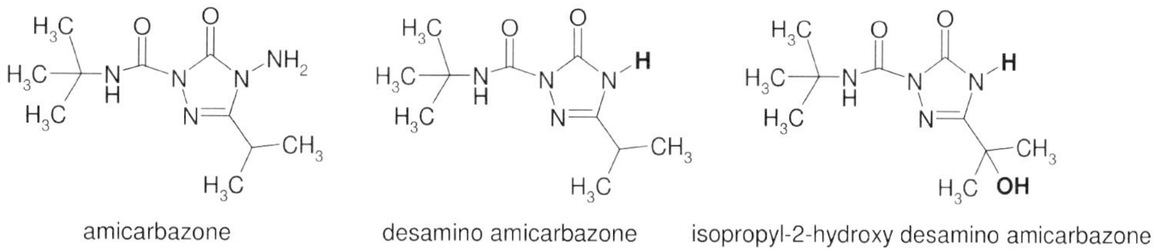

amicarbazone   desamino amicarbazone   isopropyl-2-hydroxy desamino amicarbazone

Fig. 10.8. Metabolites of aminocarbazone obtained in corn.

10.6.7
Final Remarks

Amicarbazone is the latest representative in the still economically important group of photosystem inhibitors. It belongs to the chemical class of carbamoyl triazolinones, was found and developed by Bayer AG and will be commercialized in the US in the corn market and in sugarcane growing countries by Arysta Life-Sciences.

10.7
Conclusions

PS II inhibitor herbicides were one of the most important herbicide classes. It could be shown in this chapter what influence the introduction of new herbicide classes could have on the market share of "ripened" herbicides but also the registration requirements in main markets. Nevertheless, new introductions could find a market and will be a good solution for weed control in corn and sugarcane markets.

References

1 R. R. Schmidt, HRAC Classification of Herbicides according to Mode of Action. *Brighton Crop Protection Conference – Weeds* **1997**, 1133–1140.

2 D. I. Arnon, F. R. Whatley, M. B. Allen, *Nature* **1957**, 180, 182–185.

3 A. Trebst, G. Hauska, *Naturwissenschaften* **1974**, 61, 308–316.

4 J. S. C. Wessels, R. van der Veen, *Biochem. Biophys. Acta* **1956**, 19, 548–549.

5 N. E. Good, *Plant Physiol.* **1961**, 36, 788–803.

6 D. E. Moreland, *Ann. Rev. Plant Physiol.* **1967**, 18, 365–386.

7 A. Trebst, E. Harth, *Z. Naturforsch. Teil C* **1974**, 29(5–6), 232–235.

8 B. Exer, *Experientia* **1958**, 14, 136–137.

9 D. E. Moreland, W. A. Gentner, J. J. Hilton, K. L. Hill, *Plant Physiol.* **1959**, 34, 432–435.

10 B. Exer, *Weed Res.* **1961**, 1, 233–244.

11 W. Draber, K. Dickoré, K. H. Büchel, A. Trebst, E. Pistorius, *Naturwissenschaften*, **1968**, 55, 446.

12 W. Draber, K. Dickoré, K. H. Büchel, A. Trebst, E. Pistorius in: H. Metzner, *Progress in Photosynthesis Research*, **1969**, Vol. III, p. 1789, International Union of Biological Sciences, Tübingen; see Ref. [58].

13 A. Trebst and H. Wietoska, *Z. Naturforsch. Teil C*, **1975**, 30, 499–504.

14 The Photosynthetic Process in: *Concepts in Photobiology: Photosynthesis and Photomorphogenesis*, Ed. G. S. Singhal, G. Renger, S. K. Sopory, K.-D. Irrgang and Govindjee, Narosa Publishers/New Delhi; and Kluwer Academic/Dordrecht, pp. 11–51. http://www.life.uiuc.edu/govindjee/paper/gov.html#100.

15 J. Deisenhofer, O. Epp, K. Miki, R. Huber, H. Michel, **1984**, *J. Mol. Biol.* 180, 385–398, X-ray structure analysis of a membrane protein complex. Electron density map at 3 A resolution and a model of the chromophores of the photosynthetic reaction center from *Rhodopseudomonas viridis*.

16 C. Hansch, T. Fujita, *J. Am. Chem. Soc.* **1964**, 86, 2738.

17 C. Hansch, E. W. Deutsch, *Biochem. Biophys. Acta* **1966**, 126, 117–128.

18 K. H. Büchel, W. Draber, A. Trebst, E. Pistorius, *Z. Naturforsch. Teil B* **1966**, 21, 243–254.

19 W. Draber, K. H. Büchel, H. Timmler, A. Trebst, *ACS Symposium Series* **1974**, No. 2, 100–116.

20 J. K. Seydel, *QSAR And Strategies in The Design of Bioactive Compounds*, **1985**, Paperbook, John Wiley and Sons Ltd. New York.

21 A. Leo, C. Hansch, D. Elkins, *Chem. Rev.* **1971**, 71, 525–616.

22 C. Hansch, T. E. Klein, *Methods Enzymol.* **1991**, 202, 512–543.

23 A. Gast, E. Knüsli, H. Gysin, *Experientia* **1955**, 11, 107–108.

24 A. Gast, E. Knüsli, H. Gysin, *Experientia* **1956**, 12, 146–148.

25 A. Dornow, H. Menzel, P. Marx, *Chem. Ber.* **1964**, 97, 2173–2178.

26 K. Lin, **1975**, Herbicidal 6-Amino-s-triazinediones, US 3902887 (Prio: 24. 05. 1972), E. I. Du Pont de Nemour and Comp., Wilmington, Del., USA.

27 H. C. Bucha, W. E. Cupery, J. E. Harrod, H. M. Loux, L. M. Ellis, *Science* **1962**, 137, 537–538.

28 A. Fischer, *Weed Res.* **1962**, 2, 177–184; see Ref. [37].

29 F. Reicheneder, K. Dury, A. Fischer, **1961**, Mittel zur Beeinflussung des Pflanzenwachstums, DE 1 105 232 (Prio: 21. 11. 1958), Badische Anilin- & Soda-Fabrik AG, Ludwigshafen/ Rhein, Germany.

30 H. C. Bucha, C. W. Todd, *Science* **1951**, 114, 493–494.

31 G. Hörlein, P. Langelüddeke, H. Schönowsky, **1972**, Selektive Unkrautbekämpfung mit Alkyl-phenylharnstoffderivaten, DE 20 39 041 (Prio: 06. 08. 1970), Hoechst AG, Frankfurt, Germany.

32 H. Kassebeer, *Z. Pflanz. Pflanzenschutz* **1971**, 78, 158–174.

33 W. Schäfer, R. Wegler, L. Eue, **1958**, Unkrautbekämpfungsmittel, DE 1 039 779 (Prio: 20. 04. 1957), Farbenfabriken Bayer Aktiengesell-schaft, Leverkusen, Germany.

34 K. P. Dorschner, R. L. Gates, J. R. Willard, **1963**, Selektive, herbicide Mittel, DE 1 160 236 (Prio: 11. 04. 1959), FMC Corporation, New York.

35 K. Carpenter, H. J. Cottrell, W. H. de Silva, B. J. Heywood, W. Gleeds, K. F. Rivett, M. L. Soundy, *Weed Res.* **1964**, 4, 175–195.

36 A. Zeidler, A. Fischer, G. Scheurer, *Z. Naturforsch. Teil B* **1969**, 24, 740–744.

37 A. Fischer, *Weed Res.* **1962**, 2, 177–184; see Ref. [28].

38 G. Jäger, Herbicides, page 338, in *Chemistry of Pesticides*, Ed. K. H. Büchel, transl. by G. Holmwood, John Wiley and Sons, New York, **1983**.

39 www.ficci.com/media-room/Speeches-presentations – 2006 Conference on Agrochemicals, January 12–13, 2006, Mumbai, Agrochemical Registration: A presentation by Dr. Arun Dhuri "Agrochemical Registration A Global View".

40 "A Global Approach to the Regulation of Agricultural Pesticides, A Vision for the Future" by www.epa.gov/oppfead 1/international/oecdfuture.pdf.

41 CONSLEG: 1991L0414 – 01/01/2004. http://europa.eu.int/eur-lex/en/consleg/pdf/1991/en_1991L0414 _do_001.pdf.

42 EC Directive 87/18/EEC.

43 Manual for the development and use of FAO Specifications for plant protection products. 4th Edition FAO, Via delle Terme di Caracalla, Rome, Italy.

44 Manual for the development and use of FAO Specifications for plant protection products 5th Edition FAO, Via delle Terme di Caracalla, Rome, Italy.

45 EC Directive 91/414 EEC.

46 EC Directive 94/37 EEC.

47 EC Directive 91/414 EEC Annex III Sections 7.2.1. to 7.2.3.

48 Predictive Operator Exposure Model (POEM) (SC 8001).

49 Data Requirements Handbook 28/09/04 Chapter 6-3 Environmental

Fate and Behaviour, www.Pesticides. gov.uk/aa_registration.

50 EC Directive 95/36 EC of 14 July 1995 L 172 8 22.7.1995, establishing the Data Requirements for environmental fate and behaviour in Annex II (Chapter 7) and Annex III (Chapter 9) of Directive 91/414.

51 Regulation/EC) NO 396/2995 Of The European Parliament And Of The Council of 23 February 2005, *Official J. Eur. Union* 16.03.2005, L 70/1.

52 *Official J. Eur. Union* L 379/13, 24.12.2004.

53 Enviromental Fate Behaviour, Data Requirements Handbook and Supplementary Guidance, PSD, page 86–87 under www.pesticides.gov.uk/ aa_registration.asp?id=643. EC Directive 97/57 of 21 December 1994 L 354 16 31.12.1994, establishing Annex VI (Uniform Principles) to Directive 91/414.

54 EC Directive 98/83, relating to the quality of water intended for human consumption.

55 www.bmu.de/files/pdfs/allgemein/ application/pdf/grundwasser_ richtlinie.pdf.

56 Outlooks on Pest Management Vol 15, Issue 1, 2004, *Regulatory News – February 2004*, page 5 www.researchinformation.co.uk/ pest.php/www.researchinformation. co.uk/pest/sample/sample.htm.

57 K. Westphal, W. Meiser, L. Eue, H. Hack, **1968**, Agent Herbicide, FR 1 519 180 (Prio: 16. 04. 1966), Farbenfabriken Bayer Aktiengesell-schaft, Leverkusen, Germany.

58 W. Draber, K. Dickoré, K. H. Büchel, A. Trebst, E. Pistorius, Lecture on the 1st Intern. Congress of Photosynthesis Research, Freudenstadt, Germany, June 4–8, 1968; cited in: H. Metzner (Ed.), *Progr. Photosyn. Res., Proc. Int. Congr.*, Verlag C. Lichtenstern, Mün-chen, **1969**, 1789–1795; see Ref. [12].

59 K. Dickoré, W. Draber, L. Eue, **1972**, 4-Amino-1,2,4-triazin-5-one, Verfahren zu ihrer Herstellung und ihre Verwendung als Herbizide, DE 2 107 757 (Prio: 18. 02. 1971), Bayer AG, Leverkusen, Germany.

60 A. Trebst, W. Donner, W. Draber, **1984**, *Z. Naturforsch.*, 39c, 405–411.

61 W. Draber, L. Eue, **1980**, unpublished results Bayer AG, Leverkusen, Germany.

62 H. Timmler, R. Wegler, L. Eue, H. Hack, 1,2,4-Triazin-5-one, **1971**, DE 1 670 912 (Prio: 18. 08. 1967), Farbenfabriken Bayer AG, Leverkusen, Germany, first published **1968** as ZA 68 04409.

63 K. Dickoré, K. Sasse, L. Eue, R. R. Schmidt, **1980**, 6-Substituierte 3-Dimethylamino-4-methyl-1,2,4-triazin-5-(4H)-one, Verfahren zu ihrer Herstellung und ihre Verwendung als Herbizide, DE 2 908 963 (Prio: 07. 03. 1979), Bayer AG Leverkusen, Germany.

64 K. Dickoré, K. Sasse, L. Eue, R. R. Schmidt, **1980**, 6-Cyclohexyl-3-dimethylamino-4-methyl-1,2,4-triazin-5-(4H)-on, Verfahren zu seiner Herstellung und seine Verwendung als Herbizid, DE 2 908 964 (Prio: 07. 03. 1979), Bayer AG, Leverkusen, Germany.

65 W. Draber, R. R. Schmidt, L. Eue, **1982**, 3-Dimethylamino-4-methyl-6-phenyl-1,2,4-triazin-5-one, Verfahren zu ihrer Herstellung sowie ihre Verwendung als Herbizide, DE 3 035 021 (Prio: 17. 09. 1980), Bayer AG, Leverkusen, Germany.

66 W. Draber, K. Dickoré, H. Timmler, **1973**, Verfahren zur Herstellung von 1,2,4-Triazin-5-onen, DE 2 138 031 (Prio: 29. 07. 1971), Farbenfabriken Bayer AG, Leverkusen, Germany.

67 E. Kranz, K. Findeisen, R. Schmidt, L. Eue, **1982**, Substituierte 6-Halogen-tert.-butyl-1,2,4-triazin-5-one, Verfahren zu ihrer Herstellung sowie ihre Verwendung als Herbizide, EP 49416 (Prio: 02. 10. 1980), Bayer AG, Leverkusen, Germany.

68 K. Findeisen, M. Lindig, H.-J. Santel, R. R. Schmidt, K. Lürssen, H. Strang, **1988**, Substituierte Triazolinone, EP 283876 (Prio: 24. 03. 1987), Bayer AG, Leverkusen, Germany.

69 K. Findeisen, M. Lindig, H.-J. Santel, R. R. Schmidt, **1989**, Substituierte Triazolinone, EP 305844 (Prio:

01. 09. 1987), Bayer AG, Leverkusen, Germany.

70 K. Findeisen, D. Kuhnt, K.-H. Müller, K. König, K. Lürssen, H.-J. Santel, R. R. Schmidt, **1992**, Substituierte Triazolinone, EP 503437 (Prio: 14. 03. 1991), Bayer AG, Leverkusen, Germany.

71 K. Findeisen, D. Kuhnt, K.-H. Müller, M. Haug, K. König, K. Lürssen, H.-J. Santel, R. R. Schmidt, **1992**, Substituierte Triazolinone EP 513621 (Prio: 14. 05. 1991), Bayer AG, Leverkusen, Germany.

72 K. Findeisen, K.-H. Linker, O. Schallner, K.-H. Müller, K. König, H.-J. Santel, R. R. Schmidt, **1994**, Substituierte Triazolinone, WO 1994/09012 (Prio: 12. 10. 1992), Bayer AG, Leverkusen, Germany.

73 K. Findeisen, M. Lindig, H.-J. Santel, R. R. Schmidt, K. Lürssen, **1990**, Substituierte Triazolinone, EP 398096 (Prio: 18. 05. 1989), Bayer AG, Leverkusen, Germany.

74 M. Lindig, K. Dickoré, K. Findeisen, H.-J. Santel, R. R. Schmidt, H. Strang, **1989**, Substituierte Triazolinone, EP 298371 (Prio: 10. 07. 1987), Bayer AG, Leverkusen, Germany.

75 K. Findeisen, M. Lindig, H.-J. Santel, K. Lürssen, R. R. Schmidt, **1991**, Halogen-Triazolone, DE 3 920 414 (Prio: 22. 06. 1989), Bayer AG, Leverkusen, Germany.

76 K.-H. Müller, K. König, J. Kluth, K. Lürssen, H.-J. Santel, R. R. Schmidt, **1992**, Substituierte 5-Alkoxy-1,2,4-triazol-3-(thi)one, EP 477646 (Prio: 22. 09. 1990), Bayer AG, Leverkusen, Germany.

77 M. Lindig, K. Findeisen, K.-H. Müller, H.-J. Santel, R. R. Schmidt, H. Strang, D. Feucht, **1988**, Substituierte Triazolinone, EP 294666 (Prio: 12. 06. 1987), Bayer AG, Leverkusen, Germany.

78 K.-H. Müller, M. Lindig, K. Findeisen, K. König, K. Lürssen, H.-J. Santel, R. R. Schmidt, H. Strang, **1990**, Substituierte Triazolinone, EP 370293 (Prio: 19. 11. 1988), Bayer AG, Leverkusen, Germany.

79 D. Kuhnt, K.-H. Müller, K. Findeisen, K. König, K. Lürssen, H.-J. Santel, R. R. Schmidt, **1992**, Substituierte Triazolinone, EP 511569 (Prio: 30. 04. 1991), Bayer AG, Leverkusen, Germany.

80 D. Kuhnt, K.-H. Müller, K. Findeisen, K. König, K. Lürssen, H.-J. Santel, R. R. Schmidt, **1992**, Substituted triazolinones with herbicide properties, WO 1993/04050 (Prio: 23. 08. 1991), Bayer AG, Leverkusen, Germany.

81 K. H. Müller, K. König, K. Findeisen, H.-J. Santel, K. Lürssen, R. R. Schmidt, S. Dutzmann, **1990**, Substituierte Triazolinone, EP 399294 (Prio: 24. 05. 1989), Bayer AG, Leverkusen, Germany.

82 K.-H. Müller, K. Findeisen, D. Kuhnt, K. König, K. Lürssen, H.-J. Santel, R. R. Schmidt, **1992**, Triazolinone, EP 505819 (Prio: 23. 03. 1991), Bayer AG, Leverkusen, Germany.

83 S. Iwai, M. Hatano, Y. Ishikawa, K. Kawana, **1978**, Triazoline derivatives, JP 53 135 981 (Prio: 27. 04. 1977), Nippon Soda Co., Ltd., Japan.

84 K.-H. Müller, J. Kluth, K. König, K.-R. Gassen, K. Findeisen, M. Lindig, K. Lürssen, H.-J. Santel, R. R. Schmidt, **1990**, Substituierte 4-Amino-5-alkylthio-1,2,4-triazol-3-one, EP 391187 (Prio: 07. 04. 1989), Bayer AG, Leverkusen, Germany.

85 J. Kluth, K.-H. Müller, W. Haas, K.-H. Linker, K. Findeisen, K. König, H.-J. Santel, M. Dollinger, **1994**, Substituted triazolinones and their use as herbicides, EP 625515 (Prio: 17. 05. 1993), Bayer AG, Leverkusen, Germany.

86 K.-H. Müller, K. Findeisen, M. Haug, U. Heinemann, J. Kluth, K. König, H.-J. Santel, K. Lürssen, R. R. Schmidt, **1991**, Substituierte 4,5-Diamino-1,2,4-triazol-3-(thi)one, EP 415196 (Prio: 30. 08. 1989), Bayer AG, Leverkusen, Germany.

87 D. Kuhnt, K. Findeisen, M. Haug, J. Kluth, K.-H. Müller, K. König, T. Himmler, G. Beck, H.-J. Santel, K. Lürssen, R. R. Schmidt, B. Krauskopf, **1992**, Substituierte 4,5-Diamino-1,2,4-

triazol-3-(thi)one, EP 502307 (Prio: 07. 02. 1991), Bayer AG, Leverkusen, Germany.

88 H.-J. Diehr, **1997**, Process for the preparation of substituted amino-carbonyltriazolinones, EP 757041 (Prio: 31. 07. 1995), Bayer AG, Leverkusen, Germany.

89 F. Malbec, R. Milcent, A. M. Bure, **1984**, *J. Heterocycl. Chem.* 21, 1769–1774.

90 K.-F. Kröger, L. Hummel, M. Mutscher, H. Beyer, **1965**, *Chem. Ber.* 98, 3025–3033.

91 A. Ikizler, R. Ün, **1979**, *Chim. Acta Turcica*, 7, 269–290.

92 R. Milcent, P. Vicart, A.-M. Bure, **1983**, *Eur. J. Med. Chem. – Chim. Ther.*, 18, 215–220.

93 K. König, K.-H. Müller, L. Rohe, **1990**, Verfahren zur Herstellung von 1,2,4-triazol-5-onen, EP 403889 (Prio: 21. 06. 1989), Bayer AG, Leverkusen, Germany.

94 K.-H. Müller, K. König, P. Heitkämper, **1989**, Verfahren zur Herstellung von 4-Amino-1,2,4-triazol-5-onen, EP 321833 (Prio: 22. 12. 1987), Bayer AG, Leverkusen, Germany.

95 H.-J. Diehr, K.-H. Müller, R. Lantzsch, **1997**, Process for preparing substituted aminotriazolinones, EP 759430 (Prio: 18. 08. 1995), Bayer AG, Leverkusen, Germany.

96 H.-J. Diehr, R. Lantzsch, J. Applegate, K. Jelich, **1998**, Process for the preparation of substituted oxadiazolones, EP 872480 (Prio: 15. 04. 1997), Bayer AG, Leverkusen, Germany.

97 V. C. Desai, K. Jelich, H. J. Diehr, R. Lantzsch, **1999**, Process for preparing 4-amino-1,2,4-triazolin-5-ones (Prio: 11. 12. 1998), Bayer Corporation, Pittsburgh, Pa, USA; Bayer AG, Leverkusen, Germany.

98 K. van Laak, C. Casser, M. Jautelat, M. Niehoff, **2000**, Verfahren zur Herstellung von Oxadiazolonen, DE 19 853 863 (Prio: 23. 11. 1998), Bayer AG, Leverkusen, Germany.

99 P. Dahmen, W. Thielert, K.-H. Müller, H.-J. Riebel, **1998**, Selektive Herbizide auf Basis von Carbamoyltriazolinonen, DE 19 635 060 (Prio: 30. 08. 1996), Bayer AG, Leverkusen, Germany.

100 D. Feucht, P. Dahmen, M.-W. Drewes, R. Pontzen, M. Kremer, K.-H. Müller, **2001**, Carbamoyl triazolinone based herbicides, WO 2001/37652 (Prio: 19. 11. 1999), Bayer AG, Leverkusen, Germany.

101 United States Environmental Protection Agency, Pesticide Fact Sheet Amicarbazone, October 4th, 2005, Internet Address: http://www.epa.gov/opprd001/factsheets/amicarbazone.pdf.

102 Chris Richards, President and CEO of Arysta LifeSciences, Credit Suisse/First Boston Agrochemicals Conference, 15th February 2005, Internet address: http://www.arysta.hu/feltoltes/Egyeb/Erdekes/arysta2005feb.pdf.

103 W. Thielert, G. Bohne, K.-H. Müller, **1998**, Selektive Herbizide für den Zuckerrohranbau, DE 19 635 074 (Prio: 30. 08. 1996), Bayer AG, Leverkusen, Germany.

104 B. D. Philbrook, M. Kremer, K. H. Müller, R. Deege, **1999**, *Proc. Brighton Conference – Weeds* (Vol. 1), 29–34.

105 A. E. Mathew, T. Nguyen, J. J. Murphy, Analytical residue method for MKH 3586 in plants, Abstract of papers American Chemical Society **1997**, Vol. 214, No. 1–2, AGRO 73, 214th American Chemical Society National Meeting, Las Vegas, Nevada September 7–11, 1997.

11
New Aspects of Plant Growth Regulators

Hans Ulrich Haas

Plant growth regulators (PGRs), their use, mode of action and plant-internal and -external interactions, have been the subject of intense research since they were introduced for agricultural use in the early 1930s. With new experimental results, knowledge and experience, the use and the spectrum of PGRs have increased continuously over the years, but the mode of action of growth regulators in plants is of such high complexity that even for auxins, the oldest known group of growth regulators, understanding of their mode of action is still incomplete.

This chapter gives an overview of PGRs, their current use and new developments. It is a summary of available knowledge and a tool for deeper and more intense analyses of specific items. The literature given at the end includes reviews and specialist summaries. Internet links provide detailed and up-to-date overviews, such as chemical structures, including the chemical names [1], chemistry, use and environmental aspects [2–4], and summarized overviews [5, 6].

PGRs are compounds of natural or synthetic origin used for controlling or modifying plant growth processes without apparent phytotoxic effects at the dose applied. They belong to a wide range of chemistry (Figs. 11.1 and 11.2).

Classically, there are five main categories of naturally occurring PGR, the auxins (IAA, NAA, IBA, 2,4-dichlorophenoxyacetic acid), gibberellins (GA), cytokinins (kinetin, benzyladenin, zeatin), ethylene, and growth inhibitors like abscisic acid (ABA).

Auxins were the first phytohormones detected and auxinic activity had already been observed in 1879 by J. Sachs during plant propagation. The first compounds (indole-3-acetic acid) were isolated and described by Kögl 1934 [7]. Auxins are involved in fruit ripening, phototropism, rooting, apical dominance, and cell enlargement. They are widely used as herbicides, showing activity due to an overdose and subsequent de-/regulation processes in plants. A herbicidally active auxin dose leads, for example, to an overdose of ethylene, inducing epinastic growth, tissue swelling, a stimulation of abscisic acid biosynthesis and leaf abscission [8].

Ethylene effects were firstly described by Nejebulov 1901 [9], but ethylene as a hormone could be identified only after gas chromatography was established in

						Brassinosteroids	
						Oliosaccharides	
						Uniconazole-p	
						Forchlorfenuron (CPPU)	
	2,4-DP		Ethylene			Paclobutrazol	
	2,4,5-T		Etephon			Hydrogen cyanamide	
	2,4-D		Abscisic acid (dormin)			Flurprimidol	
	Maleic hydrazide		Abscisin II			Inabenfide	Phospholipids
	Tecnazene		Chlorflurenol-methyl			Phthalimide	Trans-2-ketones
	3-CPA		Daminocide			Trinexapac-ethyl	Trans-2-aldehydes
1930	1940	1950	1960	1970	1980	1990	2000
Indole-3-acetic acid (IAA)		Kinetin		Ancymidol		Cyclanilide	
Indole-butyric acid (IBA)		Benzyladenin		Mepiquat-chloride		Prohexadione-calcium	
2-(1-naphtyl)acetic acid (NAA)		Thidiazuron		Dimethipin		1-methylcyclopropene (MCP)	
Gibberellin (GA)		Gibberellin GA3		Chlorphonium chloride		Jasmonate	
		Carbaryl		Dikegulac-sodium		Aviglycine-HCl (AVG)	
		4-CPA		Mefluidide		Monoterpenes	
		Chlorpropham (CIPC)		Ethychlozate			
		Chlormequat (CCC)		Flumetralin			

Fig. 11.1. Commercialized and new plant growth regulators and the decade of their market introduction or publication.

the 1960s [10]. Ethylene influences the balance of auxins vs. gibberellins [8], it inhibits cell division and strengthens cell walls. Furthermore, it is involved in the initiation of flowering, breaking of dormancy, abscission of parts of the plants and ripening processes. The most frequently used ethylene-based PGR is ethephon, an ethylene releaser. It breaks down in plant tissue to phosphate, chloride ions and ethylene, which acts as the PGR [11].

Gibberellins are involved in growth processes, including the elongation of internodes, flowering, dormancy, and fruit morphology. Their effects were first described by Kurosawa 1926 [12], who observed shoot elongation of rice after treatments with culture filtrates of *Fusarium moniliforme Sheld*. Yabuta was then

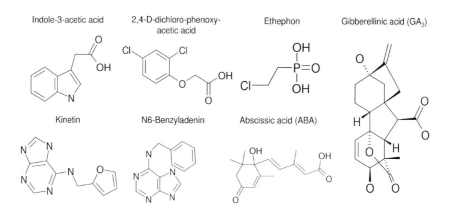

Fig. 11.2. Chemical structures of plant growth regulator compounds out of the five main PGR categories: auxins (IAA, 2,4-D), ethylene (ethephon), cytokinin (kinetin, benzyladenin), growth inhibitors (ABA), and gibberellins (GA₃).

Table 11.1 Key use areas of PGR in modern agriculture and compounds currently either in use or under development.

Factor	Example	Compounds
Plant growth	Shoot growth inhibition	Chlormequat-chloride
	Dwarfing	Ethylene, ethephon
		Mepiquat-chloride
		Mepiquat-pentaborate
		Paclobutrazol
		Prohexadione-Ca
		Promalin
		Trinexapac-ethyl
	Dormancy breaking	Ammonium nitrate
		Ca-cyanamide
		Ca-Nitrate
	Propagation	Auxins, giberellins, cytokinins
Yield and fruit quality	Flower induction	Ammonium thiosulfate
	Fruit-thinning	Ethylene, ethephon
		Gibberellinic acid
		Glutamic acid
		Paclobutrazol
		Prohexadione-Ca
	Ripening	Aviglycine HCL
	Fruit size adjustment	Benzyladenine
	Sugar accumulation	Cu-ethylenediamine
		Ethylene, ethephon
		MBTA-HCl
		Phospholipids
		Trinexapac-ethyl
Storage	Ripening	Aminoethoxyvinylglycine (AVG)
	Sprout supression	Carvone
		Chlorpropham (CIPC)
		1,2,6-DIPN
		Ethylene, ethephon
		Maleic hydrazide
		Menthol
		1-Methylcyclopropene (1-MCP)
		trans-2-Aldehyde/ketone

able to isolate a crystalline compound (5-*n*-butylpicolinic acid, fusaric acid) from the fungal culture. In 1938 Yabuta and Sumiki [13] published a first paper on the gibberellins. GA_3 was described in 1955 by Brian and Hemming [14]. Of all known gibberellins, today only gibberellic acids GA_3, GA_4 and GA_7 are of commercial importance [10].

Cytokinins act mainly through cell cycle regulation [15]. They stimulate cell division, prevent abscission, prevent rooting, enhance germination, and prevent senescence. Kinetin was firstly described in 1955 [16]. Benzyladenin was discovered by Strong in 1958 [17]. Both compounds are still the most commonly used cytokinins in plant micropropagation [7]. In addition, Strong [17] described thidiazuron, which is in use mainly to induce senescence in cotton.

Growth inhibitors like abscisic acid (dormin) and abscisin II retard growth, promote abscission, and induce dormancy. They were discovered in the 1960s [18]. They directly affect cell division and expansion and induce stomatal closure [8].

Plant growth regulators play an important role in modern agriculture. They are used to ensure and enhance quantity and quality of all parts of a crop cycle, from seed to harvest and postharvest. PGR can be grouped into main use categories, including shoot length control, yield regulation and harvest facilitation, storage control, propagation, and combined effects (Table 11.1).

Besides the auxins, which were commercialized predominantly as herbicides soon after their discovery, one of the first plant growth regulator products for shoot length control was maleic hydrazide, first described as a PGR in 1949 [19]. Since then the classical and core use of growth regulators is *shoot length control*. Growth inhibitors currently in use in cereals act mainly as gibberellin inhibitors. Chlormequat-chloride (CCC), trinexapac-ethyl, mepiquat-chloride and paclobutrazol are typical examples of such compounds (Fig. 11.3).

Rademacher [20] has summarized the chemistry of growth retardants in agronomic and orchard crops with special emphasis on gibberellin biosynthesis. Paclobutrazole and uniconazole are gibberellin inhibitors belonging to the N-containing heterocyclic triazoles. Triazoles are commonly known as fungicides, acting as demethylation inhibitors (DMI). Triazole fungicides are often reported to have growth retardant side effects [21]. Some of these PGR activities are of practical importance, such as metconazole and tebuconazole in oilseed rape.

An additional effect, which has not been given much attention, is the implication of the effects of such fungicides on the growth inhibition of weeds [22, 23]. The growth reduction of the crops and of the weeds may be about the same unless a specific weed species is selectively inhibited more than the crop.

Fruiting and growth of orchard trees is variable and dependent on climate, weather, but also plant specific factors such as alternation, the biennial fluctua-

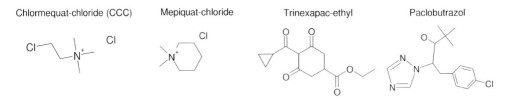

Fig. 11.3. Chemical structures of plant growth regulators mainly in use for shoot length control.

Methylcyclopropene
(1-MCP)

Norbornadiene (2,5-NBD)

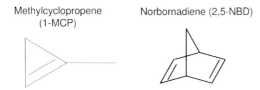

Fig. 11.4. Chemical structures of ethylene binding site inhibitors.

tion of the fruit yield in orchards. PGRs are used in this area to reduce and har-monize plant growth, to equalize and accelerate blossom and fruiting seasons (e.g. defoliation and re-growth), to precondition fruits for harvesting, and to thin fruits for better quality and fruit size and more equal yields during several years. To control tree growth in orchards but also in arable crops and flowers the gibberellin-biosynthesis inhibitors paclobutrazol [24] and prohexadione-Ca [25] are in use. Blossom thinners are ammonium thiosulfate (ATS), endothalic acid, pelargonic acid, sulfcarbamide-1-aminomethanamide, hydrogen tetraoxosulfate and hydrogen cyanamide [10]. For postbloom thinning, naphthalene acetic acid (NAA) is still important, but also a side effect of the insecticide carbaryl is used for thinning [26]. Benzyladenine (6-BA), as cytokinin, is registered for the reduc-tion of fruits, and additionally stimulates cell division in remaining fruits [27].

A further use area of PGR is the control of *storage and ripening* of fruits and plant products, e.g., cut flowers. Ethylene is broadly utilized to induce ripening. The opposite, delayed ripening for better shelf-life, is more difficult to manage. Daminocide was the first commercial compound to delay ripening of apple fruits [28]. It was replaced by the ethylene biosynthesis inhibitor aviglycine-HCl (AVG), which also had to be applied to the fruits on the tree before harvest.

A new episode of controlled ripening started with the investigation of ethylene binding site inhibitors, like 1-methylcyclopropene (1-MCP) and norbornadiene (2,5-NBD) (Fig. 11.4), and their development for market use [29]. Meanwhile 1-MCP was commercialized for postharvest treatment to delay ripening of apples [30–32].

Sprout inhibition of potatoes has mainly been driven by the use of propham, chlorpropham and maleic hydrazide during the last few years. Whilst propham (IPC) and chlorpropham (CIPC, Fig. 11.5) are applied after harvest at the begin-

Chlorpropham (CIPC)   S-(1)-Carvone   Menthol   Menthone

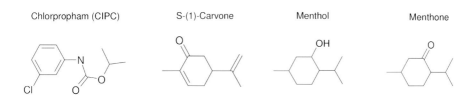

Fig. 11.5. Chemical structures of commercialized sprout suppressants.

Trans-2-aldehydes   Trans-2-ketones

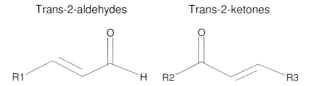

Fig. 11.6. Chemical structures of a new class of potential sprout suppressants.

ning of the storage, maleic hydrazide (MH) is applied to the potato foliage when tubers have reached a size of 40–70 mm. Tecnazene (TCNB) was also used in potato storage, but disappeared from the market due to its long degradation time. The monoterpenes S-(+)-carvone, produced from caraway seeds, has been developed commercially as a competitive product to CIPC (Fig. 11.5). Recently, menthol was commercialized for use as a sprout suppressant. Besides their sprout suppressing ability the natural terpenes also inhibit microbial growth and prevent rotting of treated potato tubers [33]. Coleman et al. [34] detected different activities of S-(+)-carvone, menthone and neomenthol, a diastereomer of menthol. The latter two showed 5–10× higher activity in suppressing tuber sprouting than S-(+)-carvone.

A patent on a new class of sprout suppressants was published recently [35]. It covers trans-2-ketones and trans-2-aldehydes (Fig. 11.6) being active as potato sprout suppressants. Known from "grass smell", trans-2-hexenal is included in this patent.

An important role of PGR is their involvement in *abiotic* and *biotic* stress defense mechanisms. Triazoles like paclobutrazole, propiconazole and tetraconazole are reported to be stress protectants [36]. Natural PGR are involved in indirect defense mechanisms of plants against herbivores [37]. Jasmonic acid, salicylic acid (Fig. 11.7) and ethylene are part of the signaling pathways of stress defense mechanisms.

Phospholipids impact the hypersensitive response and systemic acquired resistance in plants, and therefore might also be a potential new class of commercial

Jasmonates   Brassinosteroids   Salycilates   Polyamines
Jasmonic acid   Brassinolid   Salicylic acid   Spermidine

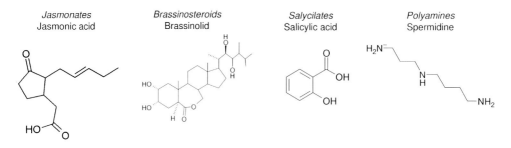

Fig. 11.7. Examples of chemicals out of potential new groups of PGR.

PGR [38]. The influence of polyamines [putrescine, spermidine (Fig. 11.7), spermine] on plant growth, including cell division, germination, till fruit development and stress response, has been reviewed by several authors [39–41]. Evans and Malmberg [42] furthermore summarized current knowledge of interactions of polyamines to commercial PGR and environmental stimuli. An interaction of polyamines with phospholipids in vesicles was described by Tadolini [43]. Polyamines may have an important role in stabilizing membranes through protection of lipid peroxidation. The metabolic link between polyamine and ethylene synthesis led to the suggestion of an impact of these PGR in abiotic and biotic interactions of the roots and the rhizosphere [44]. Romera and Alcantara [45] have summarized recent findings of ethylene involvement in the regulation of Fe-deficiency stress response. Especially in plants with iron-acquisition strategy I, the acidification of rhizosphere and subapical swelling of roots, ethylene plays a role in the regulation of stress response.

A reduction of abiotic and biotic stress in plants is also described as a side effect of quinone-outside-inhibiting (QoI)-fungicides, like the strobilurins. Wu and von Tiedemann [46, 47] reported that the strobilurin azoxystrobin and the triazole epoxiconazole exhibited strong antioxidative properties in delaying senescence and protecting barley and wheat from ozone injury. An increased resistance of tobacco against the tobacco mosaic virus *Pseudomonas syringae* pv. *tabbaci* was reported by Herms et al. [48] after a treatment of the plants with pyraclostrobin.

An abiotic stress with increasing importance in the near future is *water stress*. First reports indicate that PGRs also act as regulators under drought conditions. A positive response of kentucky bluegrass (*Poa pratensis* L.) to natural PGR was described by Schmidt in 1993 [49]. In these studies foliar application of seaweed extracts could accelerate recovery of kentucky bluegrass under serious drought conditions. Ervin and Koski [50, 51] described reduced evapotranspiration in kentucky bluegrass treated with trinexapac-ethyl. Marcum and Jiang [52] found similar effects on tall fescue (*Festuca arundinacea* S.). Zhang and Schmidt [53] and Zhang and Ervin [54] described enhanced drought tolerance of tall fescue and creeping bentgrass (*Agrostis palustris* Huds. A.) after application of humic acid or seaweed extract. In 2004, Zhang et al. [55] reported a better yield of soybeans after spray of uniconazole, brassinolide and ABA under drought conditions, compared with the untreated control. They did not detect such effects after benzyladenine (6-BA) treatment. Schubert [56] has reported effects of PGR on the yield of cereals under drought stress dependent on the PGR applied. Trinexapac-ethyl treated plants had a much higher harvest index and increased thousand-kernel weight compared with those of CCC treatments and the control.

Further new developments of commercial PGR include label extensions and mixtures of currently commercialized PGR as well as the evaluation of development of scientifically known PGR, including jasmonate [57], brassinosteroids [58], polyamines [42], phospholipids [38] and oligosaccharides [59, 60], for agricultural use.

References

1 A. Wood, www.alanwood.net/
pesticides/class_plant_growth_
regulators.html **2006**.

2 EPA, http://www.epa.gov/pesticides/
biopesticides/ingredients **2006**.

3 PMEP-Extoxnet, http://pmep.cce.
cornell.edu/profiles/extoxnet/
index.html **2006**.

4 S. Orme, S. Kegley,
www.pesticideinfo.org **2006**.

5 Wikipedia, http://en.wikipedia.org/
wiki/Plant_growth_regulator **2006**.

6 P. Sengbusch, http://www.biologie.
uni-hamburg.de/b-online/e31/31.htm
2006.

7 J.E. Preece, *HortScience* **2003**, 38(5),
1015–1025.

8 K. Grossmann, in G. Voss, G. Ramos,
Chemistry of Crop Protection, Wiley-
VCH **2003**, 131–142.

9 Jr. E. Beyer, P.W. Morgan, S.F. Yang,
in M.B. Wilkins, *Advanced Plant
Physiology*, Pitman Publishers,
London, **1984**, 111–126.

10 P.D. Petracek, F.P. Silvermann,
HortScience **2003**, 38(5), 937–942.

11 J.A. Maynard, A.C. Leopold, *Aust. J.
Chem.* **1963**, 16, 596–608.

12 E. Kurosawa, *Nat. Hist. Soc. Formosa*
1926, 16, 213–227.

13 T. Yabuta, Y. Sumiki, *J. Agric. Chem.
Soc. Jpn.* **1938**, 14, 1526.

14 P. Brian, H. Hemming, *Physiol. Plant.*
1955, 8, 669–681.

15 W. Tang, L. Harris, R.J. Newton,
J. Forestry Res. **2004**, 15(3), 227–
232.

16 C.O. Miller, F. Skoog, M.H. von
Saltza, M. Strong, *J. Am. Chem. Soc.*
1955, 77, 1329–1334.

17 F.M. Strong, *Topics in Microbial
Chemistry*, Wiley, New York, **1958**.

18 B. Milborrow, in M.B. Wilkins,
Advanced Plant Physiology, Pitman
Publishers, London, **1984**, pp. 76–
110.

19 D.L. Schöne, O.L. Hoffmann, *Science*
1949, 109, 588–590.

20 W. Rademacher, *Annu. Rev. Plant
Physiol. Plant Mol. Biol.* **2000**, 51,
501–531.

21 R.A. Fletcher, G. Hofstra, G. Jian-guo,
Plant Cell Physiol. **1986**, 27(2),
367–371.

22 J.M. Benton, A.H. Cobb, *Plant Growth
Regulation*, **1995**, 17(2), 149–155.

23 B.D. Hanson, C.A. Mallory-Smith,
B.D. Brewster, L.A. Wendling, D.C.
Thill, *Weed Technol.* **2003**, 17(4),
777–781.

24 J.D. Qinlan, *Acta Hortic.* **1980**, 114,
144–151.

25 R.J. Evans, R.R. Evans, C.L. Reguski,
W. Rademacher, *HortScience* **1999**, 34,
1200–1201.

26 D.W. Greene, *HortScience* **2002**, 37(3),
477–481.

27 P.T. Wismer, J.T.A. Proctor, D.C.
Elfving, *J. Am. Soc. Hort. Sci.* **1995**,
120, 802–807.

28 L.J. Edgerton, M.B. Hoffmann,
Proc. Am. Soc. Hort. Sci. **1965**, 57,
120–124.

29 E.C. Sisler, M. Serek, *Bot. Bull. Acad.
Sin.* **1999**, 40, 1–7.

30 M. Serek, E.C. Sisler, M.S. Reid, *Acta
Hort.* **1995**, 394, 337–346.

31 S.M. Blankenship, C.R. Unrath,
HortScience **1998**, 33, 469.

32 E.C. Sisler, M. Serek, *Plant Biol.* **2003**,
5, 473–480.

33 D. Vokou, S. Vareltzidou, P. Katinakis,
Agric., Ecosystems Environ. **1993**, 47,
223–235.

34 W.K. Coleman, G. Lonergan, P. Silk,
Am. J. Potato Res. **2001**, 78, 345–354.

35 N.R. Knowles, Knowles, L. O'Rear,
United States Patent 6855669,
published **2005**.

36 A. Gilley, R.A. Fletcher, *Plant Growth
Regulat.* **1997**, 21(3), 169–175.

37 R.M.P. Van Poecke, M. Dicke, *Plant
Biol.* **2004**, 6(4), 387–401.

38 A.K. Cowan, *Plant Growth Regulat.*
2006, 48, 97–109.

39 R.D. Slocum, R. Kaur-Sawhney, A.W.
Galston, *Arch. Biochem. Biophys.* **1984**,
235(2), 283–303.

40 T.A. Smith, *Annu. Rev. Plant Physiol.*
1985, 36, 117–143.

41 C.H. Kao, *Bot. Bull. Acad. Sin.* **1997**,
38, 141–144.

42 P.T. Evans, R.L. Malmberg, *Annu. Rev. Plant Physiol. Plant Mol. Biol.* **1989**, 40, 235–269.

43 B. Tadolini, *Biochem. J.* **1988**, 249, 33–36.

44 I. Couee, I. Hummel, C. Sulmon, G. Goueset, A. El Amrani, *Plant Cell Tissue Organ Culture*, **2004**, 76(1), 1–10.

45 F.J. Romera, E. Alcantara, *Plant Physiol.* **1994**, 105, 1133–1138.

46 Y.X. Wu, A. von Tiedemann, *Pestic. Biochem. Physiol.* **2001**, 71, 1–10.

47 Y.X. Wu, A. von Tiedemann, *Environ. Pollut.* **2002**, 116, 37–47.

48 S. Herms, K. Seehaus, H. Koehle, U. Conrath, *Plant Physiol.* **2002**, 130, 120–127.

49 R.E. Schmidt, http://www.sustane.com/pdfs/research/VA_Tech_drought_stress_in_KY_bluegrass.pdf, **1993**.

50 E.H. Ervin, A.J. Koski, *Hort. Sci* **1998**, 33(7), 1200–1202.

51 E.H. Ervin, A.J. Koski, *Crop Sci.* **1991**, 41, 247–250.

52 K.B. Marcum, H. Jiang, *J. Turfgrass Manage.* **1997**, 2(2), 13–27.

53 X. Zhang, R.E. Schmidt, *Crop Sci.* **2000**, 40, 1344–1349.

54 X. Zhang, E.H. Ervin, *Crop Sci.* **2004**, 44(5), 1737–1745.

55 M. Zhang, L. Duan, Z. Zhai, J. Li, X. Tian, B. Wang, Z. He, Z. Li, in: T. Fischer et al. *Proceedings for the 4th International Crop Science Congress*, Brisbane, Australia, **2004**.

56 S. Schubert, *DLG-Mitteilungen* **2006**, 3, 65–67.

57 R.A. Creelman, J.E. Mullet, *Proc. Natl. Acad. Sci. U.S.A.* **1995**, 92(10), 4114–4119.

58 N.B. Mandava, *Annu. Rev. Plant Physiol. Plant Mol. Biol.* **1988**, 39, 23–52.

59 P. Albersheim, B.S. Valent, *J. Cell Biol.* **1978**, 78(3), 627–643.

60 A. Darvill, C. Augur, C. Bergmann, R.W. Carlson, J.J. Cheong, S. Eberhard, M.G. Hahn, V.M. Lo, V. Marfa, B. Meyer et al., *Glycobiology* **1992**, 2(3), 181–198.